JOHNSON/EV̶̶̶̶̶DE
OUTBOARD

34 95

Volume IIA
1- and 2-Cylinder
1990-1992
Tune-Up
and
Repair Manual

SELOC
PUBLICATIONS

MARINE
MANUALS

Other books
by
Joan Lorton Coles
and/or
Clarence W. Coles

Glenn's Complete Bicycle Manual
*The Book of Harry (limited Edition)
*Glenn's Flat Rate Manual -- American Cars
Seloc's OMC Stern Drive Tune-up and Repair Manual
Seloc's Marine Jet Drive Tune-up and Repair Manual
Seloc's Mercury Volume I Tune-up and Repair Manual
Seloc's Mercury Volume II Tune-up and Repair Manual
Seloc's Mercury Volume III Tune-up and Repair Manual
Seloc's Force Outboard Tune-up and Repair Manual
Seloc's Yamaha Volume I Tune-up and Repair Manual
Seloc's Yamaha Volume II Tune-up and Repair Manual
Seloc's Yamaha Volume III Tune-up and Repair Manual
Seloc's Mariner Volume I Tune-up and Repair Manual
Seloc's Mariner Volume II Tune-up and Repair Manual
Seloc's Chrysler Outboard Tune-up and Repair Manual
Seloc's OMC Cobra Stern Drive Tune-up and Repair Manual
Seloc's MerCruiser Stern Drive Tune-up and Repair Manual
Seloc's MerCruiser Stern Drive Volume II Tune-up & Repair Manual
Seloc's Mercury/Mariner Volume I Tune-up and Repair Manual
Seloc's Mercury/Mariner Volume II Tune-up and Repair Manual
Seloc's Mercury/Mariner Volume III Tune-up and Repair Manual
Seloc's Johnson/Evinrude Volume I Tune-up and Repair Manual
Seloc's Johnson/Evinrude Volume II Tune-up and Repair Manual
Seloc's Johnson/Evinruide Volume IIA Tune-up and Repair Manual
Seloc's Johnson/Evinrude Volume III Tune-up and Repair Manual
Seloc's Johnson/Evinrude Volume IV Tune-up and Repair Manual
Seloc's Johnson/Evinrude Volume V Tune-up and Repair Manual
Seloc's Personal Watercraft Volume I Tune-up and Repair Manual
Seloc's Personal Watercraft Volume II Tune-up and Repair Manual
Seloc's Personal Watercraft Volume III Tune-up and Repair Manual
Seloc's Volvo Penta Stern Drive Volume I Tune-up and Repair Manual
Seloc's Volvo Penta Stern Drive Volume II Tune-up and Repair Manual
Seloc's Volvo Penta Stern Drive Volume III Tune-up and Repair Manual
Briggs & Stratton Single Cylinder OHV Tune-up and Repair Manual
Briggs & Stratton Single Cylinder Horizontal Crankshaft Tune-up & Repair Manual
Briggs & Stratton Single Cylinder Vertical Crankshaft Tune-up and Repair Manual

*Out of print.

SELOC'S
JOHNSON/EVINRUDE OUTBOARD

Tune-Up
and
Repair Manual

Volume IIA

1990-1995

by
Joan and Clarence Coles

Cover
An original oil painting
by
Ray Goudey II
B.F.A. Art Center
Pasadena, California

Technical Consultant
Richard L. Harrison
Lucerne Valley, California

Graphics
Barbara Flotow

This Volume IIA supercedes
Seloc's Johnson/Evinrude Volume II

Volume IIA -- Copyright © **1992**
Update Copyright © **1995**
by
Seloc Publications

Inquiries should be addressed to:

Seloc Publications
10693 Civic Center Drive
Rancho Cucamonga, California 91730

ISBN 0-89330-026-8

FOREWORD

This is a comprehensive tune-up and repair manual for all Johnson or Evinrude 1- and 2-cylinder recreational outboard units manufactured from 1990 through 1995. Competition, high-performance, and commercial units (including "after market" equipment), are not covered. The book has been designed and written for the professional mechanic, the do-it-yourselfer, and the student developing his mechanical skills.

Professional Mechanics will find it to be an additional tool for use in their daily work on Johnson/Evinrude units because of the many special helpful techniques described.

Boating Enthusiasts interested in performing their own work and in keeping their unit operating in the most efficient manner will find the step-by-step illustrated procedures used throughout the manual extremely valuable. In fact, many have said this book almost equals an experienced mechanic looking over their shoulder giving advice.

Students and **Instructors** have found the chapters divided into practical areas of interest and work. Technical trade schools, from Florida to Michigan and west to California, as well as the U.S. Navy and Coast Guard, have adapted Seloc manuals as a standard classroom text.

Troubleshooting sections have been included in many chapters to assist eh individual performing the work to quickly and accurately isolate problems to a specific area without unnecessary expense and time-consuming work. As an added aid and one of the unique features of this book, many worn parts are illustrated to identify and clarify when an item should be replaced.

Illustrations and procedural steps are so closely related and identified with matching numbers that, in most cases, captions are not used. The exploded drawings show internal parts and their relationship with each other.

Accurate comprehensive specifications and wiring diagrams are included.

TABLE OF CONTENTS

8 LOWER UNIT

9 HAND REWIND STARTER

10 TRIM/TILT SYSTEM

11 MAINTENANCE

APPENDIX

APPENDIX (CONT))

WIRE IDENTIFICATION DWGS. (CONT.)

1
SAFETY

1-1 INTRODUCTION

Today, a boat and power unit represents a sizeable investment for the owner. In order to protect this investment and to receive the maximum amount of enjoyment from the boat, it must be cared for properly while being used and when it is out of the water. Always store the boat with the bow higher than the stern and be sure to remove the transom drain plug and the inner hull drain plugs. If any type cover is used to protect the boat, plastic, canvas, whatever, be sure to allow for some movement of air through the hull. Proper ventilation will assure evaporation of any condensation due to changes in temperature and humidity.

1-2 CLEANING, WAXING, AND POLISHING

An outboard boat should be washed with clear water after each use to remove surface dirt and any salt deposits from use in salt water. Regular rinsing will extend the time between waxing and polishing. It will also give you "pride of ownership", by having a sharp looking piece of equipment. Elbow grease, a mild detergent, and a brush will be required to remove stubborn dirt, oil, and other unsightly deposits.

Stay away from harsh abrasives or strong chemical cleaners. A white buffing compound can be used to restore the original gloss to a scratched, dull, or faded area. The finish of your boat should be thoroughly cleaned, buffed, and polished at least once each season. Take care when buffing or polishing with a marine cleaner not to overheat the surface you are working, because you will burn it.

A small outboard engine mounted on an aluminum boat should be removed from the boat and stored separately. Under all circumstances, any outboard engine must **ALWAYS** be stored with the powerhead higher than the lower unit and exhaust system. This position will prevent water trapped in the lower unit from draining back through the exhaust ports into the powerhead.

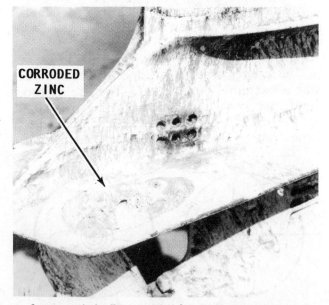

CORRODED ZINC

DRAIN PLUG

Whenever the boat is stored, for long or short periods, the bow should be slightly higher than the stern and the drain plug in the transom removed to ensure proper drainage of rain water.

Lower unit badly corroded because the zinc was not replaced. Once the zinc is destroyed, more costly parts will be damaged. Attention to the zinc condition is extremely important during boat operation in salt water.

A new zinc prior to installation. This inexpensive item will save corrosion on more valuable parts.

Most outboard engines have a flat area on the back side of the powerhead. When the engine is placed with the flat area on the powerhead and the lower unit resting on the floor, the engine will be in the proper altitude with the powerhead higher than the lower unit.

1-3 CONTROLLING CORROSION

Since man first started out on the water, corrosion on his craft has been his enemy. The first form was merely rot in the wood and then it was rust, followed by other forms of destructive corrosion in the more modern materials. One defense against corrosion is to use similar metals throughout the boat. Even though this is difficult to do in designing a new boat, particularily the undersides, similar metals should be used whenever and wherever possible.

A second defense against corrosion is to insulate dissimilar metals. This can be done by using an exterior coating of Sea Skin or by insulating them with plastic or rubber gaskets.

Using Zinc

The proper amount of zinc attached to a boat is extremely important. The use of too much zinc can cause wood burning by placing the metals close together and they become "hot". On the other hand, using too small a zinc plate will cause more rapid deterioration of the metal you are trying to protect. If in doubt, consider the fact that it is far better to replace the zincs than to replace planking or other expensive metal parts from having an excess of zinc.

When installing zinc plates, there are two routes available. One is to install many different zincs on all metal parts and thus run the risk of wood burning. Another route, is to use one large zinc on the transom of the boat and then connect this zinc to every underwater metal part through internal bonding. Of the two choices, the one zinc on the transom is the better way to go.

Small outboard engines have a zinc plate attached to the cavitation plate. Therefore, the zinc remains with the engine at all times.

1-4 PROPELLERS

As you know, the propeller is actually what moves the boat through the water. This is how it is done. The propeller operates in water in much the manner as a wood screw does in wood. The propeller "bites" into the water as it rotates. Water passes between the blades and out to the rear in the shape of a cone. The propeller "biting" through the water in much the same manner as a wood auger is what propels the boat.

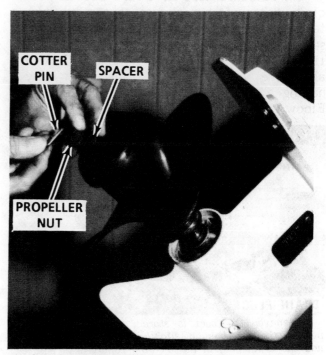

Associated parts securing the propeller on the shaft: a spacer, the propeller nut, and a cotter pin.

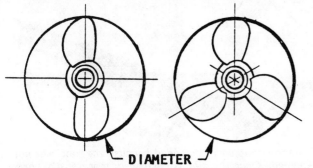

Diameter and pitch are the two basic dimensions of a propeller. The diameter is measured across the circumference of a circle scribed by the propeller blades, as shown.

Arrangement of propeller and associated parts, in order, for a small horsepower engine.

Diameter and Pitch

Only two dimensions of the propeller are of real interest to the boat owner: the diameter and the pitch. These two dimensions are stamped on the propeller hub and always appear in the same order: the diameter first and then the pitch. For instance, the number 15-19 stamped on the hub, would mean the propeller had a diameter of 15 inches with a pitch of 19.

The diameter is the measured distance from the tip of one blade to the tip of the other as shown in the accompanying illustration.

The pitch of a propeller is the angle at which the blades are attached to the hub. This figure is expressed in inches of water travel for each revolution of the propeller. In our example of a 15-19 propeller, the propeller should travel 19 inches through the water each time it revolves. If the propeller action was perfect and there was no slippage, then the pitch multiplied by the propeller rpms would be the boat speed.

Most outboard manufacturers equip their units with a standard propeller with a diameter and pitch they consider to be best suited to the engine and the boat. Such a propeller allows the engine to run as near to the rated rpm and horsepower (at full throttle) as possible for the boat design.

The blade area of the propeller determines its load-carrying capacity. A two-blade propeller is used for high-speed running under very light loads.

A rubber hub installed inside the propeller serves the same purpose as a "shear pin".

A four-blade propeller is installed in boats intended to operate at low speeds under very heavy loads such as tugs, barges, or large houseboats. The three-blade propeller is the happy medium covering the wide range between the high performance units and the load carrying workhorses.

Propeller Selection

There is no standard propeller that will do the proper job in very many cases. The list of sizes and weights of boats is almost endless. This fact coupled with the many boat-engine combinations makes the propeller selection for a specific purpose a difficult job. In fact, in many cases the propeller is changed after a few test runs. Proper selection is aided through the use of charts set up for various engines and boats. These charts should be studied and understood when buying a propeller. However, bear in mind, the charts are based on average boats

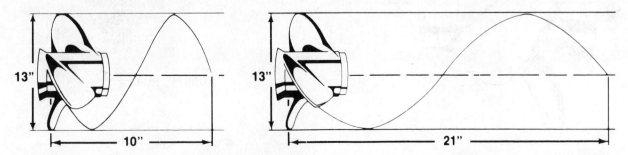

Diagram to explain the pitch dimension of a propeller. The pitch is the theoretical distance a propeller would travel through the water if there was no slippage.

with average loads, therefore, it may be necessary to make a change in size or pitch, in order to obtain the desired results for the hull design or load condition.

A wide range of pitch is available for each of the larger horsepower engines. The choice available for the smaller engines, up to about 25 hp, is restricted to one or two sizes. Remember, a low pitch takes a smaller bite of the water than the high pitch propeller. This means the low pitch propeller will travel less distance through the water per revolution. The low pitch will require less horsepower and will allow the engine to run faster and more efficiently.

It stands to reason, and it's true, that the high pitch propeller will require more horsepower, but will give faster boat speed if the engine is allowed to turn at its rated rpm.

If a higher-pitched propeller is installed on a boat, in an effort to get more speed, extra horsepower will be required. If the extra power is not available, the rpms will be reduced to a less efficient level and the actual boat speed will be less than if the lower-pitched propeller had been left installed.

All engine manufacturers design their units to operate with full throttle at, or slightly above, the rated rpm. If you run your engine at the rated rpm, you will increase spark plug life, receive better fuel economy, and obtain the best performance

from your boat and engine. Therefore, take time to make the proper propeller selection for the rated rpm of your engine at full throttle with what you consider to be an average load. Your boat will then be correctly balanced between engine and propeller throughout the entire speed range.

A reliable tachometer must be used to measure engine speed at full throttle to ensure the engine will achieve full horsepower and operate efficiently and safely. To test for the correct propeller, make your run in a body of smooth water with the lower unit in forward gear at full throttle. Observe the tachometer at full throttle.

NEVER run the engine at a high rpm when a flush attachment is installed. If the reading is above the manufacturer's recommended operating range, you must try propellers of greater pitch, until you find the one that allows the engine to operate continually within the recommended full throttle range.

If the engine is unable to deliver top performance and you feel it is properly tuned, then the propeller may not be to blame. Operating conditions have a marked effect on performance. For instance, an engine will lose rpm when run in very cold water. It will also lose rpm when run in salt water as compared with fresh water. A hot, low-barometer day will also cause your engine to lose power.

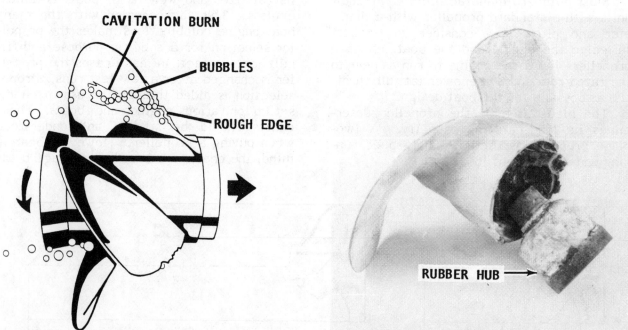

Cavitation (air bubbles) formed at the propeller. Manufacturers are constantly fighting this problem, as explained in the text.

A corroded hub on a small engine propeller. Replacement of this propeller will be less expensive than the cost of a rebuild.

Ventilation

Ventilation is the forming of voids in the water just ahead of the propeller blades. Marine propulsion designers are constantly fighting the battle against the formation of these voids due to excessive blade tip speed and engine wear. The voids may be filled with air or water vapor, or they may actually be a partial vacuum. Ventilation may be caused by installing a piece of equipment too close to the lower unit, such as the knot indicator pickup, depth sounder, or bait tank pickup.

Vibration

Your propeller should be checked regularly to be sure all blades are in good condition. If any of the blades become bent or nicked, this condition will set up vibrations in the drive unit and the motor. If the vibration becomes very serious it will cause a loss of power, efficiency, and boat performance. If the vibration is allowed to continue over a period of time it can have a damaging effect on many of the operating parts.

Vibration in boats can never be completely eliminated, but it can be reduced by keeping all parts in good working condition and through proper maintenance and lubrication. Vibration can also be reduced in some cases by increasing the number of blades. For this reason, many racers use

two-blade props and luxury cruisers have four- and five-blade props installed.

Shock Absorbers

The shock absorber in the propeller plays a very important role in protecting the shafting, gears, and engine against the shock of a blow, should the propeller strike an underwater object. The shock absorber allows the propeller to stop rotating at the instant of impact while the power train continues turning.

How much impact the propeller is able to withstand before causing the clutch hub to slip is calculated to be more than the force needed to propel the boat, but less than the amount that could damage any part of the power train. Under normal propulsion loads of moving the boat through the water, the hub will not slip. However, it will slip if the propeller strikes an object with a force that would be great enough to stop any part of the power train.

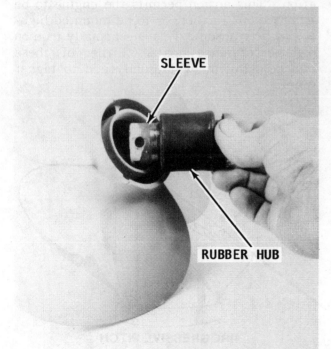

Rubber hub removed from a propeller. This hub was removed because the hub was slipping in the propeller.

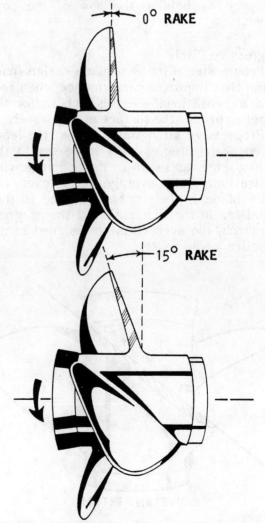

Illustration depicting the rake of a propeller, as explained in the text.

If the power train was to absorb an impact great enough to stop rotation, even for an instant, something would have to give and be damaged. If a propeller is subjected to repeated striking of underwater objects, it would eventually slip on its clutch hub under normal loads. If the propeller would start to slip, a new hub and shock absorber would have to be installed.

Propeller Rake

If a propeller blade is examined on a cut extending directly through the center of the hub, and if the blade is set vertical to the propeller hub, as shown in the accompanying illustration, the propeller is said to have a zero degree (0°) rake. As the blade slants back, the rake increases. Standard propellers have a rake angle from 0° to 15°.

A higher rake angle generally improves propeller performance in a cavitating or ventilating situation. On lighter, faster boats, higher rake often will increase performance by holding the bow of the boat higher.

Progressive Pitch

Progressive pitch is a blade design innovation that improves performance when forward and rotational speed is high and/or the propeller breaks the surface of the water.

Progressive pitch starts low at the leading edge and progressively increases to the trailing edge, as shown in the accompanying illustration. The average pitch over the entire blade is the number assigned to that propeller. In the illustration of the progressive pitch, the average pitch assigned to the propeller would be 21.

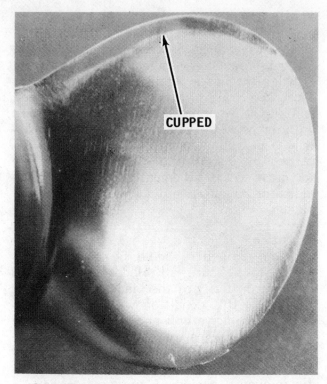

CUPPED

Propeller with a "cupped" leading edge. "Cupping" gives the propeller a better "hold" in the water.

Cupping

If the propeller is cast with a edge curl inward on the trailing edge, the blade is said to have a cup. In most cases, cupped blades improve performance. The cup helps the blades to "HOLD" and not break loose, when operating in a cavitating or ventilating situation. This action permits the engine to be trimmed out further, or to be mounted higher on the transom. This is especially true on high-performance boats. Either of these two adjustments will usually add to higher speed.

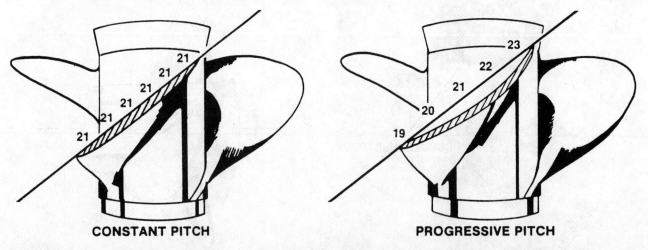

CONSTANT PITCH   PROGRESSIVE PITCH

Comparison of a constant and progressive pitch propeller. Notice how the pitch of the progressive pitch propeller, right, changes to give the blade more thrust and therefore, the boat more speed.

The cup has the effect of adding to the propeller pitch. Cupping usually will reduce full-throttle engine speed about 150 to 300 rpm below the same pitch propeller without a cup to the blade. A propeller repair shop is able to increase or decrease the cup on the blades. This change, as explained, will alter engine rpm to meet specific operating demands. Cups are rapidly becoming standard on propellers.

In order for a cup to be the most effective, the cup should be completely concave (hollowed) and finished with a sharp corner. If the cup has any convex rounding, the effectiveness of the cup will be reduced.

Rotation

Propellers are manufactured as right-hand rotation (RH), and as left-hand rotation (LH). The standard propeller for outboards is RH rotation.

A right-hand propeller can easily be identified by observing it as shown in the accompanying illustration. Observe how the blade slants from the lower left toward the upper right. The left-hand propeller slants in the opposite direction, from upper left to lower right, as shown.

When the propeller is observed rotating from astern the boat, it will be rotating clockwise when the engine is in forward gear. The left-hand propeller will rotate counterclockwise.

Propeller Modification

If poor acceleration is experienced on hard-to-plane boats, OMC suggests a slight modification be performed. The modification involves drilling three 6 mm (7/32")

holes through the outer shell in a precise pattern. The holes allow exhaust gasses to bleed onto the propeller blades causing controlled ventilation during the acceleration period. This action will allow the motor to turn at a higher rpm under acceleration, thus providing more power to plane the boat.

Layout the exact position of each hole 16 ±2mm (5/8" ±1/16") back from the inner lip and the same amount in a **CLOCKWISE** direction from the base of each blade, as shown in the accompanying illustration.

If an inner rib is located under the position of any one of the holes, another propeller must be used. If the holes are properly positioned, and the correct size is drilled, there will be no affect on top speed, maximum rpm, or ventilation in turns. Improper location or size holes will have no affect on performance, particularly in turns.

1-5 FUEL SYSTEM

With Built-in Fuel Tank

All parts of the fuel system should be selected and installed to provide maximum service and protection against leakage. Reinforced flexible sections should be installed in fuel lines where there is a lot of motion, such as at the engine connection. The flaring of copper tubing should be annealed after it is formed as a protection against hardening. **CAUTION:** Compression fittings should **NOT** be used because they are so easily overtightened, which places them under a strain and subjects them to fatigue.

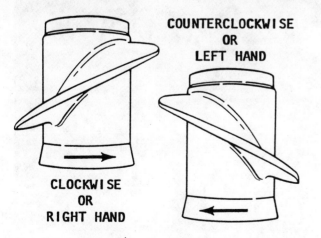

Right- and left-hand propellers showing how the angle of the blades is reversed. Right-hand propellers are by far the most popular.

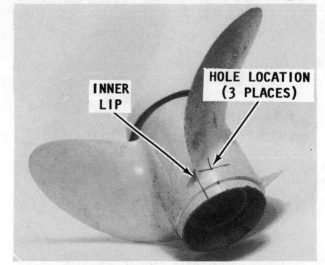

Layout for drilling three holes through the propeller shell for increased performance to provide more power to plane the boat, as explained in the text.

Such conditions will cause the fitting to leak after it is connected a second time.

The capacity of the fuel filter must be large enough to handle the demands of the engine as specified by the engine manufacturer.

A manually-operated valve should be installed if anti-siphon protection is not provided. This valve should be installed in the fuel line as close to the gas tank as possible. Such a valve will maintain anti-siphon protection between the tank and the engine.

The supporting surfaces and hold-downs must fasten the tank firmly and they should be insulated from the tank surfaces. This insulation material should be non-abrasive and non-absorbent material. Fuel tanks installed in the forward portion of the boat should be especially well secured and protected because shock loads in this area can be as high as 20 to 25 g's ("g" equals force of gravity).

Static Electricity

In very simple terms, static electricity is called frictional electricity. It is generated by two dissimilar materials moving over each other. One form is gasoline flowing through a pipe or into the air. Another form is when you brush your hair or walk across a synthetic carpet and then touch a metal object. All of these actions cause an electrical charge. In most cases, static electricity is generated during very dry weather conditions, but when you are filling the fuel tank on a boat it can happen at any time.

Fuel Tank Grounding

One area of protection against the build-up of static electricity is to have the fuel

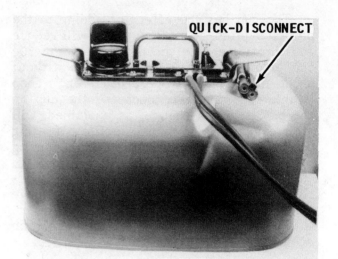

Old style pressure-type tank showing the fuel line to the engine and quick-disconnect fitting.

tank properly grounded (also known as bonding). A direct metal-to-metal contact from the fuel hose nozzle to the water in which the boat is floating. If the fill pipe is made of metal, and the fuel nozzle makes a good contact with the deck plate, then a good ground is made.

As an economy measure, some boats use rubber or plastic filler pipes because of compound bends in the pipe. Such a fill line does not give any kind of ground and if your boat has this type of installation and you do

Adding fuel to a six-gallon OMC fuel tank. Some fuel must be in the tank before oil is added to prevent the oil from accumulating on the tank bottom.

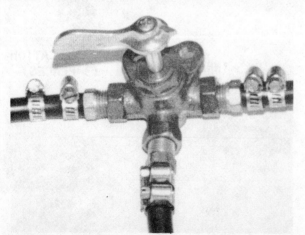

A three-position valve permits fuel to be drawn from either tank or to be shut off completely. Such an arrangement prevents accidental siphoning of fuel from the tank.

not want to replace the filler pipe with a metal one, then it is possible to connect the deck fitting to the tank with a copper wire. The wire should be 8 gauge or larger.

The fuel line from the tank to the engine should provide a continuous metal-to-metal contact for proper grounding. If any part of this line is plastic or other non-metallic material, then a copper wire must be connected to bridge the non-metal material. The power train provides a ground through the engine and drive shaft, to the propeller in the water.

Fiberglass fuel tanks pose problems of their own. One method of grounding is to run a copper wire around the tank from the fill pipe to the fuel line. However, such a wire does not ground the fuel in the tank. Manufacturers should imbed a wire in the fiberglass and it should be connected to the intake and the outlet fittings. This wire would avoid corrosion which could occur if a wire passed through the fuel. **CAUTION: It is not advisable to use a fiberglass fuel tank if a grounding wire was not installed.**

Anything you can feel as a "shock" is enough to set off an explosion. Did you know that under certain atmospheric conditions you can cause a static explosion yourself, particularly if you are wearing synthetic clothing. It is almost a certainty you could cause a static spark if you are **NOT** wearing insulated rubber-soled shoes.

As soon as the deck fitting is opened, fumes are released to the air. Therefore, to be safe you should ground yourself before opening the fill pipe deck fitting. One way to ground yourself is to dip your hand in the water overside to discharge the electricity in your body before opening the filler cap. Another method is to touch the engine block or any metal fitting on the dock which goes down into the water.

1-6 LOADING

In order to receive maximum enjoyment, with safety and performance, from your boat, take care not to exceed the load capacity given by the manufacturer. A plate attached to the hull indicates the U.S. Coast Guard capacity information in pounds for persons and gear. If the plate states the maximum person capacity to be 750 pounds and you assume each person to weigh an average of 150 lbs., then the boat could carry five persons safely. If you add another 250 lbs. for motor and gear, and the maximum weight capacity for persons and gear is 1,000 lbs. or more, then the five persons and gear would be within the limit.

Try to load the boat evenly port and starboard. If you place more weight on one side than on the other, the boat will list to the heavy side and make steering difficult. You will also get better performance by placing heavy supplies aft of the center to keep the bow light for more efficient planing.

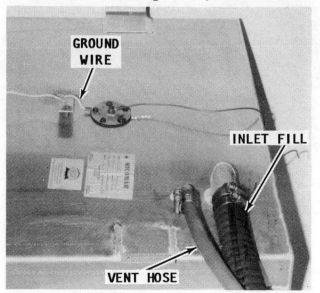

A fuel tank properly grounded to prevent static electricity. Static electricity could be extremely dangerous when taking on fuel.

U.S. Coast Guard plate affixed to all new boats. When the blanks are filled in, the plate will indicate the Coast Guard's recommendations for persons, gear, and horsepower to ensure safe operation of the boat. These recommendations should not be exceeded, as explained in the text.

Clarification

Much confusion arises from the terms, certification, requirements, approval, regulations, etc. Perhaps the following may clarify a couple of these points.

1- The Coast Guard does not approve boats in the same manner as they "Approve" life jackets. The Coast Guard applies a formula to inform the public of what is safe for a particular craft.

2- If a boat has to meet a particular regulation, it must have a Coast Guard certification plate. The public has been led to believe this indicates approval of the Coast Guard. Not so.

3- The certification plate means a willingness of the manufacturer to meet the Coast Guard regulations for that particular craft. The manufacturer may recall a boat if it fails to meet the Coast Guard requirements.

4- The Coast Guard certification plate, see accompanying illustration, may or may not be metal. The plate is a regulation for the manufacturer. It is only a warning plate and the public does not have to adhere to the restrictions set forth on it. Again, the plate sets forth information as to the Coast Guard's opinion for safety on that particular boat.

5- Coast Guard Approved equipment is equipment which has been approved by the Commandant of the U.S. Coast Guard and has been determined to be in compliance with Coast Guard specifications and regulations relating to the materials, construction, and performance of such equipment.

1-7 HORSEPOWER

The maximum horsepower engine for each individual boat should not be increased by any great amount without checking requirements from the Coast Guard in your area. The Coast Guard determines horsepower requirements based on the length, beam, and depth of the hull. **TAKE CARE NOT** to exceed the maximum horsepower listed on the plate or the warranty and possibly the insurance on the boat may become void.

1-8 FLOTATION

If your boat is less than 20 ft. overall, a Coast Guard or BIA (Boating Industry of America) now changed to NMMA (National Marine Manufacturers Association) requirement is that the boat must have buoyant material built into the hull (usually foam) to keep it from sinking if it should become swamped. Coast Guard requirements are mandatory but the NMMA is voluntary.

"Kept from sinking" is defined as the ability of the flotation material to keep the boat from sinking when filled with water

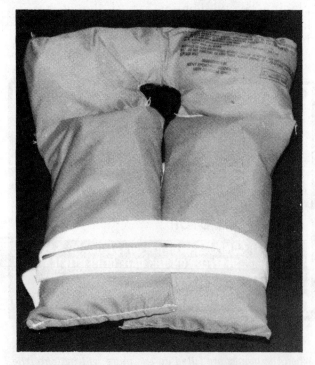

Type I PFD Coast Guard Approved life jacket. This type flotation device provides the greatest amount of buoyancy. **NEVER** *use them for cushions or other purposes.*

A Type IV PFD cushion device intended to be thrown to a person in the water. If air can be squeezed out of the cushion it is no longer fit for service as a PFD.

and with passengers clinging to the hull. One restriction is that the total weight of the motor, passengers, and equipment aboard does not exceed the maximum load capacity listed on the plate.

Life Preservers —Personal Flotation Devices (PFDs)

The Coast Guard requires at least one Coast Guard approved life-saving device be carried on board all motorboats for each person on board. Devices approved are identified by a tag indicating Coast Guard approval. Such devices may be life preservers, buoyant vests, ring buoys, or buoyant cushions. Cushions used for seating are serviceable if air cannot be squeezed out of it. Once air is released when the cushion is squeezed, it is no longer fit as a flotation device. New foam cushions dipped in a rubberized material are almost indestructible.

Life preservers have been classified by the Coast Guard into five type categories. All PFDs presently acceptable on recreational boats fall into one of these five designations. All PFDs **MUST** be U.S. Coast Guard approved, in good and serviceable condition, and of an appropriate size for the persons who intend to wear them. Wearable PFDs **MUST** be readily accessible and throwable devices **MUST** be immediately available for use.

Type I PFD has the greatest required buoyancy and is designed to turn most **UNCONSCIOUS** persons in the water from a face down position to a vertical or slightly backward position. The adult size device provides a minimum buoyancy of 22 pounds and the child size provides a minimum buoyancy of 11 pounds. The Type I PFD provides the greatest protection to its wearer and is most effective for all waters and conditions.

Type II PFD is designed to turn its wearer in a vertical or slightly backward position in the water. The turning action is not as pronounced as with a Type I. The device will not turn as many different type persons under the same conditions as the Type I. An adult size device provides a minimum buoyancy of 15½ pounds, the medium child size provides a minimum of 11 pounds, and the infant and small child sizes provide a minimum buoyancy of 7 pounds.

Type III PFD is designed to permit the wearer to place himself (herself) in a vertical or slightly backward position. The Type III device has the same buoyancy as the Type II PFD but it has little or no turning ability. Many of the Type III PFD are designed to be particularly useful when water skiing, sailing, hunting, fishing, or engaging in other water sports. Several of this type will also provide increased hypothermia protection.

Type IV PFD is designed to be thrown to a person in the water and grasped and held by the user until rescued. It is **NOT** designed to be worn. The most common Type IV PFD is a ring buoy or a buoyant cushion.

Type V PFD is any PFD approved for restricted use.

Coast Guard regulations state, in general terms, that on all boats less than 16 ft. overall, one Type I, II, III, or IV device shall be carried on board for each person in the boat. On boats over 26 ft., one Type I, II, or III device shall be carried on board for each person in the boat **plus** one Type IV device.

It is an accepted fact that most boating people own life preservers, but too few actually wear them. There is little or no excuse for not wearing one because the modern comfortable designs available today do not subtract from an individual's boating pleasure. Make a life jacket available to

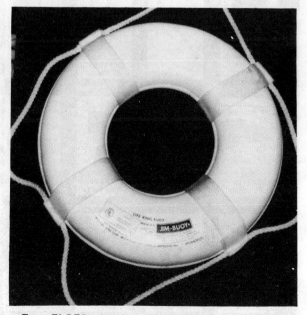

Type IV PFD ring buoy designed to be thrown. On ocean cruisers, this type device usually has a weighted pole with flag, attached to the buoy.

your crew and advise each member to wear it. If you are a crew member ask your skipper to issue you one, especially when boating in rough weather, cold water, or when running at high speed. Naturally, a life jacket should be a must for non-swimmers any time they are out on the water in a boat.

1-9 EMERGENCY EQUIPMENT

Visual Distress Signals
The Regulation

Since January 1, 1981, Coast Guard Regulations require all recreation boats when used on coastal waters, which includes the Great Lakes, the territorial seas and those waters directly connected to the Great Lakes and the territorial seas, up to a point where the waters are less than two miles wide, and boats owned in the United States when operating on the high seas to be equipped with visual distress signals.

The only exceptions are during daytime (sunrise to sunset) for:

Recreational boats less than 16 ft. (5 meters) in length.

Boats participating in organized events such as races, regattas or marine parades.

Open sailboats not equipped with propulsion machinery and less than 26 ft. (8 meters) in length.

Manually propelled boats.

The above listed boats need to carry night signals when used on these waters at night.

Pyrotechnic visual distress signaling devices **MUST** be Coast Guard Approved, in serviceable condition and stowed to be readily accessible. If they are marked with a date showing the serviceable life, this date must not have passed. Launchers, produced before Jan. 1, 1981, intended for use with approved signals are not required to be Coast Guard Approved.

USCG Approved pyrotechnic visual distress signals and associated devices include:

Pyrotechnic red flares, hand held or aerial.

Pyrotechnic orange smoke, hand held or floating.

Launchers for aerial red meteors or parachute flares.

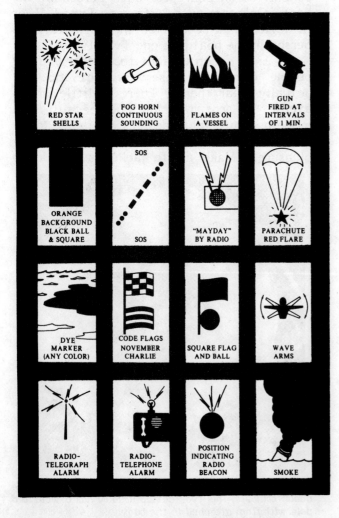

Internationally accepted distress signals.

Moisture-protected flares should be carried on board for use as a distress signal.

Non-pyrotechnic visual distress signaling devices must carry the manufacturer's certification that they meet Coast Guard requirements. They must be in serviceable condition and stowed so as to be readily accessible.

This group includes:

Orange distress flag at least 3 x 3 feet with a black square and ball on an orange background.

Electric distress light -- not a flashlight but an approved electric distress light which **MUST** automatically flash the international **SOS** distress signal (. . . - - . . .) four to six times each minute.

Types and Quantities

The following variety and combination of devices may be carried in order to meet the requirements.

1- Three hand-held red flares (day and night).

2- One electric distress light (night only).

3- One hand-held red flare and two parachute flares (day and night).

4- One hand-held orange smoke signal, two floating orange smoke signals (day) and one electric distress light (day and night).

If young children are frequently aboard your boat, careful selection and proper stowage of visual distress signals becomes especially important. If you elect to carry pyrotechnic devices, you should select those in tough packaging and not easy to ignite should the devices fall into the hands of children.

Coast Guard Approved pyrotechnic devices carry an expiration date. This date can **NOT** exceed 42 months from the date of

manufacture and at such time the device can no longer be counted toward the minimum requirements.

SPECIAL WORDS

In some states the launchers for meteors and parachute flares may be considered a firearm. Therefore, check with your state authorities before acquiring such a launcher.

First Aid Kits

The first-aid kit is similar to an insurance policy or life jacket. You hope you don't have to use it but if needed, you want it there. It is only natural to overlook this essential item because, let's face it, who likes to think of unpleasantness when planning to have only a good time. However, the prudent skipper is prepared ahead of time, and is thus able to handle the emergency without a lot of fuss.

Good commercial first-aid kits are available such as the Johnson and Johnson "Marine First-Aid Kit". With a very modest expenditure, a well-stocked and adequate kit can be prepared at home.

Any kit should include instruments, supplies, and a set of instructions for their use. Instruments should be protected in a watertight case and should include: scissors, tweezers, tourniquet, thermometer, safety

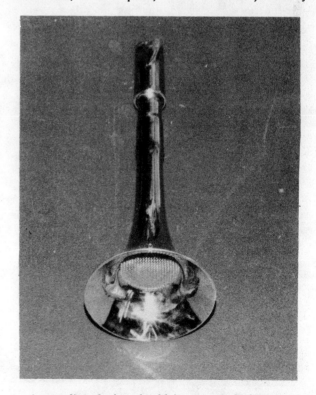

A sounding device should be mounted close to the helmsperson for use in sounding an emergency alarm.

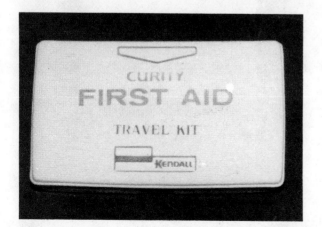

An adequately stocked first-aid kit should be on board for the safety of crew and guests.

pins, eye-washing cup, and a hot water bottle. The supplies in the kit should include: assorted bandages in addition to the various sizes of "band-aids", adhesive tape, absorbent cotton, applicators, petroleum jelly, antiseptic (liquid and ointment), local ointment, aspirin, eye ointment, antihistamine, ammonia inhalent, sea-sickness pills, antacid pills, and a laxative. You may want to consult your family physician about including antibiotics. Be sure your kit contains a first-aid manual because even though you have taken the Red Cross course, you may be the patient and have to rely on an untrained crew for care.

Fire Extinguishers

All fire extinguishers must bear Underwriters Laboratory (UL) "Marine Type" approved labels. With the UL certification, the extinguisher does not have to have a Coast Guard approval number. The Coast Guard classifies fire extinguishers according to their size and type.

Type B-I or B-II Designed for extinguishing flammable liquids. Required on all motorboats.

The Coast Guard considers a boat having one or more of the following conditions as a "boat of closed construction" subject to fire extinguisher regulations.

1- Inboard engine or engines.
2- Closed compartments under thwarts and seats wherein portable fuel tanks may be stored.
3- Double bottoms not sealed to the hull or which are not completely filled with flotation materials.
4- Closed living spaces.
5- Closed stowage compartments in which combustible or flammable material is stored.
6- Permanently installed fuel tanks.

Detailed classification of fire extinguishers is by agent and size:

B-I contains 1-1/4 gallons foam, 4 pounds carbon dioxide, 2 pounds dry chemical, and 2-1/2 pounds freon.

B-II contains 2-1/2 gallons foam, 15 pounds carbon dioxide, and 10 pounds dry chemical.

The class of motorboat dictates how many fire extinguishers are required on board. One B-II unit can be substituted for two B-I extinguishers. When the engine compartment of a motorboat is equipped with a fixed (built-in) extinguishing system, one less portable B-I unit is required.

Dry chemical fire extinguishers without

A suitable fire extinguisher should be mounted close to the helmsman for emergency use.

At least one gallon of emergency fuel should be kept on board in an approved container.

gauges or indicating devices must be weighed and tagged every 6 months. If the gross weight of a carbon dioxide (CO_2) fire extinguisher is reduced by more than 10% of the net weight, the extinguisher is not acceptable and must be recharged.

READ labels on fire extinguishers. If the extinguisher is U.L. listed, it is approved for marine use.

DOUBLE the number of fire extinguishers recommended by the Coast Guard, because their requirements are a bare **MINIMUM** for safe operation. Your boat, family, and crew, must certainly be worth much more than "bare minimum".

1-10 COMPASS

Selection

The safety of the boat and her crew may depend on her compass. In many areas weather conditions can change so rapidly that within minutes a skipper may find himself "socked-in" by a fog bank, a rain squall, or just poor visibility. Under these conditions, he may have no other means of keeping to his desired course except with the compass. When crossing an open body of water, his compass may be the only means of making an accurate landfall.

During thick weather when you can neither see nor hear the expected aids to navigation, attempting to run out the time on a given course can disrupt the pleasure of the cruise. The skipper gains little comfort in a chain of soundings that does not match those given on the chart for the expected area. Any stranding, even for a short time, can be an unnerving experience.

A pilot will not knowingly accept a cheap parachute. A good boater should not accept a bargain in lifejackets, fire extinguishers, or compass. Take the time and spend the few extra dollars to purchase a compass to fit your expected needs. Regardless of what the salesman may tell you, postpone buying until you have had the chance to check more than one make and model.

Lift each compass, tilt and turn it, simulating expected motions of the boat. The compass card should have a smooth and stable reaction.

The card of a good quality compass will come to rest without oscillations about the lubber's line. Reasonable movement in your hand, comparable to the rolling and pitching

The compass is a delicate instrument and deserves respect. It should be mounted securely and in position where it can be easily observed by the helmsman.

of the boat, should not materially affect the reading.

Installation

Proper installation of the compass does not happen by accident. Make a critical check of the proposed location to be sure compass placement will permit the helmsman to use it with comfort and accuracy. First, the compass should be placed directly in front of the helmsman and in such a position that it can be viewed without body stress as he sits or stands in a posture of relaxed alertness. The compass should be in the helmsman's zone of comfort. If the compass is too far away, he may have to bend forward to watch it; too close and he must rear backward for relief.

Do not hesitate to spend a few extra dollars for a good reliable compass. If in doubt, seek advice from fellow boaters.

Second, give some thought to comfort in heavy weather and poor visibilty conditions during the day and night. In some cases, the compass position may be partially determined by the location of the wheel, shift lever, and throttle handle.

Third, inspect the compass site to be sure the instrument will be at least two feet from any engine indicators, bilge vapor detectors, magnetic instruments, or any steel

"Innocent" objects close to the compass, such as diet coke in an aluminum can, may cause serious problems and lead to disaster, as these three photos and the accompanying text prove.

or iron objects. If the compass cannot be placed at least two feet (six feet would be better) from one of these influences, then either the compass or the other object must be moved, if first order accuracy is to be expected.

Once the compass location appears to be satisfactory, give the compass a test before installation. Hidden influences may be concealed under the cabin top, forward of the cabin aft bulkhead, within the cockpit ceiling, or in a wood-covered stanchion.

Move the compass around in the area of the proposed location. Keep an eye on the card. A magnetic influence is the only thing that will make the card turn. You can quickly find any such influence with the compass. If the influence can not be moved away or replaced by one of non-magnetic material, test to determine whether it is merely magnetic, a small piece of iron or steel, or some magnetized steel. Bring the north pole of the compass near the object, then shift and bring the south pole near it. Both the north and south poles will be attracted if the compass is demagnetized. If the object attracts one pole and repels the other, then the compass is magnetized. If your compass needs to be demagnetized, take it to a shop equipped to do the job **PROPERLY.**

After you have moved the compass around in the proposed mounting area, hold it down or tape it in position. Test everything you feel might affect the compass and cause a deviation from a true reading. Rotate the wheel from hard over to hard over. Switch on and off all the lights, radios, radio direction finder, radio telephone, depth finder and the shipboard intercom, if one is installed. Sound the electric whistle, turn on the windshield wipers, start the engine (with water circulating through the engine), work the throttle, and move the gear shift lever. If the boat has an auxiliary generator, start it.

If the card moves during any one of these tests, the compass should be relocated. Naturally, if something like the windshield wipers cause a slight deviation, it may be necessary for you to make a different deviation table to use only when certain pieces of equipment are operating. Bear in mind, following a course that is only off a degree or two for several hours can make considerable difference at the end, putting you on a reef, rock, or shoal.

Check to be sure the intended compass site is solid. Vibration will increase pivot wear.

Now, you are ready to mount the compass. To prevent an error on all courses, the line through the lubber line and the compass card pivot must be exactly parallel to the keel of the boat. You can establish the fore-and-aft line of the boat with a stout cord or string. Use care to transfer this line to the compass site. If necessary, shim the base of the compass until the stile-type lubber line (the one affixed to the case and not gimbaled) is vertical when the boat is on an even keel. Drill the holes and mount the compass.

Magnetic Items After Installation

Many times an owner will install an expensive stereo system in the cabin of his boat. It is not uncommon for the speakers to be mounted on the aft bulkhead up against the overhead (ceiling). In almost every case, this position places one of the speakers in very close proximity to the compass, mounted above the ceiling.

As we all know, a magnet is used in the operation of the speaker. Therefore, it is very likely that the speaker, mounted almost under the compass in the cabin will have a very pronounced affect on the compass accuracy.

Consider the following test and the accompanying photographs as prove of the statements made.

First, the compass was read as 190 degrees while the boat was secure in her slip.

Next a full can of diet coke in an **aluminum** can was placed on one side and the compass read as 204 degrees, a good 14 degrees off.

Next, the full can was moved to the opposite side of the compass and again a reading was observed. This time as 189 degrees, 11 degrees off from the original reading.

Finally the contents of the can were consumed, the can placed on both sides of the compass with **NO** affect on the compass reading.

Two very important conclusions can be drawn from these tests.

1- Something must have been in the contents of the can to affect the compass so drastically.

2- Keep even "innocent" things clear of the compass to avoid any possible error in the boat's heading.

REMEMBER, a boat moving through the water at 10 knots on a compass error of just 5 degrees will be almost 1.5 miles off course in only **ONE** hour. At night, or in thick weather, this could very possibly put the boat on a reef, rock, or shoal, with disastrous results.

1-11 STEERING

USCG or BIA certification of a steering system means that all materials, equipment, and installation of the steering parts meet or exceed specific standards for strength, type, and maneuverability. Avoid sharp bends when routing the cable. Check to be sure the pulleys turn freely and all fittings are secure.

1-12 ANCHORS

One of the most important pieces of equipment in the boat next to the power plant is the ground tackle carried. The engine makes the boat go and the anchor and its line are what hold it in place when the boat is not secured to a dock or on the beach.

*The weight of the anchor **MUST** be adequate to secure the boat without dragging.*

The anchor must be of suitable size, type, and weight to give the skipper peace of mind when his boat is at anchor. Under certain conditions, a second, smaller, lighter anchor may help to keep the boat in a favorable position during a non-emergency daytime situation.

In order for the anchor to hold properly, a piece of chain must be attached to the anchor and then the nylon anchor line attached to the chain. The amount of chain should equal or exceed the length of the boat. Such a piece of chain will ensure that the anchor stock will lay in an approximate horizontal position and permit the flutes to dig into the bottom and hold.

1-13 MISCELLANEOUS EQUIPMENT

In addition to the equipment you are legally required to carry in the boat and those previously mentioned, some extra items will add to your boating pleasure and safety. Practical suggestions would include: a bailing device (bucket, pump, etc.), boat

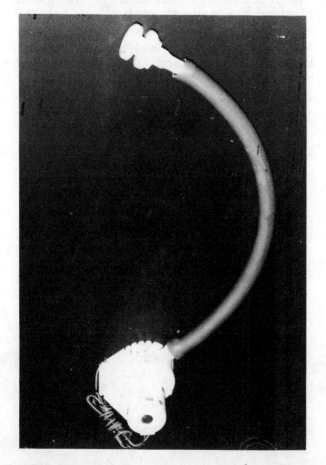

The bilge pump line must be cleaned frequently to ensure the entire bilge pump system will function properly in an emergency.

hook, fenders, spare propeller, spare engine parts, tools, an auxiliary means of propulsion (paddle or oars), spare can of gasoline, flashlight, and extra warm clothing. The area of your boating activity, weather conditions, length of stay aboard your boat, and the specific purpose will all contribute to the kind and amount of stores you put aboard. When it comes to personal gear, heed the advice of veteran boaters who say, "Decide on how little you think you can get by with, then cut it in half".

Bilge Pumps

Automatic bilge pumps should be equipped with an overriding manual switch. They should also have an indicator in the operator's position to advise the helmsman when the pump is operating. Select a pump that will stabilize its temperature within the manufacturer's specified limits when it is operated continuously. The pump motor should be a sealed or arcless type, suitable for a marine atmosphere. Place the bilge pump inlets so excess bilge water can be removed at all normal boat trims. The intakes should be properly screened to prevent the pump from sucking up debris from the bilge. Intake tubing should be of a high quality and stiff enough to resist kinking and not collapse under maximum pump suction condition if the intake becomes blocked.

To test operation of the bilge pump, operate the pump switch. If the motor does not run, disconnect the leads to the motor. Connect a voltmeter to the leads and see if voltage is indicated. If voltage is not indicated, then the problem must be in a blown fuse, defective switch, or some other area of the electrical system.

If the meter indicates voltage is present at the leads, then remove, disassemble, and inspect the bilge pump. Clean it, reassemble, connect the leads, and operate the switch again. If the motor still fails to run, the pump must be replaced.

To test the bilge pump switch, first disconnect the leads from the pump and connect them to a test light or ohmmeter. Next, hold the switch firmly against the mounting location in order to make a good ground. Now, tilt the opposite end of the switch upward until it is activated as indicated by the test light coming on or the ohmmeter showing continuity. Finally, lower the switch slowly toward the mounting position until it is deactivated. Measure the distance between the point the switch was activated and the point it was deactivated. For proper service, the switch should deactivate between 1/2-inch and 1/4-inch from the planned mounting position. **CAUTION: The switch must never be mounted lower than the bilge pump pickup.**

1-14 BOATING ACCIDENT REPORTS

New federal and state regulations require an accident report to be filed with the nearest State boating authority within 48 hours if a person is lost, disappears, or is injured to the degree of needing medical treatment beyond first aid.

Accidents involving only property or equipment damage **MUST** be reported within 10 days if the damage is in excess of $200. Some States require reporting of accidents with property damage less than $200 or total boat loss.

A **$500 PENALTY** may be asessed for failure to submit the report.

WORD OF ADVICE

Take time to make a copy of the report to keep for your records or for the insurance company. Once the report is filed, the Coast Guard will not give out a copy, even to the person who filed the report.

The report must give details of the accident and include:

1- The date, time, and exact location of the occurrence.

2- The name of each person who died, was lost, or injured.

3- The number and name of the vessel.

4- The names and addresses of the owner and operator.

If the operator cannot file the report for any reason, each person on board **MUST** notify the authorities, or determine that the report has been filed.

1-15 NAVIGATION

Buoys

In the United States, a buoyage system is used as an assist to all boaters of all size craft to navigate our coastal waters and our navigable rivers in safety. When properly read and understood, these buoys and markers will permit the boater to cruise with comparative confidence and enable

her/him to avoid reefs, rocks, shoals, and other hazards.

In the spring of 1983, the Coast Guard began making modifications to U.S. aids to navigation in support of an agreement sponsored by the International Association of Lighthouse Authorities (IALA) and signed by representatives from most of the maritime nations of the world. The primary purpose of the modifications is to improve safety by making buoyage systems around the world more alike and less confusing.

The modifications shown in the accompanying illustrations on the next page, should have been completed by 1991.

Lights

All boats are required to show lights at night. The kind and number vary with the type and size of the boat. Inland and Great Lakes Rules require that powerboats under 26 ft. carry a combination red (port) and green (starboard) 20-pt. light, visible at least a mile, and a white 32-pt. (all-around) stern light, visible for 2 miles. Powerboats 26 to 65 ft. must carry a white 32-pt. red light astern, a white 20-pt. light forward, and separate 10-pt. red and green side lights.

Hookup for testing an automatic bilge pump switch.

International rules require powerboats under 40 gross tons to carry a white 12-pt. stern light visible for 2 miles, a white 20-pt. forward light, visible at 3 miles, and a combination red-and-green 20-pt. bow light, visible for 1 mile (or separate 10-pt. red and green bow lights). When anchored, a white 32-pt. light is required.

Waterway Rules

On the water, certain basic safe-operating practices must be followed. You should learn and practice them, for to **know,** is to be able to handle your boat with confidence and safety. Knowledge of what to do and not do will add a great deal to the enjoyment you will receive from your boating investment.

In the rules that follow, the terms "port" and "starboard" are used to refer to the left and right side of the boat. One easy way to remember this basic fundamental is to consider the words "port" and "left" both have four letters and go together.

Powered boats must yield the right-of-way to all boats without motors, except when being overtaken. When meeting another boat head-on, keep to starboard, unless you are too far to port to make this practical. When overtaking another boat, the right-of-way belongs to the boat being overtaken. If your boat is being passed, you must maintain course and speed.

When two boats approach at an angle and there is danger of collision, the boat to port must give way to the boat to starboard. Always keep to starboard in a narrow channel or canal. Boats underway must stay clear of vessels fishing with nets, lines, or trawls. (Fishing boats are not allowed to fish in channels or to obstruct navigation.)

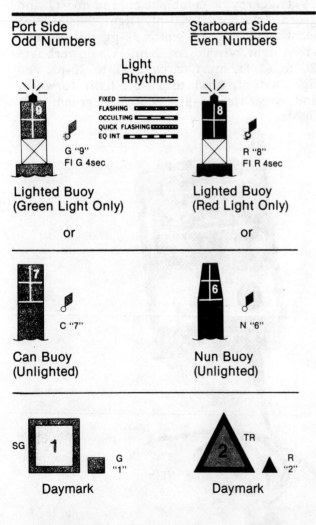

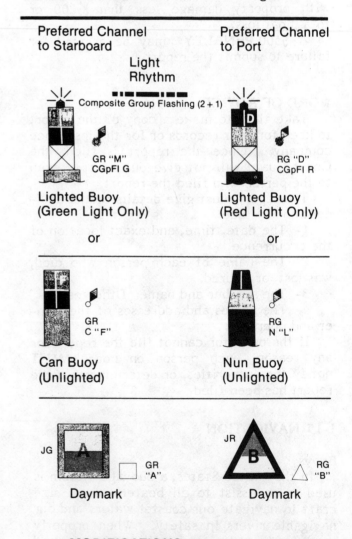

MODIFICATIONS: Port hand aids will be green with green lights. All starboard hand aids will have red lights.

MODIFICATIONS: Green will replace black. Light rhythm will be changed to Composite Gp Fl (2 + 1).

2
TUNING

2-1 INTRODUCTION

The efficiency, reliability, fuel economy and enjoyment available from engine performance are all directly dependent on having it tuned properly. The importance of performing service work in the sequence detailed in this chapter cannot be over emphasized. Before making any adjustments, check the Specifications in the Appendix. **NEVER** rely on memory when making critical adjustments.

Before beginning to tune any engine, check to be sure the engine has satisfactory compression. An engine with worn or broken piston rings, burned pistons, or badly scored cylinder walls, cannot be made to perform properly no matter how much time and expense is spent on the tune-up. Poor compression must be corrected or the tune-up will not give the desired results.

The opposite of poor compression would be to consider good compression as evidence of a satisfactory cylinder. However, this is not necessarily the case, when working on an outboard engine. As the professional mechanic has discovered, many times the compression check will indicate a satisfactory cylinder, but after the head is pulled

Damaged piston, probably caused by inaccurate fuel mixture, or improper point setting.

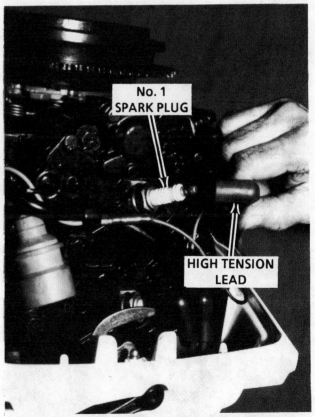

No. 1 SPARK PLUG

HIGH TENSION LEAD

Removing the high tension lead from the No. 1 spark plug in preparation to removing the plug for a compression test. The high tension lead should be grounded.

and an inspection made, the cylinder will require service.

A practical maintenance program that is followed throughout the year, is one of the best methods of ensuring the engine will give satisfactory performance at any time.

The extent of the engine tune-up is usually dependent on the time lapse since the last service. A complete tune-up of the entire engine would entail almost all of the work outlined in this manual. A logical sequence of steps will be presented in general terms. If additional information or detailed service work is required, the chapter containing the instructions will be referenced.

Each year higher compression ratios are built into modern outboard engines and the electrical systems become more complex, especially with electronic (capacitor discharge) units. Therefore, the need for reliable, authoritative, and detailed instructions becomes more critical. The information in this chapter and the referenced chapters fulfill that requirement.

2-2 TUNE-UP SEQUENCE

If twenty different mechanics were asked the question, "What constitutes a major and minor tune-up?", it is entirely possible twenty different answers would be given. As the terms are used in this manual and other Seloc outboard books, the following work is normally performed for a minor and major tune-up.

Minor Tune-up
Lubricate engine.
Drain and replace gear oil.
Adjust points.
Adjust carburetor.
Clean exterior surface of engine.
Tank test engine for fine adjustments.

Major Tune-up
Remove head.
Clean carbon from pistons and cylinders.
Clean and overhaul carburetor.
Clean and overhaul fuel pump.
Rebuild and adjust ignition system.
Lubricate engine.
Drain and replace gear oil.
Clean exterior surface of engine.
Tank test engine for fine adjustments.

During a major tune-up, a definite sequence of service work should be followed to return the engine to the maximum performance desired. This type of work should not be confused with attempting to locate problem areas of "why" the engine is not performing satisfactorily. This work is classified as "troubleshooting". In many cases, these two areas will overlap, because many times a minor or major tune-up will correct the malfunction and return the system to normal operation.

The time, effort, and expense of a tune-up will not restore an engine to satisfactory performance, if the pistons are damaged.

A boat and lower unit covered with marine growth. Such a condition is a serious hinderance to satisfactory performance.

The following list is a suggested sequence of tasks to perform during the tune-up service work. The tasks are merely listed here. Generally procedures are given in subsequent sections of this chapter. For more detailed instructions, see the referenced chapter.

1- Perform a compression check of each cylinder, see next section.

2- Inspect the spark plugs to determine their condition. Test for adequate spark at the plug, see Section 2-4.

3- Start the engine in a body of water and check the water flow through the engine. See Chapter 8.

4- Check the gear oil in the lower unit. See Chapter 8.

5- Check the carburetor adjustments and the need for an overhaul. See Chapter 4.

6- Check the fuel pump for adequate performance and delivery. See Chapter 4.

7- Make a general inspection of the ignition system. See Chapter 5.

8- Test the starter motor and the solenoid. See Chapter 7.

9- Check the internal wiring.

10- Check the timing and synchronization. See Chapter 5.

Removing the spark plugs in preparation to making a compression test. A good safety practice is to ground the high-tension leads to prevent any possibility of a spark igniting fuel or fuel vapors when the powerhead is cranked.

2-3 COMPRESSION CHECK

A compression check is extremely important, because an engine with low or uneven compression between cylinders **CANNOT** be tuned to operate satisfactorily. Therefore, it is essential that any compression problem be corrected before proceeding with the tune-up procedure. See Chapter 3.

If the powerhead shows any indication of overheating, such as discolored or scorched paint, especially in the area of the top (No. 1) cylinder, inspect the cylinders visually thru the transfer ports for possible scoring. A more thorough inspection can be made if the head is removed. It is possible for a cylinder with satisfactory compression to be scored slightly. Also, check the water pump. The overheating condition may be caused by a faulty water pump.

An overheating condition may also be caused by running the engine out of the water. For unknown reasons, many operators have formed a bad habit of running a small engine without the lower unit being submerged. Such a practice will result in an overheated condition in a matter of seconds. It is interesting to note, the same operator would never operate or allow anyone else to run a large horsepower engine without water circulating through the lower unit for cooling. Bear-in-mind, the laws governing operation and damage to a large unit **ALL** apply equally as well to the small engine.

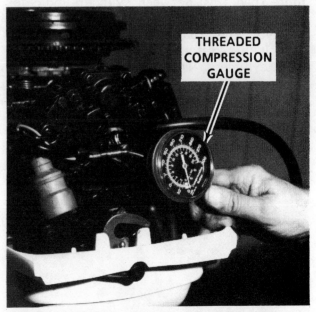

THREADED COMPRESSION GAUGE

Using a threaded compression gauge to check compression in the cylinders. The reading should be close to 100 psi (689.5 kPa). The difference between cylinders should not be over 15 psi (103.4 kPa).

Damaged spark plugs. Notice the broken electrode on the left plug. The broken part must be found and removed before returning the engine to service.

Checking Compression

Remove the spark plug wires. **ALWAYS** grasp the molded cap and pull it loose with a twisting motion to prevent damage to the connection. Remove the spark plugs and keep them in **ORDER** by cylinder for evaluation later. Ground the spark plug leads to the engine to render the ignition system inoperative while performing the compression check.

Insert a compression gauge into the No. 1, top, spark plug opening. Crank the engine with the starter, or pull on the starter· cord, thru at least 4 complete strokes with the throttle at the wide-open position, or until the highest possible reading is observed on the gauge. Record the reading. Repeat the test and record the compression for each cylinder. A variation between cylinders is far more important than the actual readings. A variation of more than 5 psi between cylinders indicates the lower compression cylinder may be defective. The problem may be worn, broken, or sticking piston rings, scored pistons or worn cylinders. These problems may only be

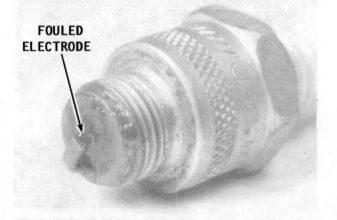

A fouled spark plug. The condition of this plug indicates problems in the air/fuel mixture or the amount of oil added to the mixture.

determined after the head has been removed. Removing the head on an outboard engine is not that big a deal and may save many hours of frustration and the cost of purchasing unnecessary parts to correct a faulty condition.

2-4 SPARK PLUG INSPECTION

Inspect each spark plug for badly worn electrodes, glazed, broken, blistered, or lead fouled insulators. Replace all of the plugs, if one shows signs of excessive wear.

Make an evaluation of the cylinder performance by comparing the spark condition with those shown in Chapter 5. Check each spark plug to be sure they are all of the same manufacturer and have the same heat range rating.

Inspect the threads in the spark plug opening of the head and clean the threads before installing the plug. If the threads are damaged, the head should be removed and and a Heli-coil insert installed. If an attempt is made to drill out the opening with the head in place, some of the filings may fall into the cylinder and cause damage to the cylinder wall during operation. Because the head is made of aluminum, the filings cannot be removed with a magnet.

When purchasing new spark plugs, **ALWAYS** ask the marine dealer if there has been a spark plug change for the engine being serviced.

Crank the engine through several revolutions to blow out any material which might have become dislodged during cleaning.

Install the spark plugs and tighten them to the proper torque value. **ALWAYS** use a new gasket and wipe the seats in the block clean. The gasket must be fully compressed

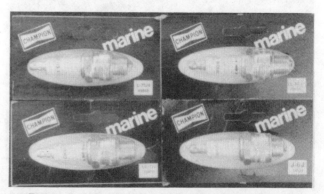

*Today, numerous type spark plugs are available for service. **ALWAYS** check with your local marine dealer to be sure you are purchasing the proper plugs for the engine being serviced.*

Worn ignition points are a common problem area contributing to poor engine performance.

on clean seats to complete the heat transfer process and to provide a gas tight seal in the cylinder. If the torque value is too high, the heat will dissipate too rapidly. Conversely, if the torque value is too low, heat will not dissipate fast enough.

2-5 IGNITION SYSTEM

Three different ignition systems are used on powerheads covered in this manual: A flywheel magneto; a capacitor discharge (CD) system with a sensor coil: and a UFI (Under Flywheel Ignition) -- a CD system with an ignition module.

If powerhead performance is less than expected, and the ignition is diagnosed as the problem area, refer to Chapter 5 for detailed service procedures. To properly synchronize the ignition system with the fuel system, see appropriate Section in Chapter 5.

Typical cam and follower arrangement. The fuel and ignition systems on any powerhead MUST be properly synchronized before maximum performance can be obtained from the unit.

Breaker Points

Powerheads equipped with the flywheel magneto system utilize breaker points. Breaker points are **NOT** used in any of the capacitor discharge (CD) ignition systems.

Rough or discolored contact surfaces are sufficient reason for replacement. The cam follower will usually have worn away by the time the points have become unsatisfactory for efficient service.

Check the resistance across the contacts. If the test indicates **ZERO** resistance, the points are serviceable. A slight resistance across the points will affect idle operation. A high resistance may cause the ignition system to malfunction and loss of spark. Therefore, if any resistance across the points is indicated, the point set should be replaced.

2-6 SYNCHRONIZING

On powerheads equipped with a CD ignition system, the timing is adjustable through synchronization, see Chapter 5. On powerheads equipped with points, the timing is controlled through adjustment of the points. If the points are adjusted too closely, the spark plugs will fire early; if adjusted with an excessive gap -- the plugs will fire too late for efficient operation.

Therefore, correct point adjustment and synchronization are essential for proper engine operation. A powerhead may be in apparent excellent mechanical condition, but perform poorly, unless the points and synchronization have been adjusted precisely, according to the Specifications in the Appendix. To synchronize the engine, see Chapter 5.

The battery MUST be located near the engine in a well-ventilated area. It must be secured in such a manner that absolutely no movement is possible in any direction under the most violent action of the boat.

2-7 BATTERY SERVICE

Many owner/operators are not fully aware of the role a battery performs with a magneto ignition system outboard engine. To clarify: With a magneto ignition system, a battery is only used to crank the engine for starting purposes. Once the engine is running properly, the battery could very well be removed without affecting engine operation. Therefore, if the battery is completely dead, the engine may be hand started with a pull cord and operate efficiently.

If a battery is used for starting, inspect and service the battery, cables and connections. Check for signs of corrosion. Inspect the battery case for cracks or bulges, dirt, acid, and electrolyte leakage. Check the electrolyte level in each cell.

Fill each cell to the proper level with distilled water or water passed thru a demineralizer.

Clean the top of the battery. The top of a 12-volt battery should be kept especially clean of acid film and dirt, because of the high voltage between the battery terminals. For best results, first wash the battery with a diluted ammonia or baking soda solution to

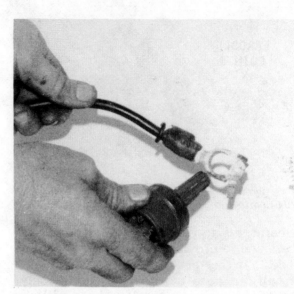

An inexpensive brush should be purchased and used to clean the battery terminals. Clean terminals will ensure a proper connection.

neutralize any acid present. Flush the solution off the battery with clean water. Keep the vent plugs tight to prevent the neutralizing solution or water from entering the cells.

Check to be sure the battery is fastened securely in position. The hold-down device should be tight enough to prevent any movement of the battery in the holder, but not so tight as to place a strain on the battery case.

A check of the electrolyte in the battery should be a regular task on the maintenance schedule on any boat.

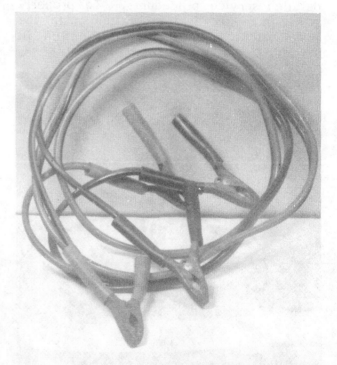

Common set of jumper cables for using a second battery to crank and start the engine. EXTREME care should be used when using a second battery, as explained in the text.

If the battery posts or cable terminals are corroded, the cables should be cleaned separately with a baking soda solution and a wire brush. Apply a thin coating of Multi-purpose Lubricant to the posts and cable clamps before making the connections. The lubricant will help to prevent corrosion.

If the battery has remained under-charged, check for high resistance in the charging circuit. If the battery appears to be using too much water, the battery may be defective, or it may be too small for the job.

Jumper Cables

If booster batteries are used for starting an engine the jumper cables must be connected correctly and in the proper sequence to prevent damage to either battery, or the alternator diodes.

ALWAYS connect a cable from the positive terminals of the dead battery to the positive terminal of the good battery **FIRST**. **NEXT**, connect one end of the other cable to the negative terminals of the good battery and the other end on the **ENGINE** for a good ground. By making the ground connection on the engine, if there is an arc when you make the connection it will not be near the battery. An arc near the battery could cause an explosion, destroying the battery and causing serious personal injury.

DISCONNECT the battery ground cable before replacing an alternator or before connecting any type of meter to the alternator.

If it is necessary to use a fast-charger on a dead battery, **ALWAYS** disconnect one of the boat cables from the battery first, to prevent burning out the diodes in the alternator.

NEVER use a fast charger as a booster to start the engine because the diodes in the alternator will be **DAMAGED**.

Alternator Charging

When the battery is partially discharged, the ammeter should change from discharge to charge between 800 to 1000rpm for all models. If the battery is fully-charged, the rpm will be a little higher.

Typical electric cranking motor installation on the powerheads covered in this manual.

Flywheel removed exposing components under and protected by the flywheel. Major parts are identified.

With the engine running, increase the throttle to approximately 5200 rpm. The ammeter reading should be approximately equal to the amperage rating of the alternator installed. With a fully-charged battery, the ammeter reading will be a bit lower because of the self-regulating characteristics of the generating systems. Before disconnecting the ammeter, reconnect the red harness lead to the positive battery terminal and install the wing nut.

2-8 CARBURETOR ADJUSTMENTS

Fuel and Fuel Tanks

Take time to check the fuel tank and all of the fuel lines, fittings, couplings, valves, flexible tank fill and vent. Turn on the fuel supply valve at the tank, if the engine is equipped with a self-contained fuel tank. If the gas was not drained at the end of the previous season, make a careful inspection for gum formation. When gasoline is allowed to stand for long periods of time, particularly in the presence of copper, gummy deposits form. This gum can clog the filters, lines, and passageway in the carburetor.

If the condition of the fuel is in doubt, drain, clean, and fill the tank with fresh fuel.

Fuel pressure at the carburetor should be checked whenever a lack of fuel volume at the carburetor is suspected.

OIL/FUEL MIXTURE

Regular Mixture

Use 50:1 oil mixture for all powerheads covered in this manaul

Break-in Mixture

Powerheads -- 2hp thru 50 hp: Use 25:1 oil mixture (two pints oil per six gallons of fuel), for first 5 hours, or using twelve gallons of fuel.

After Break-in

Use 50:1 oil mixture (one pint of oil per six gallons of fuel).

Acceptable Products

ALWAYS use OMC or BIA certified TC-WII oil lubricant.

NEVER use automotive oils, premixed fuel of unknown oil quantity, or premixed fuel richer than 50:1.

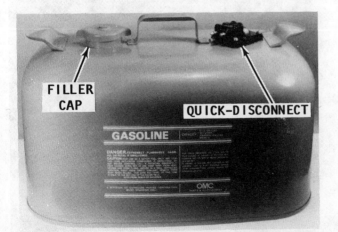

An OMC six-gallon fuel tank with the fuel line connected through a quick-disconnect fitting. Such a fitting is handy when the tank is removed from the boat for filling.

Typical fuel pump installation -- this one on a 25hp powerhead.

Idle Adjustment

Due to local conditions, it may be necessary to adjust the carburetor while the engine is running in a test tank or with the boat in a body of water. For maximum performance, the idle mixture and the idle rpm should be adjusted under actual operating conditions.

Set the idle mixture screw at the specified number of turns open from a lightly seated position. In most cases this is from 1 to 1½ turns open from close.

Start the engine and allow it to warm to operating temperature.

CAUTION: Water must circulate through the lower unit to the engine any time the engine is run to prevent damage to the water pump in the lower unit. Just five seconds without water will damage the water pump.

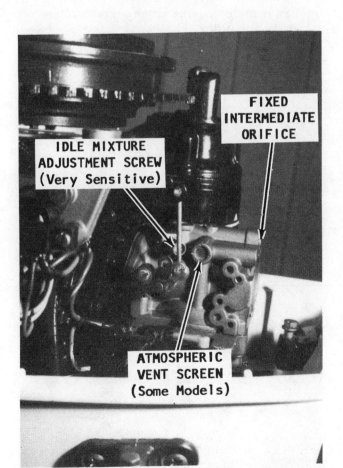

Carburetor installation on a Model 25 powerhead showing location of the fixed intermediate orifice, the idle mixture adjustment screw, and the atmospheric vent screw, which may be used on some models.

NEVER, AGAIN NEVER, operate the engine at high speed with a flush device attached. The engine, operating at high speed with such a device attached, would **RUN-A-WAY** from lack of a load on the propeller, causing extensive damage.

With the engine running in forward gear, slowly turn the idle mixture screw **COUNTERCLOCKWISE** until the affected cylinders start to load up or fire unevenly, due to an over-rich mixture. Slowly turn the idle mixture screw **CLOCKWISE** until the cylinders fire evenly and engine rpm increases. Continue to slowly turn the screw **CLOCKWISE** until too lean a mixture is obtained and the rpms fall off and the engine begins to misfire. Now, set the idle mixture screw one-quarter (1/4) turn out (counterclockwise) from the lean-out position. This adjustment will result in an approximate true setting. A too-lean setting is a major cause of hard starting a cold engine. It is better to have the adjustment on the rich side rather than on the lean side. Stating it another way, do not make the adjustment any leaner than necessary to obtain a smooth idle.

If the engine hesitates during acceleration after adjusting the idle mixture, the mixture is too lean. Enrich the mixture slightly, by turning the adjustment screw inward until the engine accelerates correctly.

With the engine running in forward gear, rotate the nylon idle adjustment screw, located on the portside of the engine, until the engine idles at the recommended rpm, as given in the Specifications in the Appendix. This idle adjustment screw is always exposed on the outside of the shroud.

Repairs and Adjustments

For detailed procedures to disassemble, clean, assemble, and adjust the carburetor, see the appropriate section in Chapter 4 for the carburetor type on the engine being serviced.

2-9 FUEL PUMPS

Many times, a defective fuel pump diaphragm is mistakenly diagnosed as a problem in the ignition system. The most common problem is a tiny pin-hole in the diaphragm. Such a small hole will permit gas to enter the crankcase and wet foul the spark plug at idle-speed. During high-speed

SCREEN

GASKET

Two of the many types of fuel pumps installed on OMC outboard engines. These fuel pumps cannot be rebuilt, as explained in the text.

operation, gas quantity is limited, the plug is not foul and will therefore fire in a satisfactory manner.

If the fuel pump fails to perform properly, an insufficient fuel supply will be delivered to the carburetor. This lack of fuel will cause the engine to run lean, lose rpm or cause piston scoring.

When a fuel pressure gauge is added to the system, it should be installed at the end of the fuel line leading to the upper carburetor. To ensure maximum performance, the fuel pressure must be 2 psi or more at full throttle.

Tune-up Task

Most fuel pumps are equipped with a fuel filter. The filter may be cleaned by first removing the cap, then the filter element, cleaning the parts and drying them with compressed air, and finally installing them in their original position.

A fuel pump pressure test should be made any time the engine fails to perform satisfactorily at high speed.

NEVER use liquid Neoprene on fuel line fittings. Always use Permatex when making fuel line connections. Permatex is available at almost all marine and hardware stores.

Only one Johnson/Evinrude fuel pump may be rebuilt, see accompanying illustration. All others pumps must be replaced as a unit. For fuel pump service, see Chapter 4.

2-10 STARTER AND SOLENOID

Starter Motor Test

Check to be sure the battery has a 70-ampere rating and is fully charged. Would you believe, many starter motors are needlessly disassembled, when the battery is actually the culprit.

Lubricate the pinion gear and screw shaft with No. 10 oil.

Commercial additives, such as Sta-Bil, may be used to keep gasoline in the fuel tank fresh. Under favorable conditions, such additives will prevent the fuel from "souring" for up to twelve full months.

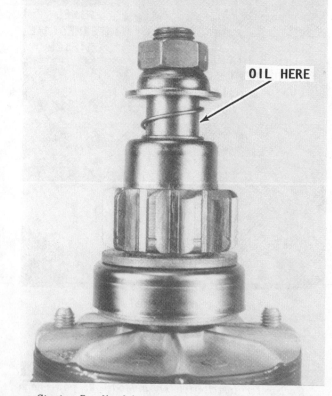

OIL HERE

Starter Bendix drive mechanism. Good maintenance practices should include just a drop of 10-weight oil periodically to the area shown.

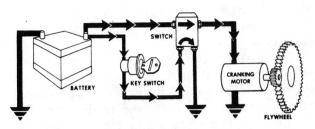

Functional diagram of a typical cranking circuit.

Connect one lead of a voltmeter to the positive terminal of the starter motor. Connect the other meter lead to a good ground on the engine. Check the battery voltage under load by turning the ignition switch to the **START** position and observing the voltmeter reading.

If the reading is 9-1/2 volts or greater, and the starter motor fails to operate, repair or replace the starter motor. See Chapter 7.

Solenoid Test

An ohmmeter is the only instrument required to effectively test a solenoid. Test the ohmmeter by connecting the red and black leads together. Adjust the pointer to the right side of the scale.

On all Johnson/Evinrude engines the case of the solenoid does **NOT** provide a suitable ground to the engine. Hundreds of solenoids

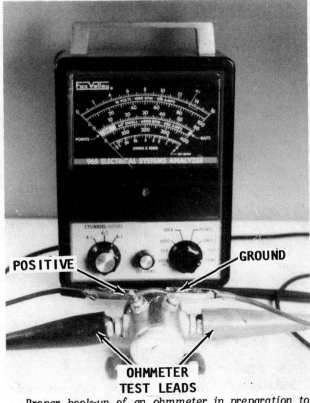

POSITIVE   GROUND

OHMMETER
TEST LEADS

Proper hook-up of an ohmmeter in preparation to testing a starter solenoid.

have been discarded because of the erroneous belief the case is providing a ground and the unit should function when 12-volts is applied. Not so! One terminal of the solenoid is connected to a 12-volt source. The other terminal is connected via a white wire to a cutout switch on top of the engine. This cutout switch provides a safety to break the ground to the solenoid in the event the engine starts at a high rpm. Therefore, the solenoid ground is made and broken by the cutout switch.

NEVER connect the battery leads to the large terminals of the solenoid, or the test meter will be damaged. Connect each lead of the test meter to each of the large terminals on the solenoid.

Using battery jumper leads, connect the positive lead from the positive terminal of the battery to the small **"S"** terminal of the solenoid. Connect the negative lead to the **"I"** terminal of the solenoid. Connect the other end of the jumper lead to the negative battery terminal. If the meter indicates continuity, the solenoid is serviceable. If the meter fails to indicate continuity, the solenoid must be replaced.

2-11 INTERNAL WIRING HARNESS

An internal wiring harness is only used on the larger horsepower engines covered in this manual. If the engine is equipped with a wiring harness, the following checks and test will apply.

Check the internal wiring harness if problems have been encountered with any of the electrical components. Check for frayed or chafed insulation and/or loose connections between wires and terminal connections.

Check the harness connector for signs of corrosion. If the harness shows any evidence of damage or corrosion, the problem must be corrected before proceeding with any harness testing.

Convince yourself a good electrical connection is being made between the harness connector and the remote control harness.

2-12 WATER PUMP CHECK

FIRST A GOOD WORD: The water pump **MUST** be in very good condition for the engine to deliver satisfactory service. The pump performs an extremely important function by supplying enough water to properly cool the engine. Therefore, in most

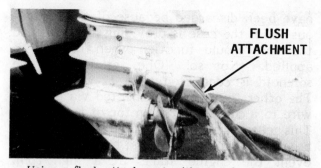

Using a flush attachment with a garden hose hookup to clean the engine water circulation system with fresh water. This arrangement may also be used while operating the engine at idle speeds to make adjustments.

cases, it is advisable to replace the complete water pump assembly at least once a year, or anytime the lower unit is disassembled for service.

Sometimes during adjustment procedures, it is necessary to run the engine with a flush device attached to the lower unit. **NEVER** operate the engine over 1000 rpm with a flush device attached, because the engine may **"RUNAWAY"** due to the no-load condition on the propeller. A "runaway" engine could be severely damaged. As the name implies, the flush device is primarily used to flush the engine after use in salt water or contaminated fresh water. Regular use of the flush device will prevent salt or silt deposits from accumulating in the water passageway. During and immediately after flushing, keep the motor in an upright position until all of the water has drained from the drive shaft housing. This will prevent water from entering the power head by way of the drive shaft housing and the exhaust ports, during the flush. It will

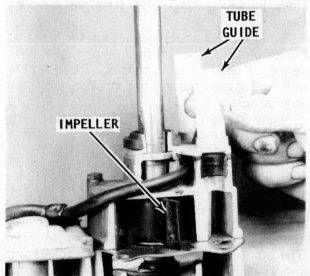

Water pump installed on a 50 hp lower unit.

also prevent residual water from being trapped in the drive shaft housing and other passageways.

To test the water pump, the lower unit **MUST** be placed in a test tank or the boat moved into a body of water. The pump must now work to supply a volume to the engine.

Lack of adequate water supply from the water pump thru the engine will cause any number of powerhead failures, such as stuck rings, scored cylinder walls, burned pistons, etc.

2-13 PROPELLER

Check the propeller blades for nicks, cracks, or bent condition. If the propeller is damaged, the local marine dealer can make repairs or send it out to a shop specializing in such work.

Remove the cotter key, propeller nut, shear pin, and the propeller from the shaft. Check the propeller shaft seal to be sure it is not leaking. Check the area just forward of the seal to be sure a fish line is not wrapped around the shaft.

When installing the propeller, **ALWAYS** use an OMC or approved seal compound on the propeller shaft splines to prevent the propeller from seizing onto the shaft.

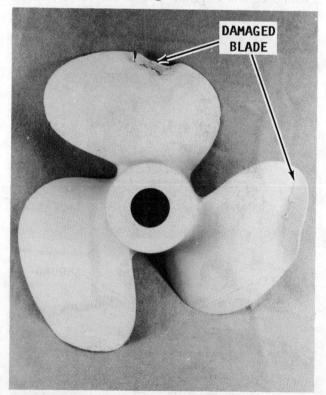

Example of a damaged propeller. This unit should have been replaced long before this amount of damage was sustained.

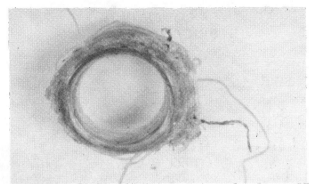

Considerable amount of fish line entangled around the propeller shaft. Some of the fish line actually melted, giving it the appearance of a washer.

Operation At Recommended RPM

Check with the local OMC dealer, or a propeller shop for the recommended size and pitch for a particular size engine, boat, and intended operation. The correct propeller should be installed on the engine to enable operation at recommended rpm.

Two rpm ranges are usually given. The lower rpm is recommended for large, heavy slow boats, or for commercial applications. The higher rpm is recommended for light, fast boats. The wide rpm range will result in greater satisfaction because of maximum performance and greater fuel economy. If the engine speed is above the recommended rpm, try a higher pitch propeller or the same pitch cupped. See Chapter 1 for explanation of propeller terms, pitch, diameter, cupped, etc.

For a dual engine installation, the next higher pitch propeller may prove the most satisfactory condition for water skiing.

2-14 LOWER UNIT

NEVER remove the vent or filler plugs when the lower unit is hot. Expanded lubricant would be released through the plug

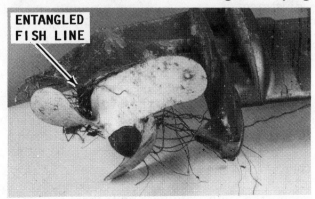

Considerable debris entangled behind the propeller. This type of maintenance neglect will seriously affect powerhead efficiency and boat performance.

New propeller ready for installation. Rebuilding a small propeller is not economical when balanced against the minor difference of purchasing a new unit.

hole. Check the lubricant level after the unit has been allowed to cool. Add only OMC approved gear lubricant. NEVER use regular automotive-type grease in the lower unit, because it expands and foams too much. Outboard lower units do not have provisions to accommodate such expansion.

If the lubricant appears milky brown, indicating the presence of water, a check should be made to determine how the water entered. If large amounts of lubricant must be added to bring the lubricant up to the full mark, a thorough inspection should be made to find the cause of the lubricant loss.

Draining Lower Unit

The fill/drain plug on Johnson/Evinrude lower units may be located towards the bottom of the unit on the port side, starboard side, or on the leading edge of the

The gear oil in the lower unit should be checked on a daily basis during the season of operation. The oil should be drained and replenished with new oil every 100 hours of operation.

lower unit. On many models a Phillips screw will be found very close to the fill/drain plug. **NEVER** remove this Phillips screw because the lower unit would then have to be disassembled in order to return the cradle for the shift dog back in place.

Remove the drain plug and then remove the vent plug located just above the anti-cavitation plate.

Filling Lower Unit

Position the drive unit approximately vertical and without a list to either port or starboard. Insert the lubricant tube into the **FILL/DRAIN** hole at the bottom plug hole, and inject lubricant until the excess begins to come out the **VENT** hole. Install the **VENT** plug first then replace the **FILL** plug with **NEW** gaskets. Check to be sure the gaskets are properly positioned to prevent water from entering the housing. Many times some of the gear lubricant is lost during installation of the plugs. Therefore, if the vent plug is removed again, and more lubricant added very **SLOWLY** using a small-spout oil can to allow air to pass out the opening, the unit will be filled to capacity.

For detailed lower unit service procedures, see Chapter 8. For lower unit lubrication capacities, see the Appendix.

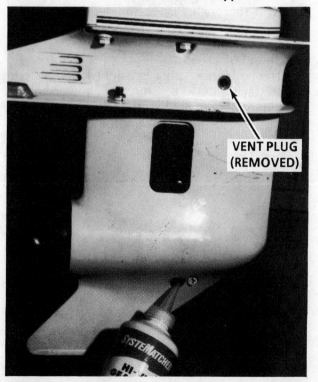

Filling the lower unit with new lubricant. Notice the unit filled through the lower drain plug opening. The vent plug MUST be removed to allow trapped air to escape as the lubricant enters.

LOWER UNIT LUBRICANT CAPACITIES

MODEL SIZE	YEAR	CAPACITY OUNCES
Colt	1990	1.2
2.3	1991 & On	3
3.3	1991 & On	3
3	1990 & On	2.7
4	1990 & On	2.7
4D	1990 & On	11
6	1990 & On	11
8	1990 & On	11
9.9	1990 & On	9
15	1990 & On	9
20	1990 & On	11
25	1990 & On	11
30	1990 & On	11
40	1990 & On	16.4
50	1990 & On	16.4

Repairs and Adjustments

For detailed procedures to disassemble, clean, assemble, and adjust the carburetor, see the appropriate section in Chapter 4 for the carburetor type on the engine being serviced.

2-15 BOAT TESTING

Hook and Rocker

Before testing the boat, check the boat bottom carefully for marine growth or evidence of a "hook" or a "rocker" in the bottom. Either one of these conditions will greatly reduce performance.

Performance

Mount the motor on the boat. Install the remote control cables and check for proper adjustment.

Check the engine rpm at full throttle. The rpm should be within the Specifications in the Appendix. All OMC engine model serial number identification plates indicate the horsepower rating and rpm range for the engine. If the rpm is not within specified range, a propeller change may be in order. A higher pitch propeller will decrease rpm, and a lower pitch propeller will increase rpm.

For maximum low speed engine performance, the idle mixture and the idle rpm should be readjusted under actual operating conditions.

3
POWERHEAD

3-1 INTRODUCTION

The carburetion and ignition principles of two-cycle engine operation **MUST** be understood in order to perform a proper tune-up on an outboard motor. Therefore, it would be well worth the time, to study the principles of two-cycle engines, as outlined in this section.

A Polaroid, or equivalent instant-type camera is an extremely useful item providing the means of accurately recording the arrangement of parts and wire connections **BEFORE** the disassembly work begins. Such a record is most valuable during the assembly work.

Tags are handy to identify wires after they are disconnected to ensure they will be connected to the same terminal from which they were removed. These tags may also be used for parts where marks or other means of identification is not possible.

THEORY OF OPERATION

The two-cycle engine differs in several ways from a conventional four-cycle (automobile) engine.

1- The method by which the fuel-air mixture is delivered to the combustion chamber.
2- The complete lubrication system.
3- In most cases, the ignition system.
4- The frequency of the power stroke.

These differences will be discussed briefly and compared with four-cycle engine operation.

Intake/Exhaust

Two-cycle engines utilize an arrangement of port openings to admit fuel to the combustion chamber and to purge the exhaust gases after burning has been completed. The ports are located in a precise pattern in order for them to be open and closed off at an exact moment by the piston as it moves up and down in the cylinder. The exhaust port is located slightly higher than the fuel intake port. This arrangement opens the exhaust port first as the piston starts downward and therefore, the exhaust phase begins a fraction of a second before the intake phase.

Actually, the intake and exhaust ports are spaced so closely together that both open almost simultaneously. For this reason, the pistons of most two-cycle engines have a deflector-type top. This design of the piston top serves two purposes very effectively.

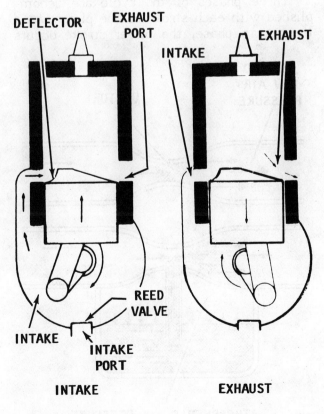

Drawing to depict the intake and exhaust cycles of a two-cycle engine.

First, it creates turbulence when the incoming charge of fuel enters the combustion chamber. This turbulence results in more complete burning of the fuel than if the piston top were flat. The second effect of the deflector-type piston crown is to force the exhaust gases from the cylinder more rapidly.

This system of intake and exhaust is in marked contrast to individual valve arrangement employed on four-cycle engines.

Lubrication

A two-cycle engine is lubricated by mixing oil with the fuel. Therefore, various parts are lubricated as the fuel mixture passes through the crankcase and the cylinder. Four-cycle engines have a crankcase containing oil. This oil is pumped through a circulating system and returned to the crankcase to begin the routing again.

Power Stroke

The combustion cycle of a two-cycle engine has four distinct phases.
1- Intake
2- Compression
3- Power
4- Exhaust

Three phases of the cycle are accomplished with each stroke of the piston, and the fourth phase, the power stroke occurs with each revolution of the crankshaft. Compare this system with a four-cycle engine. A stroke of the piston is required to accomplish each phase of the cycle and the power stroke occurs on every other revolution of the crankshaft. Stated another way, two revolutions of the four-cycle engine crankshaft are required to complete one full cycle, the four phases.

Physical Laws

The two-cycle engine is able to function because of two very simple physical laws.

One: Gases will flow from an area of high pressure to an area of lower pressure. A tire blowout is an example of this principle. The high-pressure air escapes rapidly if the tube is punctured.

Two: If a gas is compressed into a smaller area, the pressure increases, and if a gas expands into a larger area, the pressure is decreased.

If these two laws are kept in mind, the operation of the two-cycle engine will be easier understood.

Actual Operation

Beginning with the piston approaching top dead center on the compression stroke:

INDUCED LOW AIR PRESSURE

VENTURI

ATMOSPHERIC AIR PRESSURE

Air flow principle for a modern carburetor.

Adding OMC approved oil into the fuel tank.

The intake and exhaust ports are closed by the piston; the reed valve is open; the spark plug fires; the compressed fuel-air mixture is ignited; and the power stroke begins. The reed valve was open because as the piston moved upward, the crankcase volume increased, which reduced the crankcase pressure to less than the outside atmosphere.

As the piston moves downward on the power stroke, the combustion chamber is filled with burning gases. As the exhaust port is uncovered, the gases, which are under great pressure, escape rapidly through the exhaust ports. The piston continues its downward movement. Pressure within the crankcase increases, closing the reed valves against their seats. The crankcase then becomes a sealed chamber. The air-fuel mixture is compressed ready for delivery to the combustion chamber. As the piston continues to move downward, the intake port is uncovered. Fresh fuel rushes through the intake port into the combustion chamber striking the top of the piston where it is deflected along the cylinder wall. The reed valve remains closed until the piston moves upward again.

When the piston begins to move upward on the compression stroke, the reed valve opens because the crankcase volume has been increased, reducing crankcase pressure to less than the outside atmosphere. The

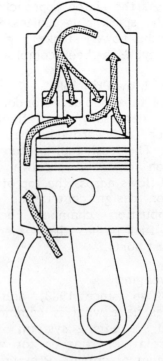

Drawing to depict fuel flow of the "loop charge" while the piston is on the down stroke.

intake and exhaust ports are closed and the fresh fuel charge is compressed inside the combustion chamber.

Pressure in the crankcase decreases as the piston moves upward and a fresh charge of air flows through the carburetor picking

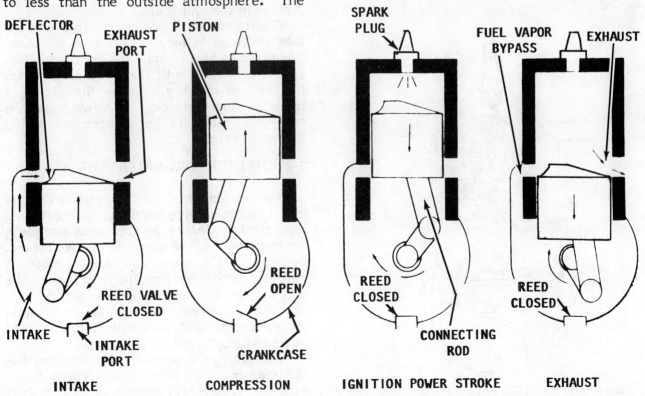

Complete piston cycle of a two-cycle engine, depicting intake, power, and exhaust.

up fuel. As the piston approaches top dead center, the spark plug ignites the air-fuel mixture, the power stroke begins and one complete cycle has been completed.

Cross Fuel Flow Principle

OMC pistons are a defector dome type. The design is necessary to deflect the fuel charge up and around the combustion chamber. The fresh fuel mixture enters the combustion chamber through the intake ports and flows across the top of the piston. The piston design contributes to clearing the combustion chamber, because the incoming fuel pushes the burned gases out the exhaust ports.

Loop Scavenging

The 40 hp (since 1983), and the 50 hp powerheads have what is commonly known as a loop scavenging system. The piston dome is relatively flat on top with just a small amount of crown. Pressurized fuel in the crankcase is forced up through the skirt of the piston and out through irregular shaped openings cut in the skirt. After the fuel is forced out the piston skirt openings it is transfered upward through long deep grooves molded into the cylinder wall. The fuel then enters the combustion portion of the cylinder and is compressed, as the piston moves upward.

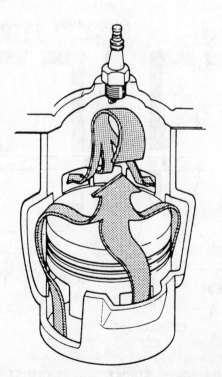

Drawing to depict the exhaust leaving the cylinder and fuel entering through the three ports in the piston.

This particular powerhead does not have intake cover plates, because the intake passage is molded into the cylinder wall as described in the previous paragraph. Therefore, if these engines are being serviced, disregard the sections covering intake cover plates.

Timing

The exact time of spark plug firing depends on engine speed. At low speed the spark is retarded -- fires later than when the piston is at or beyond top dead center. Therefore, the timing is advanced as the magneto armature plate advances.

At high speed, the spark is advanced -- fires earlier than when the piston is at top dead center.

The 40 hp (since 1983) and the 50 hp powerheads have a timing adjustment for low and high speeds. Procedures for making the timing adjustment will be found in Chapter 5.

Summary

More than one phase of the cycle occurs simultaneously during operation of a two-cycle engine. On the downward stroke, power occurs above the piston while the ports are closed. When the ports open, exhaust begins and intake follows. Below the piston, fresh air-fuel mixture is compressed in the crankcase.

On the upward stroke, exhaust and intake continue as long as the ports are open. Compression begins when the ports are closed and continues until the spark plug ignites the air-fuel mixture. Below the piston, a fresh air-fuel mixture is drawn into the crankcase ready to be compressed during the next cycle.

3-2 CHAPTER ORGANIZATION

The remainder of this chapter is divided into 13 main service sections, each covering a particular area of service with complete instructions for the work to be performed. Because of the many countless number of outboard units in the field, it would be impractical and almost impossible to give detailed procedures for removal and installation of each bolt, carburetor, starter, and other "buildup" type units.

Therefore, the sections, for the particular powerhead work to be performed, begin with the preliminary access tasks completed. As an example, disassembly of the powerhead begins with the necessary hood, cowling, and accessories removed.

The information is presented in a logical sequence for complete powerhead overhaul. The instructions can be followed generally for almost any size horsepower engine. In rare cases, where the procedures differ depending on the model being serviced, separate steps are included. One example is the three different type of crankshaft installations.

The illustrations accompanying the text are from different size units and the captions clearly identify which model is covered.

Exploded drawings, showing principle parts, for the various size powerheads are included at the end of the chapter.

Special tools may be called out in certain instances. These tools may be purchased from the local Johnson/Evinrude dealer or directly from Customer Services Department, Outboard Marine Corporation (OMC), Waukegan, Illinois, 60085.

The chapter ends with Break-in Procedures, Section 3-15, to be performed after the powerhead has been assembled, all accessories installed, and the powerhead mounted on the exhaust housing.

Torque Values

All torque values must be met when they are specified. Many of the outboard castings and other parts are made of aluminum. The torque values are given to prevent stretching the bolts, but more importantly to protect the threads in the aluminum. It is extremely important to tighten the connecting rods to the proper torque value to ensure proper service. The head bolts are probably the next most important torque value.

Powerhead Components

Service procedures for the carburetors, fuel pumps, starter, and other powerhead components are given in their respective chapters of this manual. See the Table of Contents.

Reed Installation

All reeds on Johnson/Evinrude engines covered in this manual are installed just behind the carburetor behind the intake manifold.

Cleanliness

Make a determined effort to keep parts and the work area as clean as possible.

Parts **MUST** be cleaned and thoroughly inspected before they are assembled, installed, or adjusted. Use proper lubricants, or their equivalent, whenever they are recommended.

Keep rods and rod caps together as a set to ensure they will be installed as a pair and in the proper sequence.

Needle bearings **MUST** remain as a complete set. **NEVER** mix needles from one set with another. If only one needle is damaged, the complete set **MUST** be replaced.

3-3 POWERHEAD DISASSEMBLING

Preliminary Work

Before the powerhead can be disassembled, the battery must be disconnected; fuel lines disconnected; and the carburetor, generator, starter, flywheel, and magneto, all removed. If in doubt as to how these items are to be removed, refer to the appropriate chapter.

After the accessories have been removed, remove the bolts in the front and rear of the powerhead securing the powerhead to the exhaust housing. Lift the powerhead free.

BAD NEWS

If the unit is several years old, or if it has been operated in salt water, or has not had proper maintenance, or shelter, or any number of other factors, then separating the

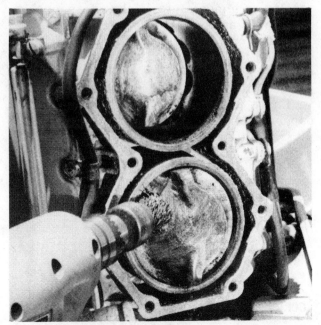

Cleaning the pistons while they remain in the powerhead. The pistons should be carefully inspected for burned areas and the cylinder walls for scoring.

powerhead from the exhaust housing may not be a simple task. An air hammer may be required on the studs to shake the corrosion loose; heat may have to be applied to the casting to expand it slightly; or other devices employed in order to remove the powerhead. One very serious condition would be the driveshaft "frozen" with the crankshaft. In this case, a circular plug-type hole must be drilled and a torch used to cut the driveshaft. Let's assume the powerhead will come free on the first attempt.

The following procedures pickup the work after these preliminary tasks have been completed.

3-4 HEAD SERVICE

Usually the head is removed and an examination of the cylinders made to determine the extent of overhaul required. However, if the head has not been removed, back out all of the head bolts and lift the head free of the powerhead.

Many, but not all, heads have a thermostat installed. In addition to the thermostat, the engine may have a thermostat bypass valve. These two items are easily removed, inspected and cleaned.

Normally, if a thermostat is not functioning properly, it is almost always stuck in the open position. An engine operating at too low a temperature is almost as much a problem as an engine running too hot.

Therefore, during a major overhaul, good shop practice dictates to replace the thermostat and eliminate this area as a possible problem at a later date.

Lay a piece of fine sandpaper or emery paper on a flat surface (such as a piece of glass) with the abrasive side facing up. With

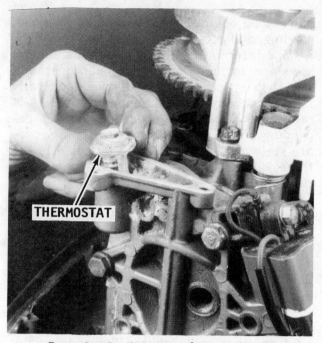

Removing the thermostat from the head.

the machined face of the head on the sandpaper, move the head in a circular motion to dress the surface. This procedure will also indicate any "high" or "low" spots.

Check the spark plug opening/s to be sure the threads are not damaged. Most marine dealers can insert a heli-coil into a spark plug opening if the threads have been damaged.

On many engines, a sending unit is installed in the head to warn the operator if the engine begins to run too hot. The light

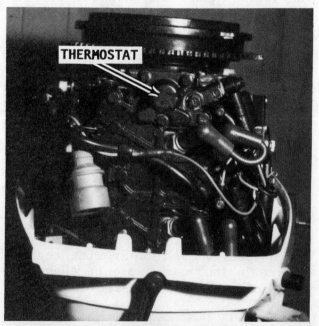

Model 25 powerhead showing convenient location of the thermostat and easy removal and installation through the three attaching bolts.

Cylinder block water passages corroded preventing proper circulation of coolant water.

on the dash can be checked by turning the ignition switch to the **ON** position, and then ground the wire to the sending unit. The light should come on. If it does not, replace the bulb and repeat the test.

3-5 REED SERVICE

DESCRIPTION

All two-cycle engines have individual ignition and fuel delivery for each cylinder. This means the cylinder is operating independently of the others. The cylinder consists of a top seal, the cylinder, the center seal, and the lower seal. This means each cylinder is completely sealed off from the others.

Therefore, with a two-cylinder powerhead, two sets of reeds are installed, one for each cylinder. These reeds may be installed on a reed plate or with a reed box, depending on the model engine. One carburetor provides fuel to both sets of reeds.

The reed arrangement operates in much the same manner as the reed in a saxophone or other wind instrument. At rest, the reed is closed and seals the opening to which it is attached. In the case of an outboard engine, this opening is between the crankcase and the carburetor. The reeds are mounted in the intake manifold, just behind the carburetor.

Actual Operation

The piston creates vacuum and pressure as it moves up and down in the cylinder. As

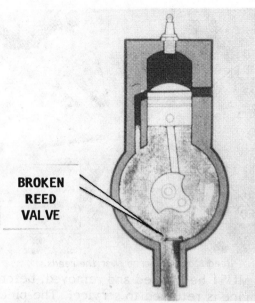

Cross-section of a cylinder to illustrate a broken reed in the crankcase.

the piston moves upward, a vacuum is created in the crankcase pulling the reed open. On the compression stroke, when the piston moves downward, the reed is forced closed.

Reed Designs

A wide range of reeds, reed plates, and reed box installations may be found on an outboard unit, due to the varying designs of the engines. All installations employ the same principle and there is no difference in their operation.

Broken Reed

A broken reed is usually caused by metal fatigue over a long period of time. The failure may also be due to the reed flexing too far because the reed stop has not been adjusted properly or the stop has become distorted. If the reed is broken, the loose

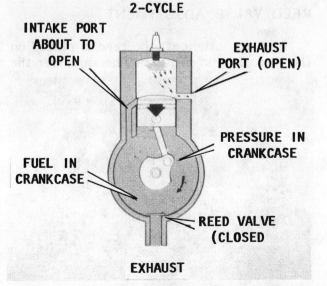

Diagram to illustrate operation of a two-cycle engine.

Reed box with a broken reed.

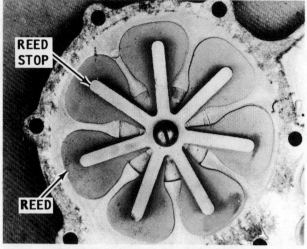

Reed stops centered over the reeds.

Close view showing the dimple on the reed plate. The reed leaves must straddle the dimple and be centered over the openings for proper operation.

piece **MUST** be located and removed, before the engine is returned to service. The piece of reed may have found its way into the crankcase, behind the bypass cover. If the broken piece cannot be located, the powerhead must be completely disassembled until it is located and removed.

The accompanying illustration depicts how a broken reed will cause a backflow through the carburetor.

An excellent check for a broken reed on an operating engine is to hold an ordinary business card in front of the carburetor. Under normal operating conditions, a very small amount of fine mist will be noticeable, but if fuel begins to appear rapidly on the card from the carburetor, one of the reeds is broken and causing the backflow through the carburetor and onto the card.

A broken reed will cause the engine to operate roughly and with a "pop" back through the carburetor.

Reed Stop to Reed Clearance

Measure the clearance between the top of the reed and the bottom of the reed stop, as shown in the accompanying illustration. The clearance should be 0.236-0.244" (6.0-6.3mm). If the clearance is not as specified,

or if the reed stops have become distorted, the most effective corrective action is to replace the stop instead of making an attempt at adjustment.

Reed to Base Plate Clearance

The specified clearance of the reed from the base plate, when the reed is at rest, is 0.010" (0.254 mm) at the tip of the reed.

An alternate method of checking the reed clearance is to hold the reed up to the sunlight and look through the back side. Some air space should be visible, but not a great amount. If in doubt, check the reed at the tip with a feeler gauge. The maximum clearance should not exceed 0.010" (0.254 mm).

The reeds must **NEVER** be turned over in an attempt to correct a problem. Such action would cause the reed to flex in the opposite direction and the reed would break in a very short time.

REED VALVE ADJUSTMENT

In many instances, the reed is placed on the reed plate in such a manner to cover the

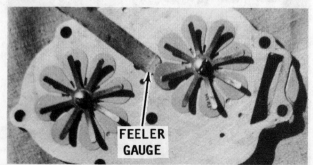

Using a feeler gauge to measure the clearance between the reed tip and the reed plate.

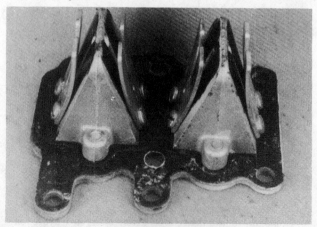

V-type reed box installed on the 9.5 hp engines.

openings in the plate. As shown in the accompanying illustration, a small indent is manufactured into the face of the plate. The leaves of the reed should be centered on this identation for proper operation. If the reed is being replaced, both reeds **AND** the reed stops should be replaced as a set.

V-Type Reed Boxes

As the name implies, these reed boxes are shaped in a "V" with a set of reeds and stops on both arms of the "V". If a problem develops with this type reed box, it is strongly recommended that the complete assembly be replaced -- reeds, box, and stops. The assembly may be purchased as a complete unit and the cost will usually not exceed the time, effort, and problems encountered in an attempt to replace only one part.

CLEANING AND SERVICE

Always handle the reeds with the utmost care. Rough treatment will result in the reeds becoming distorted and will affect their performance.

Wash the reeds in solvent, and blow them dry with compressed air from the **BACK SIDE ONLY.** Do not blow air through the reed from the front side. Such action would cause the reed to open and fly up against the reed stop. Wipe the front of the reed dry with a lint free cloth.

Clean the base plate thoroughly by removing any old gasket material.

Secure the reed blocks together with screws and nuts tightened to the torque value given in the Appendix.

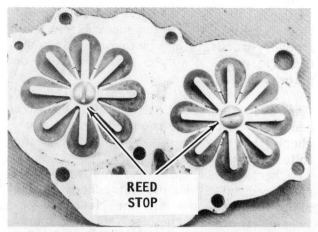

Front view of a reed plate with the two sets of reeds and reed stops in place.

Check for chipped or broken reeds. Observe that the reeds are not preloaded or standing open. Satisfactory reeds will not adhere to the reed block surface, but still there is not more than 0.010" (0.254 mm) clearance between the reed and the block surface.

DO NOT remove the reeds, unless they are to be replaced. **ALWAYS** replace reeds in sets. **NEVER** turn used reeds over to be used a second time.

Check the reed location over the reed block, or plate openings to be sure the reed is centered.

The reed assemblies are then ready for installation.

Disassembling

Disassemble the reed block by first removing the screws securing the reed stops and reeds to the reed block, and then lifting the reed stops and reeds from the block.

Front view of the reed box showing the Phillips screws that must be removed before the box can be removed from the plate.

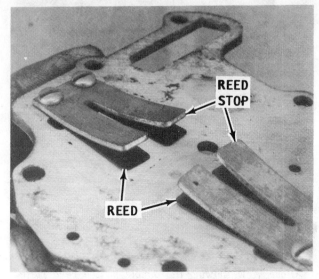

Reed stop installation with flat bars acting as the reed stop. The bars are the same width as the reed.

Clean the gasket surfaces of the reed block or plate. Check the surfaces for deep grooves, cracks, or any distortion that could cause leakage. Replace the reed block or plate if it is damaged.

After new reeds have been installed, and the reed stop and attaching screws have been tightened to the required torque value, check the new reeds as outlined in the following paragraphs.

Check to be sure the reeds are not preloaded. They should not adhere to the block or plate, and still the clearance between the reed and the block surface, should not be more than 0.010" (0.254 mm). **DO NOT** remove the reeds, unless they are to be replaced. **ALWAYS** replace reeds in sets. **NEVER** turn used reeds over to be used a second time.

Lay the reeds on a flat surface and measure all the reed stops. If there is a great difference between the stops, the entire reed stop assembly should be replaced. Any attempt to bend and get all the stops equal and level would be almost impossible.

INSTALLATION

Procedures to install the reeds to the powerhead will be found in Section 3-14, Cylinder Block Service, under Reed Box Installation.

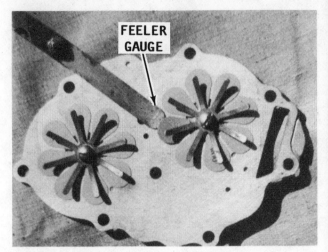

Using a feeler gauge to measure the clearance between the reed tip and the reed plate.

3-6 BYPASS COVERS

On some small horsepower units the powerhead does not contain bypass covers. The bypass cover actually covers the passageway the fuel travels from the crankcase up the side of the powerhead and into the cylinder.

Seldom does a bypass cover cause any problem. On some models, a fuel pump may be attached to one of the bypass covers.

During a normal overhaul, the bypass covers should be removed, cleaned, and new gaskets installed. Identify the covers to ensure installation in the same location from which they are removed.

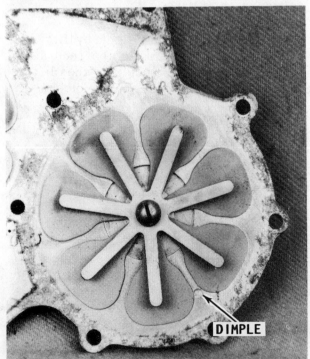

Close view showing the dimple on the reed plate. The reed leaves must straddle the dimple and be centered over the openings for proper operation.

Removing the bypass cover from a typical powerhead.

INSTALLATION

Procedures to install the bypass covers to the powerhead will be found in Section 3-14, Cylinder Block Service, under Bypass Cover and Exhaust Cover Installation.

3-7 EXHAUST COVER

The exhaust covers are one of the most neglected items on any outboard engine. Seldom are they checked and serviced. Many times an engine may be overhauled and returned to service without the exhaust covers ever having been removed.

One reason the exhaust covers are not removed is because the attaching bolts usually become corroded in place. This means they are very difficult to remove, but the work should be done. Heat applied to the bolt head and around the exhaust cover will help in removal. However, some bolts may still be broken. If the bolt is broken it must be drilled out and the hole tapped with new threads.

The exhaust covers are installed over the exhaust ports to allow the exhaust to leave the powerhead and be transferred to the exhaust housing. If the cover was the only item over the exhaust ports, they would become so hot from the exhaust gases they might cause a fire or a person would be severely burned if they came in contact with the cover.

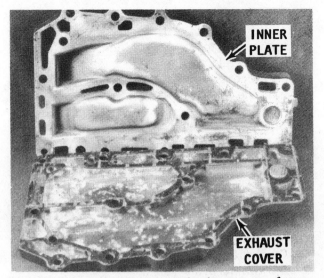

The inner exhaust plate and exhaust cover from a 40hp or 50hp powerhead.

Therefore, an inner plate is installed to help dissipate the exhaust heat. Two gaskets are installed -- one on either side of the inner plate. Water is channeled to circulate between the exhaust cover and the inner plate. This circulating water cools the exhaust cover and prevents it from becoming a hazard.

On some early model outboards, the inner plate was constructed of aluminum. Unfortunately, the aluminum would corrode through, especially in a salt water environment, and then water could enter the lower cylinder and cause a powerhead failure. The accompanying illustration clearly shows an inner plate corroded

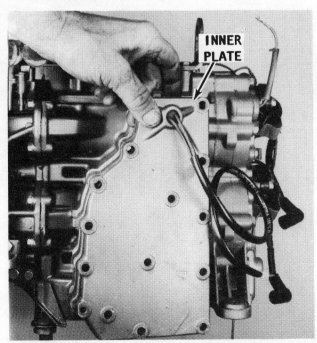

Removing the inner plate from a 40hp or 50hp powerhead.

Removing the exhaust cover from an early powerhead.

through, allowing water to enter the lower cylinder. To correct this corrosion problem, the inner plate is now made of stainless steel material.

A thorough cleaning of the inner plate behind the exhaust covers should be performed during a major engine overhaul. If the integrity of the exhaust cover assembly is in doubt, replace the complete cover including the inner plate.

On powerheads equipped with the heat/electric choke, a baffle is installed on the inside surface of the inner plate. This baffle is heated from the engine exhaust gases. Air passing through the baffle heats the choke and allows the choke to open as engine temperature rises.

CLEANING

Clean any gasket material from the cover and inner plate surfaces. Check to be sure the water passages in the cover and plate are clean to permit adequate passage of cooling water.

Inspect the inlet and outlet hole in the powerhead to be sure they are clean and free of corrosion. The openings in the powerhead may be cleaned with a small size screwdriver.

Clean the area around the exhaust ports and in the webs running up to the exhaust ports. Carbon has a habit of forming in this area.

The exhaust area of the powerhead, open for inspection and cleaning.

INSTALLATION

Procedures to install the exhaust covers will be found in Section 3-14, Cylinder Block Service, under Bypass Cover and Exhaust Cover Installation.

Removing the inner plate from an early powerhead.

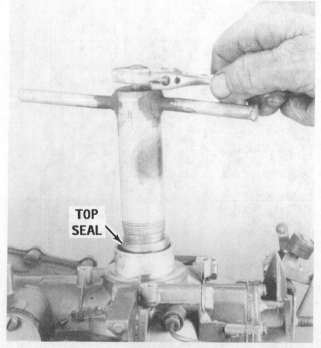

Using the proper tools to remove the top seal.

3-8 SEAL — TOP AND BOTTOM

The top seal maintains vacuum and pressure in the crankcase at the top cylinder.

REMOVAL

This seal can only be removed using one of two methods.

The first, is by using a special puller while the powerhead is still assembled. If the puller is used, thread the end of the puller into the seal. After the puller is secured to the seal, remove the seal from the powerhead by tightening the center screw on the puller. **DO NOT** attempt to use any other type of tool to remove this seal or the powerhead flanges will be damaged. If the flanges are damaged, the block must be replaced.

The second method is to remove the seal during powerhead disassembling. After the crankcase cover has been removed, the seal will be loose and can be easily removed by holding onto the bearing and prying the seal out.

INSTALLATION

To install the seal with the powerhead assembled, coat the outside diameter of the

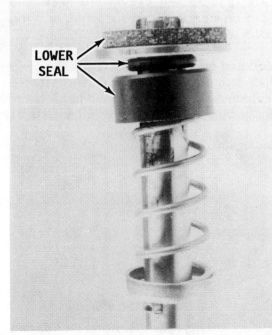

Pressing a new seal into a bearing.

seal with OMC Seal Compound. Use the special tool to tap the seal **EVENLY** into place around the crankshaft.

If the powerhead has been disassembled, a socket or other similar type tool of equal diameter as the seal may be used to tap it into the bearing. If the powerhead being

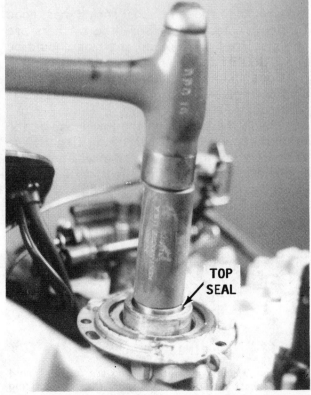

Using the proper tools to install the top seal.

Powerhead lower seal installed on the driveshaft.

serviced does not have the seal in the bearing, lay the seal in the recess of the block. When the crankcase cover is installed, the seal will be in place.

BOTTOM SEAL

The bottom seal has equal importance as the top seal. This seal is installed to maintain vacuum and pressure in the lower half of the crankcase for the lower cylinder.

The bottom seal will vary, depending on the model engine being serviced. The following procedures and accompanying illustrations cover the most common Johnson/Evinrude bottom seal installed.

Seal Mounted on the Driveshaft

When the powerhead is removed, observe around the driveshaft at the lower end, and the seal will be visible. The seal consists of a gasket, plate, an O-ring, lower seal bearing, spring, washer, and a pin. The pin is installed through the driveshaft and holds the seal upward and in place.

As the powerhead is lowered down over the driveshaft during installation, the seal is held in place and will hold the vacuum and pressure created when the engine is operating.

Removal

With the powerhead assembled, it is a simple matter to reach into the exhaust housing and remove the seal and then replace the gasket and O-ring. Check the spring, to be sure it is not distorted, and the washer for damage.

Seal Mounted on the Crankshaft

This bottom seal prevents exhaust fumes from entering the crankcase, and holds pressure and vacuum inside. The seal consists of

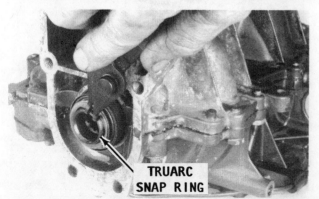

Using a pair of Truarc pliers to remove the snap ring from the crankshaft.

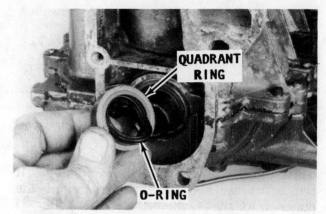

Removing the O-ring from the quadrant ring. These two items form the bottom crankshaft seal.

a quadrant ring, O-ring, retainer washer, spring, another washer, and a snap ring.

Removal

To remove this seal from the lower end of the crankshaft, use a pair of Truarc pliers and **CAREFULLY** remove the snap ring. **TAKE CARE** not to lose any of the parts due to the spring pressure against the snap ring. Notice how the quadrant O-ring fits inside the seal. This ring is also removable. Observe how the seal has a raised edge on one side. This raised edge **MUST** face upward when the seal is installed.

INSPECTION

Check to be sure the spring has good tension. Check to be sure the washers are not distorted. The quadrant ring should be **DISCARDED** and a new one installed.

Good shop practice dictates the quadrant seal be replaced each time the lower seal is serviced.

Check the groove in the lower end of the crankshaft where the truarc ring fits. If the groove is not clean, the ring will snap out and the lower sealing qualities will be lost. If the groove is badly corroded, the crankshaft must be replaced.

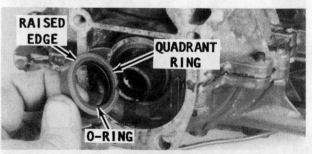

A new O-ring and quadrant ring are installed onto the crankshaft with the raised edge of the quadrant ring facing UPWARD when installed.

Remove the seal and O-ring from the cap of a 40hp or 50hp powerhead.

Seal in a Cap

Many of the larger horsepower powerheads have the seal installed in a cap. The cap is bolted to the bottom of the powerhead. The four bolts securing the cap must be removed before the crankcase cover can be removed. The cap is then removed after the cover has been removed from the cylinder block. The seal can be punched out and new ones installed without difficulty.

3-9 CENTERING PINS

All Johnson/Evinrude outboard engines have at least one, and in most cases two, centering pins installed through the crankcase cover. These pins index into matching holes in the powerhead block when the crankcase cover is installed. These pins center the crankcase cover on the powerhead block.

Cylinder block with the two centering pins installed.

Removing a centering pin from the cylinder block.

The centering pins are tapered. The pins must be carefully checked to determine how they are to be removed from the cover. In most cases the pin is removed by using a center punch and tapping the pin towards the carburetor or intake manifold side of the crankcase.

When removing a centering pin, hold the punch securely onto the pin head, then strike the punch a good hard forceful blow. **DO NOT** keep beating on the end of the pin, because such action would round the pin head until it would not be possible to drive it out of the cover.

Centering pins are the first item to be installed in the cover when replacing the crankcase cover.

3-10 MAIN BEARING BOLTS AND CRANKCASE SIDE BOLTS

The main bearing bolts are installed through the crankcase cover into the powerhead block. Most engines have two bolts installed for the top main bearing, two for the center main bearing, and two for the lower main bearing.

Removing the main bearing bolts from the powerhead.

In many cases the upper and lower main bearing bolts are **DIFFERENT** lengths. Therefore, take time to tag and identify the bolts to ensure they will be installed in the same location from which they were removed.

The crankcase side bolts are installed along the edge of the crankcase cover to secure the cover to the cylinder block. These bolts usually have a 7/16" head and all must be removed before the crankcase cover can be removed. Remove the crankcase side bolts.

Remove the main bearing bolts. Two bolts installed in the center are behind the reeds. Normally these two are not actually bolts, but Allen head screws. All six main bearing bolts must be removed before the crankcase cover can be removed.

INSTALLATION

Main bearing bolt and the crankcase side bolt installation is given in Section 3-14, Cylinder Block Service under Main Bearing Bolt and Crankcase Side Bolt Installation.

3-11 CRANKCASE COVER

REMOVAL

After all side bolts and main bearing bolts have been removed, use a soft-headed mallet and tap on the bottom side of the crankshaft. A soft, hollow sound should be heard indicating the cover has broken loose

Removing the crankcase cover from a 40hp or 50hp powerhead.

from the crankcase. If this sound is not heard, check to be sure all the side bolts and main bearing bolts have been removed. **NEVER** pry between the cover and the crankcase or the cover will surely be distorted. If the cover is distorted, it will fail to make a proper seal when it is installed.

Once the crankshaft has been tapped, as described, and the proper sound heard, the cover will be jarred loose and may be removed.

CLEANING AND INSPECTING

Wash the cover with solvent, and then dry it thoroughly. Check the mating surface to the cylinder block for damage that may affect the seal.

Cylinder block after the crankcase cover has been removed.

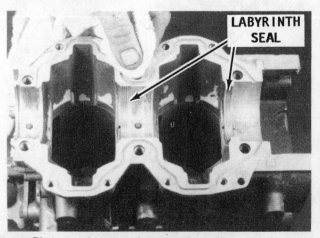
Cleaning the powerhead surface of a 40hp or 50hp powerhead. Notice the labyrinth seal at the center and bottom main bearings.

Crankcase cover with the labyrinth seal area clearly visible.

Inspect the labyrinth seal grooves at the center main bearing area to be sure they are clean and not damaged in any manner.

INSTALLATION

Installation procedures for the crankcase cover are given in Section 3-14, Cylinder Block Service, under Crankcase Cover Installation.

3-12 CONNECTING RODS AND PISTONS

The connecting rods and their rod caps are a **MATCHED set.** They absolutely **MUST** be identified, kept, and installed as a set. Under no circumstances should the connecting rod and caps be interchanged. Therefore, on a multiple piston engine, **TAKE TIME AND CARE** to tag each rod and rod cap; to keep them together as a set while they are on the bench; and to install them into the same cylinder from which they were removed as a set.

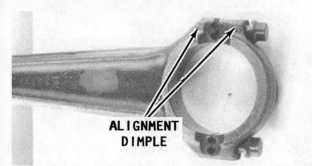

Rod and rod cap with the two alignment dimples shown.

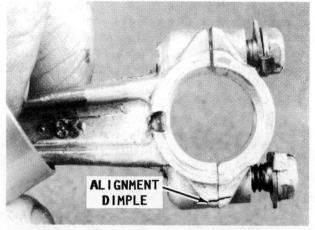

Rod and rod cap with the alignment line marks shown.

The connecting rod and its cap on 15 hp to early 40 hp engines are manufactured as a set --as a single unit. After the complete rod and cap have been made, two holes are drilled through the side of the cap and rod, and the cap is then fractured from the rod. Therefore, the cap must always be installed with its original rod. The cap half of the break can **ONLY** be matched with the other half of the break on the **ORIGINAL** rod.

The rods and caps on the smaller horsepower engines are made of aluminum with babbitt inserts. These rods and caps are manufactured as two separate items.

Inspect the rod and the rod cap before removing the cap from the crankshaft. Under normal conditions, a line or a dot is

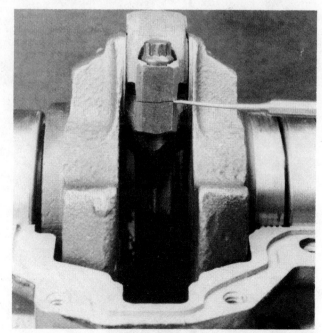

A punch points to the fractured break of a rod and its cap. The rod and cap must be matched during installation.

visible on the top side of the rod and the cap. This identification is an assist to assemble the parts together and in the proper location.

Observe into the block and notice how the rods have a "trough". Also notice the hole in the rod near where the wrist pin passes through the piston. On many rods there is also a hole in the rod at the crank end. These two holes **MUST ALWAYS** face upward during installation.

REMOVAL

To remove the rod bolts from the cap, it is recommended to loosen each bolt just a little at-a-time and alternately. This procedure will prevent one bolt from being completely removed while the other is still tightened to its recommended torque value. Such action may very likely warp the cap.

Remove the bolts as described in the previous paragraph, and then **CAREFULLY** remove the rod cap to prevent loosing the needle bearings installed under the cap, if used.

Remove the needle bearings and cages, if used, from around the crankshaft. Count the needle bearings and insert them into a separate container -- one container for each rod, with the container clearly identified to ensure they will be installed with the proper rod at the crankshaft journal from which they were removed.

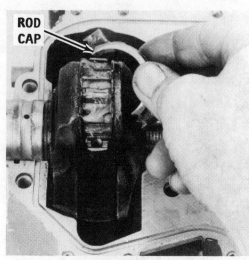

Removing the rod cap from the rod.

Tap the piston out of the cylinder from the crankshaft side. Immediately attach the proper rod cap to the rod and hold it in place with the rod bolts. The few minutes involved in securing the cap with the rod will ensure the matched cap remains with its mating rod during the cleaning and assembling work.

Identify the rod to ensure it will be installed into the cylinder from which it was removed.

Remove and identify the other rod caps, needle bearings and cages, and rods with pistons, in the same manner.

DISASSEMBLING

Before separating the piston from the rod, notice the location of the piston in relation to the rod. Observe the hole in the rod trough on one side of the rod near the wrist pin opening and another at the lower end. These holes must face toward the **TOP** of the engine during installation.

Removing the bolts from the rod cap.

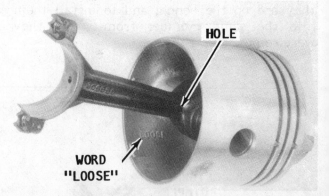

Identifying word "LOOSE" on the inside of the piston skirt and the hole in the rod at the wrist pin end. The wrist pin must be driven from the loose side of the piston out the tight side, as described in the text.

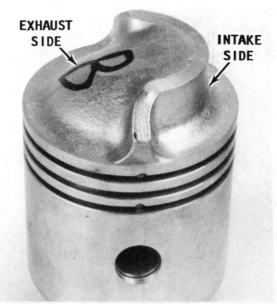

Close view of a piston with the slanted edge and sharp edges identified. The piston can only be installed one way for proper operation.

Observe the slanted edge and the sharp edge of the dome-type piston. The slanted edge **MUST** face toward the exhaust side of the cylinder and the sharp edge toward the intake side during installation.

Pistons installed in the block of a 40hp or 50hp powerhead. The word "UP" embossed on the piston must be at the top of the cylinder, as shown.

Pistons installed in the powerhead of a 40hp or 50hp unit. The word "UP" embossed on the piston must be at the top of the cylinder, as shown.

If servicing a 40 hp or 50 hp powerhead, carefully observe the hole in the rod near the wrist pin and the relationship of the irregular cutouts in the piston skirt. Only in this position will this relationship exist. The rod and piston **MUST** be assembled in this manner or the engine will run **VERY** poorly.

When the rod is installed to the piston, the relationship of the rod can only be one way. The rod holes must face upward and the piston must face as described in the previous paragraph.

Observe into the piston skirt. On most model pistons, notice the "L" stamped on the boss through which the wrist pin passes. The letter mark identifies the "loose" side of the piston and indicates side of the piston

Piston with the word "LOOSE" embossed on the inside surface of the skirt.

A broken rod, possibly caused by inadequate oil delivery or the powerhead operated in a RUNAWAY condition (excessive rpm under a "No Load" condition).

Heating a piston in hot water to expand the metal slightly.

from which the wrist pin must be driven out without damaging the piston. Some pistons may have the full word **"LOOSE"** stamped on the inside of the piston skirt.

If the piston does not have the **"L"** or the word **"LOOSE"** stamped, the wrist pin may be driven out in either direction.

It may be necessary to heat the piston in a container of boiling water in order to press the wrist pin free.

Remove the retaining clips from each end of the wrist pin. Some clips are spring wire type and may be worked free of the piston using a screwdriver. Other model pistons have a truarc snap ring. This type of ring can only be successfully removed using a pair of truarc pliers.

Place the piston in an arbor press using the **PROPER** size cradle for the piston being serviced, and with the **LOOSE** side of the piston facing **UPWARD**.

The wrist pin must be driven out **FROM** the loose side. This may not seem reasonable, but there is a very simple explanation.

By placing the piston in the arbor press cradle with the tight side down, and the arbor ram pushing from the loose side, the piston has good suport and will not be distorted. If the piston is placed in the arbor press with the loose side down, the piston would be distorted and unfit for further service.

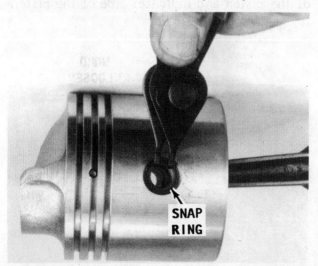

Removing the wrist pin Truarc snap ring from the piston.

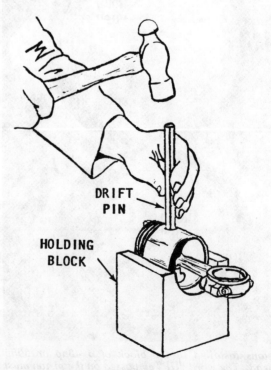

Removing the wrist pin using a holding block.

Many rods have a wrist pin bearing. Some are caged bearings and other are not. **TAKE CARE** not to lose any of the bearings when the wrist pin is driven free of the piston.

Alternate Removal Method

If an arbor press or cradle is not available, proceed as follows. Heat the piston in a container of very hot water for about ten minutes. Heating the piston will cause the metal to expand ever so slightly, but ease the task of driving the pin out. Assume a sitting position in a chair, on a box, whatever. Next, lay a couple towels over your legs. Hold your legs tightly together to form a cradle for the piston above your knees. Set the piston between your legs with the **"LOOSE"** side of the piston facing upward. Now, drive the wrist pin free using a drift pin with a shoulder. The drift pin will fit into the hole through the wrist pin and the shoulder will ride on the edge of the wrist pin. Use sharp hard "rap-type" blows

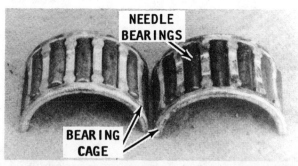

Needle bearings and cages unfit for further service.

with a hammer. Your legs will absorb the shock without damaging the piston. If this method is used on a regular basis during the busy season, your legs may develop black-and-blue areas, but no problem. The marks will disappear in a few days.

ROD INSPECTION AND SERVICE

If the rod has needle bearings, the needles should be replaced anytime a major overhaul is performed. It is not necessary to replace the cages, but a complete **NEW** set of needles should be purchased and installed.

Place each connecting rod on a surface plate and check the alignment. If light can be seen under any portion of the machined surfaces, or if the rod has a slight wobble on the plate, or if a 0.002" feeler gauge can be inserted between the machined surface and the surface plate, the rod is bent and unfit for further service.

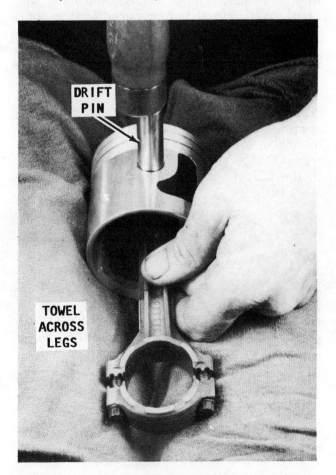

When an arbor press is not available, the piston can be supported between your legs and the wrist pin driven out using a drift pin, as described in the text.

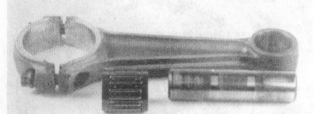

Rod, rod cap, wrist pin, and wrist pin bearing, after removal.

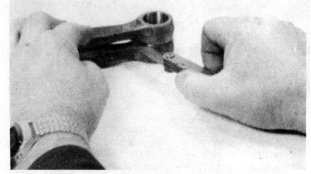

Testing two rods at the wrist pin end for warpage.

Testing two rods at the rod cap end for warpage.

SCORED AREA

Badly rusted and corroded crankshaft from a submerged engine. This crankshaft is no longer fit for service.

Inspect the connecting rod bearings for rust or signs of bearing failure. **NEVER** intermix new and used bearings. If even one bearing in a set needs to be replaced, all bearings at that location **MUST** be replaced.

Inspect the bearing surface of the rod and the rod cap for rust and pitting.

Inspect the bearing surface of the rod and the rod cap for water marks. Water marks are caused by the bearing surface being subjected to water contamination, which causes "etching". The "etching" will worsen **VERY** rapidly.

Inspect the bearing surface of the rod and rod cap for signs of spalling. Spalling is the loss of bearing surface, and resembles

flaking or chipping. The spalling condition will be most evident on the thrust portion of the connecting rod in line with the I-beam. Bearing surface damage is usually caused by improper lubrication.

Check the bearing surface of the rod and rod cap for signs of chatter marks. This condition is identified by a rough bearing surface resembling a tiny washboard. The condition is caused by a combination of low-speed low-load operation in cold water, and is aggravated by inadequate lubrication and improper fuel. Under these conditions, the crankshaft journal is hammered by the connecting rod. As ignition occurs in the cylinder, the piston pushes the connecting rod with tremendous force, and this force is transferred to the connecting rod journal.

Since there is little or no load on the crankshaft, it bounces away from the connecting rod. The crankshaft then remains immobile for a split second, until the piston travel causes the connecting rod to catch up to the waiting crankshaft journal, then hammers it.

The pitted damage to this piston crown was probably caused by a broken piston ring working its way into the combustion chamber. The little "hills" then became "hot" spots on the crown, contributing to "dieseling" after the powerhead was shut down.

*Installing the rod cap onto the rod in preparation for cleaning the inside surface. The cap **MUST** always be kept with its matching rod.*

In some instances, the connecting rod crankpin bore becomes highly polished.

While the engine is running, a "whirr" and/or "chirp" sound may be heard when the engine is accelerated rapidly from idle speed to about 1500 rpm, then quickly returned to idle. If chatter marks are discovered, the crankshaft and the connecting rods should be replaced.

Inspect the bearing surface of the rod and rod cap for signs of uneven wear and possible overheating. Uneven wear is usually caused by a bent connecting rod. Overheating is identified as a bluish bearing surface color and is caused by inadequate lubrication or operating the engine at excessive high rpm.

Inspect the needle bearings, if installed. A bluish color indicates the bearing became very hot and the complete set for the rod **MUST** be replaced, no question.

Service the connecting rod bearing surfaces according to the following procedures and precautions:

a- Align the etched marks on the knob side of the connecting rod with the etched marks on the connecting rod cap.

b- Tighten the connecting rod cap attaching bolts securely.

c- Use **ONLY** crocus cloth to clean bearing surface at the crankshaft end of the connecting rod. **NEVER** use any other type of abrasive cloth.

d- Insert the crocus cloth in a slotted 3/8" diameter shaft. Chuck the shaft in a drill press and operate the press at high speed and at the same time, keep the connecting rod at a 90° angle to the slotted shaft.

e- Clean the connecting rod **ONLY** enough to remove marks. **DO NOT** continue once the marks have disappeared.

f- Clean the piston pin end of the connecting rod using the method described in Steps d and e, but using 320 grit Carborundum cloth instead of crocus cloth.

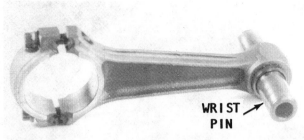

Testing the wrist pin end of the rod prior to installation.

g- Thoroughly wash the connecting rods to remove abrasive grit. After washing, check the bearing surfaces a second time.

h- If the connecting rod cannot be cleaned properly, it should be replaced.

i- Lubricate the bearing surfaces of the connecting rods with light-weight oil to prevent corrosion.

PISTON AND RING INSPECTION AND SERVICE

Inspect each piston for evidence of scoring, cracks, metal damage, cracked piston pin boss, or worn pin boss. Be especially critical during inspection if the engine has been submerged.

Carefully check each wrist pin to be sure it is not the least bit bent. If a wrist pin is bent, the pin and piston **MUST** be replaced as a set, because the pin will have damaged the boss when it was removed.

*Rod cap separated slightly from its mated rod. Notice the matching hills and valleys. The cap and rod **MUST** always be kept together -- **NEVER** interchanged.*

Piston badly scored and no longer fit for service.

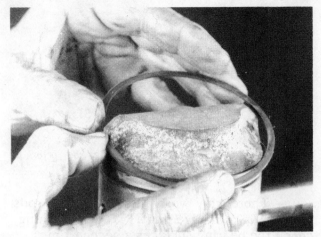

Removing the rings from the piston.

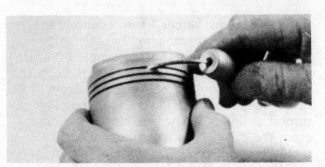

Cleaning the piston ring grooves. An automotive type ring groove cleaner should **NEVER** *be used.*

Check the wrist pin bearings. If the bearing is the pressed-in type, use your finger and determine the bearing is in good condition with no indication of binding or "rough" spots. If the wrist pin bearing is the removable type, the needle should be replaced.

Grasp each end of the ring with either a ring expander or your thumbnails, open the ring and remove it from the piston. Many times, the ring may be difficult to remove because it is "frozen" in the piston ring groove. In such a case, use a screwdriver and pry the ring free. The ring may break, but if it is difficult to remove, it **MUST** be replaced.

OBSERVE the pin in each ring groove of the piston. The ends of the ring **MUST**

straddle this pin. The pin prevents the ring from rotating while the engine is operating. This fact is the direct opposite of a four-cycle engine where the ring must rotate. In a two-cycle engine, if the ring is permitted to rotate, at one point, the opening between the ring ends would align with either the intake or exhaust port in the cylinder. At that time, the ring would expand very slightly, catch on the edge of the port, and **BREAK**.

Therefore, when checking the condition of the piston, **ALWAYS** check the pin in each groove to be sure it is tight. If one pin is the least bit loose, the piston **MUST** be replaced, without question. Never attempt to replace the pin, it is **NEVER** successful.

Check the piston ring grooves for wear, burns, distortion or loose locating pins. During an overhaul, the rings should be replaced to ensure lasting repair and proper engine performance after the work has been completed.

Clean the piston dome, ring grooves and the piston skirt. Clean the piston skirt with a crocus cloth.

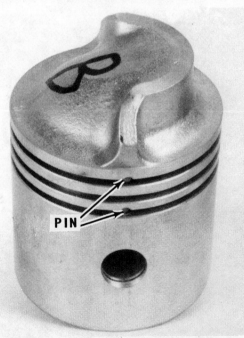

PIN

Close view of a piston showing the ring pin in the groove.

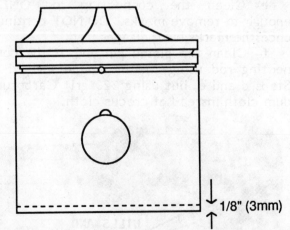

1/8" (3mm)

Use a micrometer to measure the diameter and to check the roundness of the piston. The measurements should be taken about 1/8" (3mm) up from the bottom of the skirt.

Clean carbon deposits from the top of the piston using a soft wire brush, carbon removal solution, or by sand blasting. If a wire brush is used, **TAKE CARE** not to burr or round machined edges.

Wear a pair of good gloves for protection against sharp edges, and clean the piston ring grooves using the recessed end of the proper broken ring as a tool. **NEVER** use a rectangular ring to clean the groove for a tapered ring, or use a tapered ring to clean the groove for a rectangular ring.

NEVER use an automotive-type ring groove cleaner to clean piston ring grooves, because this type of tool could loosen the piston ring locating pins. **TAKE CARE** not to burr or round the machined edges. Inspect the piston ring locating pins to be sure they are tight. There is one locating pin in each ring groove. If one locating pin is loose, the piston must be replaced. Never attempt to replace the pin, it is **NEVER** successful.

Oversize Pistons and Rings

Scored cylinder blocks can be saved for further service by reboring and installing oversize pistons and piston rings. **ONE MORE WORD:** Oversize pistons and rings are not available for all engines. At the time of this printing, the sizes listed in the Appendix were available. Check with the parts department at your local dealer for the model engine you are servicing, and to be sure the factory has not deleted a size from their stock.

ASSEMBLING

CRITICAL WORDS

Two conditions absolutely **MUST** exist when the piston and rod assembly are installed into the cylinder block.

The slanted side of the piston must face toward the exhaust side of the cylinder.

*An automotive ring compressor should **NEVER** be used to install the rings for a two-cycle engine.*

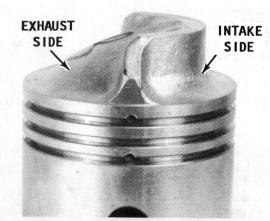

*The slanted side of the piston **MUST** face the exhaust port and the sharp edge face the intake port.*

The hole in the rod near the wrist pin opening and at the lower end of the rod must face **UPWARD**.

Therefore, the rod and piston **MUST** be assembled correctly in order for the assembly to be properly installed into the cylinder. Soak the piston in a container of very hot water for about ten minutes. Heating the piston will cause it to expand ever so slightly, but enough to allow the wrist pin to be pressed through without difficulty.

Before pressing the wrist pin into place, hold the piston and rod near the cylinder block and check to be sure both will be facing in the right direction when they are installed.

Pack the wrist pin needle bearing cage with needle bearing grease, or a good grade

Piston with the wrist pin ready for installation. Notice the ring groove pin. For proper operation and lubrication, the rod must be installed to the piston and the piston into the cylinder as described in the text.

Wrist pin entering the piston from the "LOOSE" side.

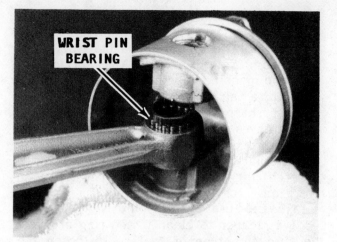

Installing the rod and wrist pin bearing into the piston.

of petroleum jelly. Load the bearing cage with needles and insert it into the end of the rod.

Slide the rod into the piston boss and check a second time to be sure the slanted side of the piston is facing toward the exhaust side of the cylinder and the hole in the rod is facing upward.

Place the piston and rod in the arbor press with the **LOOSE** or stamped **"L"** side of the piston facing **UPWARD**. Press the wrist pin through the piston and rod. Continue to press the wrist pin through until the groove in the wrist pin for the lock ring is visible on

both ends of the pin. Remove the assembly from the arbor press. Install the retaining ring onto each end of the wrist pin. Some models have a wire ring, and others have a truarc ring. Use a pair of truarc pliers to install the truarc ring.

Fill the piston skirt with a rag, towel, shop cloths, or other suitable material. The rag will prevent the rod from coming in contact with the piston skirt while it is laying on the bench. If the rod is allowed to strike the piston skirt, the skirt may become distorted.

Assemble the other pistons, rods, and wrist pins in the same manner. Fill the skirt with rags as protection until the assembly is installed.

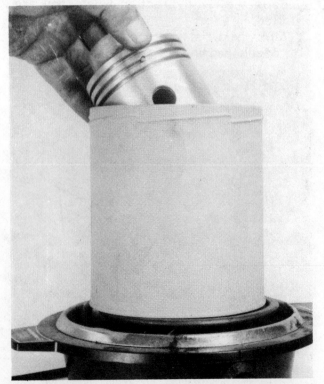

Heating the piston in hot water to expand the metal slightly as an assist to installing the wrist pin.

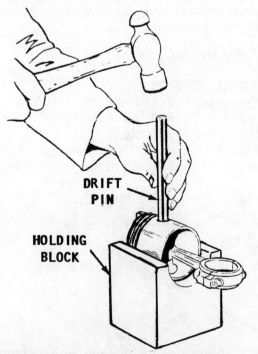

Install the wrist pin using a holding block.

Alternate Assembling Method

If an arbor press is not available, the piston may be assembled to the rod in much the same manner as described for disassembly on Page 3-21.

First, soak the piston in a container of very hot water for about ten minutes.

Before pressing the wrist pin into place, hold the piston and rod near the cylinder block and check to be sure both will be facing in the right direction when they are installed.

Pack the wrist pin needle bearing cage with needle bearing grease, or a good grade of petroleum jelly. Load the bearing cage with needles and insert it into the end of the rod.

Slide the rod into the piston boss and check a second time to be absolutely sure the slanted side of the piston is facing toward the exhaust side of the cylinder and the hole in the rod is facing upward.

Now, assume a sitting position and lay a couple towels over your lap. Hold your legs tightly together to form a cradle for piston above your knees. Set the piston between your legs with the **LOOSE** side of the piston facing upward. Drive the wrist pin through

the piston using a drift pin with a shoulder. The drift pin will fit into the hole through the wrist pin and the shoulder will ride on the end of the wrist pin. Use sharp hard "rap"-type blows with a hammer. Your legs will absorb the shock without damaging the piston. If this method is used on a regular basis during the busy season, your legs may develop black-and-blue areas, but no problem. The marks will disappear in a few days.

Continue to drive the wrist pin through the piston until the groove in the wrist pin for the lockring is visible at both ends. Install the retaining spring wire or truarc ring onto each end of the wrist pin.

Fill the piston skirt with a rag, towel, shop cloths, or other suitable material. The material will prevent the rod from coming in contact with the piston skirt while it is laying on the bench. If the rod is allowed to strike the piston skirt, the skirt may become distorted.

Assembling the other piston, rod, and wrist pin in the same manner. Fill the skirt with material as protection until the assembly is installed.

INSTALLATION

Piston and rod assembly installation procedures begin on Page 3-32.

3-13 CRANKSHAFT

REMOVAL

Lift the crankshaft assembly from the block. On some models, especially the

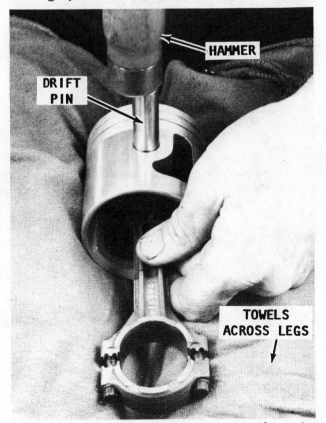

Installing the wrist pin without the use of an arbor press. The piston is supported on your lap, as described in the text.

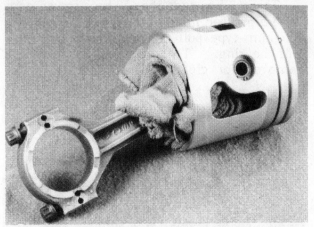

Anytime the piston is out of the cylinder, the inside should be filled with rags or shop cloths as protection against the rod striking and damaging the skirt.

Crankshaft with the upper, center, and lower main bearings ready to be removed.

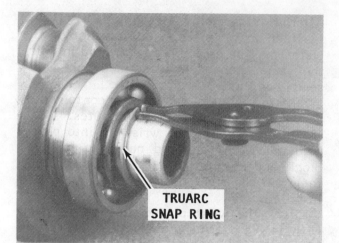

Removing the Truarc snap ring from the lower bearing of a 40hp or 50hp powerhead.

larger horsepower engines, it may be necessary to use a soft-headed mallet and tap on the bottom side of the crankshaft to jar it loose. As the crankshaft is lifted, **TAKE CARE** to work the center main bearing loose. This center bearing is a split bearing held together with a snap wire ring. On some models, the bottom half of the bearing may be stuck in the cylinder block. Therefore, the crankshaft and the center main bearing must be worked free of the block together.

If servicing a 15 hp to 30 hp powerhead, observe how the center main bearing, and the top and bottom main bearings all have a hole in the outside circumference. Notice the locating pins in the cylinder block. The purpose of this arrangement is to prevent the bearing shell from rotating. During assembling, the holes in the bearings **MUST** index with the pins in the block. Also notice the grooves in the block on one side of the center main bearing. Observe the grooves in the crankcase cover. This arrangement of grooves forms what is commonly known as a "labyrinth" seal. The grooves fill with oil and/or fuel creating a seal between the cylinders.

The 40 hp and the 50 hp, powerheads have a pressed in place lower roller bearing. A clamp-type puller is required to removed this bearing. The bearing need not be removed for cleaning and inspection. Re-

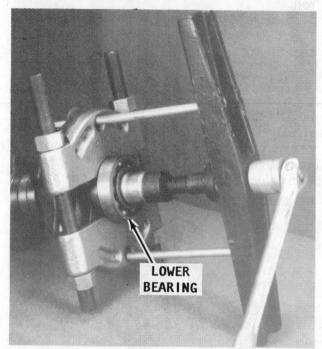

Using a puller to remove the lower bearing of a 40hp or 50hp unit.

Removing the crankshaft from the block of a 40hp or 50hp powerhead.

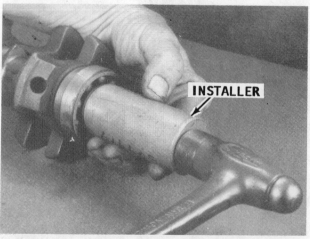

Using a sleeve-type tool over the end of the driveshaft to install the bearing.

move this bearing **ONLY** if the determination has been made that it is unfit for further service.

To install a new bearing, place the bearing onto the shaft and press it into place using an arbor press. If an arbor press is not available a socket large enough to fit over the crankshaft could be used to drive the bearing into place.

On the smaller horsepower powerheads, babbitt bearings are used for the center main bearing with needle bearings installed for the upper and lower main bearings.

CLEANING AND INSPECTION

Inspect the splines for signs of abnormal wear. Check the crankshaft for straightness. Inspect the crankshaft oil seal surfaces to be sure they are not grooved, pitted or scratched. Replace the crankshaft if it is severely damaged or worn. Check all crankshaft bearing surfaces for rust, water marks, chatter marks, uneven wear or overheating. Clean the crankshaft surfaces with crocus cloth.

Clean the crankshaft and crankshaft bearing with solvent. Dry the parts, but **NOT** the bearing, with compressed air. Check the crankshaft surfaces a second time. Replace the crankshaft if the surfaces cannot be cleaned properly for satisfactory service. If the crankshaft is to be installed for service, lubricate the surfaces with light oil.

The top and lower bearing may be easily removed from the crankshaft. The center

Badly rusted and corroded crankshaft from a submerged engine. This crankshaft is no longer fit for service.

main bearing has a spring steel wire securing the two halves together. Remove the wire, and then the outer sleeve, then the needle bearings. **TAKE CARE** not to lose any of the needles. The outer shell is a fractured break type unit. Therefore, the two halves of the shell **MUST** absolutely be kept as a set.

Check the crankshaft bearing surfaces to be sure they are not pitted or show any signs of rust or corrosion. If the bearing surfaces are pitted or rusted, the crankshaft and bearings must be replaced.

During an engine overhaul to this degree, it is a good practice to remove the seal from the top main bearing. If the same type of seal is used in the bottom main bearing, remove that seal also.

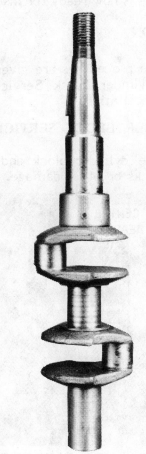
A crankshaft cleaned and ready for installation.

Crankcase cover with the labyrinth seal area clearly visible.

A crankshaft with a badly corroded "throw". This crankshaft is unfit for further service.

ASSEMBLING

Insert the proper number of needle bearings into the center main bearing cage. Install the outer sleeve over the bearing cage. Check to be sure the two halves of the outer sleeve are matched. Again, these two halves are manufactured as a single unit and then broken. Therefore, the hills and valleys of the break absolutely **MUST** match during installation.

Snap the retaining ring into place around the bearing. Slide the upper bearing onto the crankshaft journal at the upper end and the lower bearing onto the lower end. Rotate the installed bearings to be sure there is no evidence of binding or rough spots. The crankshaft is now ready for installation.

INSTALLATION

Installation procedures are given in Section 3-14, Cylinder Block Service, under Crankshaft Installation.

3-14 CYLINDER BLOCK SERVICE

Inspect the cylinder block and cylinder bores for cracks or other damage. Remove

The needles and cage of the center main bearing ready for the outside shell to be installed.

Crankshaft with the upper, center, and lower main bearings installed. Check to be sure the snap ring on the center main bearing is installed.

carbon with a fine wire brush on a shaft attached to an electric drill or use a carbon remover solution.

Use an inside micrometer or telescopic gauge and micrometer to check the cylinders for wear. Check the bore for out-of-round and/or oversize bore. If the bore is tapered, out-of-round or worn more than 0.003" - 0.004" (0.076 mm - 0.102 mm) the cylinders should be rebored and oversize pistons and rings installed.

GOOD WORDS:

Oversize piston weight is approximately the same as a standard size piston. Therefore, it is **NOT** necessary to rebore all cylinders in a block just because one cylinder requires reboring. The APBA (American Power Boat Association) accepts and permits the use of 0.015" (0.381 mm) oversize pistons.

Hone the cylinder walls lightly to seat the new piston rings, as outlined in the Honing Procedures Section in this chapter. If the cylinders have been scored, but are not out-of-round or the sleeve is rough, clean the surface of the cylinder with a cylinder hone as described in Honing Procedures, next section.

SPECIAL WORD

Cylinder sleeves may be installed on some models, but the cost is very high.

HONING PROCEDURES

To ensure satisfactory engine performance and long life following the overhaul work, the honing work should be performed with patience, skill, and in the following sequence:

a- Follow the hone manufacturer's recommendations for use of the hone and for cleaning and lubricating during the honing operation.

b- Pump a continuous flow of honing oil into the work area. If pumping is not practical, use an oil can. Apply the oil generously and frequently on both the stones and work surface.

c- Begin the stroking at the smallest diameter. Maintain a firm stone pressure against the cylinder wall to assure fast stock removal and accurate results.

d- Expand the stones as necessary to compensate for stock removal and stone wear. The best cross-hatch pattern is obtained using a stroke rate of 30 complete cycles per minute. Again, use the honing oil generously.

e- Hone the cylinder walls **ONLY** enough to de-glaze the walls.

f- After the honing operation has been completed, clean the cylinder bores with hot water and detergent. Scrub the walls with a stiff bristle brush and rinse thoroughly with hot water. The cylinders **MUST** be cleaned well as a prevention against any abrasive material remaining in the cylinder bore. Such material will cause rapid wear of new piston rings, the cylinder bore, and the bearings.

g- After cleaning, swab the bores several times with engine oil and a clean cloth, and then wipe them dry with a clean cloth. **NEVER** use kerosene or gasoline to clean the cylinders.

Checking the ring gap clearance by inserting the ring in the cylinder, as described in the text.

h- Clean the remainder of the cylinder block to remove any excess material spread during the honing operation.

WORDS OF ADVICE

If new rings are to be installed, each ring from the package **MUST** be checked in the cylinder. Errors happen. Men and machines can make mistakes. The wrong size ring can be included in a package with the proper part number.

Therefore, check **EACH** ring, one at-a-time as follows: Turn the ring sideways and lower it a couple inches into the cylinder bore. Now, turn the ring horizontal in the cylinder. It is now in its normal operating position, but without the piston. Next, use a feeler gauge and measure the distance (the gap) between the ends of the ring. The maximum and minimum allowable ring gap is listed in the Specifications in the Appendix.

Turn the piston upside down and slide it in and out of the cylinder. The piston should slide without any evidence of binding.

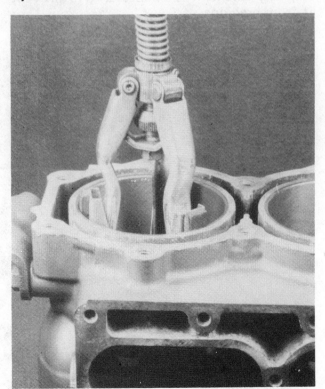

Resurfacing a cylinder wall using a honing tool.

Using a feeler gauge to check the ring end gap with the ring in the cylinder.

ASSEMBLING

SPECIAL WORD

The cylinder block assembling work should proceed quickly and without interruptions. If the work is partially completed and then left for any period of time, sealant may become hard, parts may be moved and their identity for a particular cylinder lost, or an important step may be bypassed, overlooked, or forgotten.

The following procedures pickup the work of assembling the cylinder block AFTER the various parts have been serviced and assembled. Procedures for each area are found in this chapter under separate headings.

PISTON AND ROD ASSEMBLY INSTALLATION

Several different methods are possible to install the piston and rod assembly into the cylinder. The following procedures are outlined for the do-it-yourselfer, working at home without the advantage of special tools.

First, purchase a special hose clamp with a strip of metal inside the clamp, as shown in the accompanying illustration. This piece of metal on the inside allows the outside portion of the clamp to slide on the inner strip without causing the ring to rotate.

Actually, to our knowledge, a Mercruiser dealer is the only place such an inexpensive clamp may be purchased. At the Mercruiser marine dealer, ask for an exhaust bellows hose clamp. The design of this hose clamp prevents the clamp and the piston ring from turning as the clamp is tightened. DO NOT attempt to use an ordinary hose clamp from an automotive parts house because such a clamp will cause the piston ring to rotate as the clamp is tightened. The ring MUST NOT rotate, because the ring ends must remain on either side of the dowel pin in the ring groove.

Next, coat the inside surface of the cylinder with a film of light-weight oil. Coat the exterior surface of the piston with the oil.

TAKE TIME

Take just a minute to notice how the piston rings are manufactured. Each end of the ring has a small cutout on the inside circumference. Now, visualize the ring installed in the piston groove. The ring ends must straddle the pin installed in each piston groove. As the ring is tightened around the piston, the ends will begin to come together. When the piston is installed into the cylinder bore, the two ends of the ring will come together and the cutout edge will be up against the pin. For this reason, CARE must be exercised when installing the rings onto the piston and when the piston is installed into the cylinder.

Install only the bottom ring into the bottom piston groove. Do not expand the ring any further than necessary, to prevent it from breaking.

Install the ring into the piston groove with the ends of the ring straddling the pin

Proper hose clamp to install the rings if a ring compressor is not available.

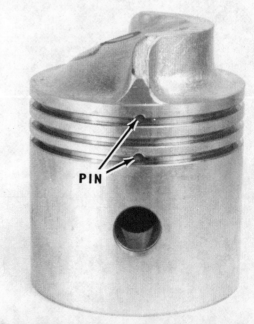

PIN

Piston ring groove pins. The ends of the ring must straddle the pin.

Installing the hose clamp over the ring prior to moving the piston further into the cylinder.

Tapping a piston into the cylinder of a 40hp or 50hp powerhead. Notice the proper type hose clamp used to compress the rings, as explained in the text.

in the groove. The ring ends **MUST** straddle the pin to prevent the ring from rotating during engine operation. In a two-cycle engine, if the ring is permitted to rotate, at one point the opening between the ring ends would align with either the intake or exhaust port in the cylinder, the ring would expand very slightly, catch on the edge of the port, and **BREAK**.

CAREFULLY insert the rod and the piston skirt down into the cylinder.

GOOD WORDS

The following four areas must be checked at this point in the assembling work.

a- The piston and rod are being installed into the same cylinder from which they were removed.

b- The hole in the rod is facing **UPWARD**.

c- The slanted side of the piston is **TOWARD** the exhaust side of the cylinder.

d- The ends of each piston ring **MUST** straddle the pin in the piston groove.

Push the piston into the cylinder until the bottom ring, just installed, is about an inch from the surface of the cylinder block.

Install the hose clamp over the piston and bottom ring. Tighten the hose clamp with one hand and at the same time rotate the clamp back-and-forth slightly with the other hand. This "rocking" motion of the clamp as it is tightened will convince you the ring ends are properly positioned on either side of the pin. Continue to tighten the clamp, and "rocking" the clamp until the clamp is against the piston skirt. At this point, the ring ends will be together and the cutout on each ring end will be against the pin.

Tap the piston with the end of a wooden tool handle until the ring enters the cylinder. Remove the hose clamp.

Install the remaining rings in the same manner, one at-a-time, making sure the ends of each ring straddle the pin in the piston groove.

Notice how the ring pins are staggered from one groove to the next, by 180°.

After the last ring has been installed and the clamp removed, tap the piston into the bore until the crown is about even with the cylinder block surface.

Tapping the piston into the cylinder with a soft-headed mallet.

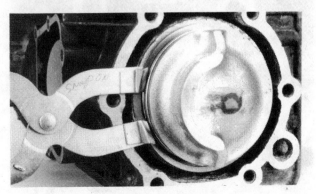

Using a ring installer to expand the ring during installation into the piston ring groove.

When installing the piston into a small horsepower powerhead, it is possible to compress the ring with the fingers of each hand, and then to push the piston into the cylinder with your thumbs.

Install the other pistons in exactly the same manner.

Turn the cylinder block upside down with the top of the block to your **LEFT.** Remove the bolts and rod caps from each rod. Set each rod cap in a definite position to ensure each will be installed onto the rod from which it was removed.

Checking the flexibility of the rings through intake port.

Checking the flexibility of the rings through the exhaust port.

Both pistons installed in the powerhead. The slanted edge of each piston is facing toward the exhaust port.

Both pistons installed into a 40hp or 50hp powerhead. The embossed word "UP" on the piston must be at the top of the cylinder, as shown.

Checking the ring tension with a small screwdriver through the exhaust port of a 40hp or 50hp powerhead.

CRANKSHAFT INSTALLATION NEEDLE MAIN AND ROD BEARINGS

The following procedures outline steps to install a crankshaft with needle upper, center, and lower main bearings. The upper and lower mains are complete bearings and cannot be disassembled. The center bearing is caged. Installation procedures for small horsepower crankshafts with babbitt upper, lower, and center main bearings and with babbitt rod bearings are given in the following sections.

Observe the pin installed in each main bearing recess. Notice the hole in each main bearing outer shell. During installation, the hole in each bearing shell **MUST** index over the pin in the cylinder block.

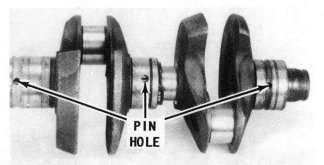

Crankshaft with the upper, lower, and center main bearings installed. Notice the hole in each bearing. Matching pins in the cylinder block must index into these holes during crankshaft installation.

Hold the crankshaft over the cylinder block with the upper end to your **LEFT**. Now, lower the crankshaft into the block, and at the same time, align the hole in each bearing to enable the pin in the block to index with the hole. Rotate each bearing slightly until all pins are properly indexed with the matching bearing hole. Once all pins are indexed, the crankshaft will be properly seated.

Apply needle bearing grease to each bearing cage. Coat the rod half of the bearing area with needle bearing grease. Needle bearing grease **MUST** be used because other types of grease will not thin out and dissipate. The grease must dissipate to allow the gasoline and oil mixture to enter and lubricate the bearing. If needle bearing grease is not available, use a good grade of petroleum jelly (Vasoline).

Insert the proper number of needle bearings into each cage. Set the bearing cage into the bottom half of the rod. With your

Bearing locating pins in the cylinder block. Each pin must index into a hole in the bearing shown in the illustration at the top of this column.

Installing the crankshaft into the block of a 40hp or 50hp powerhead. The holes in the top and center main bearings must index with the pin in the block.

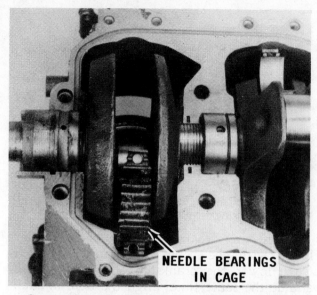

Cage and needle bearings installed into the lower portion of the rod.

fingers on each side of the rod, pull up on the rod and bring the rod up to the bottom side of the crankshaft. Put one needle bearing on each side of the crankshaft. Using needle bearing grease load the other cage and install the needle bearings into the cage. Lower the cage onto the crankshaft journal.

Install the proper rod cap to the rod with the identifying mark or dimple properly aligned to ensure the cap is being installed in the same position from which it was removed. Tighten the rod bolts fingertight, and then just a bit more.

Use a "scratchall", pick, or similar tool and move it back-and-forth on the outside surface of the rod and cap. Make the

Lowering the cage and needle bearings over the top of the crankshaft.

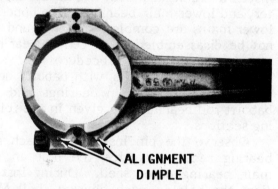

Rod and cap showing the alignment dimples.

Installing a needle bearing on each side of the crankshaft.

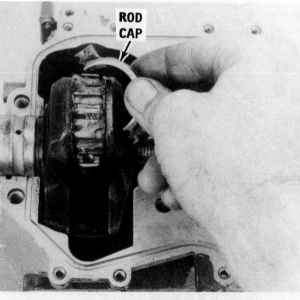

Installing the rod cap over the needle bearings and cage.

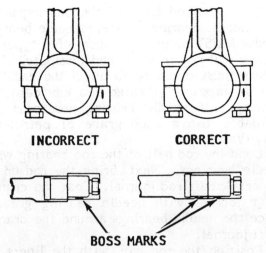

INCORRECT   CORRECT

BOSS MARKS

Correct and incorrect rod cap alignment.

Checking the flexibility of the rings through the exhaust port on a small horsepower powerhead.

movement across the mating line of the rod and cap. The tool should not catch on the rod or on the cap. The rod cap must seat squarely with the rod. If not, tap the cap until the "scratchall" will move back-and-forth on the rod and cap across the mating line without any feeling of catching. Any step on the outside will mean a step on the inside of the rod and cap. Just a whisker of a lip, will cause one of the needle bearings to catch and fail to rotate. The needle will quickly flatten, and the rod will begin to "knock". Needle bearings **MUST** rotate or the function of the bearing is lost.

Tighten the rod cap bolts alternately and evenly in three rounds to the torque value given in the Torque Table in the Appendix.

Tighten the bolts to 1/2 the torque value on the first round, to 3/4 the torque value on the second round, and to the full torque value on the third and final round. On each round, check with the pick to be sure the cap remains seated squarely.

Install the other rod cap/s in the same manner.

After the rods have been connected to the crankshaft, rotate the crankshaft until the rings on one cylinder are visible through the exhaust port. Use a screwdriver and push on each ring to be sure it has spring tension. It will be necessary to move the piston slightly, because all of the rings will not be visible at one time. If there is no spring tension, the ring was broken during installation. The piston must be removed and a new ring installed. Repeat the tension test at the intake port. Check the other cylinder/s in the same manner.

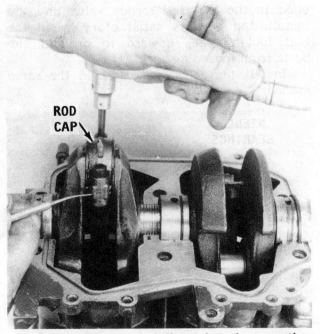

ROD CAP

Installing the rod cap bolts, and at the same time checking the cap alignment with the rod using a pick.

Check the flexibility of the rings through the exhaust port on a 40hp or 50hp powerhead, as described in the text.

CRANKSHAFT INSTALLATION
UNITS W/TOP NEEDLE MAIN BRG AND BABBITT CTR & BOTTOM
UNITS W/TOP & BOTTOM NEEDLE BRG AND CENTER BABBITT BRG
UNITS W/ALL BABBITT MAIN BRGS

This section provides detailed instructions to install a small horsepower crankshaft with any of the bearing combinations listed in the heading. Some of the powerheads covered in these paragraphs have rod liners, others do not.

The procedures pickup the work after the piston/s have been installed, as described earlier in this section.

ADVICE
Before installing the crankshaft, check to be sure each "throw" is clean and shiny. There should be no evidence of corrosion that might damage the "throw" during engine operation.

Lower the crankshaft into place in the cylinder block with the long threaded shank end at the top of the cylinder block. (It is a known fact, in more than just a few shops around the country, because of haste, the crankshaft installation work has proceeded with the short end at the top.)

The hole in the upper and lower main bearing **MUST** index into the pin in the cylinder block.

Some engines may have a lining arrangement as listed in the heading of this section. The lining is made in two parts. Install the liner half into the rod, then install the bearings as described in the next paragraph. The matching liner is to be installed into the rod cap.

Coat the rod half, of the bearing area, with needle bearing grease. Needle bearing grease **MUST** be used because other types of grease will not thin out and dissipate. The grease must dissipate to allow the gasoline and oil mixture to enter and lubricate the bearing. If needle bearing grease is not available, use a good grade of petroleum jelly (Vasoline).

Load the rod half of the rod bearing with needle bearings. Next, bring the rod up to the crankshaft rod journal. Coat the crankshaft journal with needle bearing grease. Place the needle bearings around the crankshaft journal.

Position the rod cap, with the liners (if used) over the needle bearings. Install the rod cap bolts and lockwashers. Bring the bolts up fingertight, and then just a bit more.

If the liners are used, the cap and rod automatically align properly. If liners are not used, a dowel pin is installed in the rod cap. This pin will index into a hole in the rod for proper alignment.

Tighten the rod cap bolts alternately and evenly in three rounds to the torque value given in the Specifications in the Appendix. Tighten the bolts to 1/2 the torque value on the first round, to 3/4 the torque value on the second round, and to the full torque value on the third and final round. On each round, check with the pick to be sure the cap remains seated squarely.

After the rod cap bolts have been tightened to the required torque value and the installation appears satisfactory, bend the bolt locking tabs upward to prevent the bolts from loosening.

Install the other rod cap/s in the same manner.

Rod and cap with the alignment marks visible.

Needle bearings installed in the rod cap liner and around the crankshaft.

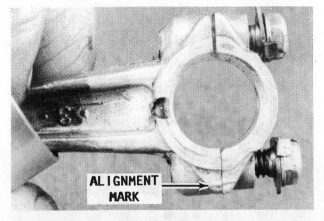

Rod and cap with the alignment marks visible.

Rod cap with liner ready for installation around the crankshaft.

After the rods have been connected to the crankshaft, rotate the crankshaft until the rings on one cylinder are visible through the exhaust port. Use a screwdriver and push on each ring to be sure it has spring tension. It will be necessary to move the piston slightly, because all of the rings will not be visible at one time. If there is no spring tension, the ring was broken during installation. The piston must be removed and a new ring installed. Repeat the tension test at the intake port. Check the other cylinder/s in the same manner.

CRANKSHAFT INSTALLATION BABBITT MAIN AND ROD BEARINGS

This section provides detailed instructions to install a small horsepower crankshaft with babbitt upper, lower, and center main bearings, and with babbitt rod bearings.

The procedures pickup the work after the piston/s have been installed, as described earlier in this section.

Lower the crankshaft into place in the cylinder block with the long threaded shank end at the top of the cylinder block. (It is a known fact, in more than just a few shops around the country, because of haste, the crankshaft installation work has proceeded with the short end at the top.)

Pull the rod up to the crankshaft journal. Position the rod cap over the crankshaft

The locking tabs must be bent upward after the rod cap bolts have been tightened to the proper torque value.

Tightening the rod cap bolts to the proper torque value.

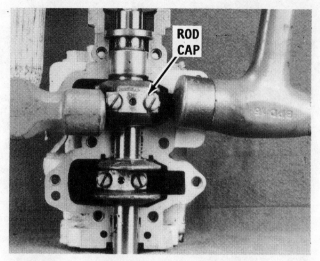

Using two hammers to fit the rod cap to the crankshaft.

journal. Install the rod cap bolts and lockwashers. Bring the bolts up fingertight, and then just a bit more.

Tighten the rod cap bolts alternately and evenly in three rounds to the torque value given in the Torque Table in the Appendix. Tighten the bolts to 1/2 the torque value on the first round, to 3/4 the torque value on the second round, and to the full torque value on the third and final round. On each round, check with the pick to be sure the cap remains seated squarely.

Repeat the procedure for the other rod and cap.

After the other rod cap has been installed and the bolts tightened to the proper torque value, hold one hammer on one side

Checking movement of the pistons and crankshaft with the flywheel temporarily installed.

of the rod and cap, and at the same time tap the other side of the rod and cap with the other hammer. Tap lightly on the top of the cap. Reverse the hammer positions and tap the opposite sides of the rod and cap. This procedure will "fit" the rod and cap to the crankshaft journal.

Repeat the "fitting" procedure for the other rod and cap.

Once the installation procedure appears satisfactory and all work has been completed, bend the bolt locking tabs upward to prevent the bolts from loosening.

CRANKCASE COVER INSTALLATION

First, check to be sure the mating surfaces of the crankcase cover and the cylinder block are clean. Pay particular attention to the labyrinth seal grooves in the center main bearing area. The mating surfaces and the seal grooves **MUST** be free of any old sealing compound or other foreign material.

CRITICAL WORDS

The remainder of the cylinder block installation work should be performed **WITHOUT** interruption. Do not begin the work if a break in the sequence is expected -- coffee, lunch, tea, whatever. Both types of

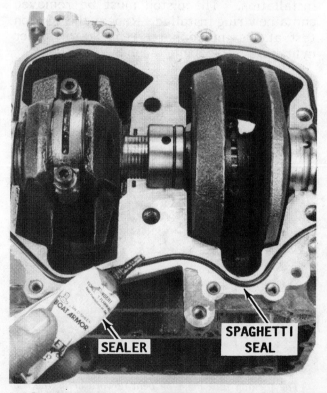

Installing sealer to the "spaghetti" seal in the crankcase cover.

Installing sealer to the cylinder block when the "spaghetti" seal is not used.

Two tapered pins installed in the cylinder block.

sealer will begin to set almost immediately, therefore, the crankcase cover installation, main bearing bolt installation and tightening, and the side bolt installation and tightening **MUST** move along **RAPIDLY**.

Apply just a small amount of 1000 Sealer into the groove in the cylinder block to hold the "spaghetti" seal in place, if used. Install a new seal into the groove. After the seal on both sides of the cylinder block has been installed, apply a light coating of 1000 Sealer to the outside edge of the "spaghetti" seal.

SPECIAL NOTE

Since 1980, the cylinder and crankcase assemblies do not have the grooves on the mating surfaces for the neoprene "spaghetti" seals. The new sealing method is to use OMC Gel-Seal (P/N 322702). This new seal is to be used on all 2-cylinder powerheads covered in this manual. **DO NOT** use similar appearing jel-type sealants, since some of them contain fillers that have a shimming affect. Such shimming could cause improper bearing location, bearing misalignment, or tight armature plate bearings.

Lay down a small bead of the Gel-Seal along one flange of the crankcase, as shown. Take care to apply sealant inside all bolt holes. Keep the sealant at least 1/4" from the labyrinth seals. If the motor is to be operated the same day, the surface opposite the one with the Gel-Seal should be sprayed with OMC LocQuic Primer. Wait several hours before starting the engine. If LocQuic Primer is not used, the assembly should sit overnight before the engine is started.

The remaining installation instructions apply to all powerheads.

Next, lower the crankcase cover into place on the cylinder block. Install the two

guide centering pins through the cover and into the block. The centering pins are tapered. Therefore, check the crankcase and notice which side has the large hole and which has the small hole. The pin must be inserted into the large hole first. If the pin is installed into the small hole first, the crankcase cover or the cylinder block will break.

MAIN BEARING BOLT AND CRANKCASE SIDE BOLT INSTALLATION

Apply a coating of 1000 Sealer to the threads of the main bearing bolts. Install and tighten the main bearing bolts finger-tight and then just a bit more.

Tighten the main bearing bolts alternately and evenly in three rounds to the torque value given in the Torque Table in

Installing the main bearing bolts through the crankcase cover into the cylinder block.

the Appendix. Be sure to check the Specifications in the Appendix for the engine being serviced.

Tighten the bolts to 1/2 the total torque value on the first round, to 3/4 the total torque value on the second round, and to the full torque value on the third and final round.

As an example: If the total torque value specified is 200 ft-lbs, the bolts should be tightened to 100 ft-lbs on the first go-around; to 150 ft-lbs on the second round; and to the full 200 ft-lbs on the third round.

Install and tighten the crankcase side bolts to the torque value given in the Appendix.

Install the Woodruff key in the crankshaft. Slide the flywheel onto the crankshaft. Rotate the flywheel through several revolutions and check to be sure all moving parts indicate smooth operation without evidence of binding or "rough" spots.

Remove the flywheel and the Woodruff key.

LOWER SEAL INSTALLATION
TYPE ATTACHED TO LOWER
END OF CRANKSHAFT

This type of seal is attached to the lower end of the crankshaft. On the smaller engines, a seal is used on the crankshaft with a spring, O-ring, and gasket. These items push up against the bottom of the powerhead to affect the seal.

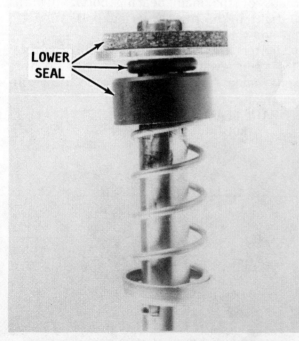

Powerhead lower seal installed on the driveshaft.

Installing the lower main seal assembly to the crankshaft.

Install the quadrant O-ring into the quadrant retainer.

Apply a small amount of light-weight oil onto the quadrant retainer and O-ring, and then slide them onto the crankshaft with the lip or raised edge of the retainer facing **UPWARD** upon installation.

Slide the large washer, spring, and small washer, onto the crankshaft, and secure them in place with the truarc snap ring.

LOWER SEAL INSTALLATION
40 HP AND 50 HP POWERHEADS

After the crankshaft and the crankcase cover have been installed, install an O-ring onto the cap. Cover the O-ring and the crankshaft with some light-weight oil. Slip the cap over the crankshaft as far as possible. Coat the threads of the attaching bolts with Loctite. Secure the cap in place with the four bolts. Tighten the bolts alternately and evenly.

EXHAUST COVER AND
BYPASS COVER INSTALLATION

Coat both sides of a **NEW** gasket with 1000 Sealer, and then place the gasket in

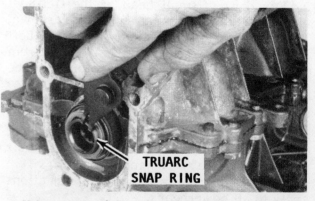

Using a pair of Truarc pliers to install the Truarc snap ring onto the crankshaft to secure the bottom seal in place.

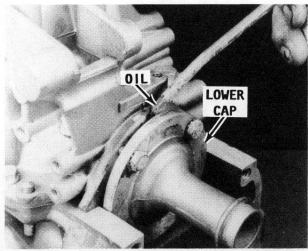

Installing the lower main bearing cap onto the crankshaft of a 40hp or 50hp powerhead.

Installing the exhaust cover.

position on the exhaust side of the cylinder block. Install the inner plate. Coat both sides of another **NEW** gasket with sealer, and then install the gasket and exhaust cover.

Secure the exhaust cover in place with the attaching hardware.

Coat both sides of a **NEW** gasket with sealer, and then place it in position on the cylinder block. Install the bypass covers and secure them in place with the attaching hardware. If a fuel pump is used, be sure the same bypass cover is installed in the position from which it was removed.

REED BOX INSTALLATION

Install the reed box and intake manifold onto the cylinder block. A gasket is usually installed on both sides of the reed box. The reeds and reed stops face inward toward the cylinder.

On 15 hp to 30 hp powerheads, a screw is installed in the center of the reed box into the cylinder block. This center screw is installed first, then the intake manifold is installed and secured in place.

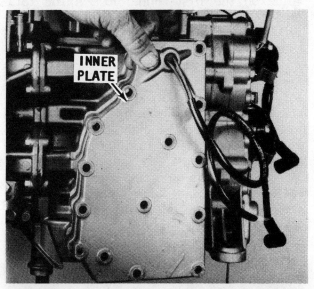

Installing the inner exhaust plate on a 40hp or 50hp powerhead.

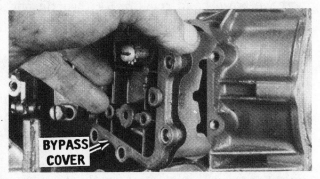

Installing the intake bypass covers. One is already in place.

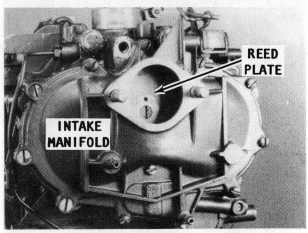

Installing the intake manifold over the reed plate.

HEAD AND POWERHEAD INSTALLATION

Place a **NEW** head gasket in place on the cylinder block. **NEVER** use automotive type head gasket sealer. The chemicals in the sealer will cause electrolytic action and eat the aluminum faster than you can get to the bank for money to buy a new cylinder block.

Install the head bolts and tighten them fingertight, then just a bit more. Tighten the bolts alternately and evenly in three rounds to the torque value specified in the Appendix. On the first round tighten the bolts to 1/2 the total torque value, on the second round to 3/4 the total torque value, and to the full torque value on the third and final round.

Install the assembled powerhead to the exhaust housing and tighten the attaching bolts alternately and evenly in three rounds to the torque value specified in the Appendix. Tighten the bolts to 1/2 the torque value on the first round, to 3/4 the total torque value on the second round, and to the full torque value on the third and final round.

Install all powerhead accessories including the flywheel, carburetor, magneto, starter, etc.. If any doubts or difficulties are encountered, follow the procedures outlined in the chapter covering the particular component. Connect the fuel lines, wiring, and battery cables.

The complete outboard unit is now ready to be started and "broke-in" according to the procedures outlined in the next section.

Tightening the head bolts to the proper torque value.

3-15 BREAK-IN PROCEDURES

Mount the engine in a test tank or body of water. If this is not possible, connect a flush attachment and garden hose to the lower unit. **NEVER** operate the engine above idle speed using the flush attachment.

If the engine is operated above an idle speed, the unit must be **IN GEAR,** preferable with a test wheel attached to the propeller shaft. If the engine is operated above an idle speed with no load on the propeller, the engine could **RUNAWAY** resulting in serious damage or destruction of the unit.

CAUTION: Water must circulate through the lower unit to the engine any time the engine is run to prevent damage to the water pump in the lower unit. Just five seconds without water will damage the water pump.

a- **ALWAYS** use OMC or BIA certified TC-W oil lubricant. Check the container label for the certification information.

b- With standard fuel tank: Use 25:1 fuel/oil mixture (16 fl oz -- 473 ml per 3-gallons of fuel), for the first 12 gallons of gasoline used, then use 50:1 mixture.

With AccuMix or VRO system: Use 50:1 fuel/oil mixture (8 fl oz -- 236 ml --per 3 gallons of fuel), in **ADDITION** to the AccuMix or VRO system for the first 12 gallons of fuel consumed.

c- Operate the powerhead at a fast idle with the unit in gear for the first 20 minutes. Check the fuel, exhaust, and water systems for leaks. Observe the water stream which should be visible at the rear -- starboard side -- of the powerhead. The stream will verify water is circulating through the powerhead.

d- During first hour of operation, do not exceed 1/2 throttle. Actually it is best to vary powerhead rpm every 10 to 15 minutes.

e- During the second hour of operation, increase powerhead rpm to 3/4 throttle and increase rpm to full throttle for short periods -- just a couple minutes. Change powerhead rpm every 10 to 15 minutes.

f- Next 8 hours of operation: Avoid full throttle for long periods of time. Change powerhead rpm every 10 to 15 minutes.

g- Check the VRO reservoir to verify oil has been used during this break-in period, before changing to clear fuel.

h- After 10 hours of operation, allow powerhead to cool, and then check torque value on the cylinder head bolts.

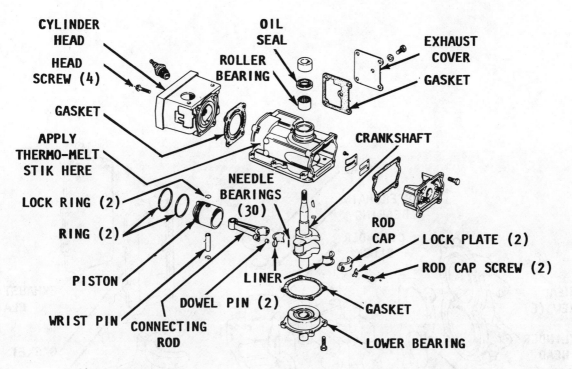

Exploded drawing of a typcial single cylinder powerhead with major parts identified.

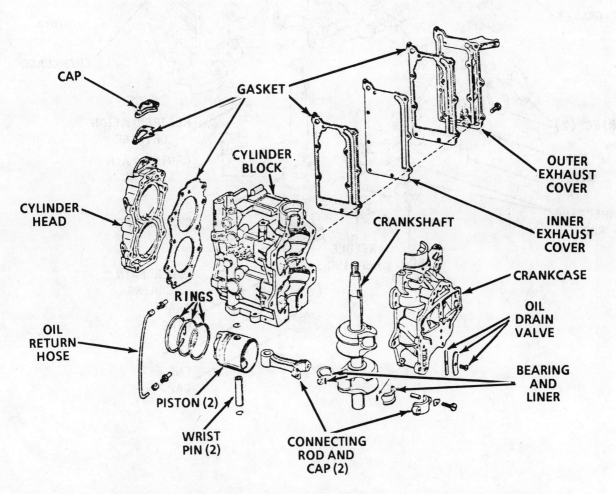

Exploded drawing of a 6hp and 8hp powerhead -- 1990 & on, with major parts identified.

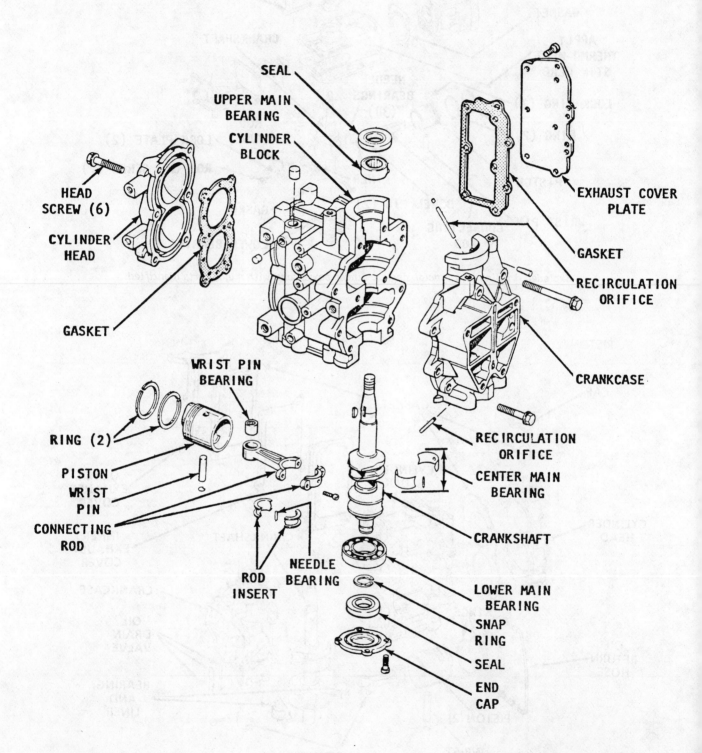

Exploded drawing of a 3hp and 4hp powerhead -- 1990 & on, with major parts identified.

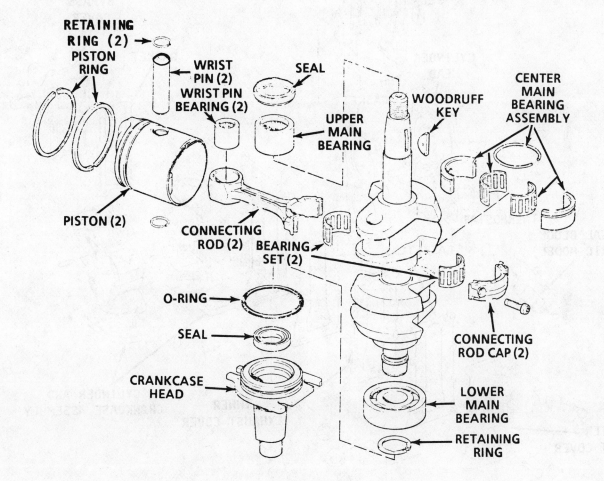

Exploded drawing of a 9.9hp and 15hp crankshaft assembly -- 1990 & on, with major parts identified. The block is shown on the next page.

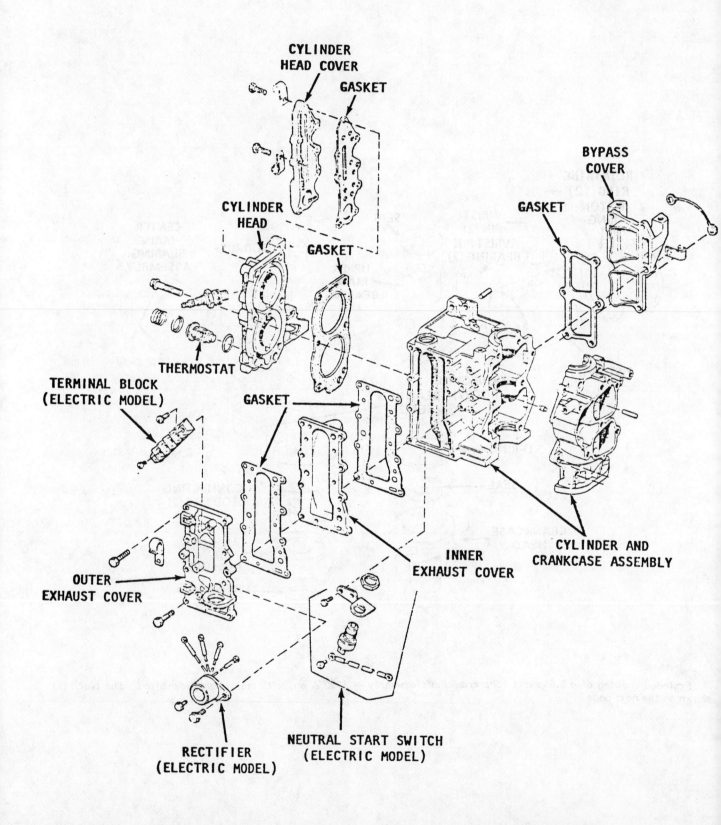

CYLINDER
HEAD COVER

GASKET

CYLINDER
HEAD

GASKET

BYPASS
COVER

GASKET

THERMOSTAT

TERMINAL BLOCK
(ELECTRIC MODEL)

GASKET

OUTER
EXHAUST COVER

INNER
EXHAUST COVER

CYLINDER AND
CRANKCASE ASSEMBLY

RECTIFIER
(ELECTRIC MODEL)

NEUTRAL START SWITCH
(ELECTRIC MODEL)

Exploded drawing of a 9.9hp and 15hp cylinder block -- 1990 and on, with major parts identified. The crankshaft assembly is shown on the previous page.

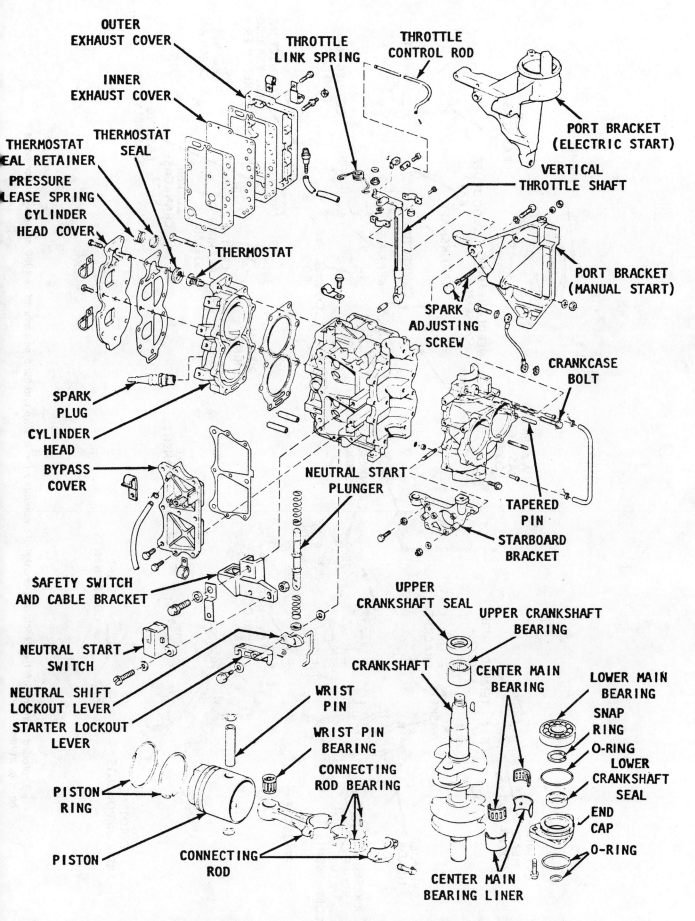

Exploded drawing of a 20hp, 25hp, and 30hp powerhead -- 1990 & on, with major parts identified.

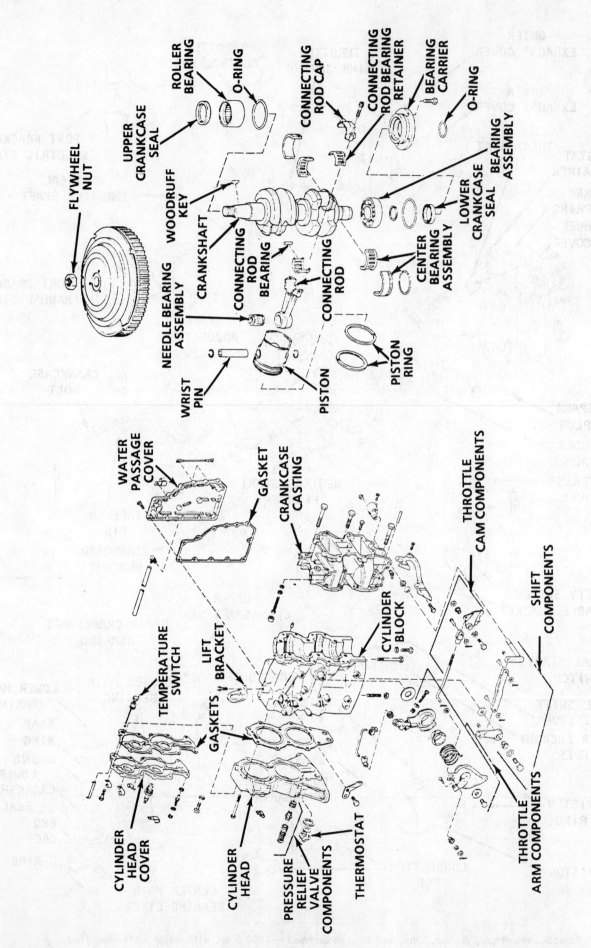

Exploded drawing of the block and crankshaft assembly for a 40hp and 50hp powerhead -- 1990 and on. Major parts have been identified.

4
FUEL

4-1 INTRODUCTION

The carburetion and ignition principles of two-cycle engine operation **MUST** be understood in order to perform a proper tune-up on an outboard motor.

If you have any doubts concerning your understanding of two-cycle engine operation, it would be best to study the operation theory section in the first portion of Chapter 3, before tackling any work on the fuel system.

Newer units (since about 1985), have been equipped with either an electric or manual primer system. The electric primer system is covered in Section 4-11. The manual primer system is covered in Section 4-12.

Since 1988, all 40hp and larger units have been equipped with an improved Variable Ratio Oil System, commonly referred to as VRO2.

Since 1986 all non-electric start 25hp and 30hp units and all 9.9hp and 15hp units have been equipped with an oil injection system known as AutoBlend. In 1987, the name was changed to AccuMix. The Accu-Mix system is available as an optional accessory on all smaller models down to the 4hp.

Both oil injection systems are covered in Section 4-13, the last section in this chapter.

4-2 GENERAL CARBURETION INFORMATION

The carburetor is merely a metering device for mixing fuel and air in the proper proportions for efficient engine operation. At idle speed, an outboard engine requires a mixture of about 8 parts air to 1 part fuel.

At high speed or under heavy duty service, the mixture may change to as much as 12 parts air to 1 part fuel.

Float Systems

A small chamber in the carburetor serves as a fuel reservoir. A float valve admits fuel into the reservoir to replace the fuel consumed by the engine.

Fuel level in each chamber is extremely critical and must be maintained accurately. Accuracy is obtained through proper adjustment of the float. This adjustment will provide a balanced metering of fuel to each cylinder at all speeds.

Following the fuel through its course, from the fuel tank to the combustion chamber of the cylinder, will provide an appreciation of exactly what is taking place. In order to start the engine, the fuel must be moved from the tank to the carburetor by a squeeze bulb installed in the fuel line.

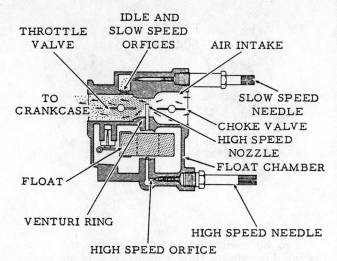

Fuel flow through the venturi, showing principle and related parts controlling intake and outflow.

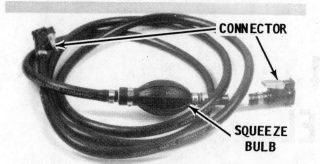

Typical fuel line with squeeze bulb and quick-disconnect fitting at each end. These items may be purchased as an assembled unit.

All powerheads covered in this manual are equipped with a manually-operated squeeze bulb in the line to transfer fuel from the tank to the powerhead until the powerhead starts.

After the powerhead starts, the fuel passes through the fuel pump to the carburetor. All systems have some type of filter installed somewhere in the line between the tank and the carburetor.

At the carburetor, the fuel passes through the inlet passage to the needle and seat, and then into the float chamber (reservoir). A float in the chamber rides up and down on the surface of the fuel. After fuel enters the chamber and the level rises to a predetermined point, a tang on the float closes the inlet needle and fuel entering the chamber is cutoff. When fuel leaves the chamber as the powerhead operates, the fuel level drops and the float tang allows the inlet needle to move off its seat and fuel once again enters the chamber. In this manner a constant reservoir of fuel is main-

tained in the chamber to satisfy the demands of the engine at all speeds.

A fuel chamber vent hole is located near the top of the carburetor body to permit atmospheric pressure to act against the fuel in each chamber. This pressure assures an adequate fuel supply to the various operating systems of the engine.

Air/Fuel Mixture

A suction effect is created each time the piston moves upward in the cylinder. This suction draws air through the throat of the carburetor. A restriction in the throat, called a venturi, controls air velocity and has the effect of reducing air pressure at this point.

The difference in air pressures at the throat and in the fuel chamber, causes the fuel to be pushed out metering jets extending down into the fuel chamber. When the fuel leaves the jets, it mixes with the air passing through the venturi. This air/fuel mixture should then be in the proper proportion for burning in the cylinder/s for maximum engine performance.

In order to obtain the proper air/fuel mixture for all engine speeds, high- and low-speed needle valves are installed. On late-model powerheads the high-speed needle

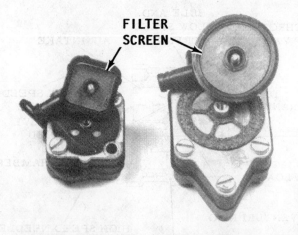

Two different type fuel pumps with the covers removed to show the filter screen. These pumps cannot be rebuilt. The only service possible is to clean the screens with solvent and then blow them dry.

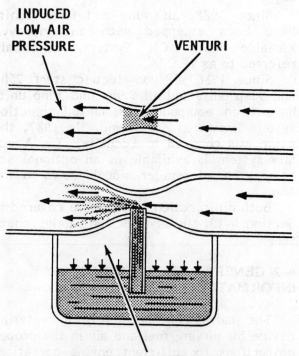

Air flow principle of a modern carburetor.

valve was replaced with a high-speed orifice. There is no adjustment with the orifice type. The needle valves are used to compensate for changing atmospheric conditions. Only 15% to 20% of the powerheads covered in this manual have an adjustable high- and low-speed needle valve.

Powerhead operation at sea level compared with performance at high altitudes is quite noticeable. A throttle valve controls the volume of air/fuel mixture drawn into the powerhead. A cold engine requires a richer fuel mixture to start and during the brief period it is warming to normal operating temperature. A choke valve is placed ahead of the metering jets and venturi to provide the extra amount of air required for start and while the engine is cold.

When this choke valve is closed, a very rich fuel mixture is drawn into the engine.

The throat of the carburetor is usually referred to as the "barrel." Carburetors installed on engines covered in this manual all have a single metering jet with a single throttle and choke plate. Single barrel carburetors are fed by one float and chamber.

4-3 FUEL SYSTEM

The fuel system includes the fuel tank, fuel pump, fuel filters, carburetor, connecting lines, with a squeeze bulb, and the associated parts to connect it all together. Regular maintenance of the fuel system to

Damaged piston, possibly caused by insufficient oil mixed with the fuel; using too-low an octane fuel; or using fuel that had "soured" (stood too long without a preservative additive).

obtain maximum performance, is limited to changing the fuel filter at regular intervals and using **FRESH** fuel. Even with the high price of fuel, removing gasoline that has been standing unused over a long period of time is still the easiest and least expensive preventive maintenance possible.

In most cases this old gas, even with some oil mixed with it, can be used without harmful effects in an automobile using regular gasoline.

Choke valve location in the carburetor venturi. The choke valve on most Johnson/Evinrude carburetors is located in front of the venturi.

Commercial additives, such as Sta-bil, may be used to prevent the fuel in the tank from "souring". Under favorable conditions, such additives will keep the fuel in a "useable" condition for up to twelve full months.

If a sudden increase in gas consumption is noticed, or if the engine does not perform properly, a carburetor overhaul, including boil-out, or replacement of the fuel pump may be required.

4-4 TROUBLESHOOTING

The following paragraphs provide an orderly sequence of tests to pinpoint problems in the system. If an outboard has not been used for some time and fuel has remained in the carburetor, it is possible that varnish may have formed. Such a condition could be the cause of hard starting, or complete failure of the powerhead to operate.

Fuel Problems

Many times fuel system troubles are caused by a plugged fuel filter, a defective fuel pump, or by a leak in the line from the fuel tank to the fuel pump. Aged fuel left in the carburetor and the formation of varnish could cause the needle to stick in its seat and prevent fuel flow into the bowl. A defective choke may also cause problems. **WOULD YOU BELIEVE,** a majority of starting troubles, which are traced to the fuel system, are the result of an empty fuel tank or aged fuel.

Fuel will begin to sour in three to four months and will cause engine starting problems. Therefore, leaving the motor setting idle with fuel in the carburetor, lines, or tank during the off-season, usually results in

Fouled spark plug, possibly caused by operator's habit of over-choking or a malfunction holding the choke closed. Either of these conditions delivered a too-rich fuel mixture to the cylinder.

very serious problems. A fuel additive such as Sta-Bil may be used to prevent gum from forming during storage or prolonged idle periods.

For many years there has been the widespread belief that simply shutting off the fuel at the tank and then running the engine until it stops is the proper procedure before storing the engine for any length of time. Right? **WRONG.**

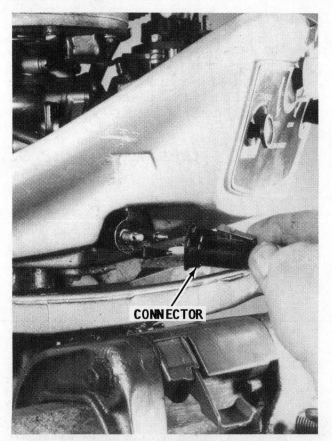

Female fuel line connector ready to be mated with the male portion of the connector.

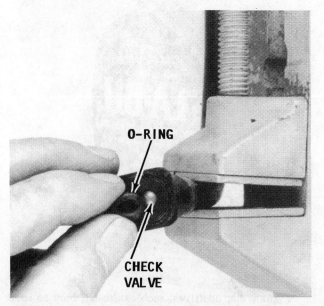

Fuel connector with the O-ring visible. These O-rings have a relative short life and must be replaced at regular intervals, as detailed in the text.

First, it is **NOT** possible to remove all fuel in the carburetor by operating the engine until it stops. Some fuel is trapped in the float chamber and other passages and in the line leading to the carburetor. The **ONLY** guaranteed method of removing **ALL** of the fuel is to take the time to remove the carburetor, and drain the fuel.

Secondly, if the engine is operated with the fuel supply shut off until it stops the fuel and oil mixture inside the engine is removed, leaving bearings, pistons, rings, and other parts with little protective lubricant, during long periods of storage.

Proper procedure involves: Shutting off the fuel supply at the tank; disconnecting the fuel line at the tank; operating the powerhead until it begins to run **ROUGH**; then stopping the engine, which will leave some fuel/oil mixture inside; and finally removing and draining the carburetor. By disconnecting the fuel supply, all **SMALL** passages are cleared of fuel even though some fuel is left in the carburetor. A light oil should be put in the combustion chamber as instructed in the Owner's Manual. On some of the carburetors covered in this manual, an orifice on the base of the fuel bowl can be removed to drain the fuel from the carburetor.

Choke Problems

When the engine is hot, the fuel system can cause starting problems. After a hot engine is shut down, the temperature inside the fuel bowl may rise to 200°F and cause the fuel to actually boil. All carburetors are vented to allow this pressure to escape to the atmosphere. However, some of the fuel may percolate over the main nozzle.

If the choke should stick in the open position, the engine will be hard to start. If the choke should stick in the closed position, the engine will flood making it very difficult to start.

In order for this raw fuel to vaporize enough to burn, considerable air must be added to lean out the mixture. Therefore, the only remedy is to remove the spark plugs; ground the leads; crank the powerhead through about ten revolutions; clean the plugs; install the plugs again; and start the engine.

If the needle valve and seat assembly is leaking, an excessive amount of fuel may enter the intake manifold in the following manner: After the powerhead is shut down, the pressure left in the fuel line will force fuel past the leaking needle valve. This extra fuel will raise the level in the fuel bowl and cause fuel to overflow into the intake manifold.

A continuous overflow of fuel into the intake manifold may be due to a sticking inlet needle or to a defective float which would cause an extra high level of fuel in the bowl and overflow into the intake manifold.

The choke plays a most important role during engine start and in controlling the amount of air entering the carburetor, under various load conditions.

The spark plug high tension leads should be disconnected and grounded any time the fuel system is opened for service or tests. An accidental spark could ignite fuel or fuel vapors causing serious personal injury.

FUEL PUMP TESTS

SAFETY CAUTION

Gasoline will be flowing during this test. Therefore, ground the high-tension lead/s to prevent an accidental spark from igniting fuel or fuel fumes.

Testing System with Squeeze Bulb

An adequate safety method, is to ground each spark plug lead. Disconnect the fuel line at the carburetor. Place a suitable container over the end of the fuel line to catch the fuel discharged. Insert a small screwdriver into the end of the line to open the check valve, and then squeeze the primer bulb and observe if there is satisfactory fuel flow from the line.

If there is no fuel discharged from the line, the check valve in the squeeze bulb may be defective, or there may be a break or obstruction in the fuel line.

If there is a good fuel flow, then crank the powerhead. If the fuel pump is operating properly, a healthy stream of fuel should pulse out of the line.

Continue cranking the powerhead and catching the fuel for about 15 pulses to determine if the amount of fuel decreases with each pulse or maintains a constant amount. A decrease in the discharge indicates a restriction in the line. If the fuel line is plugged, the fuel stream may stop. If there is fuel in the fuel tank but no fuel flows out the fuel line while the powerhead is being cranked, the problem may be in one of several areas:

Typical fuel line installation to the carburetor. This line is disconnected during certain fuel flow tests.

1- Plugged fuel line from the fuel pump to the carburetor.

2- Defective O-ring in fuel line connector into the fuel tank.

3- Defective O-ring in fuel line connector into the engine.

4- Defective fuel pump.

5- The line from the fuel tank to the fuel pump may be plugged; the line may be leaking air; or the squeeze bulb may be defective.

6- Defective fuel tank.

7- If the engine does not start even though there is adequate fuel flow from the fuel line, the fuel inlet needle valve and the seat may be gummed together and prevent adequate fuel flow.

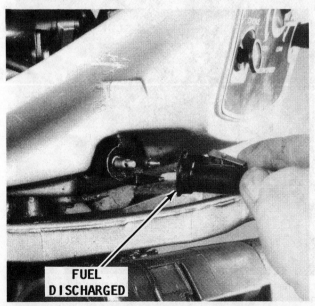

The fuel line quick-disconnect fitting at the engine separated in preparation to making a fuel flow test.

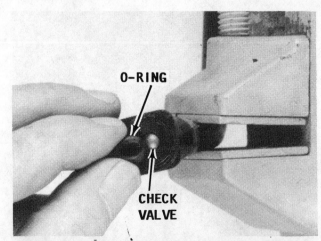

Fuel connector with the O-ring visible. These O-rings have a relative short life and may be the source of fuel problems. The O-rings should be replaced on a regular basis.

FUEL LINE TEST

On most installations, the fuel line is provided with quick-disconnect fittings at the tank and at the powerhead. If there is reason to believe the problem is at the quick-disconnects, the hose ends can be replaced as an assembly, or new O-rings may be installed. A supply of new O-rings should be carried on board for use in isolated areas where a marine store is not available. For a small additional expense, the entire fuel line can be replaced and eliminate this entire area as a problem source for many future seasons.

The primer squeeze bulb can be replaced in a short time. A squeeze bulb assembly kit, complete with the check valves installed, may be obtained from the local Johnson/Evinrude dealer. The replacement kit will also include two tie straps to secure the bulb properly in the line.

An arrow is clearly visible on the squeeze bulb to indicate the direction of fuel flow. The squeeze bulb **MUST** be instal-

A typical squeeze bulb with the directional arrow clearly visible. The squeeze bulb must be installed with the arrow pointing in the direction of fuel flow, toward the powerhead.

led correctly in the line because the check valves in each end of the bulb will allow fuel to flow in **ONLY** one direction. Therefore, if the squeeze bulb should be installed backwards, in a moment of haste to get the job done, fuel will not reach the carburetor.

To replace the bulb, first unsnap the clamps on the hose at each end of the bulb. Next, pull the hose out of the check valves at each end of the bulb. New clamps are included with a new squeeze bulb.

If the fuel line has been exposed to considerable sunlight, it may have become hardened, causing difficulty in working it over the check valve. To remedy this situation, simply immerse the ends of the hose in boiling water for a few minutes to soften the rubber. The hose will then slip onto the check valve without further problems. After the lines on both sides have been installed, snap the clamps in place to secure the line. Check a second time to be sure the arrow is pointing in the fuel flow direction, **TOWARDS** the powerhead.

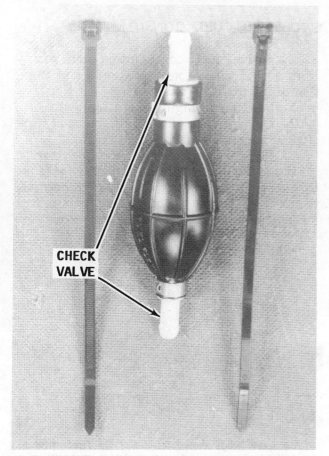

CHECK VALVE

Parts in a squeeze bulb replacement kit include the squeeze bulb, two check valves, and two tie straps to secure the bulb in the line.

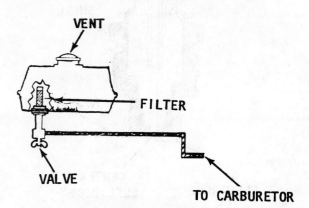

VENT

FILTER

VALVE

TO CARBURETOR

Simple line drawing of a powerhead mounted fuel tank with the filter inside the tank and an external shut-off valve.

Integral Fuel Tank System

On small powerheads equipped with an integral fuel tank above the flywheel, or else where on the powerhead, the filter in the tank may be plugged. To check the fuel flow, disconnect the fuel line at the carburetor and fuel should flow from the line if the shut-off valve is open. If fuel is not present, the usual cause is a plugged filter in the fuel tank. It is very difficult to determine if a porcelain-type filter is plugged. The general appearance that the filter is satisfactory may be a false indication. Therefore, if the filter is suspected, the best remedy is replacement. The cost is very modest and this one area is thus eliminated as a problem source.

Would you believe, many times lack of fuel at the carburetor is caused because the vent on the fuel tank was not opened.

To remove the filter, first remove the fuel tank filler cap, turn the outboard upside down and drain the tank. Next, disconnect the fuel line to the carburetor. Remove the fuel shut-off valve. In most cases, the filter will remain with the valve and can be separated from the valve by removing the fitting. If the fitting and filter do not remain with the valve when it is removed, then remove the fitting and filter from the tank.

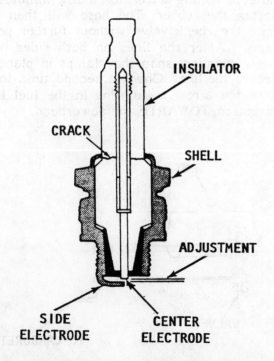

Cross-section view of a spark plug showing the principle parts with important comments for satisfactory service.

ROUGH ENGINE IDLE

If a powerhead does not idle smoothly, the most reasonable approach to the problem is to perform a tune-up to eliminate such areas as: defective points; faulty spark plugs; and synchronization out of adjustment.

Other problems that can prevent an engine from running smoothly include: An air leak in the intake manifold; uneven compression between the cylinders; and sticky or broken reeds.

Of course any problem in the carburetor affecting the air/fuel mixture will also prevent the engine from operating smoothly at idle speed. These problems usually include: Too high a fuel level in the bowl; a heavy float; leaking needle valve and seat; defective automatic choke; and improper idle and high-speed needle valve adjustments.

"Sour" fuel (fuel left in a tank without a preservative additive) will cause an engine to run rough and idle with great difficulty.

EXCESSIVE FUEL CONSUMPTION

Excessive fuel consumption can result from one of three conditions, or a combination of all three.

1- Inefficient engine operation.
2- Damaged condition of the hull, including excessive marine growth.
3- Poor boating habits of the operator.

If the fuel consumption suddenly increases over what could be considered normal, then the cause can probably be attributed to the engine or boat and not the operator.

The spark plugs should be one of the first areas to check if the powerhead is not operating properly. On the small powerheads, the point set may need attention. This set of points is unfit for service due to oxidation.

Marine growth on the hull can have a very marked effect on boat performance. This is why sail boats always try to have a haul-out as close to race time as possible. While you are checking the bottom take note of the propeller condition. A bent blade or other damage will definitely cause poor boat performance.

If the hull and propeller are in good shape, then check the fuel system for possible leaks. Check the line between the fuel pump and the carburetor while the engine is running and the line between the fuel tank and the pump when the engine is not running. A leak between the tank and the pump many times will not appear when the engine is operating, because the suction created by the pump drawing fuel will not allow the fuel to leak. Once the engine is turned off and the suction no longer exists, fuel may begin to leak.

If a minor tune-up has been performed and the spark plugs, points, and synchronization are properly adjusted, then the problem most likely is in the carburetor, indicating an overhaul is in order. Check for leaks at the needle valve and seat. Use extra care when making any adjustments affecting the fuel consumption, such as the float level or automatic choke.

ENGINE SURGE

If the engine operates as if the load on the boat is being constantly increased and decreased, even though an attempt is being made to hold a constant engine speed, the problem can most likely be attributed to the fuel pump.

Operational description and service procedures for the fuel pump are given in Section 4-9.

4-5 JOHNSON/EVINRUDE CARBURETORS

This section provides complete detailed procedures for removal, disassembly, cleaning and inspecting, assembling including bench adjustments, installation, and operating adjustments for the three type OMC carburetors installed on powerheads covered in this manual. For convenience, the carburetors are identified as **"A"**, **"B"**, and **"C"**. The carburetors within a certain type may vary as to the number of barrels and whether the low-speed and high-speed orifices are adjustable or fixed.

These carburetors are equipped with either a manual choke, or an electric choke with a manual backup. The manual backup permits the operator to operate the choke in the event the battery is dead. The following table lists the types of carburetors, the horsepower and model engines equipped with each. Many of the carburetors are used on powerheads prior to the coverage of this manual -- 1990 and on. For information prior to 1990 -- see Seloc's Johnson/Evinrude Tune-Up and Repair Manual -- Volume II.

The following table clearly indicates the size powerhead and model years for each of the type carburetors covered in this manual.

Marine growth allowed to accumulate on the lower unit will create "drag" and seriously hamper boat performance, and corrode the metal if it is not removed.

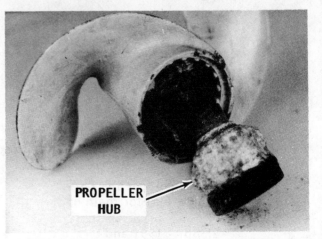

PROPELLER HUB

A corroded hub on a small engine propeller. Replacement of this propeller will be less expensive than the cost of a rebuild.

CARBURETOR INSTALLATIONS

Type "A"

Single barrel, front draft,
with adjustable low-speed
and high-speed fixed orifice
in a removable nozzle.

Colt -- 1990 Only
3hp -- 1991 and on
4hp -- 1990 and on

Type "A"

Single barrel, front draft,
with adjustable low-speed
needle valve/orifice.
High-speed is fixed orifice
in float bowl.

6hp -- 1990 and on
8hp -- 1990 and on

Type "A"

Single barrel, front draft,
with adjustable low-speed
needle valve/orifice
with special packing.
High-speed is fixed orifice
in a nozzle.

9.9hp & 15hp -- 1990 and on

Type "B"

Single barrel, front draft,
with adjustable low-speed valve/orifice
intermediate fixed orifice,
and fixed high-speed orifice.

20hp -- 1990 and on
25hp -- 1990 and on
30hp -- 1990 and on
40hp -- 1990 and on
50hp -- 1990 and on

Type "C"

Single barrel, front draft,
with adjustable low-speed
screw. High-speed is fixed
orifice in carburetor main body.

2.3hp -- 1991 and on
3.3hp -- 1991 and on

4-6 TYPE "A" CARBURETOR
(See Powerhead Listing -- This Page)

The illustrations in this section are for a typical installation. Appearances may differ, but the procedures are valid.

REMOVAL

1- Disconnect the fuel line from the powerhead or from the fuel tank at the quick-disconnect fitting.

2- Remove the hood assembly from the powerhead. Remove the choke and throttle linkage to the carburetor.

3- Remove the fuel line from the carburetor. This may be accomplished by either one of two methods: On powerheads equipped with a self-contained fuel tank, close the fuel shut-off valve, located just below the tank. Disconnect both ends, and then remove the copper fuel line between the shut-off valve and the carburetor.

4- On powerheads utilizing a separate fuel tank, remove the tie-strap or clamp securing the rubber hose connecting the fuel connector to the carburetor. Remove the rubber hose from the carburetor. Remove

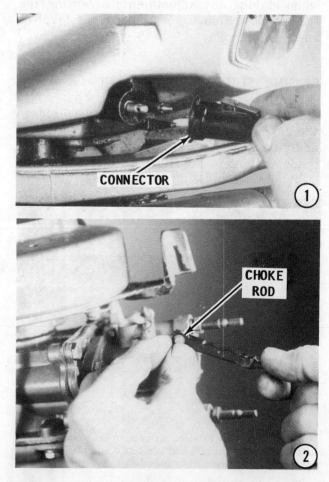

CONNECTOR

①

CHOKE ROD

②

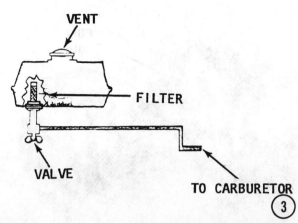

VENT

FILTER

VALVE

TO CARBURETOR

③

the nuts securing the carburetor to the crankcase. The carburetor may have to be moved slightly forward as the nuts are loosened in order to obtain enough room for the nuts to clear the studs. After the nuts are clear, lift the carburetor from the intake manifold.

5- Remove and **DISCARD** the gasket from the intake manifold or the carburetor, if the gasket adhered to the carburetor when it was removed. A new gasket is included in a carburetor repair kit.

DISASSEMBLING CARBURETOR "A"

When disassembling the carburetor, **ALWAYS** try to work on a clean work surface, such as a clean shop cloth on the bench. After the parts have been cleaned, as described in the Cleaning and Inspecting paragraphs of this section, take extra precautions to keep the parts clean to prevent contamination from entering passageways.

6- The carburetor being serviced is equipped with a fixed high-speed orifice in one of three locations, as indicated in the listing on Page 4-10. On 3hp and 4hp powerheads, 1990 and on, the high-speed orifice is located in a removable nozzle. The 6hp and 8hp powerheads, 1990 and on, have the fixed high-speed orifice in the float bowl. The 9.9hp and 15hp powerheads have the fixed high-speed orifice in a nozzle. The high-speed orifice may be located on the side of the nozzle well as shown in the illustration on the following page.

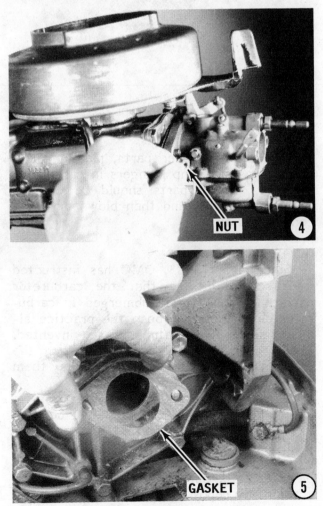

NUT   ④

GASKET   ⑤

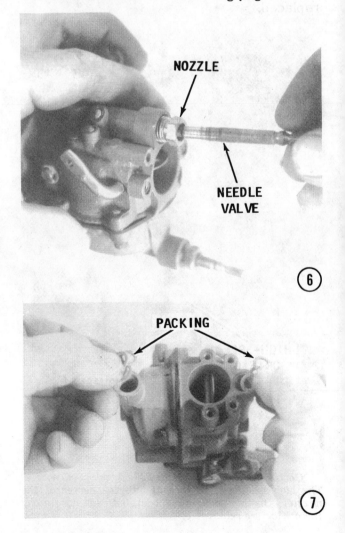

NOZZLE

NEEDLE VALVE

⑥

PACKING

⑦

7- Use a small screwdriver, and remove the packing from the needle valve cavities in the carburetor body.

8- If installed, remove the high-speed orifice, using the proper size screwdriver.

9- Turn the carburetor upside down. Remove the five attaching screws, and then lift the bowl free of the carburetor.

10- Remove the hinge pin and lift the float from the carburetor bowl cavity. As the float is lifted from the carburetor, observe the small spring attached to the needle. The needle will come out with the float assembly. Remove the needle seat from the carburetor.

GOOD WORDS

Further disassembly of the carburetor is not necessary. The nozzle in the center of the carburetor does **NOT** have to be removed in order to properly clean the carburetor. However, the nozzle can be removed with a screwdriver if it is damaged and needs to be replaced.

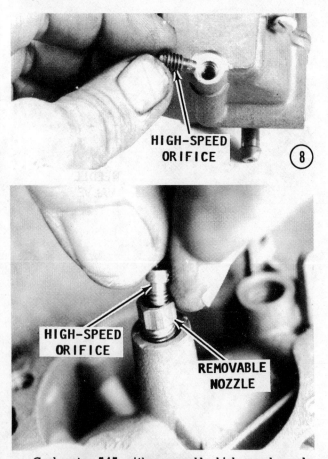

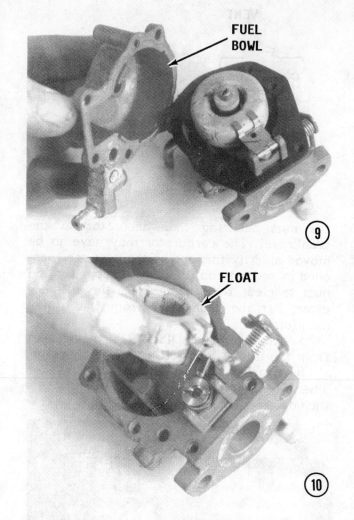

Carburetor "A" with removable high-speed nozzle. The high-speed orifice threads into the center of the nozzle. On this particular model, both the nozzle and the orifice are removable.

CLEANING AND INSPECTING

NEVER dip rubber parts, plastic parts, diaphragms, or pump plungers in carburetor cleaner. These parts should be cleaned **ONLY** in solvent, and then blown dry with compressed air.

GOOD WORDS

Since about 1983, OMC has instructed their service people that the carburetor parts should **NOT** be submerged in carburetor cleaner, as has been the practice almost since carburetors were invented. Their approved procedure is to place the parts in a shallow tray and then spray them with an aerosol carburetor cleaner.

A syringe, short section of clear plastic hose, and Isopropyl Alcohol should be used to clear passages and jets.

Blow out all passages in the castings with compressed air. Check all parts and passages to be sure they are not clogged or contain any deposits. **NEVER** use a piece of

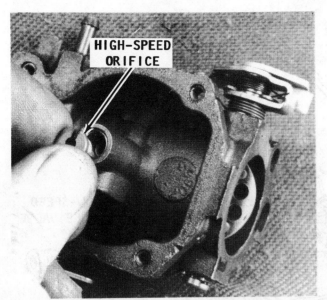

Carburetor with a non-removeable high-speed nozzle. The high-speed orifice threads into the center of the nozzle, but only the orifice is removeable.

wire or any type of pointed instrument to clean drilled passages or calibrated holes in a carburetor.

Move the throttle shaft back-and-forth to check for wear. If the shaft appears to be too loose, replace the complete throttle body because individual replacement parts are **NOT** available.

If any part of the float is damaged, the unit must be replaced. Check the float arm needle contacting surface and replace the float if this surface has a groove worn in it.

Inspect the tapered section of the idle adjusting needles and replace any that have developed a groove.

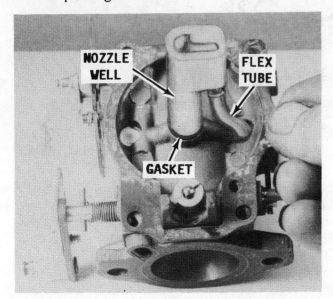

Nozzle design used on Type "A" carburetors installed on 6hp and 8hp powerheads.

If a high-speed orifice is installed on the carburetor being serviced, check the orifice for cleanliness. The orifice has a stamped number. This number represents a drill size. Check the orifice with the shank of the proper size drill to verify the proper orifice is used. The correct size orifice for the powerhead and carburetor being serviced may be obtained from the local OMC dealer.

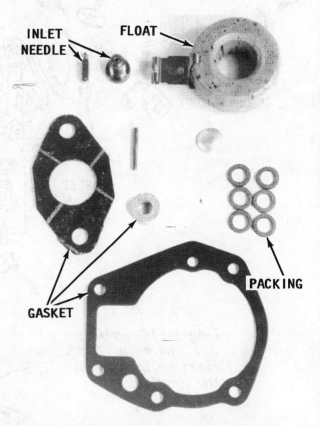

Parts necessary to properly rebuild a carburetor are included in a Johnson/Evinrude repair kit.

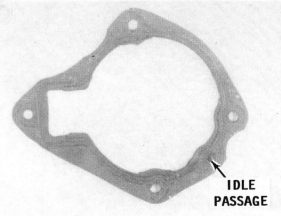

Gasket used between the bowl and the carburetor showing the opening to allow fuel to leave the bowl and enter the idle passage of the carburetor.

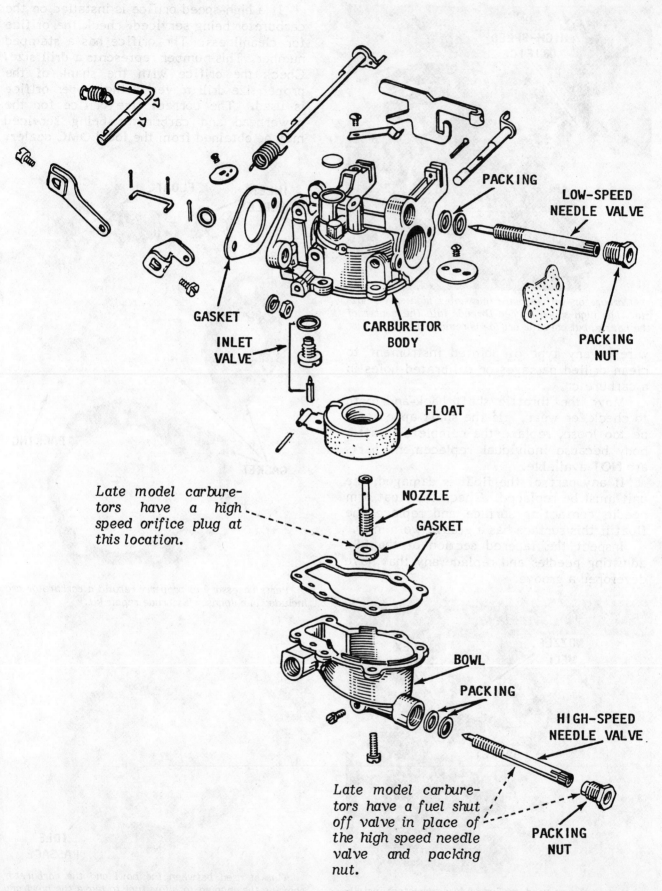

PACKING

LOW-SPEED
NEEDLE VALVE

PACKING
NUT

GASKET

INLET
VALVE

CARBURETOR
BODY

FLOAT

Late model carbure-
tors have a high
speed orifice plug at
this location.

NOZZLE

GASKET

BOWL

PACKING

HIGH-SPEED
NEEDLE VALVE

Late model carbure-
tors have a fuel shut
off valve in place of
the high speed needle
valve and packing
nut.

PACKING
NUT

Exploded drawing of the type "A" carburetor installed on small powerheads equipped with the fuel tank attached to the powerhead. The filter on these units may be in the tank or in the fuel pump.

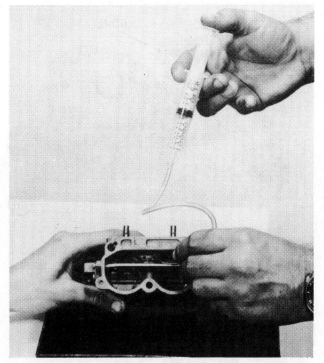

Using a Syringe, piece of clear plastic hose, and Isopropyl Alcohol to clean the idle air bleed system.

Most of the parts that should be replaced during a carburetor overhaul are included in overhaul kits available from your local marine dealer. One of these kits will contain a matched fuel inlet needle and seat. This combination should be replaced each time the carburetor is disassembled as a precaution against leakage.

ASSEMBLING TYPE "A" CARBURETOR

1- Install a **NEW** inlet nozzle seat with a **NEW** gasket into the carburetor. Slide a **NEW** gasket down over the inlet nozzle onto the surface of the carburetor.

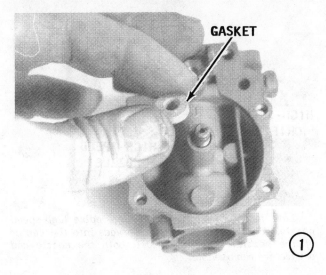

2- Attach a **NEW** inlet needle onto the spring included in the carburetor repair kit, and slip the spring over the edge of the float, as shown. Apply just a drop of oil to the inlet needle and then lower the needle into the seat. Install a **NEW** hinge pin included in the kit.

3- Hold the carburetor in a horizontal position, as shown, and observe the attitude of the float. The float must be level (parallel) with the surface of the carburetor. If the float is not level, use a pair of needle-nose pliers and **CAREFULLY** bend the float tab **SQUARELY** until the float is in the correct position (level with the carburetor). Check to be sure the float is square with the carburetor cavity (one side is not further away than the other).

Measuring Float Drop

4- Allow the float to drop under its own weight, and then measure the distance between the base of the carburetor body and the lowest edge of the float. Dimension should be as listed.

    3 and 4hp: 1-1/8" to 1-1/2" (28-33mm)
    6 and 8hp: 1" to 1-3/8" (25 to 35mm)
    9.9 and 15hp: 1-1/8" to 1-1/2" (28-33mm)

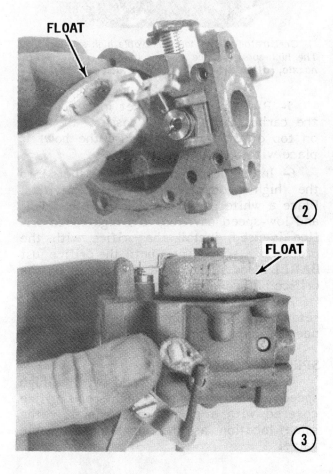

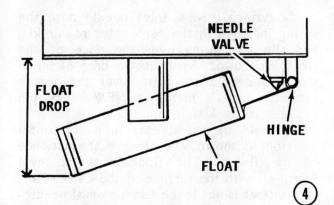

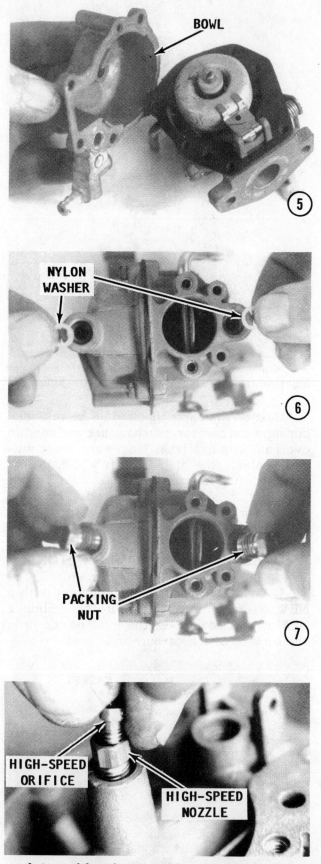

Carburetor with a non-removable high-speed nozzle.
The high-speed orifice threads into the center of the
nozzle, but only the orifice is removable.

5- Place the bowl gasket in position on
the carburetor, and then position the bowl
on top of the gasket. Secure the bowl in
place with the five screws.

6- Insert three new packing washers into
the high-speed and low-speed cavities.
Place a white plastic washer into the high-
and low-speed cavities. If a high-speed orifice is used, install the orifice with the
proper size screwdirver until the orifice just
BARELY seats. Install a new gasket on the
orifice plug, and then the plug.

7- Start the packing nuts into the carburetor, but **ONLY** far enough to allow the
needle valves to be installed.

SPECIAL WORDS

Carburetors not equipped with a high
speed needle valve have a high speed orifice
covered with a plug in approximately the
same location as the high speed needle
valve.

Late model carburetor with removeable high-speed
nozzle. The high-speed orifice threads into the center
of the nozzle. On this model, both the nozzle and
orifice are removeable.

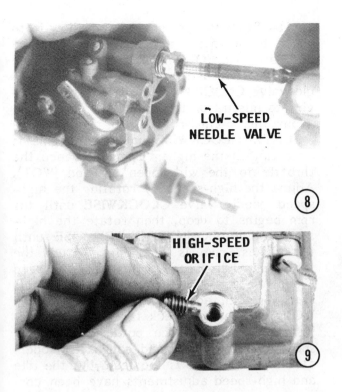

LOW-SPEED NEEDLE VALVE

⑧

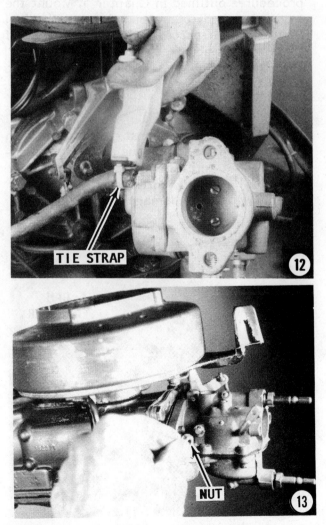

HIGH-SPEED ORIFICE

⑨

9- Install the high-speed orifice, if one is used, into the float bowl. **ALWAYS** take time to use the proper size screwdriver to prevent damaging the orifice. Tighten the orifice only until it just seats.

10- Install the high-speed orifice plug using a **NEW** gasket. Tighten the plug securely.

INSTALLATION

11- Check to be sure the surface of the intake manifold is clean and free of any old gasket material. Place a **NEW** gasket in position on the intake manifold.

12- Connect the fuel line to the carburetor, or on engines equipped with a self-contained fuel tank, replace the line between the shut-off valve and the carburetor.

13- Slide the carburetor over the intake manifold mounting studs, and then secure it in place with the two nuts. Tighten the nuts **ALTERNATELY** to avoid warping the carburetor body.

8- Install the low-speed needle valve into the upper cavity. After the valve is seated **LIGHTLY**, back it out (**COUNTER-CLOCKWISE**) 1-1/2 turns. Now, tighten the packing nut until there is drag on the needle valve, but the valve may still be rotated by hand, but with just a little difficulty.

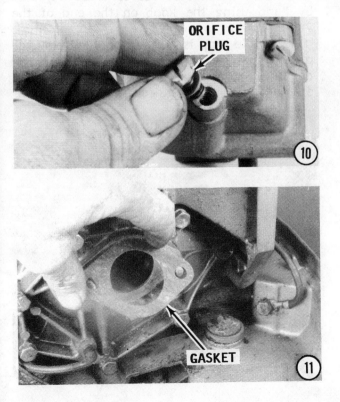

ORIFICE PLUG

⑩

GASKET

⑪

TIE STRAP

⑫

NUT

⑬

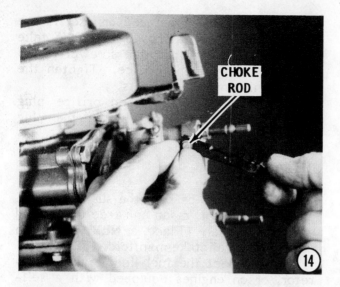

14- Connect the manual choke rod and the front cover onto the carburetor. Thread the knobs onto the low- and high-speed needle valves.

15- Check the synchronization of the fuel and ignition systems according to the procedures outlined in Chapter 5. Mount the engine in a body of water. Connect the fuel line to a fuel source. Prime the engine. If the engine is equipped with a self-contained fuel tank, open the fuel shut-off valve and allow the carburetor to fill with fuel.

CAUTION: Water must circulate through the lower unit to the engine any time the engine is run to prevent damage to the water pump in the lower unit. Just five seconds without water will damage the water pump.

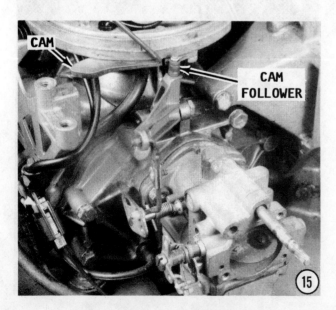

16- Start the powerhead and allow it to warm to operating temperature. Shift the outboard into **FORWARD** gear. Adjust the low-speed idle by turning the low-speed needle valve **CLOCKWISE** until the engine begins to misfire or the rpm drops noticeably. From this point, rotate the needle valve **COUNTERCLOCKWISE** until the engine is operating at the highest rpm. Advance the throttle to the wide open position (WOT). Adjust the high-speed by rotating the high-speed needle valve **CLOCKWISE** until the rpm begins to drop, then rotate the high-speed valve **COUNTERCLOCK WISE** until the highest rpm is reached. Return the throttle to idle speed. Adjust the idle speed a second time as described earlier in this step. Again, advance the throttle to the WOT position and check the high-speed adjustment. Return the powerhead to idle speed. If the powerhead coughs and operates as if the fuel is too lean, but the idle and high-speed adjustments have been correctly made, then recheck the synchronization between the fuel and ignition systems. Now, shut off the fuel supply and allow the powerhead to run until it first begins to misfire from lack of fuel. Retard the spark and shut the engine down. Tighten the sleeve nut securely to prevent the needle valves from changing position through powerhead vibration while it is operating, but still allow the needle valves to be adjusted by hand using the knob on the end of the valve.

The idle stop is located on the port side of the powerhead on the outside of the cowling. Adjust the nylon screw inward or outward to obtain the desired idle speed.

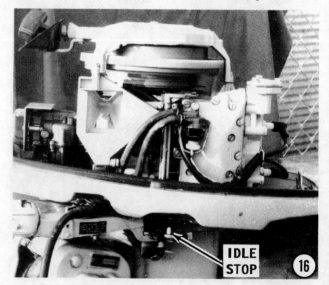

4-7 TYPE "B" CARBURETOR
(See Powerhead Listing — Page 4-10)

The illustrations in this section are for a typical Carburetor "B" installation. Appearances may differ, but the procedures are valid.

The Type **"B"** carburetor is a single barrel, front draft, with a adjustable low-speed needle, intermediate fixed orifice, and fixed high-speed orifice.

This carburetor is used on the 20hp, 25hp, 30hp, 40hp and 50hp powerheads covered in this manual.

Two carburetors are installed on each powerhead -- one for each cylinder.

The Type **"B"** carburetor utilizes an electric choke mounted on the side of the powerhead. On most models, the electric choke does not have to be removed in order to remove and service the carburetor.

REMOVAL

Preliminary Tasks

1- Disconnect the battery cables at the battery as a precaution against an accidental spark igniting the fuel or fuel fumes

present during the service work. Disconnect the fuel line from the junction on the port side of the powerhead. This is the line from the fuel pump to the carburetors. Remove the cowling. Remove the low-speed adjustment knobs at the bottom of the air silencer and remove the silencer cover.

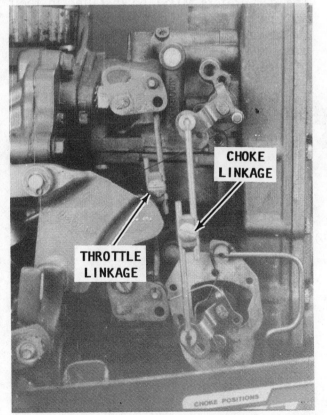

Starboard side view showing the throttle linkage and the choke linkage.

One method of disconnecting the fuel line is to remove the fuel pump cover.

2

4

2- Disconnect the hose at the bottom of the air silencer. This hose is connected to a fitting at the bottom of the crankcase, and then the air silencer.

3- Disconnect the choke and throttle linkage between the two carburetors on the port side of the engine. The linkage will snap out of a retainer on the carburetor cams. The nylon retainers **MUST** be removed before the carburetor is immersed in any type of cleaning solution. Disconnect the

choke wire from the electric choke, by first removing the O-ring and coil spring.

4- Remove the four nuts securing the carburetor to the intake manifold. Identify the top carburetor as an aid to installation. The carburetors **MUST** be installed in their original positions. The fuel connections and the choke arrangement is different for each carburetor. Both carburetors can now be removed **TOGETHER** as an assembly. The carburetor parts should be kept with the individual unit to ensure all items are installed back in their original positions.

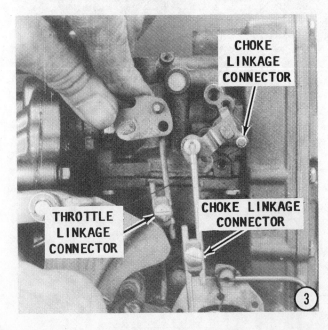

3

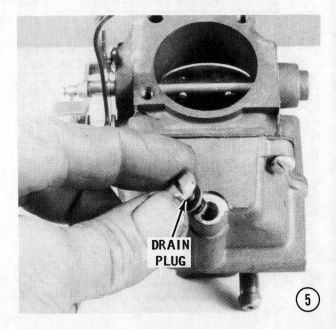

5

Only one carburetor will be rebuilt in the following procedures. Service of the second unit is to be performed in the same manner.

DISASSEMBLING

5- Disconnect the fuel hoses between the two carburetors. This is accomplished by simply cutting the tie strap or working the clip securing the hose to the carburetor fitting. Remove the drain plug or the high-speed orifice plug from the bottom of the carburetor bowl.

6- Remove the orifice from the float bowl, using the proper size screwdriver.

Remove the screw plug and washer from the top of the carburetor. Remove the orifice using the proper size screwdriver.

If the carburetor being serviced has the intermediate orifice, remove the plug and gasket from the starboard side of the carburetor. Remove the intermediate orifice using the proper size screwdriver.

7- Turn the carburetor upside-down and remove the screws securing the float bowl to the carburetor body. Lift the bowl free of the carburetor.

8- Remove and **DISCARD** the bowl gasket and the small gasket around the high-speed nozzle. The high-speed nozzle is **NOT** removable.

9- Remove the float hinge pin, and then remove the float.

10- Remove the inlet needle and seat from the carburetor body.

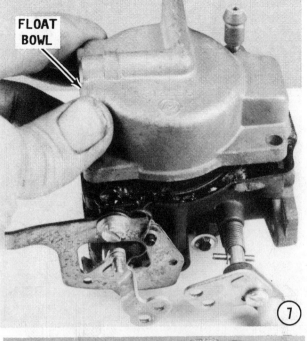

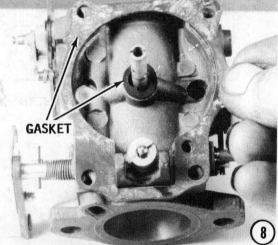

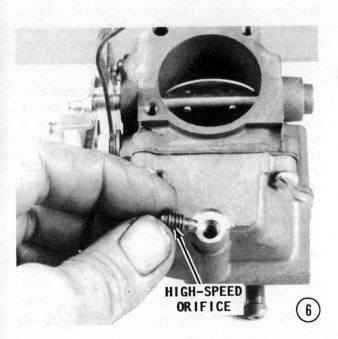

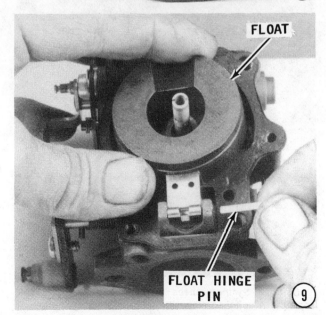

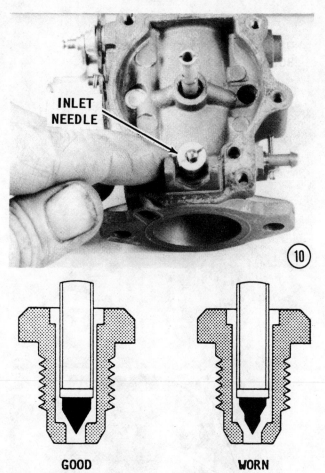

Cross-section drawing to allow comparison of a new needle and seat with one badly worn. Notice how the edges of the worn valve and the seat have become beveled.

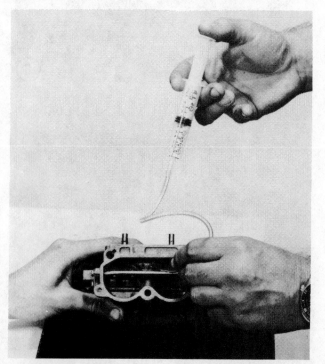

Using a Syringe, piece of clear plastic hose, and Isopropyl Alcohol to clean the idle air bleed system.

CLEANING AND INSPECTING

NEVER dip rubber parts, plastic parts, nylon parts, diaphragms, or pump plungers in carburetor cleaner. These parts should be cleaned **ONLY** in solvent, and then blown dry with compressed air.

GOOD WORDS

Since about 1983, OMC states that carburetor parts should **NOT** be submerged in carburetor cleaner, as has been the practice almost since carburetors were invented. Their approved procedure is to place the parts in a shallow tray and then spray them with an aerosol carburetor cleaner as depicted in the accompanying illustration.

A syringe, short section of clear plastic hose, and Isopropyl Alcohol should be used to clear passages and jets.

Blow out all passages in the castings with compressed air. Check all parts and passages to be sure they are not clogged or contain any deposits. **NEVER** use a piece of wire or any type of pointed instrument to clean drilled passages or calibrated holes in a carburetor.

Move the throttle shaft back-and-forth to check for wear. If the shaft appears to be too loose, replace the complete throttle body because individual replacement parts are **NOT** available.

If any part of the float or spring is damaged, the unit must be replaced. Check the float arm needle contacting surface and replace the float if this surface has a groove worn in it.

Check the low-speed and high-speed orifices for cleanliness. The orifice has a stamped number. This number represents a drill size. The local OMC dealer will be able to provide the correct size orifice for the powerhead and carburetor being serviced.

Most of the parts that should be replaced during a carburetor overhaul are included in overhaul kits available from your local marine dealer. One of these kits will contain a matched fuel inlet needle and seat. This combination should be replaced each time the carburetor is disassembled as a precaution against leakage.

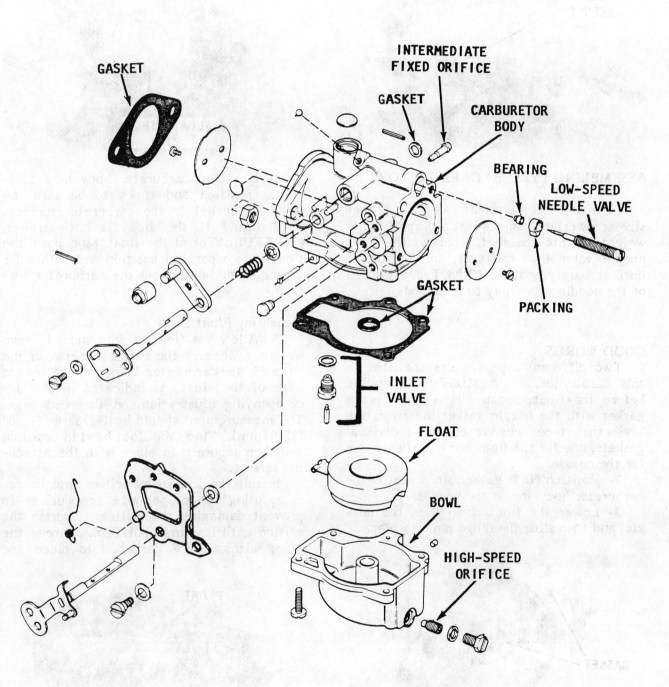

Exploded drawing of the type "B" carburetor with adjustable low-speed needle, fixed intermediate orifice, and fixed high-speed orifice. This carburetor is installed on the Model 20hp thru 50hp powerheads -- 1990 & on. Major parts are identified.

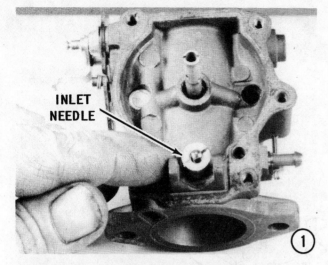

①

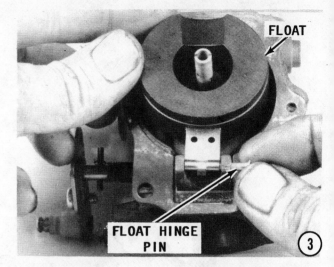

②

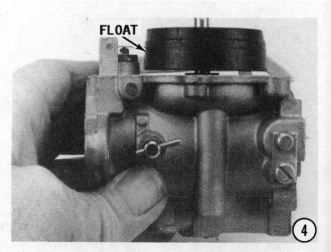

③

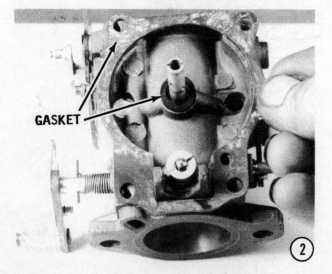

④

ASSEMBLING TYPE "B" CARBURETOR

1- Install the inlet seat using the proper size screwdriver. Inject just a drop of light-weight oil into the seat. Thread the inlet needle valve into the seat, and tighten it until it barely seats. **DO NOT** overtighten or the needle valve may be damaged.

GOOD WORDS

Two different type gaskets are used on this carburetor. One **MUST** be installed before the float, because it is a one-piece gasket with the nozzle gasket incorporated. The other type consists of two individual gaskets, one for the float and a separate one for the nozzle.

2- Position **NEW** gaskets in place on the carburetor body and on the nozzle.

3- Lower the float down over the nozzle, and then slide the hinge pin into place.

4- Hold the carburetor body in a horizontal position and check to be sure the float is parallel to the carburetor surface, as shown. If the float is not parallel, **CAREFULLY** bend the float tang until the float is in a parallel position and both sides are equal distance from the carburetor surface.

Measuring Float Drop

5- Allow the float to drop under its own weight. Measure the distance between the base of the carburetor body and the lowest edge of the float, as indicated in the accompanying illustration, on the next page. The measurement should be 1-1/8" to 1-5/8" (28-41mm). Place the float bowl in position, and then secure it in place with the attaching screws.

Install the low-speed orifice into its recess, using the proper size screwdriver to prevent damaging the orifice. Tighten the orifice until it seats **LIGHTLY**. Thread the plug, with a **NEW** gasket, into place and

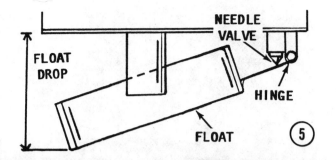

(5)

tighten it snugly. If the carburetor being serviced has the intermediate orifice, install the orifice, using the proper size screwdriver to prevent damaging the orifice. Tighten the orifice until it seats **LIGHTLY**. Thread the plug, with a **NEW** gasket, in place and tighten it snugly.

6- Install the high-speed orifice into the carburetor bowl, using the proper size screwdriver to prevent damaging the orifice.

7- Thread the plug, with a new gasket, into place and tighten it snugly.

(8)

(6)

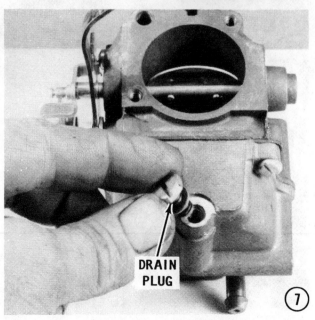

(7)

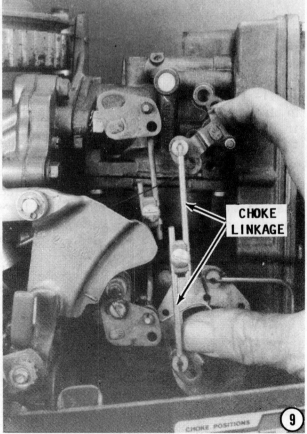

(9)

SECOND CARBURETOR

Perform Steps 1 thru 7 to assemble the second carburetor.

INSTALLATION

8- Connect the fuel line between the top and bottom carburetor. Clean the mating surface of the intake manifold. Check to be sure all old gasket material has been removed. Slide a **NEW** gasket down over the studs into place on the manifold. Install both carburetors at the same time onto the studs and secure them in place with the nuts. Tighten the nuts **ALTERNATELY** and **EVENLY**. Connect the fuel line between the fuel pump and the carburetors.

9- On the starboard side of the powerhead, install the four choke and throttle retainers, two for each carburetor. Connect the choke and throttle linkage between the two carburetors. Connect the choke coil solenoid wire onto the linkage stud, and then slide the O-ring and coil spring onto the stud to secure the wire in place. Adjust the choke butterflys by loosening the screw between the top and bottom linkage. Make the adjustment to close **BOTH** choke butterflys, then tighten the screw.

10- Adjust the throttle butterflys in the same manner. Loosen the screw between the top and bottom linkage. Make the adjustment to close **BOTH** throttle butterflys, and then tighten the screw.

11- Position a **NEW** air silencer gasket in place on the front of the engine. Connect the hose from the crankcase to the back of the air silencer. Coat the threads of the air silencer retaining screws with Loctite and then install the air silencer. Install the air silencer cover and at the same time slide the plastic adjusting knob onto the low-speed needle for each carburetor. **DO NOT** push the knobs all the way home onto the needle at this time. You may seriously consider **NOT** to install and use the piece of linkage connecting the low-speed needle valve of each carburetor.

EXPLANATION

This linkage permits adjusting both carburetors simultaneously while using only one knob. Sounds great! But when the one carburetor is adjusted, the other is also changed. A great many professional me-

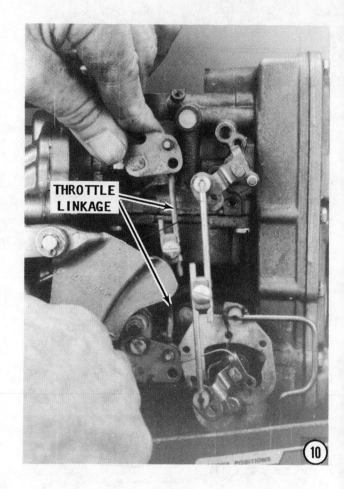

THROTTLE LINKAGE

10

AIR SILENCER

11

chanics have discovered the linkage is not required and it is far more efficient to adjust each carburetor individually.

GOOD WORDS

It is best to synchronize the fuel and ignition systems at this time. See Chapter 5. After the synchronization has been completed, proceed with the following work.

PRELIMINARY ADJUSTMENTS

12- Mount the outboard in a test tank or body of water. If this is not possible, connect a flush attachment and garden hose to the lower unit. **NEVER** operate the powerhead above idle speed using the flush attachment. If the powerhead is operated above idle speed with no load on the propeller, the powerhead could **RUNAWAY** resulting in serious damage or destruction of the unit.

CAUTION: Water must circulate through the lower unit to the engine any time the engine is run to prevent damage to the water pump in the lower unit. Just five seconds without water will damage the water pump.

Start the powerhead and allow it to warm to operating temperature. Pop the

two rubber caps, one for each carburetor, out of the front cover of the air silencer. Adjust the low-speed idle by turning the low-speed needle valve **CLOCKWISE** until the powerhead begins to misfire or the rpm drops noticeably. From this point, rotate the needle valve **COUNTERCLOCKWISE** until the powerhead is operating at the highest rpm.

If the powerhead coughs and operates as if the fuel is too lean, but the idle and high-speed adjustments have been correctly made, then recheck the synchronization between the fuel and ignition systems.

On powerheads with the flexible low-speed extension to the front of the powerhead, the retainer maintains tension on the needle and adjustment will not be lost because of vibration during operation. After the idle has been adjusted, push the idle knob retainers onto the needle until it engages the spline on the low-speed needle. Repeat the procedure for the second carburetor. Replace the rubber caps into the front cover of the air silencer.

AIR SILENCER COVER

Installing the outside cover of the air silencer. As the cover is worked into place, slide the plastic adjustments over the low-speed needle but do not lock them in place until after the adjustments have been made.

4-8 TYPE "C" CARBURETOR
(See Powerhead Listing — Page 4-10)

This section provides complete detailed procedures for removal, disassembly, cleaning and inspecting, assembling including bench adjustments, installation, and operating adjustments for the type "C" carburetor, as used originally on Models 2.3hp and 3.3hp, introduced in 1991. This carburetor is a single-barrel, float feed type with a manual choke. Fuel to the carburetor is gravity fed from a fuel tank mounted on top of the powerhead.

REMOVAL

FIRST, THESE WORDS

Good shop practice dictates a carburetor repair kit be purchased and new parts be installed any time the carburetor is disassembled.

Make an attempt to keep the work area organized and to cover parts after they have been cleaned. This practice will prevent foreign matter from entering passageways or adhering to critical parts.

Remove the spark plug lead to prevent accidental starting of the engine. Shut off the fuel supply at the fuel shut-off valve on the side of the engine. Remove the screws securing the cowling to the engine and remove the cowling.

Remove the screw securing the throttle control knob to the throttle lever and pull the knob free of the lever.

Hold the choke knob with one hand and remove the nut on the back side of the intake silencer with a wrench and the other hand. Slide the lockwasher free of the choke shaft.

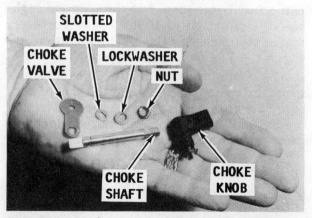

Close view of the small choke control parts used on the choke shaft.

Pull the choke knob with the choke shaft attached out through the hole in the intake silencer. The choke valve and slotted washer will come out with the shaft. Place these small items in order on the workbench as an aid during installation.

Remove the intake silencer. Loosen the bolt on the clamp securing the carburetor to the engine. Be prepared to catch a small amount of fuel with a cloth when the fuel line is disconnected. Disconnect the fuel line at the carburetor. Remove the carburetor from the reed valve housing. The clamp will come off with the carburetor.

DISASSEMBLING

1- Pry off the circlip, and then remove the pivot pin attaching the throttle control lever to the carburetor. Remove the screw, spring, washer, and nut from the carburetor top bracket.

2- Loosen, but **DO NOT** remove the top retainer brass nut on the carburetor. Hold

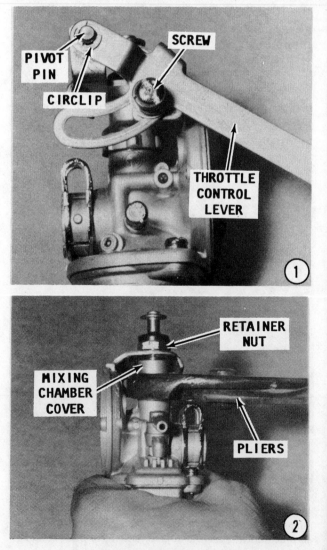

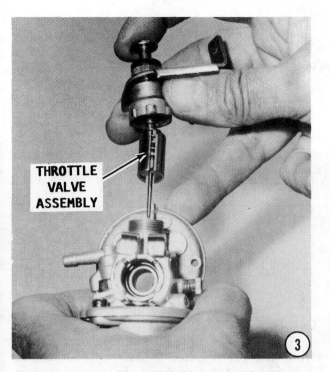

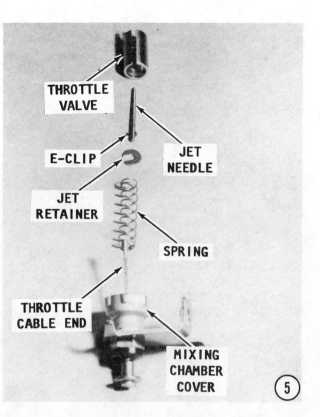

the carburetor securely with one hand and unscrew the mixing chamber cover with the other, using a pair of pliers.

3- Lift out the throttle valve assembly.

4- Compress the spring in the throttle valve assembly to allow the throttle cable end to clear the recess in the base of the throttle valve and to slide down the slot.

5- Disassemble the throttle valve consisting of the throttle valve, spring, jet needle (with an E-clip in the third groove), jet retainer, and throttle cable end.

6- Remove and discard the two screws securing the float bowl to the carburetor body. Remove the float bowl. Lift out the

float. A new pair of float bowl screws are provided in the carburetor rebuild kit.

7- Push the float pin free using a fine pointed awl.

8- Lift out the float arm and needle valve. Slide the needle valve free of the arm.

9- Unscrew the main jet from the main nozzle. Remove and **DISCARD** the float bowl gasket.

10- Count and record the number of turns required to **LIGHTLY** seat the idle adjustment screw. The number of turns will give a rough adjustment during installation.

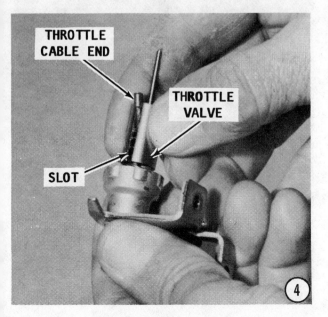

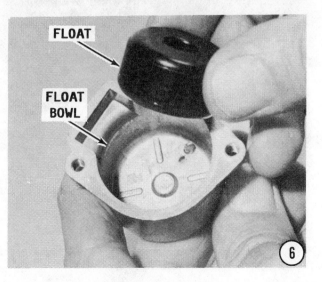

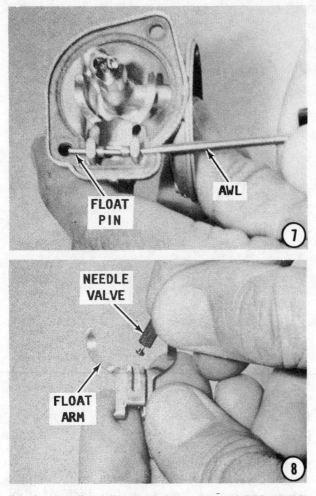

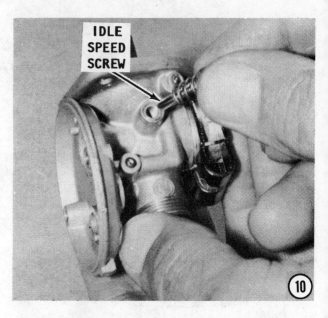

Back out the idle speed screw and **DISCARD** the screw, but **SAVE** the spring. A new screw is provided in the carburetor rebuild kit to ensure a damaged screw is not used again. This screw is most important for maximum performance.

11- Pry out and **DISCARD** the carburetor sealing ring.

GOOD WORDS

It is not necessary to remove the E-clip from the jet needle, unless replacement is required or if the engine is to be operated at a significantly different elevation.

CLEANING AND INSPECTING

NEVER dip rubber parts, plastic parts, diaphragms, or pump plungers in carburetor cleaner. These parts should be cleaned **ONLY** in solvent, and then blown dry with compressed air.

Place all metal parts in a screen-type tray and dip them in carburetor cleaner until they appear completely clean, then blow them dry with compressed air.

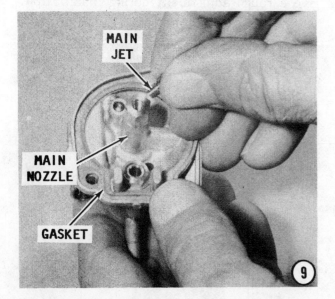

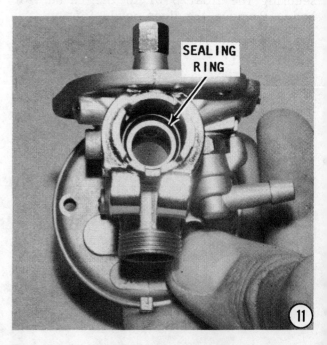

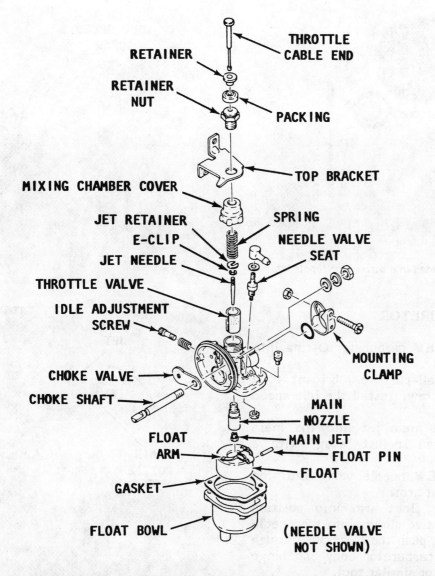

RETAINER

RETAINER
NUT

THROTTLE
CABLE END

PACKING

TOP BRACKET

MIXING CHAMBER COVER

JET RETAINER

E-CLIP

JET NEEDLE

THROTTLE VALVE

IDLE ADJUSTMENT
SCREW

SPRING

NEEDLE VALVE
SEAT

CHOKE VALVE

CHOKE SHAFT

MOUNTING
CLAMP

FLOAT
ARM

MAIN
NOZZLE

MAIN JET

FLOAT PIN

FLOAT

GASKET

FLOAT BOWL

(NEEDLE VALVE
NOT SHOWN)

Exploded drawing of the type "C" carburetor with adjustable low-speed screw. High-speed is a fixed orifice in the carburetor main body. This carburetor is installed on the 2.3hp and 3.3hp powerheads -- 1991 & on.

Blow out all passages in the castings with compressed air. Check all parts and passages to be sure they are not clogged or contain any deposits. **NEVER** use a piece of wire or any type of pointed instrument to clean drilled passages or calibrated holes in a carburetor.

Move the throttle shaft back and forth to check for wear. If the shaft appears to be too loose, replace the complete throttle body because individual replacement parts are **NOT** available.

Inspect the main body, airhorn, and venturi cluster gasket surfaces for cracks and burrs which might cause a leak. Check the float for deterioration. Check to be sure

the float spring has not been stretched. If any part of the float is damaged, the unit must be replaced. Check the float arm needle contacting surface and replace the float if this surface has a groove worn in it.

Inspect the tapered section of the idle adjusting needles and replace any that have developed a groove.

As previously mentioned, most of the parts which should be replaced during a carburetor overhaul are included in overhaul kits available from your local marine dealer. One of these kits will contain a matched fuel inlet needle and seat. This combination should be replaced each time the carburetor is disassembled as a precaution against leakage.

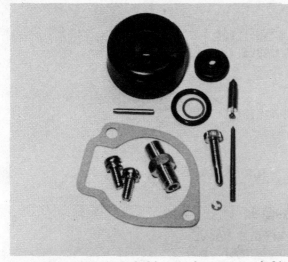

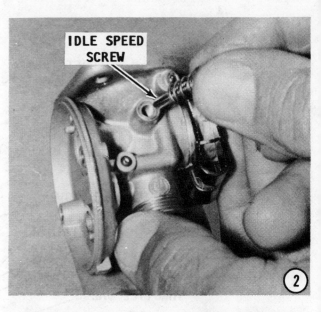

Parts usually included in a carburetor repair kit.

ASSEMBLING
TYPE "C" CARBURETOR

1- Install a **NEW** carburetor O-ring into the caburetor body.

2- Apply an all-purpose lubricant to a **NEW** idle speed screw. Install the idle speed screw and spring.

3- Install the main jet into the main nozzle and tighten it just "snug" with a screwdriver.

4- Slide a **NEW** needle valve into the groove of the float arm.

5- Lower the float arm into position with the needle valve sliding into the needle valve seat. Now, push the float pin through the holes in the carburetor body and hinge using a small awl or similar tool.

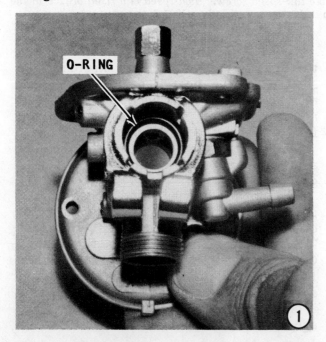

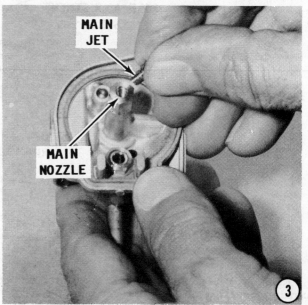

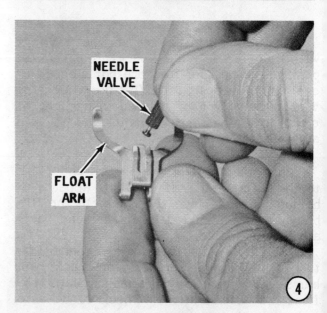

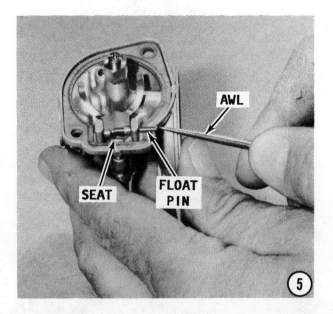

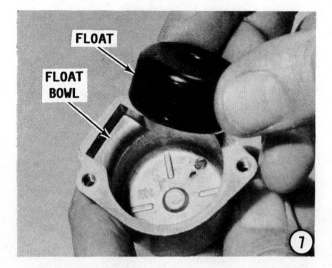

6- Hold the carburetor body in a perfect upright position. Check the float hinge adjustment. The vertical distance between the top of the hinge and the top of the gasket should be 0.090" (2.3mm). **CAREFULLY,** bend the hinge, if necessary, to achieve the required measurement.

7- Position a new float bowl gasket in place on the carburetor body. Install the float into the float bowl. Place the float bowl in position on the carburetor body, and then secure it with the two Phillips head screws.

Throttle Jet Needle Adjustment

If the E-clip on the jet needle is lowered, the carburetor will cause the engine to operate "rich". Raising the E-clip will cause the engine to operate "lean".

If the mixture is too rich, the engine will experience a decrease in rpm as the throttle

is advanced beyond 2/3 of full throttle. If the mixture is too lean, the engine temperature will rapidly increase. Since there are no temperature monitoring devices on these small hp engines, the operator must be aware of operating temperature at all times.

If the outboard is to be operated at higher altitude -- raise E-clip to compensate for rarefied air.

8- With an E-clip in the third groove from the flat end of the jet needle, begin to assemble the throttle valve components by inserting the E-clip end of the jet needle into the throttle valve (the end with the

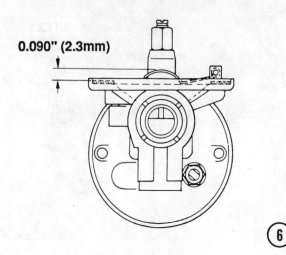

0.090" (2.3mm)

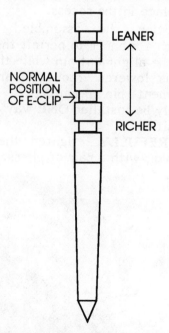

Normally, the E-clip is in the groove shown. If the clip is moved up into a higher groove, the air/fuel mixture will become leaner. If the clip is moved down to a lower groove, the mixture will richen.

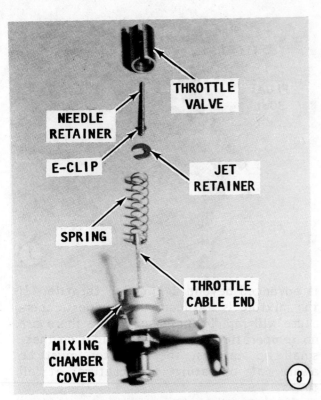

recess for the throttle cable end). Next, place the needle retainer into the throttle valve over the E-clip and align the retainer slot with the slot in the throttle valve.

9- Thread the spring over the end of the throttle cable and insert the cable into the retainer end of the throttle valve. Compress the spring and at the same time, guide the cable end through the slot until the end locks into place in the recess.

10- Position the assembled throttle valve in such a manner to permit the slot to slide over the alignment pin while the throttle valve is lowered into the carburetor. This alignment pin permits the throttle valve to only be installed **ONE** way -- in the correct position.

11- **CAREFULLY** tighten the mixing chamber cover with a pair of pliers.

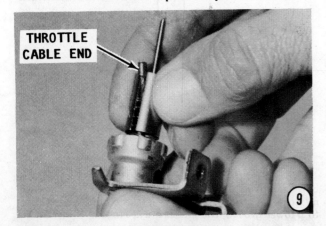

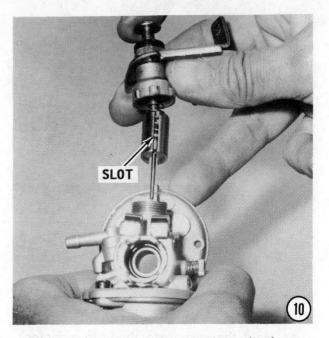

12- Slide the pivot pin through the top carburetor bracket and then through the upper hole in the throttle control lever. Secure the pin with the Circlip. Slide the spring, then the washer over the screw and align the slot in the control lever with the threaded hole in the top carburetor bracket. Install the screw until the lever moves freely with just enough friction to provide the operator with a feel of the throttle opening.

INSTALLATION

Position the carburetor clamp over the carburetor mounting collar, with the locking bolt and nut in place in the clamp. Slide the carburetor into place on the reed valve housing. Secure the carburetor in place by tightening the bolt and nut securely.

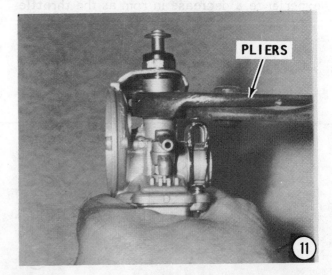

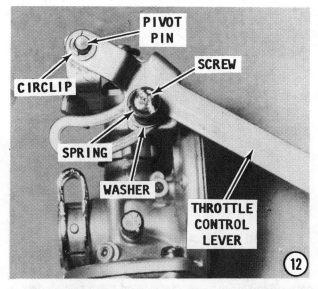

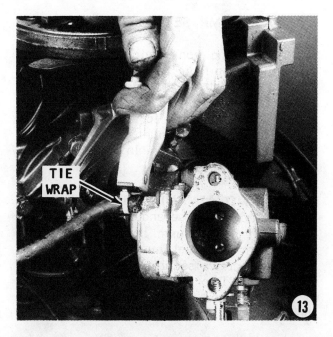

Position the intake silencer against the carburetor.

Push the choke knob onto the slotted end of the choke shaft. Place the threaded end through the intake silencer and slide the choke valve, followed by the slotted washer, onto the protruding threaded end past the threads onto the slotted part of the shaft. Secure the intake silencer to the carburetor with the two Phillips head screws. Hold the choke knob. Slide the lock washer onto the end of the choke shaft, and then thread on the nut.

WORDS FROM EXPERIENCE

Starting this nut is not an easy task. Holding the nut with an alligator clip may prove helpful. Tighten the nut just **SNUG-LY**.

13- Connect the fuel hose to the carburetor using a **NEW** tie wrap. Install the throttle control knob onto the throttle lever and secure it with the Phillips head screw.

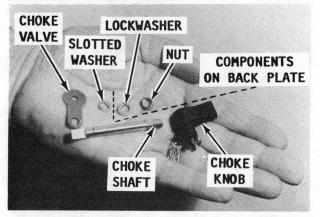

Close view of the small choke control parts used on the choke shaft.

Preliminary Idle Speed Adjustment

SLOWLY tighten the idle speed screw until it **BARELY** seats, then back it out the same number of turns recorded during disassembling. If the number of turns was not recorded, back the screw out 1-3/4 turns as a rough adjustment. A finer adjustment will be made with the engine in the water at the end of the overhaul procedures.

Closing Tasks

Mount the outboard in a test tank, on the boat in a body of water, or connect a flush attachment and hose to the lower unit. Connect a tachometer to the engine.

NEVER, AGAIN, NEVER operate the powerhead at high speed with a flush device attached. The powerhead operating at high speed with such a device attached, would **RUNAWAY** from lack of a load on the propeller, causing extensive damage. Start the powerhead and check the completed work.

CAUTION: Water must circulate through the lower unit to the engine any time the engine is run to prevent damage to the water pump in the lower unit. Just five seconds without water will damage the water pump.

Run the powerhead at half open throttle until it reaches normal operating temperature. Adjust the idle speed screw until the powerhead idles between 1100 and 1300 rpm. Allow approximately 15 seconds for the powerhead to respond to the new adjustment.

4-9 FUEL PUMP SERVICE

To test the non-serviceable pump, remove the pump from the powerhead operate the squeeze bulb until it is firm, and then carefully observe the vacuum hole in the back side of the pump for any indication of fuel. The smallest amount of fuel indicates a damaged diaphragm. In this case, the pump **MUST** be replaced.

PUMP REMOVAL AND INSTALLATION

Identify each hose and its location, then disconnect the vacuum hose and two fuel hoses from the fuel pump. Remove the attaching screws securing the pump to the powerhead. Two screws are visible on top of the pump and the third is hidden behind the fuel inlet nipple.

CLEANING AND INSPECTING

The pump cannot be disassembled, therefore, the only maintenance is to remove the cover; clean the filter screen; and install new gaskets during installation. If the pump is defective, it must be replaced as a unit.

Install the cover, and then place the pump in position on the powerhead with a **NEW** gasket. Secure the pump with the attaching screws. Connect the vacuum hose, if used, and the two fuel hoses to the pump in the same position from which they were removed. Actually, each fitting is identified by word designation embossed on the pump. The vacuum line is connected to

Different fuel pump installed on the powerheads covered in this manual. Service on these two models is limited to removing the cover and cleaning the screen.

the vacuum fitting; the inlet fuel line from the fuel tank to the inlet fitting on the cover; and the outlet line to the remaining fitting.

4-10 FUEL TANK SERVICE

Modern fuel tanks are not pressurized. A squeeze bulb is used to move fuel from the tank to the carburetor until the powerhead is operating. Once the engine starts, the fuel pump, mounted on the powerhead transfers fuel from the tank to the carburetor. The pickup unit in the tank is sold as a complete unit, but without the gauge and float.

1- To replace the pickup unit, first remove the four screws securing the unit in the tank. Next, lift the pickup unit up out of the tank.

Side of the powerhead with the vacuum opening shown. The fuel pump is mounted over this opening to receive the vacuum/pressure from the crankcase for operation.

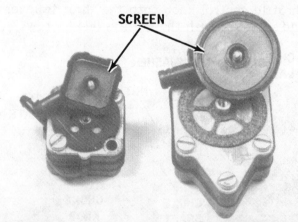

The same two fuel pumps shown at the top of this page, with the covers removed exposing the screen. The screen and the gasket are the only replaceable items.

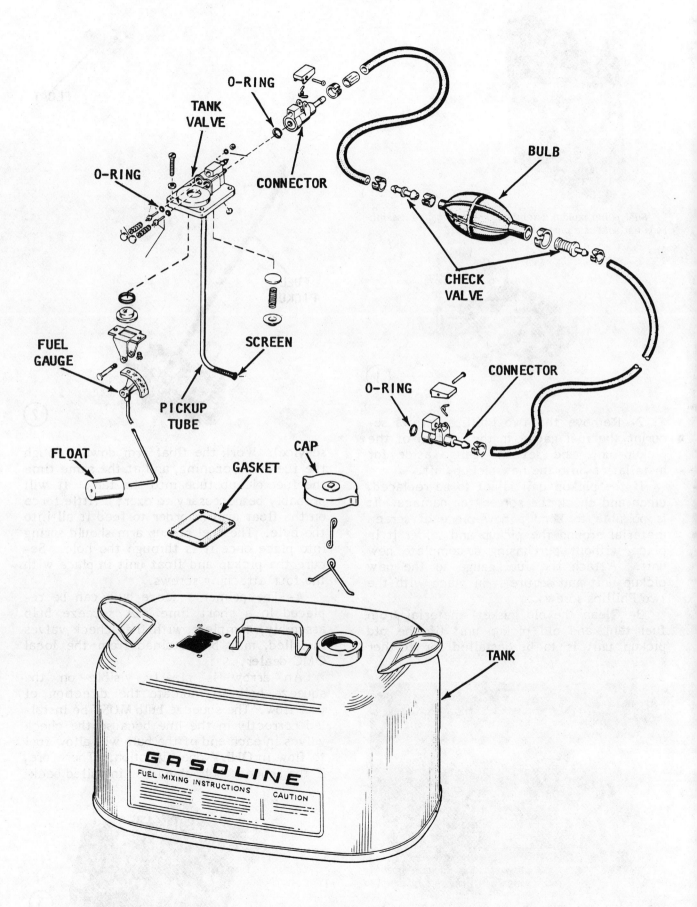

Exploded view of a modern non-pressure type fuel tank using a squeeze bulb in the fuel line.

Fuel pump pickup assembly. The fuel gauge assembly is not sold as a part of the pickup unit.

①

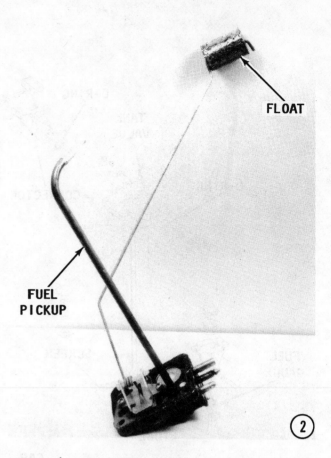

FLOAT

FUEL PICKUP

②

2- Remove the two Phillips screws securing the fuel gauge to the bottom of the pickup unit and set the gauge aside for installation onto the new pickup unit.

If the pickup unit is not to be replaced, clean and check the screen for damage. It is possible to bend a new piece of screen material around the pickup and solder it in place without purchasing a complete new unit. Attach the fuel gauge to the new pickup unit and secure it in place with the two Phillips screws.

3- Clean the old gasket material from fuel tank and old pickup unit (if the old pickup unit is to be installed for further

service). Work the float arm down through the fuel tank opening, and at the same time the fuel pickup tube into the tank. It will probably be necessary to exert a little force on the float arm in order to feed it all into the hole. The fuel pickup arm should spring into place once it is through the hole. Secure the pickup and float unit in place with the four attaching screws.

4- The primer squeeze bulb can be replaced in a short time. A squeeze bulb assembly, complete with the check valves installed, may be obtained from the local OMC dealer.

An arrow is clearly visible on the squeeze bulb to indicate the direction of fuel flow. The squeeze bulb **MUST** be installed correctly in the line because the check valves in each end of the bulb will allow fuel to flow in **ONLY** one direction. Therefore, if the squeeze bulb should be installed back-

Typical Johnson/Evinrude non-pressurized fuel tank.

③

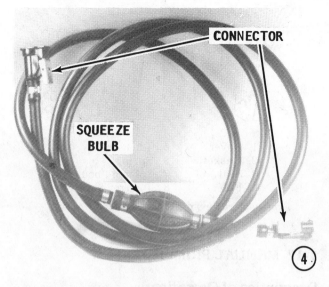

CONNECTOR

SQUEEZE
BULB

4

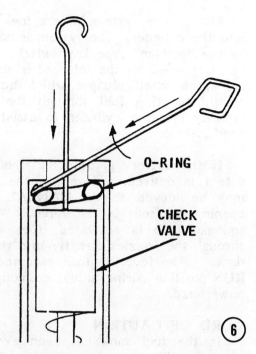

O-RING

CHECK
VALVE

6

wards (in a moment of haste to get the job done), fuel will not reach the carburetor.

5- To replace the bulb, first unsnap the clamps on the hose at each end of the bulb. Next, pull the hose out of the check valves at each end of the bulb. New clamps are included with a new squeeze bulb. If the fuel line has been exposed to considerable sunlight, it may have become hardened, causing difficulty in working it over the check valve. To remedy this situation, simply immerse the ends of the hose in boiling water for a few minutes to soften the rubber and the hose will then slip onto the check valve without further problems. After the lines on both sides have been installed, snap the clamps in place to secure the line. Check a second time to be sure the arrow is pointing in the fuel flow direction, **TOWARDS** the engine.

6- Use two ice picks or similar tool, and push down the check valve of the connector and work the O-ring out of the hole.

7- Apply just a drop of oil into the hole of the connector. Apply a thin coating of

oil to the surface of the O-ring. Pinch the O-ring together and work it into the hole while simultaneously using a punch to depress the check valve inside the connector.

4-11 ELECTRIC PRIMER CHOKE SYSTEM

The electric primer system consists of a solenoid valve, distribution lines, and injection nozzles. The nozzles are tapped into the bypass covers. During powerhead cranking when the choke system is operating, fuel is injected through metered holes in the nozzles directly into the cylinders, instead of being routed in the usual manner through the crankcase.

During engine operation, from the fuel tank, the fuel passes through the fuel line, to the fuel pump, and into the carburetor. From the carburetor the fuel and air mixture passes through the crankcase and into the cylinder.

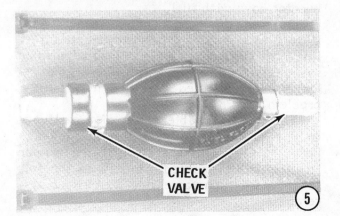

CHECK
VALVE

5

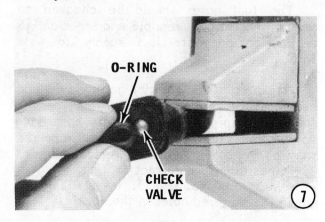

O-RING

CHECK
VALVE

7

The primer system injects fuel directly into the cylinder. The system is controlled by the "push-in" type key switch. As the key is pushed in, the solenoid is activated, moving a small plunger which acts as a pump, injecting fuel through the nozzles directly into the cylinder to assist power-head startup.

If the battery is dead and the choking effect is desired, a lever on the solenoid may be moved to the **MANUAL** position opening the seat in the valve. When the squeeze bulb is activated, fuel will pass through the nozzles directly into the cylinders. The lever is then returned to the **RUN** position during actual cranking of the powerhead.

WORD OF CAUTION

If the fuel tank has been exposed to direct sunlight, pressure may have developed inside the tank. Therefore, when the solenoid lever is moved to the **MANUAL** position, an excessive amount of fuel may be forced into the cylinders. As a safety precaution, under possible fuel tank pressure conditions, the fuel tank cap should be opened slightly to allow the pressure to escape before attempting to start the engine.

SOLENOID TESTING

Connect an ohmmeter to the solenoid between the blue/white stripe lead and the black (ground) wire. Observe the reading. The ohmmeter should indicate 5.5 + 1.5 ohms. If the reading is not within the prescribed range, the solenoid is defective and must be replaced.

The accompanying illustration will be helpful in ordering and replacing parts of the primer choke system.

The fuel hoses should be checked to ensure they remain flexible and are clear to permit an adequate fuel supply to pass through.

Pay particular attention to any evidence of a crack in a fuel line which may permit fuel to escape and cause a very hazardous condition. The diagrams in the Appendix may be very helpful during replacement of fuel lines to ensure correct routing.

OMC tool No. 326623, is available to clean the metered holes in the nozzles.

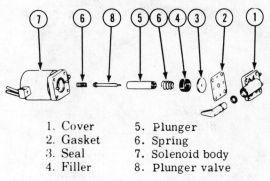

1. Cover	5. Plunger
2. Gasket	6. Spring
3. Seal	7. Solenoid body
4. Filler	8. Plunger valve

Exploded drawing of a primer solenoid valve.

4-12 MANUAL PRIMER SYSTEM

Description of Operation

During powerhead operation: from the fuel tank, the fuel passes through the fuel line, to the fuel pump, and into the carburetor where it is mixed with ambient air. From the carburetor, the fuel/air mixture passes through the crankcase and into the cylinder.

The manual primer system injects fuel directly into the cylinder. The system is activated by the "push-in" type choke lever. As the lever is pushed in, a small plunger moves inside the cylinder and acts as a pump, injecting fuel through the nozzles directly into the cylinder to assist power-head startup.

TROUBLESHOOTING

If the manual primer system is suspected of not functioning correctly, remove the fuel line from the primer at the carburetor fitting.

Place the end of the fuel line just removed into a suitable container. Squeeze the fuel tank primer bulb to fill the carburetor bowl with fuel.

Operate the primer choke lever twice. If fuel squirts from the disconnected fuel line into the container, the manual primer system is functioning correctly. If not, a kinked or restricted fuel line may be the problem.

The most probable cause of a malfunctioning primer system is internal leakage past the O-rings. Therefore if the primer itself is still suspected, proceed to the following paragraph to service the primer system.

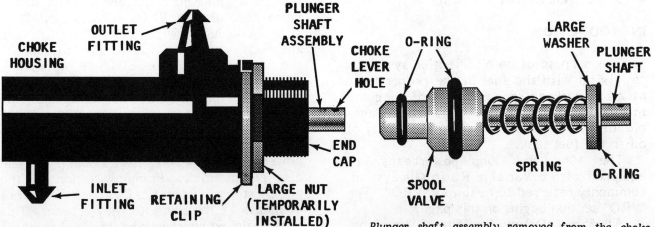

The primer choke valve removed from the power-head. The large nut is temporarily installed onto the threads of the end cap for safe keeping.

Plunger shaft assembly removed from the choke housing with major parts identified.

SERVICING MANUAL PRIMER

Removal

Disconnect and plug the inlet and outlet fuel lines to prevent loss of fuel and contamination. Remove the choke lever from the plunger. Back off the large nut securing the choke assembly to the lower cowling and lift the assembly free of the powerhead.

Pry the retaining clip from the choke body housing. Pull out the end cap, plunger, and spool valve assembly. Slide the end cap from the plunger. Remove and discard the O-ring around the end cap.

SPECIAL WORDS

Observe the three small O-rings, two on the spool valve and one around the plunger shaft. These three O-rings are made from a special material and **MUST** be replaced with a genuine OMC replacement part. Matching O-rings will **NOT** work!

Remove and discard the three O-rings. Remove the large washer and spring from the plunger shaft.

Cleaning and Inspecting

Inspect the grooves of the spool valve and the shaft of the plunger for any scratches or burrs. Polish away any imperfections using crocus cloth. If a smooth finish cannot be obtained without removing excessive material, replace the spool valve and plunger assembly.

Inspect the condition of the plunger spring, replace as required.

Two one-way valves, one at each fuel fitting, can be tested by blowing through them in turn. Each valve is functioning correctly if it allows air to pass one direction, but not in the other direction. If a valve allows air to be drawn both in and out, the valve is defective. Individual valves are not serviceable. The primer body must be replaced.

Assembling

Install the two new O-rings around the spool valve. Slide the spring, followed by the large washer and the third O-ring, over the plunger. Install a new O-ring over the end cap and place the end cap over the plunger end. Insert the assembly into the primer housing and install the retaining clip to secure everything together.

Slide the assembled primer into the opening in the lower cowling and thread the large nut over the protruding threads. Tighten the nut securely.

Install the fuel lines to the appropriate fittings and snap the choke lever into the vertical hole in the plunger.

4-13 OIL INJECTION SYSTEMS

INTRODUCTION

The purpose of an oil injection system is to mix oil with the fuel in the proper ratio at all powerhead speeds to ensure adequate lubrication. The system replaces the age-old method of manually adding a quantity of oil to the fuel tank.

The 40hp and 50hp powerheads are equipped with a Variable Ratio Oil System, commonly referred to as simply "VRO". The "VRO" section begins on this page.

All 3hp, 4 Delux, 6hp, 8hp, 9.9hp, 15hp, 20hp, 25hp and 30hp powerheads covered in this manual are equipped with an oil injection system known as AccuMix. Actually, this is the same system introduced in 1986, under the trade name "AutoBlend". The name was changed to AccuMix in 1987.

ACCUMIX DESCRIPTION

The AccuMix system is located entirely inside the portable fuel tank. A 1-1/2 quart reservoir cannister contains enough oil for almost five tank fulls of fuel. An oil metering pump is located at the base of the cannister. This oil pump is activated by pulses from the fuel pump installed on the powerhead. The oil metering pump automatically blends fuel from the fuel pickup in the portable tank with oil in the reservoir cannister. The oil/fuel mixture passes through a built-in filter also located inside the cannister.

A low-oil warning indicator activates a warning horn when the level of oil falls below one pint. If the operator sustains powerhead operation after the warning horn sounds, the fuel supply is automatically cut off to shutdown the powerhead. The powerhead cannot be restarted until oil has been added to the cannister.

Procedures to service the AccuMix oil injection system begin Page 4-47.

VARIABLE RATIO OIL SYSTEM DESCRIPTION

The VRO system consists of an oil reservoir (tank), a VRO oil line primer, a pump to move the oil from the tank to the power-

head, a warning horn, a spark arrestor in the pulse hose to the VRO pump, an oil inlet filter, a vacuum switch in the fuel line and the necessary hoses and fittings to connect the various items for efficient operation. All connections in the system **MUST** be airtight to prevent serious damage to the powerhead.

As the name implies, the VRO pump moves oil from the oil reservoir to the powerhead. However, it is a **dual** pump and also pumps fuel. Pumping action of the pump stops automatically if fuel is not available at the pump for any reason. This automatic pump shutdown feature prevents the carburetors from filling with oil.

The warning horn, located in the control box serves two functions.

First, as a low oil level warning; The horn will sound for 1/2 second every 20 seconds if the oil tank level reaches 1/4 of the tank's capacity. The low oil warning circuit consists of a sending unit in the oil reservoir, a ground wire to the engine, and a wire to the warning horn through the key switch.

Secondly, the warning horn will sound for 1/2 second every 1/2 second to produce a very urgent warning signal. To continue powerhead operation after the no oil warning horn sounds would almost certainly invite serious damage to internal moving parts and powerhead seizure!

The spark arrestor on all models is installed in the pulse hose to the VRO pump. This flame arrestor prevents a backfire flame from entering the VRO pump. A clamp positioned on the hose prevents the spark arrestor from migrating up the hose to the pump.

Variable Ratio Oil (VRO) system installed on a late model powerhead.

GOOD WORDS

Anytime the pulse hose is disconnected at the powerhead, **TAKE CARE** to be sure the spark arrestor remains in the hose. The spark arrestor **MUST** be properly positioned in the pulse hose to prevent serious and permanent damage to the VRO pump.

The oil inlet filter, located in the VRO reservoir oil pickup line, prevents any dirt or foreign material from entering the pump. If the filter should need cleaning, the hose assembly should be removed and reverse flushed using clean solvent. **DO NOT** attempt to remove the filter because the filter and hose are serviced and replaced as an assembly.

NEW POWERHEAD BREAK-IN PROCEDURE

A complete new outboard unit, a new powerhead, or a rebuilt powerhead, must have oil mixed in the fuel tank **IN ADDITION** to the VRO system. See Chapter 3, Page 3-44 for complete detailed instructions for the break-in period for all new or rebuilt powerheads.

CRITICAL WORDS

To be convinced the VRO system is working properly, the operator should observe a drop in the oil supply in the VRO oil reservoir during the 10-hour break-in period.

At the end of the 10 hour period, the powerhead mounted fuel filter should be inspected. Remove any dirt or foreign matter collected in the filter.

TROUBLESHOOTING VRO SYSTEM

This short section list a few of the probable problems that might occur in the system with suggested corrective action.

The next section -- **SERVICING** outlines in detail how the tests and service work is to be performed.

Warning Horn Sounds

a- Oil level in the oil reservoir is below 1/4 full. Add oil to the reservoir.

b- Disc on the pickup unit may not be positioned properly. Remove the pickup unit from the reservoir and correct as required.

SERVICING VRO SYSTEM

Servicing consists of making simple tests and checks. A vacuum gauge, pressure gauge, a "T" fitting, a short section of clear plastic hose, a source of low-pressure compressed air, and a couple of normal shop tools are all that is required to service the VRO system.

Check Fuel and Oil Circuits

1- Verify there is more than 1/4 tank of oil in the reservoir. Disconnect the oil inlet hose from the VRO pump. Be prepared to catch oil as it is ejected from the end of the hose. Squeeze the bulb and verify oil is ejected from the open end of the hose. The

presence of oil verifies a clear line from the tank to this point. If no oil is ejected, clean the line from the tank and repeat the test.

2- Disconnect the mixed fuel outlet hose from the VRO pump.

3- Insert a "T" fitting into the end of the hose. Insert a drop of oil into each end of a short section of clear plastic hose. Connect one end of the clear piece of plastic hose to one arm of the "T" and the other end of the hose to the VRO pump. Connect a 0-15 psi pressure gauge to the leg of the "T". Secure the connections with tie straps or hose clamps. Mount the engine in an adequate size test tank or move the boat to a body of water.

NEVER operate the engine using a flush attachment for this test. If the engine is operated above idle speed with no load on the propeller, the engine could **RUNAWAY** resulting in serious damage or destruction of the unit.

CAUTION: Water must circulate through the lower unit to the engine any time the engine is run to prevent damage to the water pump in the lower unit. Just five seconds without water will damage the water pump.

Start the engine and shift the unit into gear. Advance the throttle to the near wide open position. Check the pressure gauge. The gauge should indicate 34 kPa (5 psi) to 103 kPa (15 psi) pressure at near full throttle. Each time the pump pulses a small squirt of oil, in addition to the fuel passing through, should be observed in the clear plastic hose discharging from the pump.

Also, the fuel pressure will drop approximately 6.8 kPa (1 psi) to 13.7 kPa (2 psi) and a "click" sound may be heard each time the pump pulses and discharges oil.

Results

Fuel pressure satisfactory and oil is observed discharging from the pump through the clear plastic hose: Fuel and oil systems are verified satisfactory.

No fuel pressure: Check quantity of fuel in the tank to be sure fuel level reaches the pickup. Add fuel if required. Check fuel line to be sure it is not pinched or kinked restricting fuel flow. Check engine pulse hose to be sure it is not pinched, leaking or disconnected. If all above conditions are satisfactory, the VRO pump is defective and **MUST** be replaced as a unit.

Low fuel pressure: Check for restricted fuel filter at the engine. Check the engine pulse hose to be sure it is not pinched, kinked, leaking, or restricting fuel flow. Squeeze the fuel primer bulb a few times to force a possible fuel vapor-lock out of the system.

Warning Horn Check

4- Slide the "boot" down the wire to clear the knife disconnect between the temperature switch and the horn lead. This is not an easy task, but with a pair of needle nose pliers and some patience, it can be done. After the "boot" is clear, disconnect

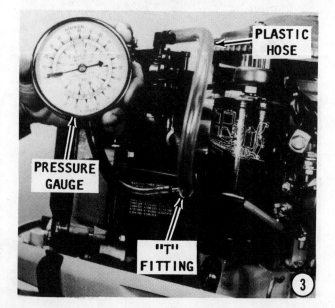

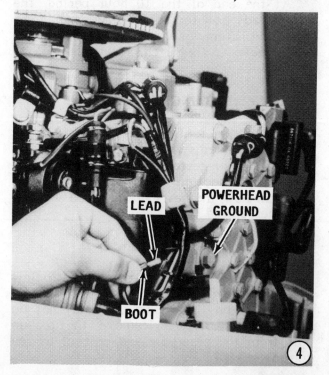

VACUUM
PULSE
HOSE

CLAMP

5

clean solvent. **TAKE CARE** to ensure the spark arrestor installed in the hose stays in the hose. A clamp is positioned on the hose to prevent the spark arrestor from migrating up the hose to the pump. If the hose is damaged and requires replacement, the hose and flame arrestor are purchased as an assembly. The flame arrestor cannot be purchased separately.

VRO Pump

If troubleshooting and service work indicates the VRO pump to be defective, it must be replaced. The pump cannot be serviced or repaired. The pump is removed by first disconnecting the hoses and then removing the three mounting bolts. Lift the VRO pump free.

Powerhead Mounted Fuel Filter

6- If a fuel inlet filter is mounted on the powerhead, the filter can be separated and inspected without removing the hoses. The filter should be inspected at the end of the 10-hour break-in period and at regular intervals as part of normal engine maintenance and service.

GOOD WORDS

Use **ONLY** OMC approved clamps on the connections on the inlet side of the VRO

the fitting. Turn the key switch to the **ON** position. Now, make contact with the lead onto a "clean" place on the powerhead. The horn should sound. If the horn does not sound, there is a problem between the knife disconnect and the remote control box.

Clean Vacuum Pulse Hose

5- Remove the vacuum pulse hose. The hose may be cleaned by back-flushing with

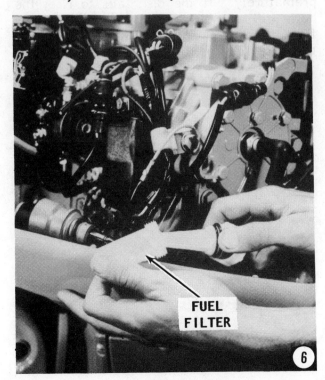

FUEL
FILTER

6

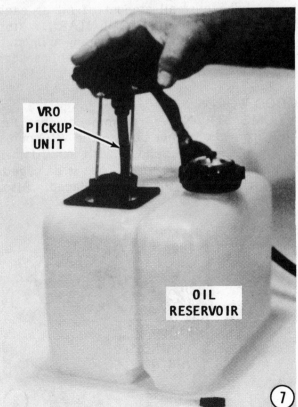

VRO
PICKUP
UNIT

OIL
RESERVOIR

7

pump. A screw type hose clamp will very likely pinch or break the hose causing a suction leak. A tie-wrap will not tighten down to the degree required to prevent a vacuum leak.

Vacuum Hose Check

This check will verify the system in good condition from the oil pickup to the end of the hose.

7- Remove the VRO pickup unit from the oil reservoir by first removing the four mounting screws and then lifting the pickup straight up and out of the reservoir. Inspect the oil pickup filter, and clean it, if necessary.

8- Purge the system of oil and any small particles of foreign matter by using low-pressure compressed air through the VRO pickup end.

9- Connect a vacuum gauge to the end of the hose. Insert the plug-type nipple that is shipped with the engine (snapped to the fuel hose at the engine) into the VRO pickup. If the plug has been lost, one can be easily made out of suitable material. Secure the plug with a clamp. Pump the gauge until 17.7 cm (7") of mercury is indicated. The system should hold the vacuum reading.

10- If the system fails to hold the required vacuum reading, check each connection by applying a small amount of oil at each fitting. The oil will **MOMENTARILY** stop the leak and the vacuum reading will stop dropping. Carefully inspect the hose for damage.

GOOD WORDS

OMC **STRONGLY** recommends that the hose from the primer bulb to the VRO pump be one continuous hose with no fittings between. Therefore, if the hose is damaged, the entire length should be replaced. Also,

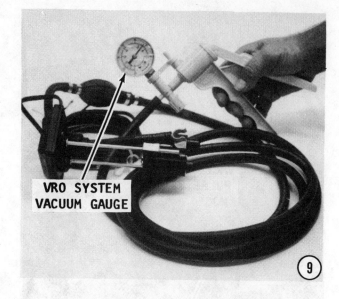

VRO SYSTEM VACUUM GAUGE

if a dual engine installation is used, **DO NOT** "T" into the line from the primer bulb. Use a separate oil reservoir, with separate primer bulb and hose to the VRO pump on the second engine -- a completely separate system for each engine.

11- If the horn sounds indicating low oil or no oil, the contacts on the pickup may not be positioned properly. A disc rises slightly when oil is added into the reservoir and lowers slightly as the oil level drops. Check to be sure the contact surface on both sides of the disc is riding equally with the other side to prevent sounding the horn prematurely. If the disc fails to rise, the horn will sound continuously. Clean the float chamber in solvent and the disc should then rise clear of the contacts.

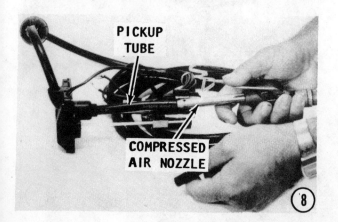

PICKUP TUBE

COMPRESSED AIR NOZZLE

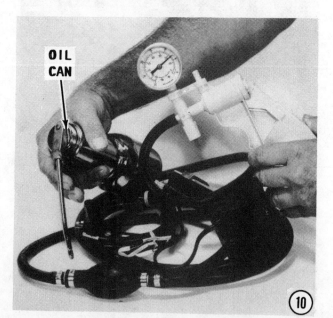

OIL CAN

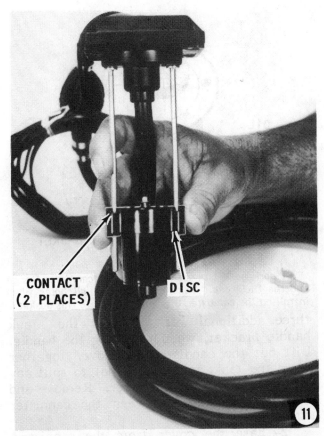

CONTACT
(2 PLACES)   DISC

11

Fuel Line Vacuum Test

12- Connect a "T" fitting and vacuum gauge to the outlet hose from the fuel tank, as shown. Check to be sure all fittings and connections are tight.

NEVER operate the engine above idle speed using a flush attachment for this test.

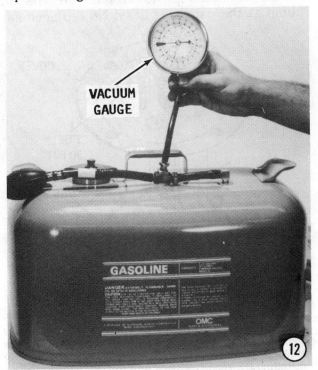

VACUUM
GAUGE

GASOLINE

12

If the engine is operated above idle speed with no load on the propeller, the engine could **RUNAWAY** resulting in serious damage or destruction of the unit.

CAUTION: Water must circulate through the lower unit to the engine any time the engine is run to prevent damage to the water pump in the lower unit. Just five seconds without water will damage the water pump.

Start the powerhead and operate it at idle speed. Observe the vacuum gauge. The gauge should indicate no more than 17.7 cm (7") of mercury.

ACCUMIX
SYSTEM DESCRIPTION

The AccuMix oil injection system is designed to provide a fuel/oil ratio of 50/1 regardless of powerhead rpm.

The AccuMix system is located entirely inside the portable fuel tank. A 1-1/2 quart reservoir cannister contains enough oil for almost five tankfulls of fuel. An oil metering pump is located at the base of the cannister. This oil pump is activated by pulses from the fuel pump installed on the powerhead. The oil metering pump automatically blends fuel from the fuel pickup in the portable tank with oil in the reservoir cannister. The oil/fuel mixture passes through a built-in filter also located inside the cannister.

A low-oil warning indicator activates a warning horn when the level of oil falls below one pint. If the operator sustains powerhead operation after the warning horn sounds, the fuel supply is automatically cut off to shutdown the powerhead.

A visual low oil level indicator is located on the reservoir cannister cover. This indicator is of the "glass eye" type, similar to those found on newer style automotive batteries.

SPECIAL WORDS
ON TROUBLESHOOTING

Due to the inherent design of this oil injection system, very few individual parts can be repaired or replaced, if found to be defective. As an example: if the low oil level float is found to be defective, unfortunately, the entire system must be replaced. Therefore, no "troubleshooting" procedures are given, because no evaluating tests have been provided by the manufacturer.

If any problem is encountered with the delivery of oil, follow the procedures outlined in the following paragraphs. The instructions deal mostly with cleaning out the system. If the problem persists, the only remedy is a new system.

GENERAL MAINTENANCE

Maintenace of this type oil injection system is limited to draining and flushing the oil reservoir cannister each season. To properly clean the cannister and the integral oil filter, it should be removed from the fuel tank and flushed with fresh gasoline or solvent.

Procedures for reservoir cannister service will be found in the following paragraphs.

When the reservoir cannister has been removed from the fuel tank, the fuel line pickup screen can also be serviced.

Clean the vent screw and the area around the screw each time the fuel tank is filled. This vent screw **MUST** be **FULLY OPEN** to allow powerhead operation and to permit air to enter the tank and take the place of the consumed fuel. The vent screw and fuel tank cap cannot be serviced, other than cleaning. If defective for any reason, the assembly must be replaced as a unit.

OIL RESERVOIR CANNISTER SERVICE

Removal

1- Disconnect the electrical harness and the fuel line connector from the top of the cannister. Remove the eight slotted head retaining screws, securing the cover and cannister to the fuel tank. Observe the

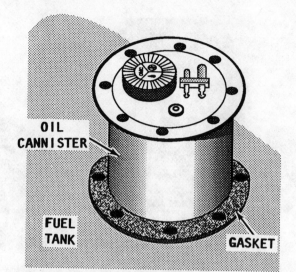

A cork gasket is used between the cannister flange and the fuel tank. Take care to align all holes when installing the cannister to avoid damage to this gasket.

three additional washers under the tank handle bracket, when removing the handle. Lift out the cannister and cover together from the tank, taking care not to spill any oil remaining in the cannister. Remove and discard the gasket between the cannister flange and the tank.

2- Ease the cover from the cannister. The fuel tube must be disengaged from the cavity in the cover. Remove and discard the O-ring between the cover and the cannister.

Good Words

Do not attempt to disassemble the oil pump at the cannister base. No replacement

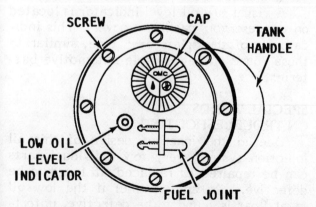

A top view of the AccuMix oil injection system. Eight screws secure the handle and the cannister to the fuel tank.

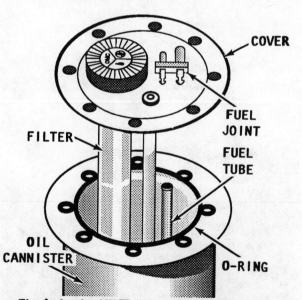

The fuel tube MUST index with the cavity directly under the fuel joint and the eight holes in the O-ring MUST align with the cannister and cover holes prior to installing the cannister back into the fuel tank.

parts are available. If defective the pump must be replaced as an assembly.

CLEANING AND INSPECTING

Drain any oil from the cannister. Obtain a container of solvent and "dunk" the oil filter a couple of times. Take care not to alter the low oil level indicator float. Use a shop towel moistened with solvent and wipe down the surfaces of the cover and the length of the low oil level indicator tube. Blow the cover assembly and filter dry with compressed air.

Pour some solvent into the cannister to rinse any residue from the cannister walls. Make certain no debris obstructs the oil pump pickup or the low oil cutoff float mounted on the pump.

Inspect and service the fuel line pickup screen, as necessary.

Assembling

1- Secure the fuel pickup line into the clip at the base of the cannister. Place a new gasket on the fuel tank surface and lower the cannister into the tank. Align the cannister flange holes with the gasket holes.

2- Apply a light coat of OMC Triple-Guard grease on both sides of the O-ring, and then position the ring onto the cannister flange. Align the holes in the O-ring with the flange holes.

3- Lower the cover over the cannister and make sure the fuel tube indexes with the cavity on the underside of the cover.

4- Position one washer at each of the three holes for the fuel tank handle bracket, and then hold the handle in place while these screws are started. Install the remaining screws fingertight. Tighten all eight screws alternately and evenly to a torque value of 10 in lb (1.1Nm).

Connect the electrical harness and the fuel connector to the fittings on top of the cannister.

5
IGNITION

5-1 INTRODUCTION

The less an outboard engine is operated, the more care it needs. Allowing an outboard engine to remain idle will do more harm than if it is used regularly. To maintain the engine in top shape and always ready for efficient operation at any time, the engine should be operated every 3 to 4 weeks throughout the year.

The carburetion and ignition principles of two-cycle engine operation **MUST** be understood in order to perform a proper tune-up on an outboard motor.

If you have any doubts concerning your understanding of two-cycle engine operation, it would be best to study the operation theory section in the first portion of Chapter 3, before tackling any work on the ignition system.

Three different ignition systems are used on powerheads covered in this manual. One is a flywheel magneto system used on the single cylinder powerheads. The second is a CD (Capacitor Discharge) system with external sensor coil (not mounted under the flywheel), used on some 2-cylinder powerheads up to about the middle of 1990 or early 1991. The third is a CD system with UFI (Under Flywheel Ignition).

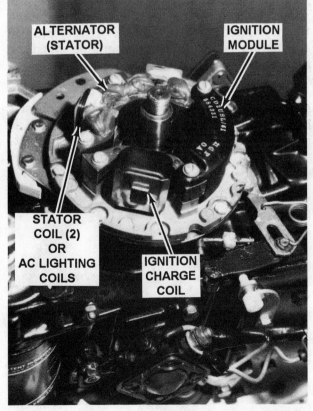

Typical UFI system installed on a 25hp powerhead with alternator and AC lighting coils. If a rectifier is used, it is mounted elsewhere on the powerhead, not under the flywheel.

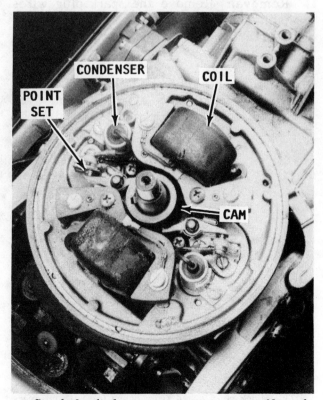

Simple 2-cylinder magneto ignition system. Naturally, a single cylinder powerhead would have only one point set, one coil, and one condenser.

As the name implies, the UFI system is housed entirely under the flywheel. During the production year -- 1990, the manufacturer phased out CDII capacitor discharge ignition with sensor except on the very small Model 2.3, 3, 3.3 and 4 powerheads. **HOWEVER**, powerheads of various sizes leaving the factory in 1990 and possibly early 1991, could be equipped with either ignition system.

Easy Identification

The easiest way to determine which ignition system is installed on the 2-cylinder powerhead being serviced, is to trace back the main harness from the ignition coil mounted on the powerhead. If the leads can be traced back under the flywheel, the powerhead is equipped with UFI. If the harness is traced back to a power pack (black box), the powerhead is equipped with CD ignition with external sensor.

5-2 PRELIMINARY CHECKS ANY SYSTEM

SPARK PLUG EVALUATION

Removal: Remove the spark plug wires by pulling and twisting on only the molded cap. **NEVER** pull on the wire or the connection inside the cap may become separated or the boot damaged. Remove the spark plugs and keep them in order. **TAKE CARE** not to tilt the socket as you remove the plug or the insulator may be cracked.

*Damaged spark plugs. Notice the broken electrode on the left plug. The broken part **MUST** be found and removed before returning the engine to service.*

Examine: Line the plugs in order of removal and carefully examine them to determine the firing conditions in each cylinder. If the side electrode is bent down onto the center electrode, the piston is traveling too far upward in the cylinder and striking the spark plug. Such damage indicates the wrist pin or the rod bearing is worn excessively. In all cases, an engine overhaul is required to correct the condition. To verify the cause of the problem, turn the engine over by hand. As the piston moves to the full up position, push on the piston crown with a screwdriver inserted through the spark plug hole, and at the same time rock the flywheel back-and-forth. If any play in the piston is detected, the engine must be rebuilt.

Correct Color: A proper firing plug should be dry and powdery. Hard deposits inside the shell indicate too much oil is being mixed with the fuel. The most important evidence is the light gray color of the

This spark plug is foul from operating with an over-rich condition, possibly an improper carburetor adjustment.

This spark plug has been operating too-cool, because it is rated with a too-low heat range for the engine.

Today, numerous type spark plugs are available for service. ALWAYS check with your local marine dealer to be sure you are purchasing the proper plug for the engine being serviced.

porcelain, which is an indication this plug has been running at the correct temperature. This means the plug is one with the correct heat range and also that the air-fuel mixture is correct.

Rich Mixture: A black, sooty condition on both the spark plug shell and the porcelain is caused by an excessively rich air-fuel mixture, both at low and high speeds. The rich mixture lowers the combustion temperature so the spark plug does not run hot enough to burn off the deposits.

Deposits formed only on the shell is an indication the low-speed air-fuel mixture is too rich. At high speeds with the correct mixture, the temperature in the combustion chamber is high enough to burn off the deposits on the insulator.

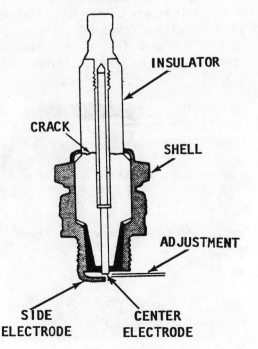

Cut-a-way drawing showing major spark plug parts.

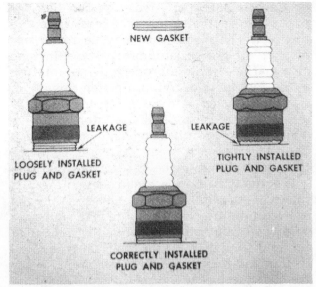

Drawing to illustrate a spark plug properly installed, center, and other plugs, left and right, improperly installed.

Too Cool: A dark insulator, with very few deposits, indicates the plug is running too cool. This condition can be caused by low compression or by using a spark plug of an incorrect heat range. If this condition shows on only one plug it is most usually caused by low compression in that cylinder. If all of the plugs have this appearance, then it is probably due to the plugs having a too-low heat range.

Fouled: A fouled spark plug may be caused by the wet oily deposits on the insulator shorting the high-tension current to ground inside the shell. The condition may also be caused by ignition problems which prevent a high-tension pulse being delivered to the spark plug.

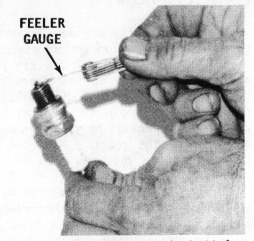

The spark plug gap should always be checked before installing new or used spark plugs.

Carbon Deposits: Heavy carbon-like deposits are an indication of excessive oil in the fuel. This condition may be the result of poor oil grade, (automotive-type instead of a marine-type); improper oil-fuel mixture in the fuel tank; or by worn piston rings.

Overheating: A dead white or gray insulator, which is generally blistered, is an indication of overheating and pre-ignition. The electrode gap wear rate will be more than normal and in the case of pre-ignition, will actually cause the electrodes to melt as shown in this illustration. Overheating and pre-ignition are usually caused by improper point gap adjustment; detonation from using too-low an octane rating fuel; an excessively lean air-fuel mixture; or problems in the cooling system.

Electrode Wear: Electrode wear results in a wide gap and if the electrode becomes carbonized it will form a high-resistance path for the spark to jump across. Such a condition will cause the engine to misfire during acceleration. If all plugs are in this condition, it can cause an increase in fuel consumption and very poor performance during high-speed operation. The solution is to replace the spark plugs with a rating in the proper heat range and gapped to specification.

Red rust-colored deposits on the entire firing end of a spark plug can be caused by water in the cylinder combustion chamber. This can be the first evidence of water entering the cylinders through the exhaust manifold because of scale accumulation. This condition **MUST** be corrected at the first opportunity. Refer to Chapter 3, Powerhead Service.

POLARITY CHECK

Coil polarity is extremely important for proper battery ignition system operation. If a coil is connected with reverse polarity, the spark plugs may demand from 30 to 40 percent more voltage to fire. Under such demand conditions, in a very short time the coil would be unable to supply enough voltage to fire the plugs. Any one of the following three methods may be used to quickly determine coil polarity.

1- The polarity of the coil can be checked using an ordinary D.C. voltmeter. Connect the positive lead to a good ground. With the engine running, momentarily touch the negative lead to a spark plug terminal.

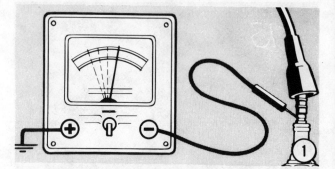

The needle should swing upscale. If the needle swings downscale, the polarity is reversed.

2- If a voltmeter is not available, a pencil may be used in the following manner: Disconnect a spark plug wire and hold the metal connector at the end of the cable about 1/4" from the spark plug terminal.

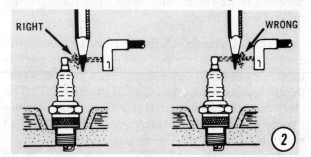

Now, insert an ordinary pencil tip between the terminal and the connector. Crank the engine with the ignition switch ON. If the spark feathers on the plug side and has a slight orange tinge, the polarity is correct. If the spark feathers on the cable connector side, the polarity is reversed.

3- The firing end of a used spark plug can give a clue to coil polarity. If the

"DISHED" AREA

Typical magneto ignition system installed on a single cylinder powerhead.

ground electrode is "dished", it may mean polarity is reversed.

WIRING HARNESS

CRITICAL WORDS: These next two paragraphs may well be the most important words in this chapter. Misuse of the wiring harness is the most single cause of electrical problems with outboard power plants.

A wiring harness is used between the key switch and the powerhead. This harness seldom contains wire of sufficient size to allow connecting accessories. Therefore, anytime a new accessory is installed, **NEW** wiring should be used between the battery and the accessory. A separate fuse panel **MUST** be installed on the control panel. To connect the fuse panel, use one red and one black No. 10 gauge wire from the battery. If a small amount of 12-volt current should be accidently attached to the magneto system, the coil may be damaged or **DESTROYED**. Such a mistake in wiring can easily happen if the source for the 12-volt accessory is taken from the key switch. Therefore, again let it be said, **NEVER** connect accessories through the key switch.

5-3 FLYWHEEL MAGNETO IGNITION
Colt 1990 Only
2.3HP 1991 and ON
3HP 1990 and ON
3.3HP 1991 and ON
4HP 1990 Only

DESCRIPTION

READ AND BELIEVE. A battery installed to crank the powerhead **DOES NOT** mean

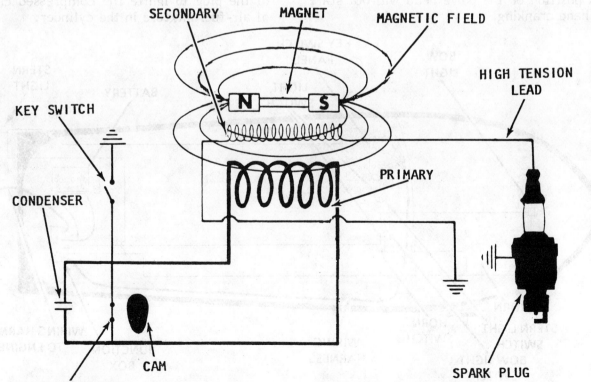

Schematic diagram of a simple single cylinder magneto ignition system with principle parts identified.

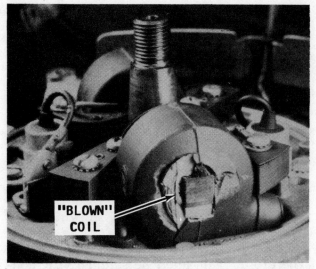

A coil *DESTROYED* when 12-volts was connected into the magneto wiring system. *Mechanics report in 85% of the cases, the damage occurs when an accessory is connected through the key switch.*

the engine is equipped with a battery-type ignition system. A magneto system uses the battery only to crank the powerhead. Once the powerhead is running, the battery has absolutely no effect on engine operation. Therefore, if the battery is low and fails to crank the powerhead properly for starting, the powerhead may be cranked manually, started, and operated. Under these conditions, the key switch must be turned to the **ON** position or the powerhead will not start by hand cranking.

A magneto system is a self-contained unit. The unit does not require assistance from an outside source for starting or continued operation. Therefore, as previously mentioned, if the battery is dead, the engine may be cranked manually and the powerhead started.

The flywheel-type magneto unit consists of an armature plate and a permanent magnet built into the flywheel. The ignition coil, condenser and breaker points are mounted on the armature plate.

As the pole pieces of the magnet pass over the heels of the coil, a magnetic field is built up about the coil, causing a current to flow through the primary winding.

Now, at the proper time, the breaker points are separated by action of a cam, and the primary circuit is broken. When the circuit is broken, the flow of primary current stops and causes the magnetic field about the coil to break down instantly. At this precise moment, an electrical current of extremely high voltage is induced in the fine secondary windings of the coil. This high voltage is conducted to the spark plug where it jumps the gap between the points of the plug to ignite the compressed charge of air-fuel mixture in the cylinder.

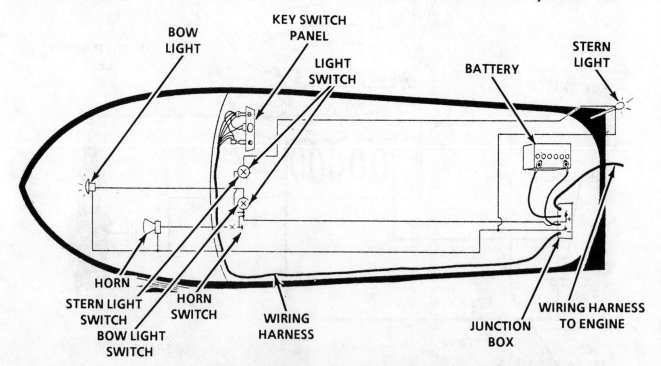

*Functional diagram to illustrate proper hookup of accessories through a junction box. If a junction is not installed on the boat, connect accessories directly to the battery. **NEVER** connect accessories through the key switch.*

TROUBLESHOOTING

Always attempt to proceed with the troubleshooting in an orderly manner. The "shotgun" approach will only result in wasted time, incorrect diagnosis, replacement of unnecessary parts, and frustration.

Begin the ignition system troubleshooting with the spark plug/s and continue through the system until the source of trouble is located.

Remember, a magneto system is a self-contained unit. Therefore, if the engine has a key switch and wire harness, remove them from the powerhead and then make a test for spark. If a good spark is obtained with these two items disconnected, but no spark is available at the plug when they are connected, then the trouble is in the harness or the key switch. If a test is made for spark at the plug with the harness and switch connected, check to be sure the key switch is turned to the **ON** position.

WIRING HARNESS

CRITICAL WORDS: These next two paragraphs may well be the most important words in this chapter. Misuse of the wiring harness is the most single cause of electrical problems with outboard powerplants.

A wiring harness is used between the key switch and the powerhead. This harness seldom contains wire of sufficient size to allow connecting accessories. Therefore, anytime a new accessory is installed, **NEW** wiring should be used between the battery and the accessory. A separate fuse panel **MUST** be installed on the control panel. To connect the fuse panel, use one red and one black No. 10 gauge wires from the battery. If a small amount of 12-volt current should be accidently attached to the magneto system, the coil will be damaged or **DESTROYED.** Such a mistake in wiring can easily happen if the source for the 12-volt accessory is taken from the key switch. Therefore, again let it be said, **NEVER** connect accessories through the key switch.

Key Switch

A magneto key switch operates in **REVERSE** of any other type key switch. When the key is moved to the **OFF** position, the circuit is **CLOSED** between the magneto and ground. In some cases, when the key is turned to the **OFF** position the points are grounded. For this reason, an automotive-type switch **MUST NEVER** be used, because the circuit would be opened and closed in reverse, and if 12-volts should reach the coil, the coil will be **DESTROYED.**

Spark Plugs

1- Check the plug wires to be sure they are properly connected. Check the entire length of the wire/s from the plug/s to the magneto under the armature plate. If the wire is to be removed from the spark plug, **ALWAYS** use a pulling and twisting motion as a precaution against damaging the connection.

2- Attempt to remove the spark plug/s by hand. This is a rough test to determine if the plug is tightened properly. You should not be able to remove the plug without using the proper socket size tool. Remove the spark plug/s and identify from which cylinder they were removed. Examine each plug and evaluate its condition as described in Section 5-2.

If the spark plugs have been removed and the problem cannot be determined, but the plug appears to be in satisfactory condition, electrodes, etc., then replace the plugs in the spark plug openings.

A conclusive spark plug test should always be performed with the spark plugs

SPARK PLUG LEAD

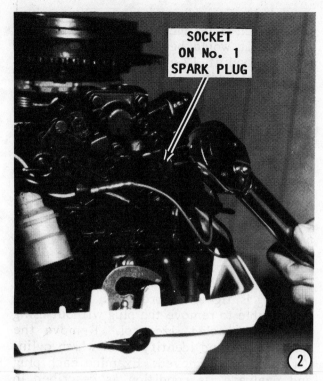

SOCKET
ON No. 1
SPARK PLUG

②

installed. A plug may indicate satisfactory spark when it is removed and tested but under a compression condition may fail. An example would be the possibility of a person being able to jump a given distance on the ground, but if a strong wind is blowing, his distance may be reduced by half. The same is true with the spark plug. Under good compression in the cylinder, the spark may be too weak to ignite the fuel properly.

Therefore, to test the spark plug under compression, replace it in the engine and tighten it to the proper torque value. A-

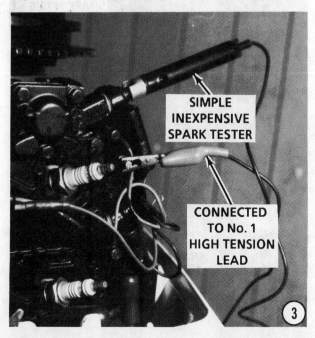

SIMPLE
INEXPENSIVE
SPARK TESTER

CONNECTED
TO No. 1
HIGH TENSION
LEAD

③

nother reason for testing for spark with the plugs installed is to duplicate actual operating conditions regarding flywheel speed. If the flywheel is rotated with the pull cord with the plugs removed, the flywheel will rotate much faster because of the no-compression condition in the cylinder, giving the **FALSE** indication of satisfactory spark.

3- Use a spark tester and check for spark at each cylinder. If a spark tester is not available, hold the plug wire about 1/4-inch from the powerhead. Turn the flywheel with a pull starter or electrical starter and check for spark. A strong spark over a wide gap must be observed when testing in this manner, because under compression a strong spark is necessary in order to ignite the air-fuel mixture in the cylinder. This means it is possible to think you have a strong spark, when in reality the spark will be too weak when the plug is installed. If there is no spark, or if the spark is weak, the trouble is most likely under the flywheel in the magneto.

ONE MORE WORD: Each cylinder has its own ignition system in a flywheel-type ignition system. This means if a strong spark is observed on one cylinder and not at another, only the weak system is at fault. However, it is always a good idea to check and service all systems while the flywheel is removed.

Compression

A compression check is extremely important, because an engine with low or uneven compression between cylinders **CANNOT** be tuned to operate satisfactorily. Therefore, it is essential that any compression problem be corrected before proceeding with the tune-up procedure. Powerheads covered in this manual -- single and twins -- should have close to 100 psi cylinder pressure. The difference between cylinders should not be much over 15 psi

If the powerhead shows any indication of overheating, such as discolored or scorched paint, especially in the area of the top (No. 1) cylinder, inspect the cylinders visually thru the transfer ports for possible scoring. A more thorough inspection can be made if the head is removed. It is possible for a cylinder with satisfactory compression to be scored slightly. Also, check the water pump. The overheating condition may be caused by a faulty water pump.

An overheating condition may also be caused by running the engine out of the water. For unknown reasons, many operators have formed a bad habit of running a small engine without the lower unit being submerged. Such a practice will result in an overheated condition in a matter of seconds. It is interesting to note, the same operator would never operate or allow anyone else to run a large horsepower engine without water circulating through the lower unit to the powerhead for cooling. Bear in mind, the laws governing operation and damage to a large unit **ALL** apply equally as well to the small engine.

Checking Compression

4- Remove the spark plug wires. **ALWAYS** grasp the molded cap and pull it loose with a twisting motion to prevent damage to the connection. Remove the spark plugs and keep them in **ORDER** by cylinder for evaluation later. Ground the spark plug leads to the engine to render the ignition system inoperative while performing the compression check.

Insert a compression gauge into the No. 1 (top), spark plug opening. Crank the engine with the starter, or pull on the

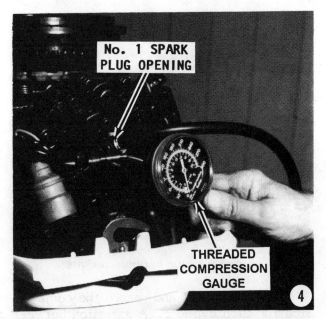

starter cord, through at least 4 complete piston strokes with the throttle at the wide-open position, or until the highest possible reading is observed on the gauge. Record the reading.

An acceptable pressure reading for a powerhead covered in this manual would be about 100 psi (689.5 kPa) or more. The cylinders should register within -- say 10%, or 10 psi (69kPa), of each other.

The preferred method of checking the cylinder walls and rings is to pull the head and make an inspection. This method will also reveal the piston condition in each cylinder.

Checking the rings and cylinder walls through the opening on the exhaust side of the engine to be sure the walls are not scored and the rings are not stuck in the piston (fail to expand properly).

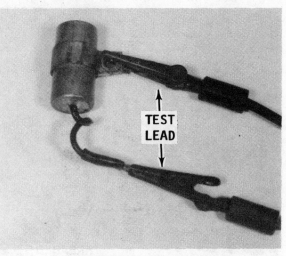

Proper hookup to test a condenser.

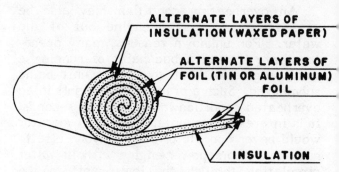

Rough sketch to illustrate how the waxed paper, aluminum foil, and insulation are rolled in a typical condenser.

Repeat the test and record the compression for each cylinder. A variation between cylinders is far more important than the actual readings. A variation of more than 10% between cylinders indicates the lower compression cylinder may be defective. The problem may be worn, broken, or sticking piston rings, scored pistons or worn cylinders. These problems may only be determined after the head has been removed. Removing the head on an outboard powerhead is not that big a deal, and may save many hours of frustration and the cost of purchasing unnecessary parts to correct a faulty condition.

Condenser

In simple terms, a condenser is composed of two sheets of tin or aluminum foil laid one on top of the other, but separated by a sheet of insulating material such as waxed paper, etc. The sheets are rolled into a cylinder to conserve space and then inserted into a metal case for protection and to permit easy assembly.

The purpose of the condenser is to absorb or store the secondary current built up in the primary winding at the instant the breaker points are separated. By absorbing or storing this current, the condenser prevents excessive arcing and the useful life of the breaker points is extended. The condenser also gives added force to the charge produced in the secondary winding as the condenser discharges.

Modern condensers seldom cause problems, therefore, it is not necessary to install a new one each time the points are replaced. However, if the points show evidence of arcing, the condenser may be at fault and should be replaced. A faulty condenser may not be detected without the use of special test equipment. The modest cost of a new condenser justifies its purchase and installation to eliminate this item as a source of trouble.

Breaker Points

The breaker points on an outboard powerhead are an extremely important part of the ignition system. A set of points may appear to be in good condition, but they may be the source of hard starting, misfiring, or poor powerhead performance. The rules and knowledge gained from association with 4-cycle engines does not necessarily apply to a 2-cycle engine. The points should be replaced every 100 hours of operation or at least once a year. **REMEMBER**, the less an outboard engine is operated, the more care it needs. Allowing an outboard engine to remain idle will do more harm than if it is used regularly.

A breaker point set consists of two points. One is attached to a stationary bracket and does not move. The other point is attached to a moveable mount. A spring is used to keep the points in contact with

Worn and corroded breaker points unfit for further service.

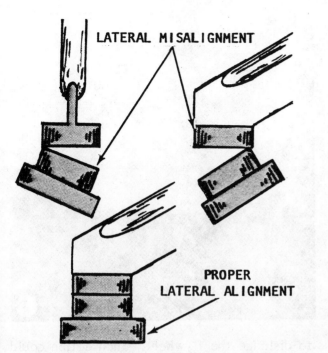

Drawing to illustrate proper point alignment, bottom set, compared with exaggerated misalignment of the other two.

each other, except when they are separated by the action of a cam built into the flywheel or machined on the crankshaft. Both points are constructed with a steel base and a tungsten cap fused to the base.

To properly diagnose magneto (spark) problems, the theory of electricity flow must be understood. The flow of electricity

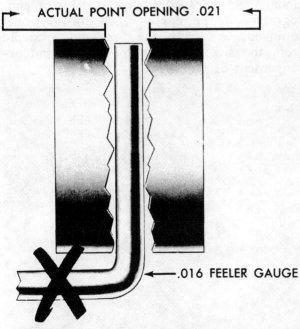

Drawing to depict how a 0.016" feeler gauge may be inserted between a badly worn set of points and the actual opening is 0.021". The point set must be in good condition to obtain an accurate adjustment.

through a wire may be compared with the flow of water through a pipe. Consider the voltage in the wire as the water pressure in the pipe and the amperes as the volume of water. Now, if the water pipe is broken, the water does not reach the end of the pipe. In a similar manner if the wire is broken the flow of electricity is broken. If the pipe springs a leak, the amount of water reaching the end of the pipe is reduced. Same with the wire. If the installation is defective or the wire becomes grounded, the amount of electricity (amperes) reaching the end of the wire is reduced.

Check the wiring carefully, inspect the points closely and adjust them accurately. The point setting for **ALL** powerheads covered in this section (except the 2.3hp 1991 and on), is 0.020" (0.5mm). The point gap for the 2.3hp powerhead is 0.014" (0.35mm). Even though the 3.3hp powerhead is very similar in almost every detail to the 2.3hp, and was introduced in the same year, the point gap for the 3.3hp engine is 0.020" (0.5mm).

SERVICING FLYWHEEL MAGNETO IGNITION SYSTEM

General Information

Magnetos installed on outboard engines will usually operate over extremely long periods of time without requiring adjustment or repair. However, if ignition system problems are encountered, and the usual corrective actions such as replacement of spark plugs does not correct the problem, the magneto output should be checked to determine if the unit is functioning properly.

PHOTOGRAPHS for the illustrations in this section were taken of single and twin cylinder powerheads. Naturally, the single cylinder powerheads will have only one set of breaker points, one coil, etc.

REMOVAL

1- Remove the hood or enough of the powerhead cover to expose the flywheel. Disconnect the battery connections from the battery terminals, if a battery is used to crank the powerhead. If a hand starter is installed, remove the attaching hardware from the legs of the starter assembly and lift the starter free.

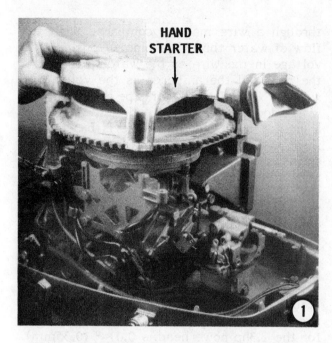

ROUND RATCHET PLATE

HAND STARTER

FLYWHEEL NUT

2- On hand rewind starter models, a round ratchet plate is attached to the flywheel to allow the hand starter to engage in the ratchet and thus rotate the flywheel. This plate must be removed before the flywheel nut is removed.

3- Remove the nut securing the flywheel to the crankshaft. It may be necessary to use some type of flywheel strap to prevent the flywheel from turning as the nut is loosened.

4- Install the proper flywheel puller using the same screw holes in the flywheel that are used to secure the ratchet plate removed in Step 2. **NEVER** attempt to use a puller which pulls on the outside edge of the flywheel or the flywheel may be damaged. After the puller is installed, tighten the center screw onto the end of the crankshaft. Continue tightening the screw until the flywheel is released from the crankshaft. Remove the flywheel. **DO NOT** strike the puller center bolt with a hammer in an attempt

to dislodge the flywheel. Such action could seriously damage the lower seal and/or lower bearing.

5- **STOP**, and carefully observe the magneto and associated wiring layout. Study how the magneto is assembled. **TAKE TIME** to make notes on the wire routing. Observe how the heels of the laminated core, with the coil attached, is flush with the boss on the armature plate. These items must be replaced in their proper positions. You may elect to follow the practice of many professional mechanics by taking a series of photographs of the engine with the flywheel removed: one from the top, and a couple from the sides showing the wiring and arrangement of parts.

FLYWHEEL PULLER

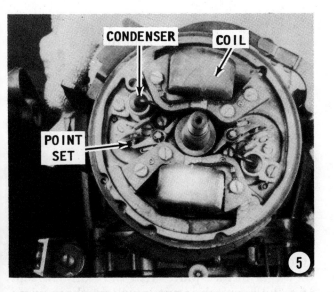

remove the flat retainer holding the set of points together.

9- Lift the moveable side of the points free of the other half of the set.

10- Remove the hold-down screw securing the non-moveable half of the point set to the armature plate.

Breaker Points/Condenser Service

The armature plate does not have to be removed to service the magneto. If it is necessary to remove the plate for other service work, such as to replace the coil or to replace the top seal, see Step 12.

For simplicity and clarity, the following procedures and accompanying illustrations cover a one-cylinder ignition system. If larger than one-cylinder is being serviced, repeat the procedures for each coil and breaker point assembly.

6- Remove the screw attaching the wires from the coil and condenser to one set of points. On engines equipped with a key switch, "kill" button, or "runaway" switch, a ground wire is also connected to this screw.

7- Using a pair of needle-nose pliers, remove the wire clip from the post protruding through the center of the points.

8- Again, with the needle-nose pliers,

After the flywheel has been removed it should be placed on the bench with the magnets facing upward. This position will help prevent small particles from becoming attached to the magnets.

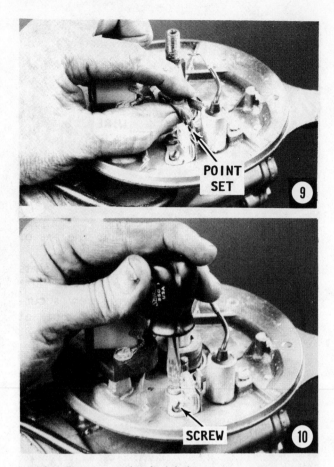

POINT SET ⑨

SCREW ⑩

11- Remove the hold down screw securing the condenser to the armature plate. Observe how the condenser sets into a recess in the armature plate.

Repeat the procedure for the other set of points.

Armature Plate Removal

First, These Words: It is not necessary to remove the armature plate unless the top seal or the coil is to be replaced.

12- Disconnect the advance arm connecting the armature plate with the power

CONDENSER ⑪

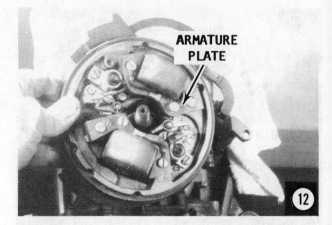

ARMATURE PLATE ⑫

shaft on the side of the powerhead. Next, remove the wires connecting the underside of the armature plate with the "kill" switch. If a "kill" switch is not installed, these wires are connected to the wiring harness plug. The wires of most units have a quick-disconnect fitting. Remove the wires from the vacuum (runaway) switch, if one is installed.

13- Observe the four screws in a square pattern through the armature plate. Two of these screws pass through the laminated core and the armature plate into the powerhead retainer. The other two pass just through the plate. Loosen these four screws. After the screws are loose, lift the armature plate up the crankshaft and clear of the engine. If any oil is present on top of the armature plate, or on the points, the top seal **MUST** be replaced.

Top Seal Replacement

Replacement of the top seal on a Johnson/Evinrude engine is **NOT** a difficult task,

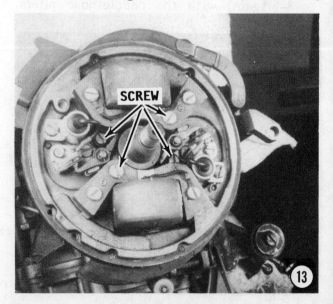

SCREW ⑬

with the proper tools: a seal remover and seal installer. **NEVER** attempt to remove the seal with screwdrivers, punch, pick, or other similar tool. Such action will most likely damage the collars in the powerhead.

Obtain OMC Seal Remover P/N 387780. A 1-1/8" open end wrench is needed to hold the remover portion of the tool, while a 3/8" open end wrench is used on the top bolt.

14- To remove the seal, first, work the point cam up and free of the driveshaft. Next, remove the Woodruff key from the crankshaft. A pair of side-cutters is a handy tool for this job. Grasp the Woodruff key with the side-cutters and use the leverage of the pliers against the crankshaft to remove the key.

15- Work the special tool into the seal. Observe how the special tool is tapered and has threads. Continue working and turning the tool until it has a firm grip on the inside of the seal. Now, tighten the center screw of the puller against the end of the crankshaft and the seal will begin to lift from the collars. Continue turning this center screw until the seal can be raised manually from the crankshaft.

16- To install the new seal: Coat the inside diameter of the seal with a thin layer of oil. Apply OMC sealer to the outside diameter of the seal. Slide the seal down the crankshaft and start it into the recess of the powerhead. Use the special tool and work the seal completely into place in the recess.

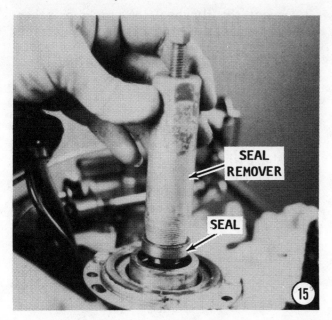

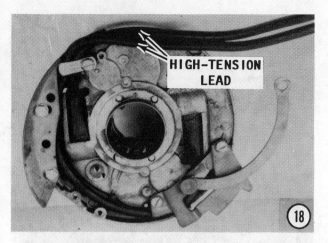

HIGH-TENSION LEAD

18

17- Install the Woodruff key into the crankshaft. On some models, a pin was used to locate the cam for the points. If the pin was used, install it at this time. Oberve the difference to the sides of the cam. On almost all cams, the word **TOP** is stamped on one side. Also, on some cams, the groove does not go all the way through. Therefore, it is very difficult to install the cam incorrectly, with the wrong side up.

Slide the cam down the crankshaft with the word **TOP** facing upward. Continue working the cam down the crankshaft until it is in place over the Woodruff key or pin.

If the coil is **not** to be removed, proceed directly to Step 7. To remove the coil, perform the procedures in the following section.

Coil Removal from the Armature Plate

The armature plate must be removed as described earlier in this section, Step 12 and Step 13. Notice how the coil has a laminat-

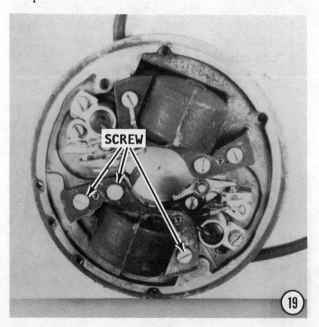

SCREW

19

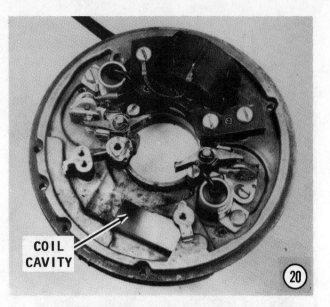

COIL CAVITY

20

ed core. The coil cannot be separated, that is, the laminations from the core.

18- Turn the armature plate over and notice how the high-tension leads are installed on the plate in a recess. The routing of the wires is misleading. The wire to the No. 1 spark plug is **NOT** connected to the No. 1 coil as might be expected.

19- Remove the three screws attaching the coils to the armature plate.

20- Hold the armature plate and separate the coils from the plate. As the coil is separated from the plate, observe the high-tension lead to the spark plug inside the coil. Work the small boot, if used, and the high-tension lead from the coil.

CLEANING AND INSPECTING

Inspect the flywheel for cracks or other damage, especially around the inside of the center hub. Check to be sure metal parts have not become attached to the magnets. Verify each magnet has good magnetism by using a screwdriver or other tool.

Thoroughly clean the inside taper of the flywheel and the taper on the crankshaft to prevent the flywheel from "walking" on the crankshaft while the engine is running.

Check the top seal around the crankshaft to be sure no oil has been leaking onto the armature plate. If there is **ANY** evidence the seal has been leaking, it **MUST** be replaced, as outlined earlier in this section.

Test the armature plate to verify it is not loose. Attempt to lift each side of the plate. There should be little or no evidence of movement.

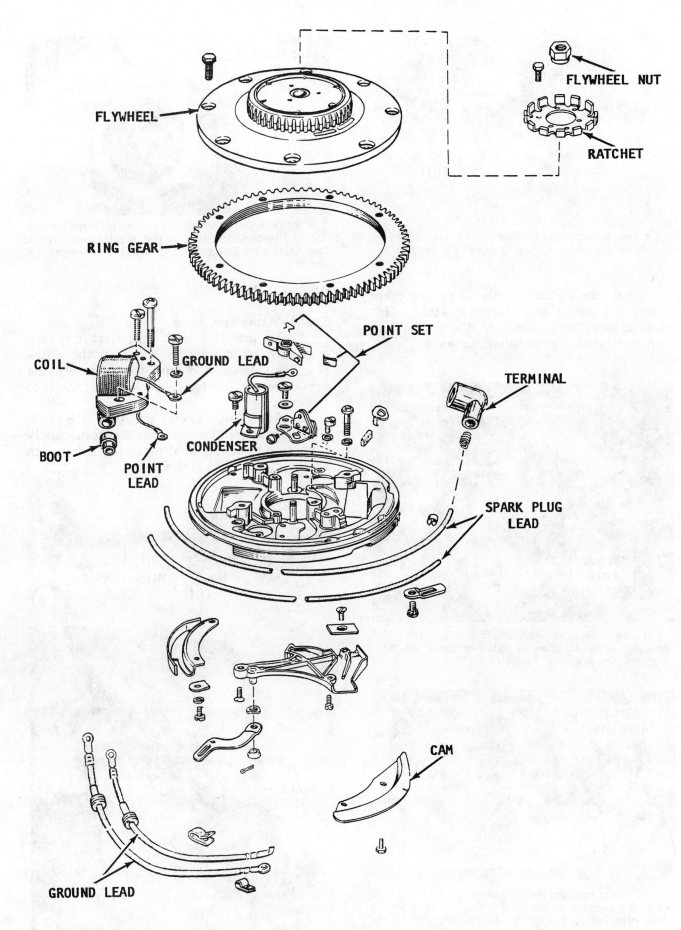

Exploded drawing of a typical magneto system. Only one coil and set of points are shown.

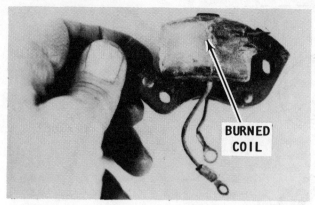

A coil burned where the high-tension lead enters the coil on the bottom side. Arcing caused the damage.

Clean the surface of the armature plate where the points and condenser attach. Install a new condenser into the recess and secure it with the hold-down screw.

A coil destroyed when the side blew out. This damage was caused when 12-volts was connected to the magneto circuit at the key switch.

A broken crankshaft and cracked flywheel damaged when the engine was operated at a high rpm with a flush attachment and garden hose connected to the lower unit.

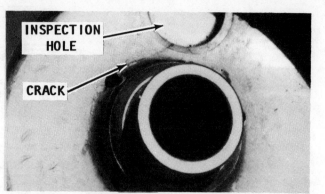

Cracks in the flywheel hub caused by metal fatigue due to flywheel construction and the inspection hole. This hole is no longer incorporated in late-model flywheels.

ASSEMBLING

Coil to Armature Plate

1- To install a new coil, first turn the armature plate over, and loosen the spark plug lead wires, and push them through the armature plate. Now, work the leads into the coil.

2- After the leads are into the coil, work the small boot up onto the coil. Apply a coating of rubber seal material underneath the boot, if a boot is used.

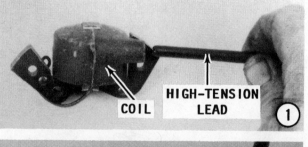

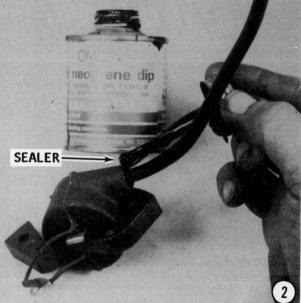

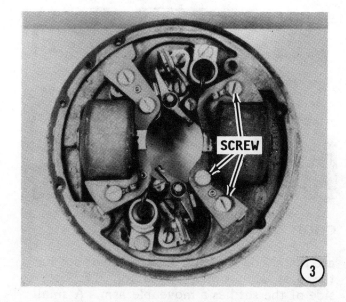

3- Start the three screws through the laminated core into the armature plate, but **DO NOT** tighten them. If the powerhead being serviced has a second coil, install the other coil in the same manner.

4- Check to be sure the spark plug (high-tension) leads are properly positioned in the coil and are securely attached to the bottom side of the armature plate.

5- To adjust the coil: A special ring tool is required that fits down over the armature plate. This tool will properly locate the coil in relation to the flywheel. Install this special tool over the armature plate. Push outward on the coil and secure the two outer screws.

6- If a special ring tool is not available, and in an emergency, hold a straight edge against the boss on the armature plate and bring the heel of the laminated core out square against the edge of the boss on the

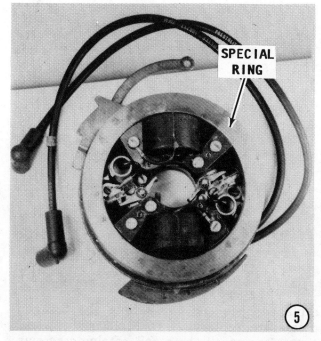

armature plate. The ground wire for the coil should be attached under the head of the top screw passing through the laminated core.

Wick Replacement

7- The wick, mounted in a bracket under the coil, can be replaced without removing the armature plate. The wick **SHOULD** be

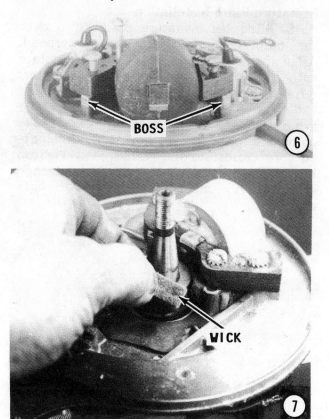

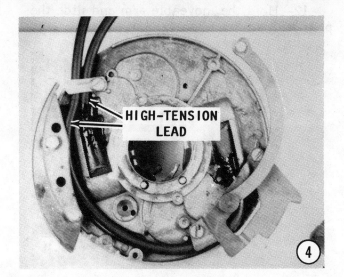

ARMATURE PLATE

⑧

replaced each and every time the breaker points are replaced. To replace the wick, simply loosen all three coil retaining screws and remove the one screw through the wick holder. Lift the coil slightly and remove the wick and wick holder. Slide the new wick into the holder; install the holder and wick under the coil; and secure it in place with the retaining screw. Adjust the coil as described earlier in this section, Steps 5 and 6, and tighten the three screws.

Armature Plate Installation

8- Slide the armature plate down over the crankshaft and onto the powerhead. Align the screw holes in the armature plate with the holes in the powerhead retainer. After the armature plate is in place, install and tighten the two screws securing the armature plate to the retainer. Now, take up on the three screws through the laminated core closest to the crankshaft. Tighten the screws securely. Attach the advance arm from the magneto to the tower shaft arm.

CLIP

⑨

CONDENSER   ⑩

GOOD WORDS

The points **MUST** be assembled as they are installed. One side of each point set has the base and is non-moveable. The other side of the set has a moveable arm. A small wire clip and a flat retainer are included in each point set package.

9- Hold the base side of the points and the flat retainer. Notice how the base has a bar at right angle to the points. Observe the hole in the bar. Observe the flat retainer. Notice that one side has a slight indentation. When the points are installed, this indentation will slip into the hole in the base bar.

Point Set

10- Install the condensers and secure them in place with their hold-down screws.

11- Hold the base side of the points and slide it down over the anchor pin onto the armature plate. Install the wavy washer and hold-down screw to secure the point base to the armature plate. Tighten the hold-down screw securely.

12- Hold the moveable arm and slide the points down over post, and at the same

SCREW

POINT SET BASE SIDE   ⑪

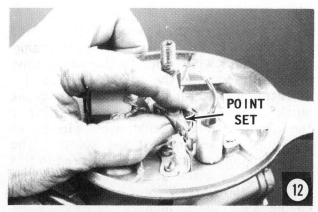

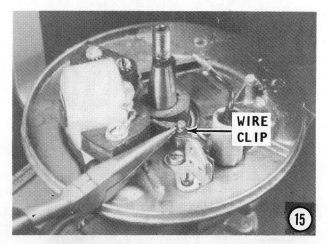

time, hold back on the points and work the spring arm to the inside of the post of the base points. Continue to work the points on down into the base.

13- Observe the points. The points should be together and the spring part of the moveable arm on the inside of the flat post.

14- Install the flat retainer onto the flat bar of the base points. Check to be sure the flat spring from the other side of the points is on the inside of the retainer. Push the retainer inward until the indentation slips into the hole in the base. The retainer **MUST** be horizontal with the armature plate.

15- Install the wire clip into the groove of the post.

CRITICAL WORDS

As the coil, condenser, and "kill" switch wire are being attached to the point set, take the following precautions and adjustments:

a- The wire between the coil and the points should be tucked back under the coil and as far away from the crankshaft as possible.

b- The condenser wire leaving the top of the condenser and connected to the point set, should be bent downward to prevent the flywheel from making contact with the wire. A countless number of installations have been made only to have the flywheel rub against the condenser wire and cause failure of the ignition system.

c- Check to be sure all wires connected to the point set are bent downward toward the armature plate. The wires **MUST NOT**

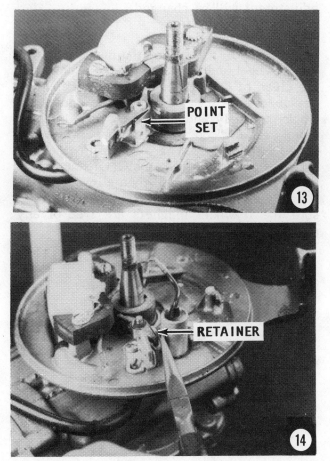

CORRECT ALIGNMENT

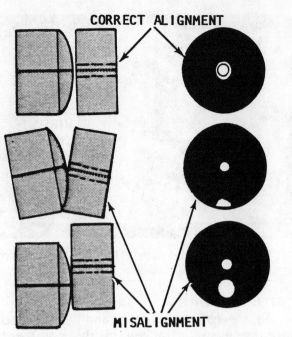

MISALIGNMENT

Before setting the breaker point gap, the points must be properly aligned (top). ALWAYS bend the stationary point, NEVER the breaker lever. Attempting to adjust an old worn set of points is not practical because oxidation and pitting of the points will always give a false reading.

touch the plate. If any of the wires make contact with the armature plate, the ignition system will be grounded and the engine will fail to start.

16- Connect the wire leads to the set of points, with the attaching screw.

GOOD WORDS

The point spring tension is predetermined at the factory and does not require adjustment. Once the point set is properly installed, all should be well. In most cases, breaker contact and alignment will not be necessary. If a slight alignment adjustment should be required, **CAREFULLY** bend the insulated part of the point set.

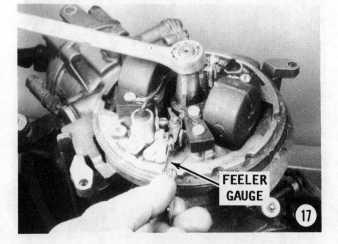

FEELER GAUGE

17

Point Adjustment

17- Install the flywheel nut onto the end of the crankshaft. Now, rotate the crankshaft **CLOCKWISE** and at the same time observe the cam on the crankshaft. Continue rotating the crankshaft until the rubbing block of the point set is at the high point of the cam. At this position, use a wire gauge or feeler gauge and set the points. The point setting for **ALL** engines covered in this section (except the 2.3hp 1991 and on), is 0.020" (0.5mm). The point gap for the 2.3hp engine is 0.014" (0.35mm).

A wire gauge will always give a more accurate adjustment than a feeler gauge. Work the gauge between the points and, at the same time, turn the eccentric on the armature plate until the proper adjustment is obtained. Rotate the crankshaft a complete revolution and again check the gap adjustment. After the crankshaft has been turned and the points are on the high point of the cam, check to be sure the hold-down screw is tight against the base. There is enough clearance to allow the eccentric on the base points to turn. If the hold-down screw is tightened **AFTER** the point adjustment has been made, it is very likely the adjustment will be changed.

Flywheel Installation

18- Check to be sure the inside taper of the flywheel and the taper on the crankshaft are clean of dirt or oil. These surfaces must be clean and dry to permit the tapers to "lock" properly and to prevent the flywheel from "walking" on the crankshaft while the powerhead is operating. To ensure the two surfaces are clean, use OMC Cleaning Solvent, or equivalent, and then allow time for the solvent to dry thoroughly. Check to be sure the flywheel magnets are free of any metal parts.

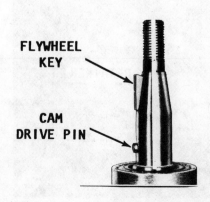

FLYWHEEL KEY

CAM DRIVE PIN

18

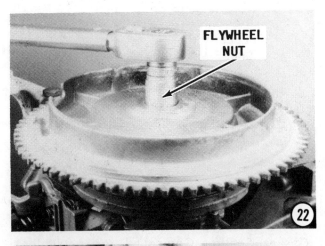

This particular engine differs from the text procedures because a washer is installed under the flywheel nut.

20- Rotate the flywheel **CLOCKWISE** and check to be sure the flywheel does not contact any part of the magneto or the wiring.

21- Place the ratchet for the starter on top of the flywheel and install the three 7/16" screws. On some model powerheads, a plate retainer covers these screws.

22- Coat the flywheel nut with OMC

Model Colt & Junior

Place the key in the crankshaft keyway, with the outer edge parallel to the centerline of the crankshaft, as indicated in the accompanying illustration. Apply OMC Nut Lock, or equivalent, to the cam drive pan, and then install the pin into the crankshaft. Install the cam with the side marked **TOP** facing **UP**.

19- Slide the flywheel down over the crankshaft with the keyway in the flywheel aligned with the key on the crankshaft.

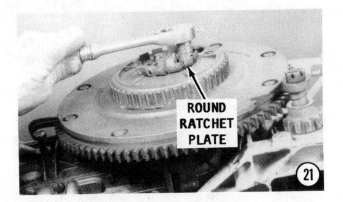

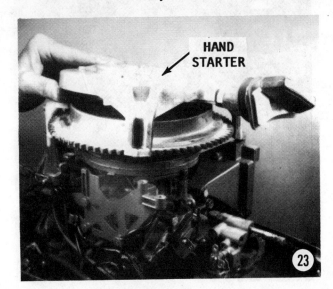

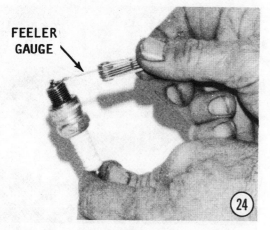

Gasket Sealing Compound, or equivalent, and then thread the flywheel nut onto the crankshaft and tighten it to the following torque value:

Colt & Junior	22-25ft lb (30-40Nm)
2.3 & 3.3	29-33ft lb (40-45Nm)
3, 4 & 4D	30-40ft lb (40-54Nm)
5 thru 8	40-50ft lb (54-70Nm)

23‑ After the ratchet and flywheel nut have been installed, install the hand starter over the flywheel, if one is used. Check to be sure the ratchet engages the flywheel properly.

24‑ Set the gap on each spark plug at 0.030".

25‑ Install the spark plugs and tighten them to the torque value of 210-246 in lbs. Connect the battery leads to the battery terminals, if a battery is used with a starter motor to crank the powerhead.

To synchronize the powerhead, proceed directly to Section 5-6.

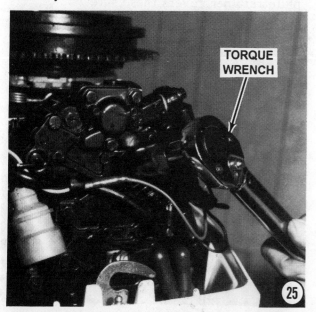

A good grade of OMC approved oil should always be used.

5-4 CDII IGNITION WITH SENSOR COIL

During the production year -- 1990, the manufacturer phased out CDII capacitor discharge ignition with sensor on most of the 2-cylinder powerheads covered in this manual. Model 3 and Model 4 powerheads retained the CDII system. The 1-cylinder powerheads still have the straight magneto ignition system, covered in Section 5-3, of this chapter.

Powerheads of various sizes leaving the factory in early 1990 could be equipped with either the CD with sensor coil, covered in this section, or with the UFI system, covered in the next section, 5-5.

The Appendix contains wire identification drawings for several versions of the CDII system.

Easy Identification
The easiest way to determine which ignition system is installed on the 2-cylinder powerhead being serviced, is to trace back the main harness from the ignition coil mounted on the powerhead. If the leads can be traced back under the flywheel, the powerhead is equipped with UFI. If the harness is traced back to a power pack ("black box") or to an external ignition module, the powerhead is equipped with CD ignition with sensor.

DESCRIPTION

READ AND BELIEVE. A battery installed to crank the powerhead **DOES NOT** mean the powerhead is equipped with a battery-type ignition system. A CD magneto system uses the battery only to crank the powerhead. Once the powerhead is running, the battery has absolutely no affect on powerhead operation. Therefore, if the battery is low and fails to crank the powerhead properly for starting, the powerhead may be cranked manually, started, and operated. Under these conditions, the key switch must be turned to the **ON** position or the powerhead will not start by hand cranking.

A CD magneto system is a self-contained unit. The unit does not require assistance from an outside source for starting or continued operation. Therefore, as previously mentioned, if the battery is dead, the powerhead may be cranked manually and the engine started.

This capacitor discharge (CD) ignition system consists of the following assemblies: flywheel, charge coil, sensor, power pack, and ignition coils.

The flywheel assembly contains two integral magnets installed 180° apart. These magnets are charged with opposite polarity.

The charge coil assembly is composed of a large coil of wire which generates alternating current (ac). The current is fed into the Power Pack.

The sensor assembly consists of a small coil of wire which generates the triggering voltage necessary for the Power Pack.

In simple terms, the Power Pack contains the electronic circuits required to produce ignition at the proper time.

The ignition coils generate approximately 30,000 volts which is fed to the spark plugs to ignite the fuel/air mixture in the cylinders.

Before performing maintenance work on the system, it would be well to take time to read and understand the introduction information presented in Section 5-1 at the beginning of this chapter, the Description at the start of this section, and the Theory of Operation in the following paragraphs.

THEORY OF OPERATION

This system generates approximately 30,000 volts which is fed to the spark plugs

without the use of a point set or an outside voltage source.

To understand how high voltage current is generated and reaches a spark plug, it is well to understand a couple of basic terms used throughout the following paragraphs and troubleshooting procedures.

Diode -- a small electrical part that allows current to flow in only one direction. On the accompanying diagrams the current flow is indicated by the direction of the arrow.

Silicon Controlled Rectifier -- commonly referred to as a SCR -- a small electrical part in which current can flow through in only one direction, and in which the current can flow only when a "Gate" in the part receives a positive (+) input to trigger or open the gate in the SCR. On the accompanying diagrams the direction of current flow is indicated by an arrow.

Capacitor -- a part that stores voltage. One end of the capacitor is positive (+), and the other end is negative (-). The voltage is stored in the capacitor until the trigger circuit is activated.

While studying the following circuit explanations, imagine the flywheel rotating very, very slowly.

Capacitor, Charge Circuit

The terms and letter designations are keyed to reference illustration **"A"**.

As the flywheel rotates, the magnetic field of a magnet passes through the charge coil winding. As a result, the charge coil produces a small amount of alternating current (AC). This current flows from the

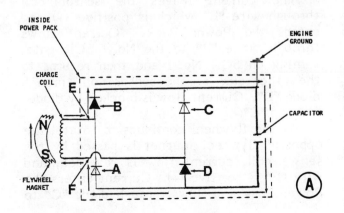

charge coil through wire **"E"** which is positive (+), then enters the Power Pack. It then flows through diode **"B"** which applies a positive (+) charge to the ground side of the capacitor. Current flow is blocked by diodes **"A"** and **"C"**. On the return path, current flows from the capacitor to the charge coil through diode **"D"** and wire **"F"**.

Alternating current from the charge coil has been changed (rectified) into direct current (DC) for capacitor charge by the four diodes. The diodes maintain a positive (+) charge on the ground side of the capacitor, regardless of the constantly changing charge coil output.

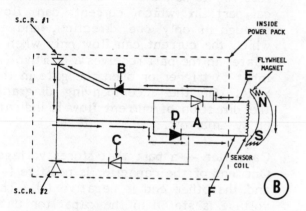

Triggering Circuit

The terms and letter designations are keyed to reference illustration **"B"**.

Purpose of the triggering circuit is to control which silicon controlled rectifier (SCR) will turn on and allow the capacitor to discharge. A sensor coil installed under the flywheel produces the signal necessary to turn on the SCR which controls the ignition timing.

After the capacitor is charged, the flywheel continues to rotate so the magnetic field of a magnet passes through the sensor coil winding, producing alternating current.

The current leaves the sensor coil through wire **"E"** which is positive (+) and enters the Power Pack. Current flows through diode **"B"** to the No. 1 SCR gate, turning on SCR No. 1 and then returns to the other side of the sensor coil through diode **"D"**. Current flow is blocked by diodes **"A"** and **"C"**.

As the flywheel continues to rotate, the opposite flywheel magnet is passed by the sensor coil, current flow is reversed, and wire **"F"** is positive (+). Current enters the Power Pack and flows through diode **"C"** to

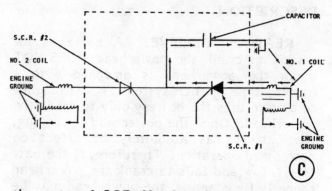

the gate of SCR No. 2 and returns to the other side of the sensor coil through diode **"A"**. Current flow is blocked by diode **"B"** and bypasses diode **"D"**.

The four diodes in the triggering circuit direct current flow to the SCR gates and back to the sensor.

Capacitor Discharge (CD) Circuit

The terms and letter designations are keyed to reference illustration **"C"**.

When SCR No. 1 is triggered, the positive (+) charge stored in the capacitor flows to powerhead ground, enters the No. 1 ignition coil through the grounded primary winding wire, and then flows through SCR No. 1 to the other side of the capacitor.

During capacitor discharge, current flows through the ignition coil primary winding and SCR No. 1 until the capacitor is discharged and SCR No. 1 turns itself off. The capacitor can now be recharged as described in the capacitor charge circuit paragraphs.

Current flowing through the No. 1 ignition coil primary winding produces a large magnetic field surrounding the secondary winding and produces the high voltage to the spark plug to ignite the fuel/air mixture in the cylinder.

When SCR No. 2 is triggered, the capacitor is discharged through the No. 2 ignition coil primary winding. Current flows through the SCR No. 2 and returns to the other side of the capacitor. The No. 2 ignition coil then produces the high voltage to the spark plug to ignite the fuel/air mixture in the next cylinder.

Ignition Stop/Kill Circuit

The terms and letter designations are keyed to reference illustration **"D"**.

The capacitor is the heart of the ignition system. Therefore, the stop button (or key switch) prevents capacitor charge. One end

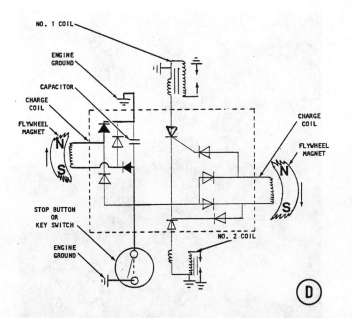

NO. 1 COIL
ENGINE GROUND
CAPACITOR
CHARGE COIL
FLYWHEEL MAGNET
STOP BUTTON OR KEY SWITCH
ENGINE GROUND
CHARGE COIL
FLYWHEEL MAGNET
NO. 2 COIL

of the capacitor is connected to powerhead ground. When the stop button (or key switch) is closed, the capacitor is grounded. When both ends of the capacitor are connected to powerhead ground, current flow from the charge coil will by-pass the capacitor as shown in reference illustration "D". If the capacitor is not charged, ignition cannot occur.

TROUBLESHOOTING

Always attempt to proceed with the troubleshooting in an orderly manner. The "shotgun" approach will only result in wasted time, incorrect diagnosis, replacement of unnecessary parts, and frustration.

Begin the ignition system troubleshooting with the spark plug/s and continue through the system until the source of trouble is located.

Compression

A compression check is extremely important, because an engine with low or uneven compression between cylinders CANNOT be tuned to operate satisfactorily. Cylinder compression for the 2-cylinder powerheads covered in the manual should be close to 100 psi. The difference between cylinders should not be much over 15 psi. It is essential -- any compression problem be corrected before proceeding with the tune-up procedure. See Chapter 3.

If the powerhead shows any indication of overheating, such as discolored or scorched paint, especially in the area of the top (No.

1) cylinder, inspect the cylinders visually thru the transfer ports for possible scoring. A more thorough inspection can be made if the head is removed. It is possible for a cylinder with satisfactory compression to be scored slightly. Also, check the water pump. The overheating condition may be caused by a faulty water pump.

An overheating condition may also be caused by running the engine out of the water. For unknown reasons, many operators have formed a bad habit of running a small horsepower outboard without the lower unit being submerged. Such a practice will result in an overheated condition in a matter of seconds. It is interesting to note, the same operator would never operate or allow anyone else to run a large horsepower engine without water circulating through the lower unit to the powerhead for cooling. Bear in mind, the laws governing operation and damage to a large unit ALL apply equally as well to the small engine.

Checking Compression

1- Remove the spark plug wires. ALWAYS grasp the molded cap and pull it loose with a twisting motion to prevent damage to the connection.

2- Remove the spark plugs and keep them in ORDER by cylinder for evaluation

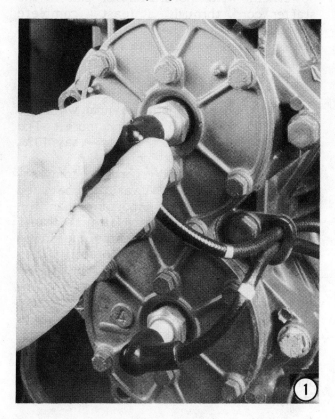

later. Ground the spark plug leads to the engine to render the ignition system inoperative while performing the compression check and to prevent damage to the ignition coil. If a high tension lead is not grounded, the coil will attempt to match the demand created by the spark trying to jump from the electrode to the nearest ground.

Insert a compression gauge into the No. 1 (top) spark plug opening. Crank the powerhead with the starter, or pull on the starter cord, through at least 4 complete piston strokes with the throttle at the wide-open position, or until the highest possible reading is observed on the gauge. Record the reading.

An acceptable pressure reading for a powerhead covered in this manual would be about 100 psi (689.5 kPa) or more. The cylinders should register within -- say 10%, or 10 psi (69kPa), of each other.

Repeat the test and record the compression for each cylinder. A variation between cylinders is far more important than the actual readings. A variation of more than 5 psi between cylinders indicates the lower compression cylinder may be defective. The problem may be worn, broken, or sticking piston rings, scored pistons or worn cylinders. These problems may only be determined after the head has been removed. Removing the head on an outboard powerhead is not that big a deal, and may save many hours of frustration and the cost of purchasing unnecessary parts to correct a faulty condition.

CHECKING CD IGNITION WITH SENSOR COIL

The following equipment is required to test the CD ignition system with sensor coil.

There is no way on this green earth to properly or accurately check the system without the following test equipment:

Spark Tester -- with the air gap adjusted to 1/2 inch (12.7 mm). A Stevens or Electro Specialties CD77 Voltmeter Tester will do the job.
Ohmmeter -- capable of indicating low ohms (Rx1) and high ohms (Rx1,000).
Peak Reading Voltmeter -- meter must be capable of registering peak voltage.
Jumper Wires -- four required. These jumper wires may be made using a piece of No. 16 solid wire about 8 inches long with the insulation stripped back about an inch from each end.

Spark Plugs
1- Check the spark plug high tension leads to be sure they are properly connected. Check the entire length of the leads from the plugs to the coils. If the lead is to be removed from the spark plug, **ALWAYS** use a pulling and twisting motion as a precaution against damaging the connection.

2- Attempt to remove the spark plugs by hand. This is a rough test to determine if the plug is tightened properly. You should

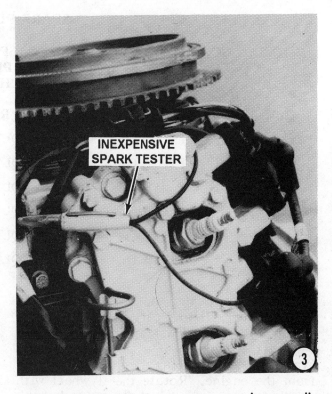

not be able to remove the plug without using the proper socket size tool. Remove the spark plugs and identify from which cylinder they were removed. Examine each plug and evaluate its condition as described in Section 5-2.

If the spark plugs have been removed and the problem cannot be determined, but the plug appears to be in satisfactory condition, electrodes, etc., then replace the plugs in the spark plug openings.

A conclusive spark plug test should always be performed with the spark plugs installed. A plug may indicate satisfactory spark when it is removed and tested, but under a compression condition may fail. An example would be the possibility of a person being able to jump a given distance on the ground, but if a strong wind is blowing, his distance may be reduced by half. The same is true with the spark plug. Under good compression in the cylinder, the spark may be too weak to ignite the fuel properly.

Therefore, to test the spark plug under compression, replace it in the powerhead and tighten it to the proper torque value -- 18-21 ft lbs (24-27Nm). Another reason for spark testing with the plugs installed is to duplicate actual operating conditions regarding flywheel speed. If the flywheel is rotated with the pull cord with the plugs removed, the flywheel will rotate much

faster because of the no-compression condition in the cylinder, giving the **FALSE** indication of satisfactory spark.

3- A neon test light or a spark tester capable of testing for spark while cranking and also while the powerhead is operating, can be purchased from almost any automotive parts supply store.

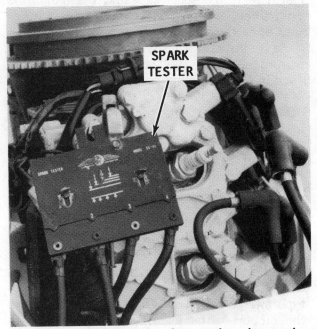

A more expensive and sophisticated spark tester than the tester shown at the top of this column. Such a tester can be connected to more than one high tension lead at a time.

SPARK PLUG TABLE

MODEL	NO. CYL.	YEAR	IGN. TYPE	SPARK PLUG CHAMP.	PLUG GAP INCHES	PLUG GAP mm	TORQUE VALUE IN. LBS	TORQUE VALUE FT LBS
Colt	1	1990 & On	Magneto	RJ6C	0.030	8mm	216-240	18-21
2.3	1	1991 & On	Magneto	QL77JC4	0.030	8mm	216-240	18-21
3.3	1	1991 & On	Magneto	QL77JC4	0.030	8mm	216-240	18-21
3	1	1990	Magneto	QL77JC4	0.030	8mm	216-240	18-21
3	1	1991 & On	CDII	QL77JC4	0.030	8mm	216-240	18-21
4	1	1990	Magneto	QL77JC4	0.030	8mm	216-240	18-21
4	1	1991 & On	CDII	QL77JC4	0.030	8mm	216-240	18-21

Use a spark tester and check for spark at each cylinder. If the spark tester is used, it should be held at least 2" from any metal part of the engine. If a spark tester is not available, hold the plug wire about 1/4-inch from the engine. Rotate the flywheel with a pull starter or electrical starter and check for spark. A strong spark over a wide gap must be observed when testing in this manner, because under compression a strong spark is necessary in order to ignite the air-fuel mixture in the cylinder. This means it is possible to think you have a strong spark, when in reality the spark will be too weak when the plug is installed. If there is no spark, or if the spark is weak, the trouble is most likely under the flywheel in the magneto or in the key switch.

If a satisfactory spark cannot be obtained, proceed with the troubleshooting as outlined in the following paragraphs.

GOOD WORDS

The leads from the ignition module and sensor coil have a common 5- or 6-pin connector. Numerous references to this connector will be made in the following tests.

KEY SWITCH WORDS

The key switch operates in **REVERSE** of any other type key switch. When the key is moved to the **OFF** position, the circuit is **CLOSED**. For this reason, an automotive-type switch **MUST NEVER** be used, because the circuit would be opened and closed in reverse, and if 12-volts should reach the coil, the coil will be **DESTROYED**, as shown on Page 5-6 and Page 5-18.

SAFETY WORDS

DO NOT touch the terminal ends of the jumper wires while making connections to the ignition module or while cranking the powerhead. **NEVER** come in contact with ignition coils or the spark tester while the powerhead is being cranked.

Ignition coils may be tested with the coil installed on the powerhead. However, if the coil is removed, test **MUST** be made on a wooden or insulated bench top to prevent leakage or possible shock.

Do not come in contact with the tester leads or the ignition coil during an output test.

When the powerhead is being cranked, be **CAREFUL** not to touch the high tension leads or the ignition coils to prevent shock and possible serious personal injury.

MONEY SAVING WORDS

To take testing of electrical components one step further: An erroneous reading could be obtained if the hand cranking speed of the powerhead was not fast enough to produce the required voltage. This condition would occur if using an undercharged battery.

The powerhead can be hand cranked at a higher rpm with the spark plugs removed. Therefore before replacing an expensive electrical, non-returnable component, remove the spark plugs; crank the powerhead with the hand rewind starter; and check for a better meter reading.

As a last resort, take the suspected faulty ignition part to the local dealer and have him give it his full testing treatment before making the new purchase.

TESTING IGNITION COMPONENTS
CDII WITH SENSOR COIL
AND POWER PACK

GOOD WORDS

At press time, OMC was using the following color code system. We list as "always", but add -- "subject to change".

Red -- always a "hot" lead when outboard cables are connected to the battery.

Purple -- always a hot lead when key switch is turned to the ON or START position.

Black -- always a ground lead.

Tan -- always a warning system --low oil or excessive temperature.

Gray -- always a tachometer lead.

Black/Yellow -- always the "Kill" circuit.

Yellow/Red -- always the start circuit.

Purple/White -- always the primer solenoid lead.

Blue/Green, Blue/White, or Green/White -- always the trim/tilt circuits. Lead with Green is DOWN circuit; lead with Blue is UP circuit.

Yellow, Yellow/Blue, Yellow/Gray -- always the charging circuit.

WIRING DIAGRAM

The wiring diagram from the Appendix is reproduced here in this section on Page 5-33, as an aid to performing tests on the various ignition components.

Test No. 1
Ignition Output

Remove the high-tension leads from both spark plugs. Adjust the gap on the spark tester to 1/2" (12mm). Connect the tester leads to the spark plug leads. Attach the tester clip to a good ground on the powerhead. Install the emergency stop switch clip and lanyard assembly, if the unit is so equipped.

Crank the powerhead. A good spark should jump each tester gap alternately.

If the results are good for both cylinders, proceed directly to Test No. 14.

If the test results are satisfactory for one cylinder, but not for the other, proceed directly to Test No. 13.

If both cylinders fail the test, proceed with the next test, No. 2.

Test No. 2
Stop Circuit Test

With the spark tester still connected, as for Test No.1, disconnect the 5-pin connector between the armature plate and the power pack -- see wiring diagram on Page 5-33. Insert jumper wires between the "A", "B", "C", and "D" terminals of the connector.

Crank the powerhead and observe the results on the spark tester.

If spark is now observed jumping alternately at each gap, the problem is in the stop circuit. Proceed to Test No. 3, Stop Button Ohm Test, or to Test No. 4, Key Switch Ohm Test.

If there is spark for one cylinder, but not the other, proceed directly to Test No. 13.

If there is no spark for either cylinder, proceed directly to Test No. 5.

Test No. 3
Stop Button — Ohms

Most production steering handles have a combination safety stop button/emergency stop device. A slip and lanyard assembly is installed behind a stop button on the steering handle. When the clip/lanyard is installed the stop device is in the RUN position. If the lanyard is removed for any reason, the device is immediately placed in the STOP position and the powerhead is shut down. To stop the powerhead when the lanyard is in place the button is held in the depressed position until the powerhead shuts down.

The powerhead cannot be started without the clip/lanyard in place behind the button.

Begin this test by installing the clip/lanyard behind the stop button. Set an ohmmeter to the highest ohm scale. Connect the Red meter lead to the "E" terminal on the armature plate side of the connector -- see wiring diagram on Page 5-33. Connect the Black meter lead to a good ground on the powerhead. The meter must indicate a very high ohm reading.

Momentarily depress the stop button. The meter should indicate a very low reading.

Remove the clip/lanyard. The meter must indicate a very low reading again.

If the test results are not satisfactory, the stop button must be replaced as a very necessary safety measure.

Test No. 4
Key Switch — Ohms

Install the clip/lanyard behind the stop button. Set the ohmmeter to indicate low and high ohms. Connect the Red meter lead to the "E" terminal of the armature plate side of the connector -- see wiring diagram on the next page. Connect the Black meter lead to a good ground on the powerhead.

With the key in the **OFF** position, the meter must indicate a low reading.

Turn the key switch to the **ON** position. The meter must indicate a high reading. If the reading is low, disconnect the harness Black/Yellow leads from the key switch terminal "M". If the meter indicates a high reading, test the key switch, see Chapter 6 -- Electrical. If the meter indicates a high reading, proceed directly to Test No. 9.

Separate the emergency stop switch lead from the harness Black/Yellow lead at the key switch.

If the meter now indicates a high reading, test the emergency stop switch.

If the meter indicates a low reading, test the wiring harness.

If the original problem is failure of the powerhead to shut down, test for an open in the Black/Yellow lead, damage to the key switch, or a damaged power pack.

Connect all the circuits disconnected during this test -- No. 4.

Test No. 5
Charge Coil — Grounded

Obtain a peak-reading voltmeter and set the meter to "NEG" and "500". Disconnect the 5-pin connector between the armature plate and the power pack.

Connect the Red meter lead to the "A" terminal on the armature plate side of the connector -- see wiring diagram on the next page. Connect the Black meter lead to a good ground on the powerhead. Crank the powerhead and observe the meter for any indication of voltage. Leave the Black meter connected to the powerhead ground and move the Red meter lead to the "D" terminal on the armature plate side of the connector. Again, crank the powerhead.

If the meter indicates any amount of voltage during either test, the charge coil or the leads are grounded. Replace the lead/s as required, or replace the charge coil.

If the meter indicated no voltage was present during either test, proceed to the next test -- No. 6.

Test No. 6
Charge Coil — Output

Set the peak-reading voltmeter to "NEG" and "500". Connect the Black voltmeter lead to the "A" terminal on the armature plate side of the disconnected 5-pin connector -- see wiring diagram on the next page. Connect the Red meter lead to the "D" terminal on the armature plate side of the connector. Crank the powerhead and observe the meter.

If the meter indicates 230 volts or more, proceed directly to Test No. 9 -- Sensor Coil Test.

If the meter indicates less than 230 volts, make a careful inspection of the wiring and the connectors.

If the wiring and connectors are in good condition, proceed to the next test -- No. 7.

Test No. 7
Charge Coil — Ohms

Obtain an ohmmeter and set the scale for a 500 plus reading. Connect the ohmmeter leads to the "A" and "D" terminals on the armature plate side of the 5-pin connector -- see wiring diagram on the following page.

The meter should indicate 515-635 ohms.

Test No. 8
Charge Coil — Ohms — Grounded

Connect the Black meter lead to a good ground on the powerhead. Make contact with the Red meter leads alternately to the "A" and "D" terminals on the armature plate side of the disconnected 5-pin connector -- see wiring diagram on the following page.

If the meter registers any movement at all, the charge coil or the charge coil leads are grounded.

Locate and repair the grounded lead/s or replace the charge coil.

Connect all disconnected leads during this test.

Test No. 9
Sensor Coil — Grounded

Disconnect the 5-pin connector between the armature plate and the power pack --see wiring diagram on the following page. Obtain a peak-reading voltmeter and set the meter to "NEG" and "5".

Connect the Black meter lead to a good powerhead ground. Crank the powerhead and alternately make contact with the Red meter lead to the "B" and "C" terminals on

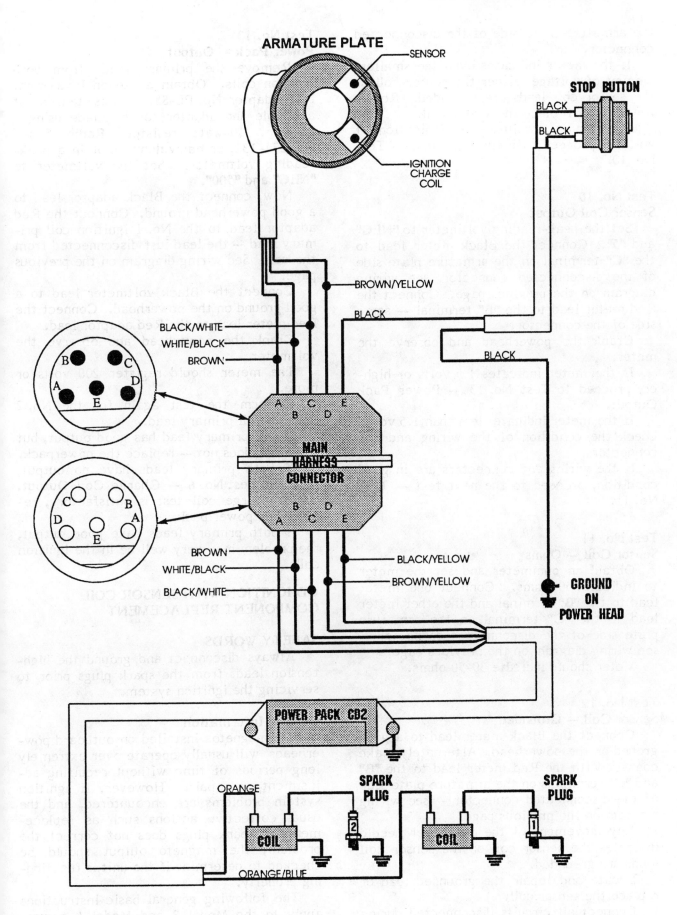

ARMATURE PLATE

SENSOR

IGNITION
CHARGE
COIL

STOP BUTTON

BLACK

BLACK

BROWN/YELLOW

BLACK

BLACK

BLACK/WHITE
WHITE/BLACK
BROWN

B
C
A
D
E

A   C   E
B   D

**MAIN
HARNESS
CONNECTOR**

B   D
A   C   E

C
D   B
E   A

BROWN
WHITE/BLACK
BLACK/WHITE

BLACK/YELLOW
BROWN/YELLOW

**GROUND
ON
POWER HEAD**

POWER PACK CB2

ORANGE

**SPARK
PLUG**

**SPARK
PLUG**

COIL

2

COIL

1

ORANGE/BLUE

Wire identification drawing from the Appendix to assist in conducting the tests outlined in this section.

the armature plate side of the disconnected connector.

If the meter indicates even the smallest amount of voltage, either the sensor coil or the sensor coil leads are grounded. Repair the leads or replace the sensor coil.

If there is absolutely no indication of voltage, proceed with the next test -- Test No. 10.

Test No. 10
Sensor Coil Output

Set the peak-reading voltmeter to "NEG" and "5". Connect the Black meter lead to the "C" terminal on the armature plate side of the disconnected connector, see wiring diagram on the previous page. Connect the Red meter lead to the "B" terminal -- same side of the connector.

Crank the powerhead and observe the meter.

If the meter indicates 1.5 volts or higher, proceed to Test No. 13 -- Power Pack Output.

If the meter indicates less than 1.5 volts, check the condition of the wiring and the connectors.

If the wiring and connectors are in good condition, proceed to the next test -- Test No. 11.

Test No. 11
Sensor Coil — Ohms

Obtain an ohmmeter and set the meter to indicate 50 ohms. Connect one meter lead to the "B" terminal and the other meter lead to the "C" terminal on the armature plate side of the disconnected connector -- see wiring diagram on the previous page.

Meter should indicate 30-50 ohms.

Test No. 12
Sensor Coil — Grounded

Connect the Black meter lead to a good ground on the powerhead. Alternately make contact with the Red meter lead to the "B" and "C" terminals on the armature plate side of the disconnected connector -- see wiring diagram on the previous page.

Any movement of the ohmmeter needle indicates the sensor coil or the sensor coil leads are grounded.

Locate and repair the grounded lead or replace the sensor coil.

Connect all circuits disconnected during this test.

Test No. 13
Power Pack — Output

Remove the primary leads from both ignition coils. Obtain a Steven Instrument load adapter No. PL-88. If this item is not available and adaptor can be made using a 10-ohm, 10-watt resistor, Radio Shack No. 271-132, or equivalent. Obtain a peak-reading voltmeter. Set the voltmeter to "NEG" and "500".

Now, connect the Black adaptor lead to a good powerhead ground. Connect the Red adaptor lead to the No. 1 ignition coil primary lead -- the lead just disconnected from the coil. See wiring diagram on the previous page.

Connect the Black voltmeter lead to a good ground on the powerhead. Connect the Red meter lead to the Red adaptor lead.

Crank the powerhead and observe the voltmeter.

The meter should register 200 volts or more.

Perform the test again for the No. 2 ignition coil primary lead.

If one primary lead has good output, but the other does not -- replace the powerpack.

If both primary leads have no output, perform Test No. 6 -- Charge Coil Output. If the charge coil test is satisfactory, replace the power pack.

If both primary leads have good output, the problem may very well be in the ignition coils.

CD IGNITION WITH SENSOR COIL COMPONENT REPLACEMENT

SAFETY WORDS

Always disconnect and ground the high-tension leads from the spark plugs prior to servicing the ignition system.

General Information

CD magnetos installed on outboard powerheads will usually operate over extremely long periods of time without requiring adjustment or repair. However, if ignition system problems are encountered, and the usual corrective actions such as replacement of spark plugs does not correct the problem, the magneto output should be checked to determine if the unit is functioning properly.

The following general basic instructions apply to the Model 3 and Model 4 powerheads using CDII ignition system.

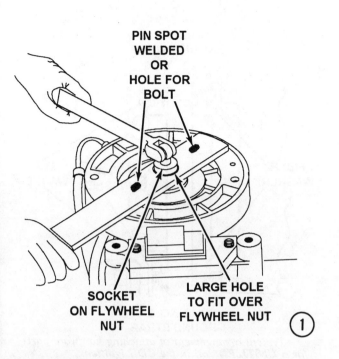

PIN SPOT
WELDED
OR
HOLE FOR
BOLT

SOCKET
ON FLYWHEEL
NUT

LARGE HOLE
TO FIT OVER
FLYWHEEL NUT

①

TYPICAL &
ACCEPTABLE
FLYWHEEL
PULLER

②

FLYWHEEL REMOVAL

1- Remove the hood or enough of the powerhead cover to expose the flywheel. Disconnect the battery connections from the battery terminals, if a battery is used to crank the powerhead. If a hand rewind starter is installed, remove the attaching hardware from the legs of the starter assembly and lift the starter free. Remove the nut securing the flywheel to the crankshaft. It may be necessary to use some type of flywheel holding device to prevent the flywheel from turning as the nut is loosened. A flat bar with a large hole in the center for access to the flywheel nut and two pins spot welded to the bar to fit into the two holes in the upper surface of the flywheel, will serve the purpose, as indicated in the accompanying illustration. If welding is not available, make a careful measurement and drill a couple holes in the bar to allow bolts to pass through and index into the flywheel holes.

2- Install the proper flywheel puller using the screw holes in the flywheel. **NEVER** attempt to use a puller which pulls on the outside edge of the flywheel or the flywheel may be damaged. After the puller is installed, tighten the center screw onto the end of the crankshaft. Continue tightening the screw until the flywheel is released from the crankshaft. Remove the flywheel. **DO NOT** strike the puller center bolt with a hammer in an attempt to dislodge the flywheel. Such action could seriously damage the lower seal and/or lower bearing.

3- **STOP,** and carefully observe the ignition components and associated wiring layout. Study how the parts are arranged. **TAKE TIME** to make notes on the wire routing. The ignition components and the wire routing **MUST** be replaced back in their original positions.

NOTE

The accompanying illustration shows the arrangement of parts under the flywheel for

OBSERVE LAYOUT
AND WIRING
BEFORE SERVICING

③

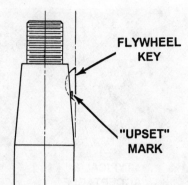

Flywheel key alignment in the crankshaft for the Model 3 thru 8hp powerheads.

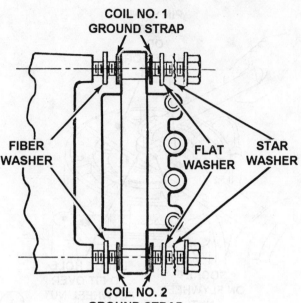

Typical arrangement of attaching hardware parts for a UNITIZED coil in the CDII ignition.

a Model 25 equipped with UFI. However, the importance of this step is to emphasize how critical the layout of parts and the routing of wiring can be for any powerhead.

You may elect to follow the practice of many professional mechanics by taking a series of photographs of the powerhead with the flywheel removed: one from the top, and a couple from the sides showing the wiring and arrangement of parts.

FLYWHEEL INSTALLATION

Clean the crankshaft with solvent and blow it dry with compressed air.

Check to be sure the flywheel magnets are free of any metal parts.

If the key is not already in place, position the key in the crankshaft keyway, with the "upset" mark facing **DOWN**. Check to be sure the inside taper of the flywheel and the taper on the crankshaft are clean of dirt or oil, to prevent the flywheel from "walking" on the crankshaft while the powerhead is operating. Slide the flywheel down over the crankshaft with the keyway in the flywheel aligned with the key on the crankshaft.

Rotate the flywheel **CLOCKWISE** and check to be sure the flywheel does not contact any part of the ignition parts or the wiring.

Thread the flywheel nut onto the crankshaft. Use the same arrangement employed during removal of the flywheel nut to prevent the crankshaft from rotating and tighten the nut to a torque value 30-40ft lb (40-54Nm).

Carefully align the five-pin connector. Assemble the two halves. Just a drop of acetone or denatured alcohol will permit the

pins of the male half to index with the cavities in the other half. Secure the connector with the retainer.

If servicing a manual start model, install the starter assembly onto the powerhead, see Chapter 9.

IGNITION COIL REPLACEMENT

Removal

Note the screw and washer placement arrangement, **BEFORE** disconnecting the leads. Disconnect the primary and secondary leads from the ignition coil terminals.

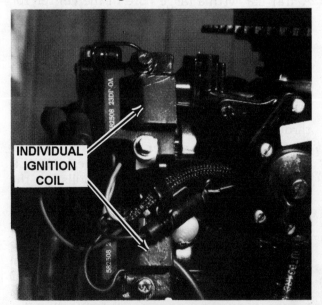

Typical individual coil installation on a 2-cylinder powerhead. Some powerheads have both coils encased in a single unitized sealed unit.

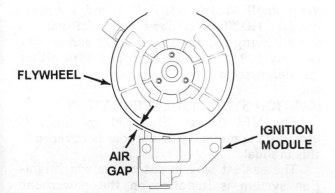

Simple line drawing to illustrate the air gap between the ignition module and the powerhead, described in the test.

Remove the attaching hardware, as indicated in the accompanying illustration, and then lift the ignition coil free of the powerhead.

Installation
Place the ignition coil in position on the powerhead and secure it with the attaching hardware, paying close attention to the order of parts, as indicated in the illustration.

SPECIAL WORDS
Apply a light film of OMC Triple-Guard grease to the ribbed portion of the spark plug ceramics and also to the openings of each spark plug lead cover. Install the spark plug leads.

Apply a light film of OMC Triple-Guard grease to the opening of each primary lead cover, and then install the covers.

The Orange/Blue primary lead **MUST** be connected to the No. 1 ignition coil and the Orange primary lead to the No. 2 coil.

Connect the primary lead to the coil **FIRST**, and then the secondary lead, with the screw and washer arrangement noted during removal.

IGNITION MODULE REPLACEMENT UNITS WITH EXTERNAL MODULE
(Not Installed Under Flywheel)

Removal
Separate the ignition module connector. Remove the attaching screws and remove the ignition module. Clean the module mounting bosses to ensure a good ground when the module is installed.

Installation
Place the ignition module in position on the powerhead bosses. Apply a light coating

of OMC Gasket Sealing Compound, or equivalent, to the threads of the mounting screws. Tighten the mounting screws just fingertight and then back the screw off slightly.

A non-metallic gauge is usually supplied with a new ignition module. Set the air gap between the ignition module and the flywheel to 0.013-0.017" (0.33-0.43mm), as indicated in the accompanying line drawing.

After the air gap has been adjusted, tighten the mounting screws to a torque value of 60-84 in lb (7-9Nm).

Restore the ignition module connector.

POWER PACK REPLACEMENT

Removal
Separate the connector between the power pack and the ignition coils.

GOOD WORDS
Bear in mind and **REMEMBER** the routing of the power pack leads and the location of the clamps to ensure proper lead routing and positioning during installation.

Remove the attaching hardware securing the power pack to the powerhead.

MORE GOOD WORDS
Except on rare occasions, when noted, always install lockwashers on electrical connectors. Place the washer on the powerhead side of the connector to provide a good ground.

Installation
The leads and half of the connectors will be installed on a new power pack. The assembly is completely ready for installation when purchased.

Connect the power pack ground wire, and route the wire to prevent any pull when the pack is installed. Pull on this ground wire could very easily result in ignition failure.

Mount the power pack and tighten the attaching screws to a torque value of 60-84 in lbs (7-9Nm).

Restore the connection between the power pack and the ignition coils. Carefully align the connector halves. Assemble the two halves. Just a drop of acetone or denatured alcohol will permit the pins of the male half to index more easily in the cavities of the other half. Secure each connector with the retainer.

SENSOR COIL (PLATE) REPLACEMENT

Removal
Remove the flywheel as described in the previous section.

If the sensor coil is to be replaced, separate the 6-prong ignition module connector.

Remove the three screws securing the sensor plate, and then lift the plate free of the retainer plate. **MAKE A NOTE** of the color code for the sensor leads and identify the color and terminal position in the connector to **ENSURE** the lead terminals are installed in the proper position in the connector.

Disconnect the sensor coil leads from the connector. This is accomplished by first inserting a stiff piece of wire, such as a straightened paper clip, into the slot next to the terminal of the lead to be disconnected. The wire moves a tang allowing the terminal to be removed. Now, gently pull the wire back and slide the terminal out of the connector body.

Installation
Slide the sensor lead terminals into the connector in the same position from which they were removed. The terminal will "snap" into the connector and will be held in place by the tang.

Position the sensor plate in place on the retainer plate and secure it with the three attaching screws.

Install the flywheel and tighten the flywheel nut to a torque value of 30-40 ft lb (40-54Nm)

WORK CHECK
Check the completed work by mounting the outboard in a test tank, in a body of water, or with a flush attachment connected to the lower unit. If a flush attahment is used **NEVER** operate the powerhead above an idle speed, because the no-load condition on the propeller would allow the powerhead to **RUNAWAY** resulting in serious damage or destruction of the powerhead.

5-5 UFI SYSTEM (UNDER FLYWHEEL IGNITION)

As the name implies, this ignition system is housed entirely under the flywheel. During the production year --1990, the manufacturer phased out CDII capacitor discharge ignition with sensor except on the very small Model 2.3, 3, 3.3 and 4 powerheads. **HOWEVER**, powerheads of various sizes leaving the factory in 1990 and possibly early 1991, could be equipped with either ignition system.

IGNITION SYSTEM IDENTIFICATION
The UFI (Under Flywheel Ignition), is used on the majority of outboards covered in this manual.

The easiest way to determine which ignition system is installed on the powerhead being serviced, is to trace the main harness from the ignition coil mounted on the powerhead. If the leads are routed back under the flywheel, the powerhead is equipped with UFI. If the harness is routed to a power pack (black box), the powerhead is equipped with CD ignition with sensor -- covered in the previous section of this chapter.

DESCRIPTION AND OPERATION
Four major components are used in the UFI system:

A flywheel with two permanent magnets, a charge coil, an ignition module and an ignition coil for each cylinder.

Two other coils -- the stator or alternator coils -- are also mounted under the flywheel, on the stator plate opposite the ignition components. These two coils provide current to charge the battery and are covered in this section because of their close proximity to the ignition components. The alternator charging system is also covered in more detail in Chapter 6 -- Electrical.

Permanent Magnets
Two permanent magnets are attached to the underside of the flywheel. As the flywheel rotates, the lines of magnetic flux set up by the magnets, cut through the coils installed on the stator plate to induce voltage in those circuits. Loose or damaged (cracked) magnets will lead to incorrect ignition timing and insufficient coil output.

Charge Coil
The charge coil is a short coil consisting of many windings of fine wire wrapped around an "E" shaped metal laminated core. As each flywheel magnet passes by the charge coil, a voltage is induced in the coil. As the flywheel continues to rotate and the coil is no longer influenced by the magnet, the magnetic field collapses. Therefore, an

alternating current is produced at the charge coil. This AC current is changed to DC by a diode inside the ignition module.

Ignition Module

The ignition module is a curved "black box" housed under the flywheel, on top of the stator plate and has several functions:

a- The module is used to convert AC from the charge coil, to DC current.

b- The module stores the DC in an internal capacitor, then releases it and automatically controls the timing.

20hp thru 50hp Powerheads

c- The S.L.O.W. (Speed Limiting Overheat Warning) system is an integral part of the ignition module. If the powerhead temperature exceeds 180°F (82°C), the S.L.O.W. circuitry will automatically decrease powerhead speed to approximately 2,000 rpm. The powerhead must cool to 162°F (72°C) and powerhead speed must return to idle before normal powerhead operation may be resumed.

d- An rpm Limiting Circuit is also an integral part of the ignition module. This circuit limits powerhead speed and helps protect the powerhead from damage caused by prolonged operation at high rpm. Excessive operation at high rpm can be caused by: Improper powerhead installation, worn propeller blades, cavitation at the propeller, or improper propeller selection.

e- The ignition coil -- one for each cylinder -- is the only component of the UFI system which is not housed under the flywheel. Both coils may be housed in a unitized sealed unit, or they may be separate coils, depending on the powerhead. Each coil consists of two sets of windings separated from each other by an insulating material. The ignition coil has primary and secondary circuit connections. The individual coils and the unitized coils are mounted on and grounded to the powerhead.

The coil boosts the DC voltage instantly to as much as 40,000 volts to the spark plugs.

Charge Coil Circuit

With each revolution of the flywheel, at the point in time when the ignition timing marks align, an alternating current is induced in the charge coil. This current flows to a series of SCRs -- electronic switches.

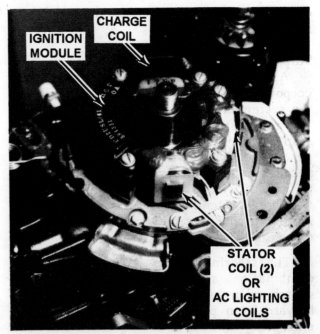

Typical UFI system installed under the flywheel on a Model 25hp powerhead.

These switches select the correct ignition coil primary winding to energize, by sensing the polarity of the charging coil output.

In this manner, a spark at the plug may be accurately timed by the timing marks on the flywheel relative to the magnets on the flywheel. As many as 100 sparks per second may be provided when the powerhead is operating at 6000rpm.

Charging System

The charging system consists of the flywheel magnets, two stator (AC Lighting) coils, a rectifier, and the battery. The charging system is mentioned in this chapter because of its close proximity to the ignition components under the flywheel. Detailed testing, servicing, and replacement procedures are presented in Chapter 6 -- Electrical.

As the flywheel magnets rotate, voltage is induced in the stator coils. The two stator coils are mounted on a large "E" shaped metal laminated core, mounted on the stator plate, across from the ignition module and the charging coil. This alternating current passes to the rectifier.

The rectifier converts the alternating current into DC, which may then be stored in the battery. The rectifier is a sealed unit mounted separately -- not under the flywheel -- and contains a number of diodes.

By definition, the diodes allow current to pass in only one direction. If one of the diodes is defective, the entire unit **MUST** be replaced.

Stop Switch Circuit

As many as three components are incorporated in the ignition system to ground the system and prevent the spark plugs from firing -- shutting down the powerhead. The stop button, the key switch, and the emergency stop switch with lanyard, are all connected to the ignition module through the powerhead wiring harness. When any one of the three switching devices is activated, the ignition module output is directed to ground, disabling the ignition system.

The steering handle is equipped with a combination stop switch/emergency stop switch with lanyard. When the clip and lanyard is in place behind the switch, the emergency switch is in the **RUN** position. When the clip and lanyard is removed from the emergency stop switch, the ignition is grounded in the **STOP** position. The powerhead may be shut down without removing the clip and lanyard by depressing the stop switch in the center of the emergency stop switch.

Any defect in these three switches will prevent the powerhead from starting. Therefore, if the powerhead will not start, the stop switch circuit components should be the first parts tested before testing any ignition components.

TROUBLESHOOTING UFI AND CHARGING SYSTEMS

Always attempt to proceed with the troubleshooting in an orderly manner. The "shot in the dark" approach will only result in wasted time, incorrect diagnosis, replacement of unnecessary parts, and frustration.

Begin the ignition system troubleshooting with the spark plugs and continue through the system until the source of trouble is located.

Spark Plugs

1- Check the plug wires to be sure they are properly connected. Check the entire length of the wires from the plugs to the magneto under the stator plate. If the wires are to be removed from the spark plugs, **ALWAYS** use a pulling and twisting motion as a precaution against damaging the connection.

HIGH TENSION LEAD (TWIST & PULL TO REMOVE)

①

Attempt to remove the spark plugs by hand. This is a rough test to determine if the plug is tightened properly. The attempt to loosen the plug by hand should fail. The plug should be tight and require the proper socket size tool. Remove each spark plug and evaluate its condition as described in Section 5-2.

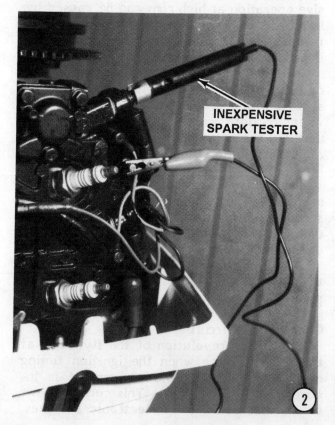

INEXPENSIVE SPARK TESTER

②

2- Use a spark tester and check for spark. If a spark tester is not available, hold the plug wire about 1/4" (6.4mm) from the powerhead. Rotate the flywheel with the hand pull starter or crank the powerhead with the cranking motor and check for spark. A strong spark over a wide gap must be observed when testing in this manner, because under compression a strong spark is necessary in order to ignite the air/fuel mixture in the cylinder. This means it is possible to think a strong spark is present, when in reality the spark will be too weak when the plug is installed. If there is no spark, or if the spark is weak, the trouble is most likely under the flywheel in the ignition components.

Compression

3- Before spending too much time and money attempting to trace a problem to the ignition system, a compression check of the cylinders should be made. If a cylinder does not have adequate compression, troubleshooting and attempted service of the ignition or fuel system will fail to give the desired results of satisfactory powerhead performance.

Remove the spark plug high tension leads by pulling and twisting ONLY on the molded cap. NEVER pull on the lead because the connection inside the cap may be separated or the boot may be damaged. Remove the spark plugs. Insert a compression gauge into one cylinder spark plug opening. Ground the spark plug leads to prevent damage to the ignition coil. If a lead is not grounded, the coil will attempt to match the demand cre-

ated by the spark trying to jump from the electrode to the nearest ground.

Crank the powerhead through several revolutions by pulling rapidly on the hand rewind starter pull rope, or with the electric cranking motor, if the powerhead is so equipped. Note the compression reading.

An acceptable pressure reading for a powerhead covered in this manual would be about 100 psi (689.5 kPa) or more. The cylinders should register within -- say 10%, or 10 psi (69kPa), of each other.

Erratic Firing Condition

A faulty ignition module can cause a continuous or erratic spark. To determine if each cylinder is firing at the correct time and only once per crankshaft revolution, the flywheel position must be checked.

First, remove the spark plug high tension leads by pulling and twisting ONLY on the molded cap. NEVER pull on the lead because the connection inside the cap may be separated or the boot may be damaged. Remove the spark plugs.

Next, insert the eraser end of a new pencil into the No. 1 spark plug opening. Slowly rotate the flywheel by hand in a CLOCKWISE direction until the greatest length of the pencil is exposed out of the spark plug opening. This position indicates the piston is approximately at top dead center (TDC). Make a mark on the flywheel in line with the timing pointer. Identify this mark as "No. 1" if not already marked.

Repeat the same procedure to establish a "No. 2" mark on the flywheel with the pencil inserted into the No. 2 spark plug opening.

Now, remove the pencil and install the spark plugs and high tension leads.

Mount the outboard in a test tank or body of water. NEVER attempt to perform this test with a flush attachment connected to the lower unit. The no-load condition on the propeller would cause the powerhead to RUNAWAY resulting in serious damage or destruction of the unit.

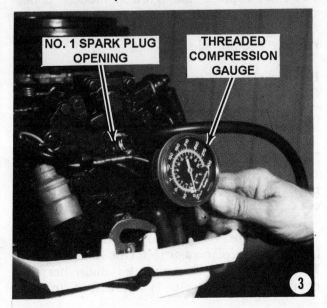

NO. 1 SPARK PLUG OPENING

THREADED COMPRESSION GAUGE

CAUTION: Water must circulate through the lower unit to the engine any time the engine is run to prevent damage to the water pump in the lower unit. Just five seconds without water will damage the water pump.

Start the powerhead and allow it to reach operating temperature. Operate the powerhead under load at the speed at which the ignition problem exists.

Obtain an induction type timing light. Connect the light to a 12V power source. Connect the timing light pickup around the No. 1 high tension lead. Aim the light at the timing pointer. Only the "No. 1" mark should align with the timing pointer. If the "No. 2" mark appears close to the pointer, or if the "No. 1" mark appears continuously around the flywheel, the ignition coil primary leads are connected incorrectly, or the ignition module is defective.

S.L.O.W. Service
20hp thru 50hp Powerheads Only

The S.L.O.W. (Speed Limiting Overheat Warning) system is an integral part of the ignition module. If the powerhead temperature exceeds 180°F (82°C), the S.L.O.W. circuitry will automatically decrease powerhead speed to approximately 2,000rpm. The powerhead must cool to approximately 155°F (68°C) and powerhead speed must return to idle before normal powerhead operation is possible.

The ignition module receives input signals from the temperature switch located in the cylinder head. A warning horn sounds when powerhead temperature exceeds a preset limit. On remote electric start units, the S.L.O.W warning system is isolated from other warning horn signals by a blocking diode in the powerhead wiring harness.

Temperature Switch Operational Test

Install the correct test propeller and start the powerhead.

The outboard **MUST** be mounted in a test tank or body of water to test the switch. **NEVER** attempt to perform this test with a flush attachment connected to the lower unit. The no-load condition on the propeller would cause the powerhead to **RUNAWAY** resulting in serious damage or destruction of the unit.

CAUTION: Water must circulate through the lower unit to the engine any time the engine is run to prevent damage to the water pump in the lower unit. Just five seconds without water will damage the water pump.

Allow the powerhead to reach operating temperature.

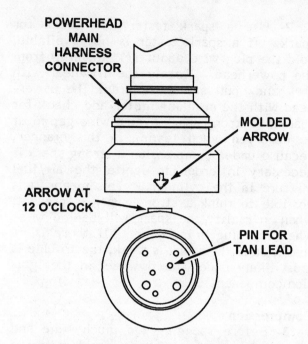

Tan pin identification at the quick-disconnect fitting for testing the blocking diode, as described in the text.

Advance the throttle to 3500rpm. Disconnect the Tan lead from the temperature switch in the cylinder head and attach the disconnected lead to a good powerhead ground. If powerhead speed drops to approximately 2000rpm, the wiring and module functions are satisfactory. The problem is in the temperature switch. Refer to the heading "Temperature Switch Resistance Test". If the switch proves defective, it must be replaced.

If powerhead speed does not drop to approximately 2000rpm, when the lead is grounded on the powerhead, check the continuity of the Tan lead. If the lead checks out fine, replace the ignition module under the flywheel. Refer to the procedures given on Page 5-48 to "pull" the flywheel and replace the module.

Blocking Diode Test
Remote Electric Models Only

If the powerhead harness blocking diode is defective, the S.L.O.W. warning system is activated when the warning horn receives a "no oil" or "low oil" signal. To test the diode: First, disconnect the main powerhead harness connector lead and the temperature switch Tan lead from the switch.

Next, obtain an ohmmeter and select the Rx1000 scale. Now, measure the resistance between the disconnected Tan lead and the Tan pin in the disconnected main harness connector. This pin is identified in the accompanying illustration. Make a note of the reading. Reverse the meter connections and again, note the reading. The meter should register a "high" reading -- almost infinity, in one direction, and a "low" reading -- almost zero, in the other direction. If the meter reading is not as expected, replace the powerhead harness blocking diode.

Temperature Switch Resistance Test

Disconnect the Tan lead from the temperature switch lead located in the cylinder head. Pull the switch from the head. Obtain an ohmmeter and select the Rx1000 scale. Connect the meter across the switch -- one meter lead to the switch lead and the other meter lead to the metal portion of the switch. Obtain about a cup of automotive motor oil with a flashpoint of above 300°F (150°C). The manufacturer recommends the use of OMC Cobra 4-Cycle Motor Oil. Boil some water and obtain a thermometer. Place the switch and thermometer into the cup of oil. Place the cup of oil, the switch and the thermometer into a larger container with about an inch of tepid water. Slowly add the boiling water to the larger container to increase the temperature of the oil.

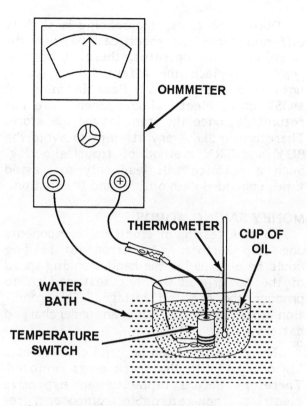

Setup using an oil bath to check operation -- opening and closing -- of the temperature switch.

SAFETY WORDS
DO NOT USE AN OPEN FLAME AS A HEAT SOURCE TO WARM THE OIL.

Continue to add the boiling water and stir the oil with the thermometer. The temperature switch is functioning correctly if the meter indicates the switch is open at room temperature -- meter indicates an infinite resistance. As the temperature of the oil rises, the switch should close at approximately 180°F (82°C) -- meter indicates almost zero resistance.

Lift out the cup of oil from the boiling water and allow it to cool. As the oil temperature drops, the switch will remain closed until the temperature lowers to approximately 155°F (68°C) -- meter indicates almost zero resistance. When the temperature drops below this value, the switch will once again open -- meter will indicate an infinite resistance.

If the meter readings are not as expected, replace the temperature switch.

TESTING UFI COMPONENTS
GENERAL INFORMATION

Due to the complex nature of the circuitry and high energy output pulses of microsecond duration, conventional testing devices such as a Volt/Ohm/Ammeter will **NOT** measure electrical output with the degree of accuracy required. An electronic ignition tester, as used by the local dealer is an electrical device capable of measuring the peak energy output of capacitor discharge ignition units. This instrument was designed to troubleshoot ignition systems such, as the CDII and UFI. Unfortunately this tester is not easily available. Therefore, this chapter includes only those tests which may be performed with common standard instruments.

If a component, must be replaced, usually the hand rewind starter, if so equipped, must be removed and the flywheel "pulled" to gain access. See Chapter 9 to remove the hand rewind starter, then to Page 5-48 to "pull" the flywheel

WORDS FROM EXPERIENCE

During the tests, if a reading is slightly different from the specifications, but the powerhead still operates, there is no real need to replace the affected component, until it actually fails. Bear in mind, in **MOST** cases electrical components are not returnable, once the item leaves the store. Therefore, make every attempt to avoid the **BUY** and **TRY** method of troubleshooting. Such a practice will lead only to wasted time, unneeded cash outlay, and frustration.

MONEY SAVING WORDS

To take testing of electrical components one step further: An erroneous reading could be obtained if the hand cranking speed of the powerhead was not fast enough to produce the required voltage. This condition would occur if using an undercharged battery.

The powerhead can be hand cranked at a higher rpm with the spark plugs removed. Therefore before replacing an expensive electrical, non-returnable component, remove the spark plugs; crank the powerhead with the hand rewind starter; and check for a better meter reading.

As a last resort, take the suspected faulty ignition part to the local dealer and have him give it his full testing treatment before making the new purchase.

SPARK VERIFICATION

Use a spark tester and check for spark. If a spark tester is not available, hold the high tension lead about 1/4" (6.4mm) from the powerhead. Rotate the flywheel with the hand pull starter or crank the powerhead with the starter motor and check for spark. A strong spark over a wide gap must be observed when testing in this manner.

If no spark is observed, the problem probably lies in the stop circuit. Perform the tests under the heading "Stop Circuit Resistance Tests".

If there is spark at one cylinder, but not at the other, the problem is probably in the ignition components on the stator plate. Perform the tests under the heading "Stator Plate Output Tests".

If there is spark at both spark plugs, and the powerhead still fails to start, perform the tests under the heading "Running Output Tests".

Stop Circuit Resistance Tests

The following tests are performed with the powerhead **NOT** operating.

Install the lanyard and clip behind the combination stop button/emergency stop switch. Disconnect the lead to the stop button at the 1-pin connector. Obtain an ohmmeter. Set the meter to the Rx1000 ohm scale. Measure the resistance between the disconnected stop button lead and a good powerhead ground. The meter should register a "high" resistance -- approaching infinite resistance.

Momentarily depress the stop button. The meter should register a "low" resistance -- appproaching zero ohms. Now, remove the clip and lanyard from behind the stop button. The meter should again register a "low" reading. If the meter readings are as expected, the stop circuit is functioning properly. If the meter readings are not as expected, replace the combination stop button/emergency stop switch.

If the model being serviced is equipped with a key switch, continue with the following tests:

Install the clip and lanyard behind the emergency stop switch. Disconnect the wire harness from the key switch to the powerhead. With the ohmmeter set to the Rx1000 scale, measure the resistance between the powerhead harness Black/Yellow lead and a good powerhead ground. With the key in the **OFF** position, the meter should register a "low" resistance. With the key in the **ON** position, the meter should register a "high" resistance.

If the meter readings are as expected, proceed to the next heading and continue to test the ignition components.

If the meter readings are not as expected, disconnect the emergency stop switch lead from the harness Black/Yellow lead at the key switch. Measure the resistance between the disconnected lead and ground. If the meter registers a "high" resistance, replace the emergency stop switch. If the meter registers a "low" resistance, the problem is probably not in the stop switch -- check the wiring harness.

Ignition Plate Output Tests

Remove the spark plug high tension lead by pulling and twisting **ONLY** on the molded cap. **NEVER** pull on the lead because the connection inside the cap may be separated

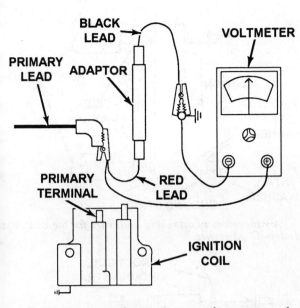

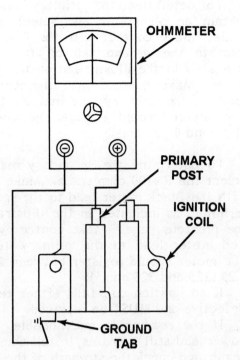

Setup using a voltmeter to test the ignition plate output.

or the boot may be damaged. Ground both leads to prevent the powerhead from accidentally starting. Twist and remove the primary lead from the ignition coil. Obtain Stevens Instrument load adaptor P/N PL-88 or Radio Shack 10 ohm 10 watt resistor P/N 271-132.

Connect the No. 1 ignition coil primary lead to the Red adaptor lead. Connect the Black adaptor lead to a good ground on the powerhead. Obtain a voltmeter capable of reading at least 200 volts DC. Make contact with the Red meter lead to the Red adaptor lead, and the Black meter lead to the ground connection.

Crank the powerhead with the hand rewind starter or cranking motor and observe the meter. The meter should register a minimum of 175 volts.

Repeat the test for the No. 2 ignition coil primary lead.

If both leads have an acceptable output voltage, but the powerhead still fails to start, the ignition coil must be tested. Proceed to the heading "Ignition Coil Testing".

If one primary lead has no output, but the other lead has an acceptable output, replace the ignition module located under the flywheel. Proceed to Page 5-48 to "pull" the flywheel and remove the module.

If both primary leads have no output, proceed to the heading "Charge Coil Test".

Connect the two primary leads back in their original locations. Refer to the wiring diagrams in the Appendix for lead identification.

Setup using an ohmmeter to measure the primary resistance of the ignition coil.

Ignition Coil Testing

The ignition coil/s may remain mounted on the powerhead during testing.

Using a twisting motion, remove the high tension lead and the primary lead from the coil.

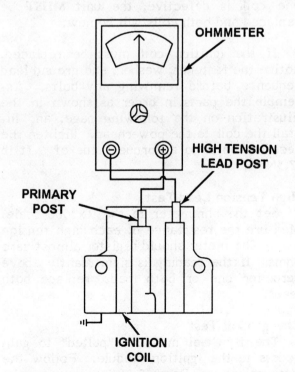

Setup, using an ohmmeter, to measure the secondary resistance of the ignition coil.

To determine the primary resistance: Obtain an ohmmeter and select the Rx1 scale. Make contact with the Black meter lead to the ground tab on the coil, as indicated in the illustration on the previous page. Make contact with the Red meter lead to the primary terminal on the coil. The meter should register between 0.05 ohms and 0.15 ohms.

To determine the secondary resistance: Select the Rx100 ohm scale. Make contact with the Black meter lead to the spark plug terminal, as indicated in the illustration on the previous page. Make contact with the Red meter lead to the primary terminal. The meter should register between 2.25 and 3.25 (225 and 325 ohms).

If an ignition coil fails either test, it is defective and **MUST** be replaced.

If the reading is questionable, but the powerhead still operates, install a spark plug tester and check the strength of the plug.

If the spark is acceptable, it is not necessary to replace the ignition coil. However, keep this area in mind for possible future trouble.

If the spark is weak, it will get weaker. Therefore, replace the ignition coil. Some models have a "unitized" coil unit -- both coils sealed in single package. Therefore, if one coil is defective, the unit **MUST** be replaced and both coils will be new.

If the ignition coil must be replaced, notice the fastener, washer, and ground lead sequence before removing any bolts. Assemble the parts in order as shown in the illustration on the following page, and install the coil to the powerhead. Tighten the securing bolts to a torque value of 6 ft lb (7.5Nm).

High Tension Lead Test

Set the ohmmeter to the Rx1000 scale. Measure the resistance of each high tension lead. The meter should register almost zero ohms. If the reading is significantly above zero for one or both leads, replace both leads.

Charge Coil Test

The flywheel must be "pulled" to gain access to the ignition module. Follow the instructions on Page 5-48 to remove the flywheel.

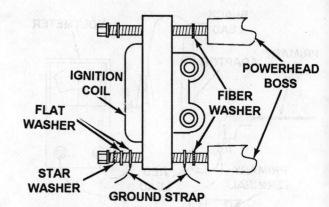

Arrangement of attaching hardware for the coil on a Model 4D powerhead.

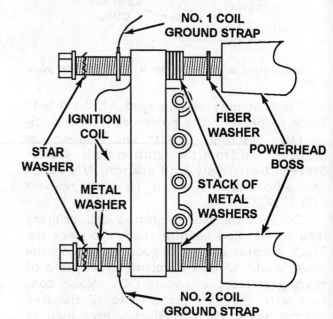

Arrangement of attaching hardware for the coil on a Model 5hp thru Model 8hp powerhead.

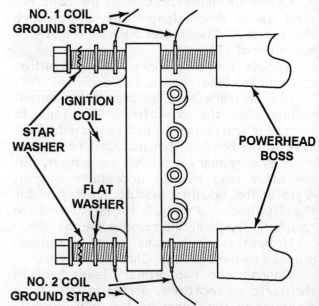

Arrangement of attaching hardware for the coil on a Model 9.9hp and larger powerhead.

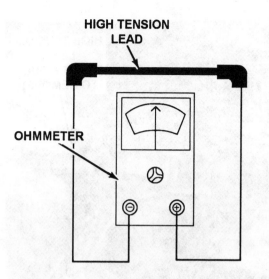

HIGH TENSION
LEAD

OHMMETER

Setup, using an ohmmeter, to measure resistance in a high tension lead.

Remove the screw securing the ignition module to the charge coil laminated core. Loosen, but do **NOT** remove the screw securing the ignition module to the stator coil laminated core. Swing the module away from the charge coil laminated core. Remove the screw securing the charge coil to the stator plate. Separate the Brown and Brown/Yellow leads under the coil at their quick disconnect fittings.

Obtain an ohmmeter and set the meter to read a maximum of 1000 ohms. Measure the resistance across the two disconnected leads from the coil. The meter should register 800 to 1000 ohms.

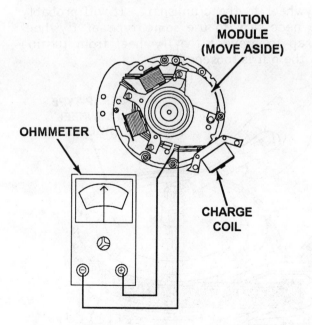

IGNITION
MODULE
(MOVE ASIDE)

OHMMETER

CHARGE
COIL

Setup, using an ohmmeter, to measure charge coil resistance.

Next, set the meter to the highest resistance scale and make contact with the Black meter lead to a good ground on the powerhead. Make contact with the Red meter to the disconnected Brown lead from the coil. The meter should register a "high" reading -- approaching infinity. Repeat the test for the Brown/Yellow lead. If both leads give an acceptable reading, but the powerhead still will not start, replace the ignition module. If any of the three resistance tests performed on the charge coil do not give acceptable results, replace the charge coil. Refer to the instructions on Pages 5-50 and 5-51 to replace the charge coil or ignition module and install the flywheel.

Running Output Tests

Remove the No. 1 primary lead from the No. 1 ignition coil. Obtain a Stevens Instrument P/N TS-77 terminal extender or equivalent. Install the extender onto the ignition coil primary post using a **CLOCKWISE** motion. Then, install the primary lead over the terminal extender. Obtain a voltmeter capable of reading about 250 volts DC.

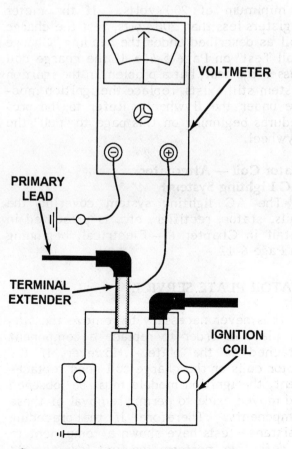

VOLTMETER

PRIMARY
LEAD

TERMINAL
EXTENDER

IGNITION
COIL

Setup, using a voltmeter, to check the running output, as described in the text.

The outboard **MUST** be mounted in a test tank or body of water to determine the output of the powerhead. **NEVER** attempt to perform this test with a flush attachment connected to the lower unit. The no-load condition on the propeller would cause the powerhead to **RUNAWAY** resulting in serious damage or destruction of the unit.

CAUTION: Water must circulate through the lower unit to the engine any time the engine is run to prevent damage to the water pump in the lower unit. Just five seconds without water will damage the water pump.

Start the powerhead and allow it to warm to operating temperature. Operate the powerhead under load, at the speed at which the ignition problem exists. If the outboard is in a body of water, obtain the services of an assistant to help with the testing.

Make contact with the Black meter lead to a good powerhead ground. Then, **MOMENTARILY**, make contact with the Red meter lead to the exposed portion of the terminal extender on the primary post of the ignition coil. The meter should register a minimum of 200 volts. If the meter registers less than 200 volts, test the charge coil as described under the heading "Charge Coil Test" on Page 5-46. If the charge coil passes the test, but a problem in the ignition system still exists, replace the ignition module under the flywheel. Refer to the procedures beginning on this page to "pull" the flywheel.

Stator Coil — Alternator (AC Lighting System)

The AC lighting system covering the coils, stator, rectifier, etc. is presented in detail in Chapter 6 -- Electrical, beginning on Page 6-12

STATOR PLATE SERVICE

It is never necessary to remove the stator plate in order to replace a component attached to the plate. However, if the stator coils or the charge coil need replacement, the ignition module must be loosened and moved aside to permit removal of these components. Therefore, if the preceding resistance tests have shown a component to be defective, perform the first few steps to "pull" the flywheel from the powerhead and

HIGH TENSION LEAD (TWIST & PULL TO REMOVE)

then proceed to the appropriate heading for further work.

FLYWHEEL REMOVAL

1- Remove the spark plug high tension leads. **ALWAYS** grasp the molded cap and pull it loose with a twisting motion to prevent damage to the connection.

2- Remove the cowling or enough of the powerhead cover to expose the flywheel. Disconnect the battery leads from the battery terminals, if a battery is used to crank the powerhead. If a hand starter is installed, remove the attaching hardware from the legs of the starter assembly and lift the starter free. Remove the nut securing the flywheel to the crankshaft. It will probably be necessary to use some type of flywheel strap to prevent the flywheel from turning as the nut is loosened.

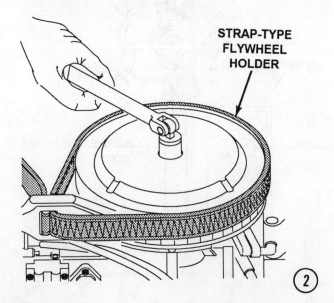

STRAP-TYPE FLYWHEEL HOLDER

TYPICAL & ACCEPTABLE FLYWHEEL PULLER

3

3- Install the proper flywheel puller using the screw holes in the flywheel. **NEVER** attempt to use a puller which pulls on the outside edge of the flywheel. Such a device will surely damage the flywheel. After the puller is installed, tighten the center screw onto the end of the crankshaft. Continue tightening the screw until the flywheel is released from the crankshaft. Remove the flywheel. **DO NOT** strike the puller center bolt with a hammer in an attempt to dislodge the flywheel. Such action could seriously damage the lower seal and/or lower bearing. Striking the flywheel will weaken the magnets housed under the flywheel and significantly reduce the ignition module output.

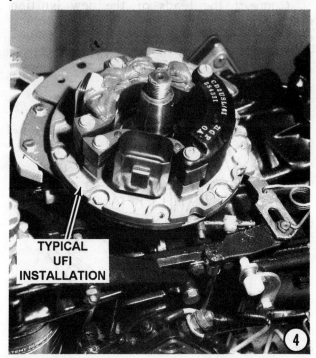

TYPICAL UFI INSTALLATION

4

4- **STOP**, and carefully observe the stator plate and associated wiring layout. **TAKE TIME** to make notes on the wire routing. Observe how the heels of the laminated cores, with the stator coils and charge coil attached, are flush with the boss on the plate. These items must be replaced in their proper positions. You may elect to follow the practice of many professional mechanics by taking a series of "Polaroid" type photographs of the powerhead with the flywheel removed -- one from the top, and a couple from the sides showing the wiring and arrangement of parts.

GOOD WORDS

If the only work to be done on the stator plate is replacement of an ignition or charging component, then proceed to the appropriate heading. If the plate is being removed because of a powerhead overhaul, then proceed with the following work.

5- Pry off the spark advance link. Loosen the five Philips head screws securing the stator plate to the support and retainer plates. Lift off the stator plate, with wire harness attached, up and over the crankshaft.

Remove the four bolts securing the retainer and support plates to the powerhead. Lift off the support and retainer plates and bearing assembly.

CLEANING AND INSPECTING

Inspect the flywheel for cracks or other damage, especially around the inside of the center hub. A damaged flywheel can be **EXTREMELY** dangerous. The forces exerted on the flywheel while it is rotating at high rpm are sufficient to shatter an unbalanced

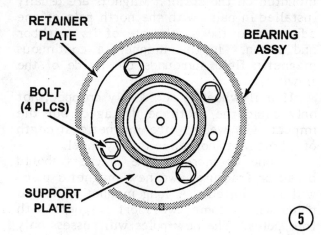

RETAINER PLATE

BEARING ASSY

BOLT (4 PLCS)

SUPPORT PLATE

5

damaged flywheel. Inspect the magnets to see if they are held tightly around the outer circumference of the flywheel rim. Check the surface of the magnets for fractures and grooves.

Grooves around the inner perimeter would indicate a component mounted on the stator plate is not aligned properly or the component is loose. Fractured or damaged magnets cannot be replaced. Both the magnets and flywheel share the same part number. Therefore attempting to remove the magnets would be a pointless task. If magnets or the flywheel is defective, the entire flywheel **MUST** be replaced.

Check to be sure small metal particles have not become attached to the magnets. Verify each magnet has good magnetism by using a screwdriver or other suitable tool.

Thoroughly clean the inside taper of the flywheel and the taper on the crankshaft to prevent the flywheel from "walking" on the crankshaft during operation.

Test the stator plate assembly to verify it is not loose. Attempt to lift each component on the plate. There should be little or no evidence of movement.

SPECIAL WORDS ON
FLYWHEEL MAGNETS

The outer ends of any magnet are called poles. One end is the north pole and the other end is the south pole. The magnetic field surrounding a magnet is concentrated around these two poles.

Some flywheel magnets are fairly long and curved around the outer perimeter of the underside of the flywheel. Others are short and are mounted around the center hub, depending on the location of the coils mounted on the stator. Magnets are usually installed in pairs with the north pole of one adjacent to the south pole of its neighbor and so on. In this manner, a continuous magnetic field surrounds the inside of the flywheel.

If a flywheel is accidently dropped, not only could the teeth be damaged, but the impact will weaken the magnetic strength of all magnets housed in the flywheel.

If one or more of the magnets should break or fracture, two new magnetic poles will be created. A long magnet with two poles will become two short magnets with four poles. The new poles will possess only

half the magnetic strength of an original pole. The overall magnetic field will be altered. The new field of the shorter magnets will not extend to cover the area of the flywheel.

Serious consequences apply to a CD type ignition system in the event of a flywheel magnet fracture. The ignition module and coil are energized by the concentrated magnetic field at the magnetic poles. If new poles are suddenly created, the coil and module will receive conflicting signals from the magnets and may even attempt to fire the cylinder twice in one revolution.

All these reasons require the flywheel to be handled with **CARE**.

If there is evidence of oil on the stator plate assembly, the oil seal/s at the forward main bearing inside the crankcase may have failed. Refer to Chapter 3, Powerhead.

IGNITION MODULE, CHARGE COIL, OR STATOR COIL SERVICE

Remove the two screws securing the wire harness retainer to the stator plate and then remove the retainer.

Ignition Module Replacement

Remove the two screws securing the module to the stator coil laminated core and the charge coil laminated core. Separate the leads under the module at their quick-disconnect fittings.

Connect the leads of the new ignition module to the wire harness leads, matching color-to-color. Tuck the connections underneath, and check to be sure the leads pass under the retainer plate in **ONE** layer and no

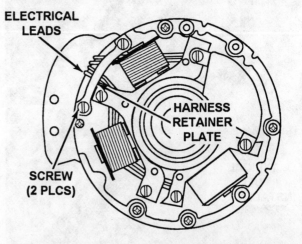

Harness retainer plate secures all the leads from the components mounted on the stator plate.

lead passes over another lead. Position the module over the two mounting holes in the laminated cores. Install the two screws and tighten them just handtight. The screws will be tightened to the specified torque value after the module is aligned to the stator plate.

The machined surfaces of the stator plate governs location of the ignition module. The outside edge of the ignition module **MUST** be flush with the machined surfaces on the stator plate to prevent contact with the flywheel magnets.

Proper location also assures maximum output from the module.

Special tool, Coil Locating Ring, OMC P/N 334994, will simplify alignment of the module. This tool is machined to fit over the stator plate bosses. Place the locating ring in position, push the ring toward the module, and at the same time, pull the module toward the ring, and then tighten the module retaining screws to a torque value of 30-40 in lb (3.4-4.5Nm).

ALTERNATE METHOD

If the special locating ring is not available to properly locate the ignition module the following procedure has been used successfully. Position the ignition module in place and just barely tighten the mounting screws with the outside edge of the ignition module flush with the machined surface of the stator plate.

Next, apply about three thicknesses (approximately 0.025" (0.64mm) of masking tape across the magnets on the flywheel. Slip the flywheel down over the crankshaft

and then rotate the flywheel slowly **CLOCKWISE**. Remove the flywheel and inspect the masking tape. If the magnets have dug into the masking tape, the module must be moved inward slightly. If there is absolutely no sign of the magnets making even the slightest contact with the tape, the module must be moved outward just a few thousands of an inch (mm). Once some evidence of the tape making the smallest amount of contact with the protrusion on the module is achieved, consider the module adjusted and secure the mounting screws securely to prevent shifting during powerhead operation. Make a final check to ensure the module did not shift during final tightening of the screws.

Charge Coil Replacement

Remove the screw securing the ignition module to the charge coil laminated core. Loosen, but do not remove the screw securing the ignition module to the stator coil laminated core. Swing the module clear of the charge coil laminated core. Remove the screws securing the charge coil to the stator plate. Separate the leads under the coil at their quick disconnect fittings.

Connect the leads of the new charge coil to the wire harness leads, matching color-to-color. Tuck the connections underneath and position the coil over the two mounting holes in the stator plate. Swing the ignition module back into position over the charge coil.

Install the two screws and tighten them just handtight. The screws will be tightened

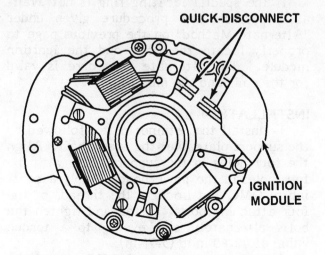

Removing or installing the ignition module mounted under the flywheel.

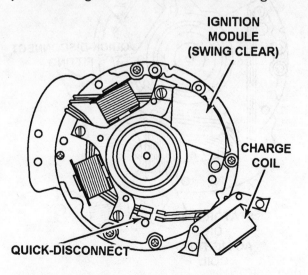

Removing or installing the ignition charge coil mounted on the stator plate under the flywheel.

to the specified torque value after the coil and module are aligned to the stator plate.

The machined surfaces of the stator plate governs location of the charge coil and the ignition module. The outside edges of the coil and module **MUST** be flush with the machined surfaces on the stator plate to prevent contact with the flywheel magnets.

Proper location also assures maximum output from the coil and module.

Special tool, Coil Locating Ring, OMC P/N 334994, will simplify alignment of the coil and module. This tool is machined to fit over the stator plate bosses. Place the locating ring in position, push the ring toward the coil and module, and at the same time, pull the coil and module toward the ring, and then tighten the three coil and module retaining screws to a torque value of 30-40 in lb (3.4-4.5Nm).

GOOD WORDS

If the special locating ring is not available, perform the procedure given under "Alternate Method" on the previous page to properly locate the charge coil and the ignition module. The alternate procedure is valid for both the coil and module.

Stator Coil Replacement

Remove the screw securing the ignition module to the stator coil laminated core. Loosen, but do **NOT** remove the screw se-

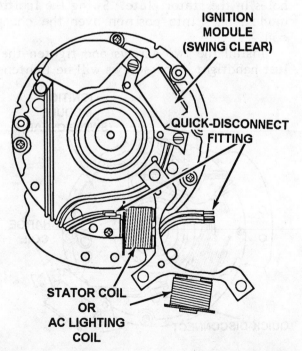

Removing or installing the stator (AC lighting) coils, as described in the text.

curing the ignition module to the charge coil laminated core. Swing the module away from the stator coil laminated core. Remove the other two screws securing the stator coils to the stator plate. Separate the leads under the coils at their quick disconnect fittings.

Connect the leads of the new stator coils to the wire harness leads, matching color-to-color. Tuck the connections underneath and position the coils over the three mounting holes in the stator plate. Swing the ignition module back into position over the stator coil laminated core.

Install the three screws and tighten them just handtight. The screws will be tightened to the specified torque value after the coil and module are aligned to the stator plate.

The machined surfaces of the stator plate governs location of the stator coils and the ignition module. The outside edges of the coils and module **MUST** be flush with the machined surfaces on the stator plate to prevent contact with the flywheel magnets.

Proper location also assures maximum output from the coils and module.

Special tool, Coil Locating Ring, OMC P/N 334994, will simplify alignment of the coils and module. This tool is machined to fit over the stator plate bosses. Place the locating ring in position, push the ring toward the coils and module, and at the same time, pull the coils and module toward the ring, and then tighten the four coil and module retaining screws to a torque value of 30-40 in lb (3.4-4.5Nm).

GOOD WORDS

If the special locating ring is not available, perform the procedure given under "Alternate Method" on the previous page to properly locate the coils and the ignition module. The alternate procedure is valid for the coils and the module.

INSTALLATION

1- Install the retainer plate, followed by the support plate over the crankshaft. Align the four holes in the support plate with the four holes in the powerhead. Apply a coating of OMC Nut Lock to the threads of the four attaching bolts. Install and tighten the bolts alternately and evenly to a torque value of 48-60 in lb (5-7Nm).

Apply an even coat of OMC Moly Lube to the crankcase boss area, as shown. Apply

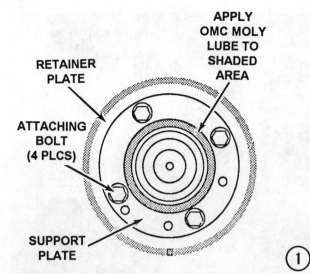

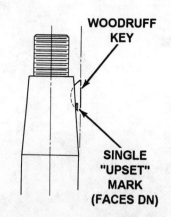

Woodruff key alignment in the crankshaft for the Model 3hp thru 8hp powerhead.

the same lubricant to the inside groove of the bearing assembly.

2- Apply OMC Nut Lock to the threads of the five stator plate Phillips head screws. Position the stator plate above the retainer plate with mounting holes aligned. Index the jaws of a pair of needlenose pliers into the two slots in the bearing assembly around the retainer plate. Compress the bearing with the pliers, and at the same time guide the stator plate down over the retainer plate and bearing. Check to be sure the holes in the stator plate align with those in the retainer plate. Install and tighten the five screws to a torque value of 25-35 in lb (2.8-4.0Nm).

Connect the spark advance link. Move the stator plate to the full advance position and make sure there is sufficient slack in the ignition wire harness to allow for full advance.

Flywheel Installation

3- Clean the crankshaft with solvent and blow it dry with compressed air.

Check to be sure the flywheel magnets are free of any metal particles.

If the Woodruff key is not already in place, position the key in the crankshaft keyway. Refer to the accompanying illustration for the correct positioning of the key for the model powerhead being serviced. Check to be sure the inside taper of the flywheel and the taper on the crankshaft are clean of dirt or oil, to prevent the flywheel from "walking" on the crankshaft while the powerhead is operating. Slide the flywheel down over the crankshaft with the keyway in the flywheel aligned with the key on the crankshaft.

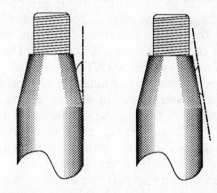

Alignment of the Woodruff key in the crankshaft for the Model 9.9 thru 15hp (right), and the Model 20 thru 55hp (left).

Rotate the flywheel **CLOCKWISE** and check to be sure the flywheel does not contact any component of the UFI system or the wiring.

Thread the flywheel nut onto the crankshaft and tighten it to the torque value given in the listing below. Use the same holding device to prevent the flywheel from rotating as was used during disassembling.

3hp	30-40 ft lb	(40-54Nm)
4hp & 4D	30-40 ft lb	(40-54Nm)
5-8hp	40-50 ft lb	(54-74Nm)
9.9-15hp	45-50 ft lb	(60-70Nm)
20-55hp	100-105 ft lb	(135-140Nm)

If servicing a manual start model, install the starter assembly on top of the powerhead -- see Chapter 9.

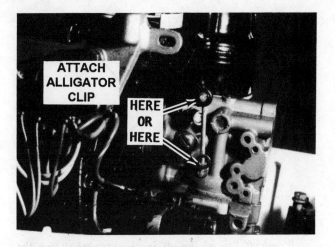

5-6 SYNCHRONIZATION
FUEL AND IGNITION SYSTEMS

Timing is **NOT** adjustable with a flywheel CD magneto system. The timing is controlled by the point setting. The correct point setting for **ALL** powerheads, with points, covered in this manual is 0.020". The fuel and ignition systems **MUST** be carefully synchronized to achieve maximum performance from the powerhead. In simple terms, synchronization is timing the carburetion to the ignition. This means, as the throttle is advanced to increase powerhead rpm, the carburetor and ignition systems are both advanced equally and at the same rate.

Therefore, any time the fuel system or the ignition system is serviced to replace a faulty part, or any adjustments are made for any reason, the powerhead synchronization must be carefully checked and verified.

Before making any adjustments with the synchronization, the ignition system should be thoroughly checked according to the procedures outlined in this chapter and the fuel checked according to the procedures outlined in Chapter 4.

GOOD WORDS

When making the synchronization adjustment, it is well to know and understand exactly what to look for and why. The critical time when the throttle shaft in the carburetor begins to move is of the utmost importance. First, realize that the time the cam follower makes contact with the cam is not the time the throttle shaft starts to move. Instead, the critical time is when the follower hits the designated position (as described in the next five paragraphs) and the throttle shaft **AT THE CARBURETOR** begins to move.

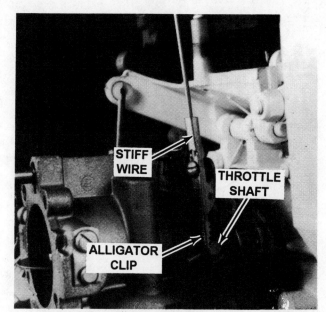

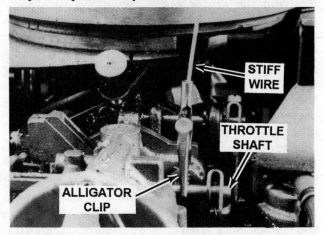

Above, left and right, and below, illustrate attachment of the alligator clip and piece of stiff wire to the throttle shaft to emphasize shaft movement.

A considerable amount of play exists between the follower at the top of the carburetor through the linkage to the actual throttle shaft. Therefore, the most important consideration is to watch for movement of the **THROTTLE SHAFT**, and not the follower. Movement of the shaft can be exaggerated by attaching a short piece of stiff wire to an alligator clip; grinding down the teeth on one side of the clip; and then attaching the clip to the throttle shaft, as shown in the examples on this page. The wire jiggling will instantly indicate movement of the shaft.

The line drawings accompanying the text have purposely been kept as simple as possible. Unnecessary details have been eliminated so as not to clutter the illustration and to reveal to the reader only what is essential when making a particular adjustment.

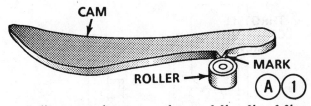

Roller set at the cam mark on a 2.3hp, 3hp, 3.3hp, and 4hp powerhead.

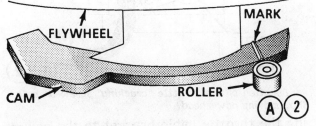

Roller set at the cam mark on a 4D powerhead.

2.3hp, 3hp, 3.3hp, 4hp, and 4D Powerheads Reference — "A" Series Illustrations

Movement of the throttle shaft can be exaggerated by attaching a short piece of stiff wire to an alligator clip; grinding down the teeth on one side of the clip; and then attaching the clip to the throttle shaft, as shown in the accompanying illustrations, on the previous page. Attach the clip on the **PORT** side for 3hp and 4hp models, -- on the **STARBOARD** side for 4D models. The wire jiggling will instantly indicate movement of the shaft.

Perform the following procedures in the sequence given for correct timing and carburetor synchronization.

Advance the throttle until the tip of the wire begins to move. When movement occurs, the center of the roller must align with the single mark on the cam. If not, back out the cam follower adjustment screw on the **PORT** side on the throttle shaft, with OMC Ball Hex Drive P/N 327622, until the throttle valve is completely closed. Now, rotate the screw **CLOCKWISE** until the throttle shaft just starts to rotate.

To adjust the carburetor: First, adjustment is made with the boat in a body of water, with the correct propeller installed. Start the powerhead and allow it to reach operating temperature.

To make the low speed adjustment: Loosen the two throttle cable nuts, one on each side of the throttle cable support bracket. Back out the idle speed adjustment screw, located at the rear of the powerhead just under the flywheel, -- to allow the stator plate to contact the stop cast into the powerhead.

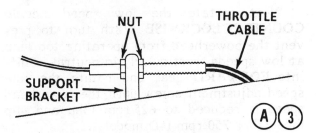

Correct tension on the throttle cable is very important to proper carburetor synchronization.

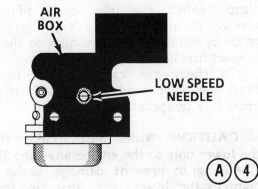

Location of the low speed needle on a 3hp and 4hp powerhead.

CAUTION: Water must circulate through the lower unit to the engine any time the engine is run to prevent damage to the water pump in the lower unit. Just five seconds without water will damage the water pump.

Start the powerhead and rotate the idle speed adjustment screw until the powerhead idles between 700 and 800 rpm. Adjust the low speed needle until the highest consistent powerhead speed is attained. Allow about 15 seconds between each adjustment for the engine to stabilize.

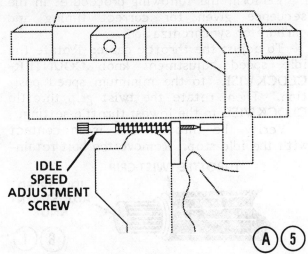

Location of the idle speed adjustment screw on a 3hp and 4hp powerhead.

Next, rotate the low speed needle COUNTERCLOCKWISE 1/8th turn to prevent the powerhead from operating too lean at low speeds, particularly in neutral. Shift into FORWARD gear and rotate the idle speed adjustment screw until the powerhead speed is reduced to 625 rpm (3hp and 4hp models) or 750 rpm (4D models).

To adjust the throttle cable: Shut down the powerhead. Rotate the throttle twist grip to the idle position. Move the stator plate against the idle speed adjustment screw. Eliminate the slack in the throttle cable by adjusting the two nuts on the cable support bracket.

To check the carburetor adjustments: Start the powerhead and allow it to reach operating temperature.

CAUTION: Water must circulate through the lower unit to the engine any time the engine is run to prevent damage to the water pump in the lower unit. Just five seconds without water will damage the water pump.

Operate the outboard in FORWARD gear for a full minute. Reduce powerhead speed to between 700 and 800 rpm and shift into NEUTRAL. The powerhead should continue to operate smoothly.

If the powerhead pops or stalls, the air/fuel mixture is probably too lean. Rotate the low speed needle 1/16th turn COUNTER CLOCKWISE, allowing about 15 seconds between adjustments, until the powerhead responds as expected.

5hp, 6hp, and 8hp Powerheads
Reference — "B" Series Illustrations

Perform the following procedures in the sequence given for correct timing and carburetor synchronization.

To adjust the throttle cable: Rotate the idle speed adjustment knob COUNTER-CLOCKWISE to the minimum speed position. Then, rotate the twist grip throttle CLOCKWISE to the lowest throttle setting.

Verify the throttle cam is in contact with the idle stop. Remove the bolt retain-

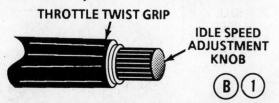

THROTTLE TWIST GRIP

IDLE SPEED ADJUSTMENT KNOB

Ⓑ ①

Simple drawing showing the idle speed adjustment knob on a tiller model unit.

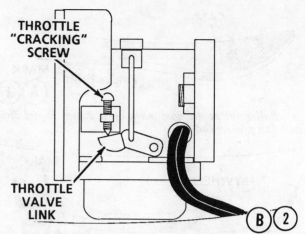

THROTTLE "CRACKING" SCREW

THROTTLE VALVE LINK

Ⓑ ②

Location of the throttle "cracking" screw on a 5hp, 6hp, and 8hp powerhead.

ing the throttle cable bracket to the powerhead. Rotate the throttle cable bracket COUNTERCLOCKWISE through one revolution to preload the throttle in the closed position and attach the bracket to the powerhead with the single bolt. Tighten the bolt securely.

To adjust the carburetor: Before starting the engine -- obtain OMC Ball Hex Screwdriver P/N 327622. Insert the screwdriver through the opening in the air box and back out the throttle cam follower screw until the follower no longer contacts the cam. If necessary, advance the throttle to gain access to the throttle cam screw. Back off the cam screw to the lowest setting and check to be sure the follower clears the cam.

Rotate the low speed needle, located on the STARBOARD side just above the throttle shaft, CLOCKWISE until the needle is lightly seated. Back the needle out exactly 2-1/2 turns. Loosen the throttle "cracking" screw, identified in the accompanying illustration. Rotate the screw CLOCKWISE through 4 complete turns after the screw tip makes contact with the throttle valve link.

Start the powerhead and allow it to reach operating temperature.

CAUTION: Water must circulate through the lower unit to the engine any time the engine is run to prevent damage to the water pump in the lower unit. Just five seconds without water will damage the water pump.

Shift the outboard into FORWARD gear and operate the powerhead at the lowest throttle setting. Adjust the low speed needle until the highest consistent rpm is attained. Allow about 15 seconds between each adjustment for the engine to stabilize, and then, rotate the low speed needle through 1/8th turn **COUNTERCLOCKWISE** to prevent a lean condition at idle speeds.

Rotate the "cracking" screw until the powerhead idles at 675 rpm.

Use the ball hex screwdriver to adjust the throttle cam follower until the follower contacts the cam **AND** powerhead rpm starts to increase. Back off the low speed needle 1/8th turn.

If powerhead rpm decreases as the cam follower makes contact with the cam, the low speed needle adjustment is too lean. Rotate the low speed needle through 1/8th turn **COUNTERCLOCKWISE** to correct the condition and then adjust the position of the cam follower as described. Shut down the engine.

To adjust the wide open throttle stop: With the powerhead **NOT** operating, set the throttle to the wide open position. Verify the position of the throttle shaft roll pin. The roll pin should be vertical. If not, use the ball hex screwdriver to adjust the throttle cam position until the roll pin is exactly vertical.

To check the carburetor adjustments, Start the powerhead and allow it to reach operating temperature.

CAUTION: Water must circulate through the lower unit to the engine any time the engine is run to prevent damage to the water pump in the lower unit. Just five seconds without water will damage the water pump.

Operate the outboard in **FORWARD** gear for a full minute. Reduce powerhead speed to between 700 and 800 rpm and shift into **NEUTRAL**. The powerhead should continue to operate smoothly.

If the powerhead pops or stalls, the air/fuel mixture is probably too lean. Rotate the low speed needle 1/16th turn **COUNTERCLOCKWISE**, allowing about 15 seconds between adjustments, until the powerhead responds as expected.

9.9hp and 15hp Powerheads
Reference — "C" Series Illustrations

Perform the following procedures in the sequence given for correct timing and carburetor synchronization. Remove both remote control cables from the powerhead before commencing these adjustments.

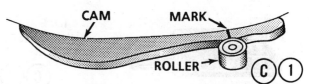

Roller set at the cam mark on the 9.9hp and 15hp powerheads.

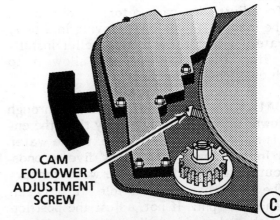

Location of the cam follower adjustment screw for the cam follower pickup point.

Movement of the throttle shaft can be exaggerated by attaching a short piece of stiff wire to an alligator clip; grinding down the teeth on one side of the clip; and then attaching the clip to the starboard throttle shaft. The wire jiggling will instantly indicate movement of the shaft.

To determine the cam follower pickup point: Advance the throttle until the tip of the wire begins to move. When movement occurs, the center of the roller must align with the single mark on the cam. If not, back out the cam follower adjustment screw, identified in the accompanying illustration, with OMC Ball Hex Drive P/N 327622, until the throttle valve is completely closed, and then, rotate the screw **CLOCKWISE** until the throttle shaft just starts to rotate.

To adjust the wide open throttle stop: With the powerhead **NOT** operating, set the throttle to the wide open position. Verify the position of the throttle shaft roll pin. The roll pin should be vertical. If not, use the ball hex screwdriver to adjust the throttle cam position until the roll pin is exactly vertical.

To adjust the idle speed: With the powerhead **NOT** operating, lightly seat the low speed needle valve. Back out the needle valve one full turn (9.9hp) or 7/8 turn (15hp). If servicing a tiller model, rotate the idle speed knob **COUNTERCLOCKWISE** to the full **SLOW** position.

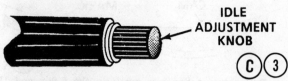

Simple drawing showing the idle speed adjustment knob on a tiller model unit.

To adjust the carburetor: Adjustment will be made with the boat moving in a body of water, with the correct propeller installed. Start the powerhead and allow it to reach operating temperature.

CAUTION: Water must circulate through the lower unit to the engine any time the engine is run to prevent damage to the water pump in the lower unit. Just five seconds without water will damage the water pump.

The powerhead should idle at between 650 and 700 rpm. If not, adjust the position of the throttle cable ball -- tiller model, or adjust the idle screw -- remote model, to correct the idle speed.

Connect the remote control cables.

To check the carburetor adjustments: Start the powerhead and allow it to reach operating temperature.

CAUTION: Water must circulate through the lower unit to the engine any time the engine is run to prevent damage to the water pump in the lower unit. Just five seconds without water will damage the water pump.

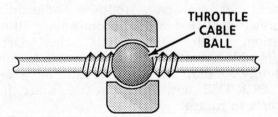

Simple drawing showing the throttle cable ball on a tiller model unit.

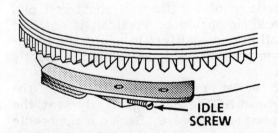

The idle speed on a tiller model is made by moving the ball on the throttle cable. On a remote model this adjustment is made with the idle screw shown in this drawing.

Operate the outboard in FORWARD gear for a full minute. Reduce powerhead speed to between 700 and 800rpm and shift into NEUTRAL. The powerhead should continue to operate smoothly.

If the powerhead pops or stalls, the air-/fuel mixture is probably too lean. Rotate the low speed needle 1/16th turn COUNTERCLOCKWISE, allowing about 15 seconds between adjustments, until the powerhead responds as expected.

20hp Powerhead
Reference -- "D" Series Illustrations

Perform the following procedures in the sequence given for correct timing and carburetor synchronization.

Movement of the throttle shaft can be exaggerated by attaching a short piece of stiff wire to an alligator clip; grinding down the teeth on one side of the clip; and then attaching the clip to the starboard throttle shaft. The wire jiggling will instantly indicate movement of the shaft.

To determine the cam follower pickup point: Advance the throttle by moving the throttle control lever, until the tip of the wire begins to move. When movement occurs, the center of the roller must align between the two marks on the cam. If the roller does not align, loosen the screw securing the adjustment lever to the throttle shaft. Move the adjustment lever up or down to make the adjustment.

To check the maximum spark advance: The outboard MUST be operated with the proper test wheel. DO NOT operate the powerhead with a propeller, or flush adaptor while performing this adjustment.

Start the powerhead and allow it to reach operating temperature. Boat movement must be restrained -- secured to the dock.

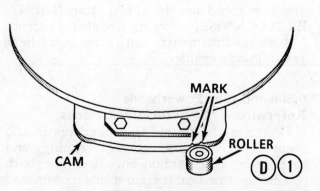

Roller at the cam mark on a 20hp powerhead.

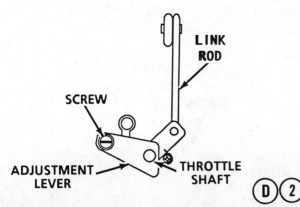

If the roller does not align between the two marks on the cam, the adjustment lever must be repositioned until the roller does align.

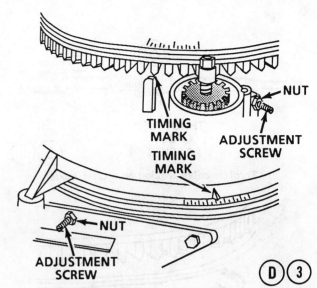

Location of the adjustment to check the maximum spark advance on an electric start model (top), and a manual start (bottom).

CAUTION: Water must circulate through the lower unit to the engine any time the engine is run to prevent damage to the water pump in the lower unit. Just five seconds without water will damage the water pump.

Connect a timing light to the top cylinder. Start the powerhead and allow it to reach operating temperature. Operate the powerhead at W.O.T (wide open throttle), with the outboard in **FORWARD** gear. Aim the timing light at the timing marks on the flywheel perimeter. On electric start models, the mark on the timing pointer should align between the 33° and 35° mark on the flywheel. On manual start models, the triangular pointer embossed on the manual starter housing should align between the 33° and 35° mark on the flywheel.

If the marks do not align, shut down the engine. Loosen the nut on the timing adjustment screw, identified in the accompanying illustrations. Turn the adjustment screw to correct the timing. Rotating the screw through one revolution **CLOCKWISE**, retards the timing one degree. Rotating the screw through one revolution **COUNTERCLOCKWISE**, advances the timing one degree. Tighten the nut on the timing adjustment screw to hold this new adjustment.

To adjust the idle speed: First, adjustment is made with the boat in a body of water, with the correct propeller installed. Start the engine and allow it to reach operating temperature.

CAUTION: Water must circulate through the lower unit to the engine any time the engine is run to prevent damage to the water

pump in the lower unit. Just five seconds without water will damage the water pump.

If servicing a tiller model, rotate the idle adjustment knob **COUNTERCLOCKWISE** to the full **SLOW** position.

All models: The engine should idle at between 650 and 700 rpm in **FORWARD** gear.

If not, adjust the idle speed on tiller models by moving the throttle cable bracket. If working on a remote model, adjust the throttle arm stop screw to correct the idle speed.

To check the carburetor adjustments: Start the engine and allow it to reach operating temperature.

CAUTION: Water must circulate through the lower unit to the engine any time the engine is run to prevent damage to the water pump in the lower unit. Just five seconds without water will damage the water pump.

Operate the powerhead with the outboard in **FORWARD** gear for a full minute. Reduce powerhead speed to between 700

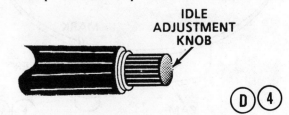

*Rotate the idle speed knob to the full **SLOW** position before making the idle speed adjustment.*

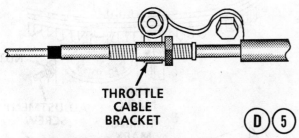

THROTTLE CABLE BRACKET

D 5

Throttle cable bracket on a tiller model.

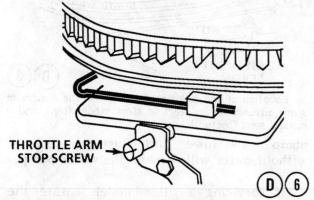

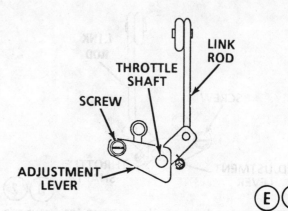

LINK ROD

THROTTLE SHAFT

SCREW

ADJUSTMENT LEVER

E 2

If the roller on the cam does not align between the two marks on the cam, the adjustment lever must be repositioned until the roller does align.

Movement of the throttle shaft can be exaggerated by attaching a short piece of stiff wire to an alligator clip; grinding down the teeth on one side of the clip; and then attaching the clip to the starboard throttle shaft. The wire jiggling will instantly indicate movement of the shaft.

To determine the cam follower pickup point: Advance the throttle until the tip of the wire begins to move. When movement occurs, the center of the roller must align between the two marks on the cam. If not, loosen the screw securing the adjustment lever to the throttle shaft. Move the adjustment lever up or down to make the adjustment.

THROTTLE ARM STOP SCREW

D 6

Location of the idle speed adjustment on a tiller model or remote model -- 20hp unit.

and 800rpm and shift into **NEUTRAL**. The powerhead should continue to operate smoothly.

If the powerhead pops or stalls, the air/fuel mixture is probably too lean. Rotate the low speed needle 1/16th turn **COUNTERCLOCKWISE**, allowing about 15 seconds between adjustments, until the powerhead responds as expected.

25hp and 30hp Powerheads
Reference -- "E" Series Illustrations

Perform the following procedures in the sequence given for correct timing and carburetor synchronization. Remove both remote control cables from the engine before commencing these adjustments.

To adjust the throttle control rod: Before making any adjustment to the throttle control rod, be sure the offset on the pivot block faces toward the control rod collar.

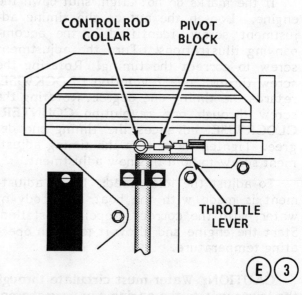

CONTROL ROD COLLAR

PIVOT BLOCK

THROTTLE LEVER

E 3

Details of the throttle control rod for making an adjustment, as explained in the text.

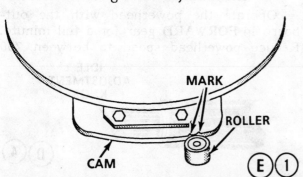

MARK

ROLLER

CAM

E 1

Roller at the cam mark on a 25hp and 30hp powerhead.

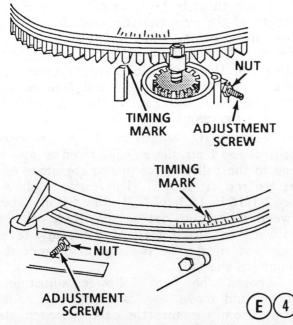

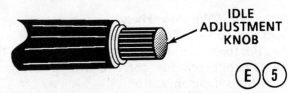

Location of the adjustment to check the maximum spark advance on an electric start unit (top), and a manual start unit (bottom).

Rotate the propeller shaft and at the same time, move the shift lever into the **FORWARD** position. Loosen the control rod collar screw. Advance the throttle lever until the lever makes contact with the stop cast into the cylinder block. Now, move the throttle control rod forward until the carburetor throttle plate is horizontal. Slide the control rod collar backwards until it makes contact with the pivot block. Tighten the screw on the collar to hold this new adjustment.

To check the maximum spark advance, the outboard **MUST** be operated with the proper test wheel. **DO NOT** operate the powerhead with a propeller, or flush adaptor while performing this adjustment.

Start the engine and allow it to reach operating temperature. Boat movement must be restrained -- secured to the dock.

CAUTION: Water must circulate through the lower unit to the engine any time the engine is run to prevent damage to the water pump in the lower unit. Just five seconds without water will damage the water pump.

Connect a timing light to the top cylinder. Start the powerhead and allow it to reach operating temperature. Operate the powerhead at W.O.T. (wide open throttle), with the outboard in **FORWARD** gear. Aim the timing light at the timing marks on the

flywheel perimeter. On electric start models, the mark on the timing pointer should align between the 33° and 35° mark on the flywheel. On manual start models, the triangular pointer embossed on the manual starter housing should align between the 29° and 31° mark on the flywheel.

If the marks do not align, shut down the engine. Loosen the nut on the timing adjustment screw, identified in the accompanying illustrations. Turn the adjustment screw to correct the timing. Rotating the screw through one revolution **CLOCKWISE**, retards the timing one degree. Rotating the screw through one revolution **COUNTER-CLOCKWISE**, advances the timing one degree. Tighten the nut on the timing adjustment screw to hold this new adjustment.

To adjust the idle speed: First, adjustment is made with the boat in a body of water, with the correct propeller installed. Start the engine and allow it to reach operating temperature.

CAUTION: Water must circulate through the lower unit to the engine any time the engine is run to prevent damage to the water pump in the lower unit. Just five seconds without water will damage the water pump.

If servicing a tiller model, rotate the idle adjustment knob **COUNTERCLOCK-WISE** to the full **SLOW** position.

All models: The powerhead should idle at between 650 and 700 rpm with the outboard in **FORWARD** gear.

If not, adjust the idle speed on tiller models by moving the throttle cable bracket. If working on a remote model, adjust the throttle arm stop screw to correct the idle speed.

SPECIAL CARBURETOR WORDS

The carburetor installed on the 25hp and 30hp engines is equipped with an idle mixture needle valve. This needle valve controls the idle mixture and is preset at the factory. The needle valve **MUST NOT** be

*Rotate the idle speed knob to the full **SLOW** position before making the idle speed adjustment.*

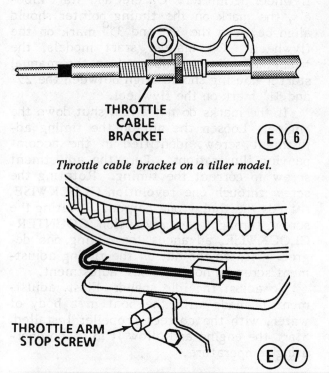

THROTTLE CABLE BRACKET E 6

Throttle cable bracket on a tiller model.

THROTTLE ARM STOP SCREW E 7

Location of the idle speed adjustment on a tiller model or remote model -- 20hp unit.

disturbed from the factory setting except for rebuilding or cleaning purposes.

To check the carburetor adjustments: Start the powerhead and allow it to reach operating temperature.

CAUTION: Water must circulate through the lower unit to the engine any time the engine is run to prevent damage to the water pump in the lower unit. Just five seconds without water will damage the water pump.

Operate the outboard in **FORWARD** gear for a full minute. Reduce powerhead speed to between 700 and 800 rpm and shift into **NEUTRAL.** The powerhead should continue to operate smoothly.

If the powerhead pops or stalls, the air/fuel mixture is probably too lean. Rotate the low speed needle 1/16th turn **COUNTERCLOCKWISE,** allowing about 15 seconds between adjustments, until the powerhead responds as expected.

40hp and 50hp Powerheads
Reference — "F" Series Illustrations

Perform the following procedures in the sequence given for correct timing and carburetor synchronization.

Make the following preliminary linkage adjustments: Disconnect the throttle cable

from the throttle lever arm. Measure the length of the long throttle control rod from the pivot point center to the ball joint center. This distance should be 7-13/16 in (19.8cm). If the distance between centers is not as specified, pry the rod free of the throttle cam and rotate the rod end to correct the length of the rod.

Measure the length of the short spark control rod from the ball joint center at one end to the ball joint center at the other end of the rod, as shown. This distance should be 2-1/16 in (5.3cm). If the distance between centers is not as specified, pry the forward ball joint free of the spark lever cam and rotate the rod end to correct the length of the rod.

Loosen the cam follower adjustment screw and move the cam follower roller away from the throttle cam. Loosen the locknut on the idle speed screw and adjust the screw until the roller on the spark lever cam is between the two alignment marks on the curved slot of the throttle lever. Tighten the locknut on the screw when the adjustment is completed. Move the cam follower roller to make contact with the throttle cam and tighten the adjustment screw.

To synchronize the throttle cam: Loosen the carburetor lever adjustment screw on the top carburetor. Rotate both throttle shafts slightly and permit them to snap shut. Apply a gentle pressure on the adjustment tab, as shown, and tighten the adjustment screw. Check to be sure both throttle shafts begin to rotate at exactly the same time.

THROTTLE CONTROL ROD   **CONTROL ROD COLLAR**   **PIVOT BLOCK**

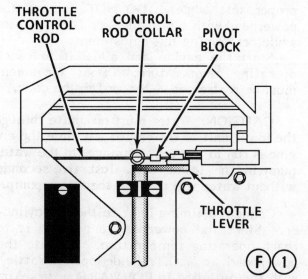

THROTTLE LEVER

F 1

Details of the throttle control rod for making an adjustment, as explained in the text.

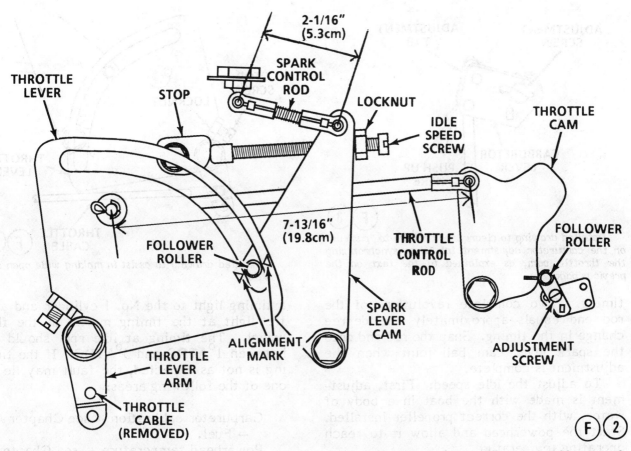

Detailed drawing of the throttle linkage at idle speed, showing the given length of the spark control rod between the centers of the two sockets and the given distance between the two sockets of the throttle control rod.

To adjust the cam pickup point: Verify both carburetor throttle plates are closed. Check to be sure the throttle cam follower roller makes contact with the throttle cam at the embossed mark on the cam. The mark must align with the center of the roller. If not, loosen the adjustment screw and reposition the roller until it makes contact at the mark on the cam. Tighten the screw to hold the new adjustment.

Hold the idle speed screw against its stop on the engine. Check to be sure the spark lever follower roller is centered between the two marks on the slot in the throttle lever. Use a feeler gauge to measure the clearance between the throttle cam and the throttle cam follower roller. A clearance of 0.010 in (0.25mm) must exist between the cam and roller. If not, adjust the length of the throttle control rod until the correct clearance is obtained: Pry the rod free of the throttle cam and rotate the rod end to correct the length of the rod.

Install the throttle cable into the upper of the two holes in the throttle lever.

To check the maximum spark advance, the outboard MUST be operated with the proper test wheel. DO NOT operate the powerhead with a propeller, or flush adaptor while performing this adjustment.

CAUTION: Water must circulate through the lower unit to the engine any time the engine is run to prevent damage to the water pump in the lower unit. Just five seconds without water will damage the water pump.

Connect a timing light to the top cylinder. Start the powerhead and allow it to reach operating temperature. Operate the powerhead at a minimum of 5000 rpm with the outboard in FORWARD gear. Aim the timing light at the timing marks on the flywheel perimeter. The mark on the timing pointer should align between the 18° and 20° mark on the flywheel.

If the timing needs adjustment, shut down the powerhead and alter the length of the spark control lever to correct the timing, as follows: Pry off the forward ball joint on the spark control rod from the ball joint on the spark lever cam. Rotate the rod end CLOCKWISE to advance the timing, or COUNTERCLOCKWISE to retard the

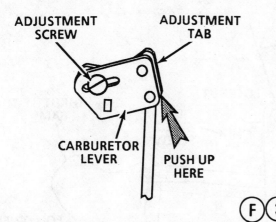

ADJUSTMENT
SCREW

ADJUSTMENT
TAB

CARBURETOR
LEVER

PUSH UP
HERE

F 3

Detailed drawing to clearly show where to "push up" on the carburetor adjustment tab when synchronizing the throttle cam, as explained in the text on the previous page.

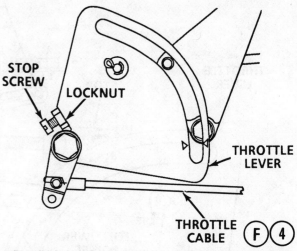

STOP
SCREW

LOCKNUT

THROTTLE
LEVER

THROTTLE
CABLE

F 4

Detailed drawing to assist in making wide open stop position.

timing. Two complete revolutions of the rod end equals approximately one degree change in the timing. Snap the rod end onto the spark lever cam ball joint when the adjustment is complete.

To adjust the idle speed: First, adjustment is made with the boat in a body of water, with the correct propeller installed. Start the powerhead and allow it to reach operating temperature.

CAUTION: Water must circulate through the lower unit to the engine any time the engine is run to prevent damage to the water pump in the lower unit. Just five seconds without water will damage the water pump.

Shift the lower unit into **FORWARD** gear. The powerhead should idle at between 725 and 775 rpm with the idle speed screw resting against the stop. If not, loosen the locknut on the idle speed screw and reset the clearance between the throttle cam and the throttle cam follower roller at 0.010in (0.25mm) as directed in the previous paragraphs, and then make the necessary adjustment using the idle speed screw. Tighten the locknut to hold the new adjustment.

With the engine still operating, connect

a timing light to the No. 1 cylinder and aim the light at the timing marks on the flywheel. The timing at idle rpm should be between 1°ATDC and 5°ATDC. If the timing is not as expected, the fault may lie in one of the following areas:

Carburetor calibration -- see Chapter 4 -- Fuel.

Powerhead temperature -- see Chapter 5 -- Ignition.

Powerhead condition -- see Chapter 3 -- Powerhead.

Flywheel keyway condition -- see Chapter 5 -- Ignition.

Woodruff key condition -- see Chapter 5 -- Ignition.

To check the W.O.T. (wide open throttle) stop position: With the powerhead **NOT** operating, advance the throttle lever to the wide open position. At this point the carburetor throttle shaft roll pins should be in the vertical position indicating the throttle valves are wide open. If not, loosen the locknut on the stop screw at the throttle lever, identified in the accompanying illustration. Rotate the screw as necessary to bring the roll pins into vertical alignment. Tighten the locknut on the stop screw to hold this new adjustment.

6
ELECTRICAL

6-1 INTRODUCTION

The battery, gauges, and horns, charging system, and the cranking system are all considered subsystems of the electrical system. Each of these units or subsystems will be covered in detail in this chapter beginning with the battery.

Ignition for powerheads covered in this manual are: A conventional magneto; a Type II CD system with sensor coil; or a second Type II CD system with UFI -- Under Flywheel Ignition, see Chapter 5 -- Ignition.

A battery is not required to operate the powerhead. Most of the larger horsepower units use a cranking motor for starting and the battery is only used to supply power for this motor.

The starting circuit consists of a cranking motor and a cranking motor-engaging mechanism. A solenoid is used as a heavy-duty switch to carry the heavy current from the battery to the cranking motor. The solenoid is actuated by turning the ignition key to the **START** position. On some models, a pushbutton is used to actuate the solenoid.

These powerheads are also equipped with a hand starter for use when the electric cranking motor system is inoperative.

6-2 BATTERIES

The battery is one of the most important parts of the electrical system. In addition to providing electrical power to start the engine, it also provides power for operation of the running lights, radio, electrical accessories, and possibly the pump for a bait tank.

Because of its job and the consequences, (failure to perform in an emergency) the best advice is to purchase a well-known brand, with an extended warranty period, from a reputable dealer.

The usual warranty covers a prorated replacement policy, which means you would be entitled to a consideration for the time left on the warranty period if the battery should prove defective before its time.

Do not consider a battery of less than 70-ampere hour capacity. If in doubt as to how large your boat requires, make a liberal estimate and then purchase the one with the next higher ampere rating.

MARINE BATTERIES

Because marine batteries are required to perform under much more rigorous conditions than automotive batteries, they are constructed much differently than those used in automobiles or trucks. Therefore, a marine battery should always be the No. 1 unit for the boat and other types of batteries used only in an emergency.

A fully charged battery, filled to the proper level with electrolyte, is the heart of the ignition system. Engine starting and efficient performance can never be obtained if the battery is below a fully charged rating.

Marine batteries have a much heavier exterior case to withstand the violent pounding and shocks imposed on it as the boat moves through rough water and in extremely tight turns.

The plates in marine batteries are thicker than in automotive batteries and each plate is securely anchored within the battery case to ensure extended life.

The caps of marine batteries are "spill proof" to prevent acid from spilling into the bilges when the boat heels to one side in a tight turn, or is moving through rough water.

Because of these features, the marine battery will recover from a low charge condition and give satisfactory service over a much longer period of time than any type of automotive-type unit.

BATTERY CONSTRUCTION

A battery consists of a number of positive and negative plates immersed in a solution of diluted sulfuric acid. The plates contain dissimilar active materials and are kept apart by separators. The plates are grouped into what are termed elements. Plate straps on top of each element connect all of the positive plates and all of the negative plates into groups.

The battery is divided into cells which hold a number of the elements apart from the others. The entire arrangement is contained within a hard-rubber case. The top is a one-piece cover and contains the filler caps for each cell. The terminal posts

protrude through the top where the battery connections for the boat are made. Each of the cells is connected to its neighbor in a positive-to-negative manner with a heavy strap called the cell connector.

BATTERY RATINGS

Two ratings are used to classify batteries: One is a 20-hour rating at 80°F and the other is a cold rating at 0°F. This second figure indicates the cranking load capacity and is referred to as the Peak Watt Rating of a battery. This Peak Watt Rating (PWR) has been developed to measure the cold-cranking ability of the battery. The numerical rating is embossed on each battery case at the base and is determined by multiplying the maximum current by the maximum voltage.

The ampere-hour rating of a battery is its capacity to furnish a given amount of amperes over a period of time at a cell voltage of 1.5. Therefore, a battery with a capacity of maintaining 3 amperes for 20 hours at 1.5 volts would be classified as a 60-ampere hour battery.

Do not confuse the ampere-hour rating with the PWR, because they are two unrelated figures used for different purposes.

A replacement battery should have a power rating equal or as close to the old unit as possible.

BATTERY LOCATION

Every battery installed in a boat must be secured in a well-protected ventilated area. If the battery area is not well ventilated, hydrogen gas which is given off during charging could become very explosive if the gas is concentrated and confined. Because of its size, weight, and acid content, the battery must be well-secured. If the battery should break loose during rough boat maneuvers, considerable damage could be done, including damage to the hull.

BATTERY SERVICE

The battery requires periodic servicing and a definite maintenance program to ensure extended life. If the battery should test satisfactorily, but still fails to perform properly, one of five problems could be the cause.

The battery MUST be located near the engine in a well-ventilated area. It must be secured in such a manner that absolutely no movement is possible in any direction under the most violent actions of the boat.

1- An accessory might have accidently been left on overnight or for a long period during the day. Such an oversight would result in a discharged battery.

2- Slow speed engine operation for long periods of time resulting in an undercharged condition.

3- Using more electrical power than the generator can replace will result in an undercharged condition.

4- A defect in the charging system. A faulty generator, defective regulator, defective alternator or diodes on units with an alternator, or high resistance somewhere in the system could cause the battery to become undercharged.

5- Failure to maintain the battery in good order. This might include a low level of electrolyte in the cells; loose or dirty cable connections at the battery terminals; or possibly an excessive dirty battery top.

Electrolyte Level

The most common practice of checking the electrolyte level in a battery is to remove the cell cap and visually observe the level in the vent well. The bottom of each vent well has a split vent which will cause

One of the most effective means of cleaning the battery terminals is to use a wire brush designed for this specific purpose.

the surface of the electrolyte to appear distorted when it makes contact. When the distortion first appears at the bottom of the split vent, the electrolyte level is correct.

Some late model batteries have an electrolyte level indicator installed which operates in the following manner:

A transparent rod extends through the center of one of the cell caps. The lower tip of the rod is immersed in the electrolyte when the level is correct. If the level should drop below normal, the lower tip of the rod is exposed and the upper end glows as a warning to add water. Such a device is only necessary on one cell cap because if the electrolyte is low in one cell it is also low in the other cells. **BE SURE** to replace the cap with the indicator onto the second cell from the positive terminal.

During hot weather and periods of heavy use, the electrolyte level should be checked more often than during normal operation. Add colorless, odorless, drinking water to bring the level of electrolyte in each cell to the proper level. **TAKE CARE** not to overfill, because adding an excessive amount of water will cause loss of electrolyte and any loss will result in poor performance, short battery life, and will contribute quickly to corrosion. **NEVER** add electrolyte from another battery. Use only clean pure water.

Cleaning

Dirt and corrosion should be cleaned from the battery just as soon as it is discovered. Any accumulation of acid film or dirt will permit current to flow between the terminals. Such a current flow will drain the battery over a period of time.

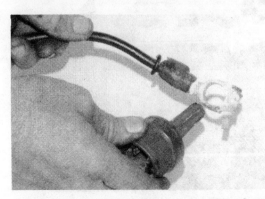

An inexpensive brush can be purchased and used to clean battery lead connectors to ensure a proper connection.

Clean the exterior of the battery with a solution of diluted ammonia or a soda solution to neutralize any acid which may be present. Flush the cleaning solution off with clean water. **TAKE CARE** to prevent any of the neutralizing solution from entering the cells, by keeping the caps tight.

A poor contact at the terminals will add resistance to the charging circuit. This resistance will cause the voltage regulator to register a fully charged battery, and thus cut down on the alternator output adding to the low battery charge problem.

Scrape the battery posts clean with a suitable tool or with a stiff wire brush. Clean the inside of the cable clamps to be sure they do not cause any resistance in the circuit.

Battery Testing

A hydrometer is a device to measure the percentage of sulfuric acid in the battery electrolyte in terms of specific gravity. When the condition of the battery drops from fully charged to discharged, the acid leaves the solution and enters the plates, causing the specific gravity of the electrolyte to drop.

The following six points should be observed when using a hydrometer:

1- NEVER attempt to take a reading immediately after adding water to the battery. Allow at least 1/4 hour of charging at a high rate to thoroughly mix the electrolyte with the new water and to cause vigorous gassing.

2- ALWAYS be sure the hydrometer is clean inside and out as a precaution against contaminating the electrolyte.

3- If a thermometer is an integral part of the hydrometer, draw liquid into it several times to ensure the correct temperature before taking a reading.

4- BE SURE to hold the hydrometer vertically and suck up liquid only until the float is free and floating.

5- ALWAYS hold the hydrometer at eye level and take the reading at the surface of the liquid with the float free and floating.

Disregard the light curvature appearing where the liquid rises against the float stem. This phenomenon is due to surface tension.

6- DO NOT drop any of the battery fluid on the boat or on your clothing, because it is extremely caustic. Use water and baking soda to neutralize any battery

A check of the electrolyte in the battery should be on the maintenance schedule for any boat. A hydrometer reading of 1.300 or in the green band, indicates the battery is in satisfactory condition. If the reading is 1.150 or in the red band, the battery needs to be charged. Observe the six safety points given in the text when using a hydrometer.

A pair of pliers should be used to tighten the wingnuts, when they are used. Securing the wingnuts by hand is not adequate, the connections will vibrate loose.

liquid that does accidently drop.

After withdrawing electrolyte from the battery cell until the float is barely free, note the level of the liquid inside the hydrometer. If the level is within the green band range, the condition of the battery is satisfactory. If the level is within the white band, the battery is in fair condition, and if the level is in the red band, it needs charging badly or is dead and should be replaced. If the level fails to rise above the red band after charging, the only answer is to replace the battery.

JUMPER CABLES

If booster batteries are used for starting an engine the jumper cables must be connected correctly and in the proper sequence to prevent damage to either battery, or to the alternator diodes.

ALWAYS connect a cable from the positive terminal of the dead battery to the positive terminal of the good battery **FIRST**. **NEXT**, connect one end of the other cable to the negative terminal of the good battery and the other end to a good ground on the powerhead. By making the ground connection on the powerhead, if there is an arc when you make the connection it will not be near the battery. An arc near the battery could cause an explosion, destroying the battery and causing serious personal **INJURY**.

DISCONNECT the battery ground cable before replacing an alternator or before connecting any type of meter to the alternator charging circuit.

If it is necessary to use a fast-charger on a dead battery, **ALWAYS** disconnect one of the boat cables from the battery **FIRST**, to prevent burning out the diodes in the rectifier. **NEVER** use a fast-charger as a booster to start the engine because the voltage regulator may be **DAMAGED**.

STORAGE

If the boat is to be laid up for the winter or for more than a few weeks, special attention must be given to the battery to prevent complete discharge or possible damage to the terminals and wiring. Before putting the boat in storage, disconnect and remove the batteries. Clean them thoroughly of any dirt or corrosion, and then charge them to full specific gravity reading. After they are fully charged, store them in a clean cool dry place where they will not be damaged or knocked over.

NEVER store the battery with anything on top of it or cover the battery in such a manner as to prevent air from circulating around the filler caps. All batteries, new and old, will discharge during periods of storage, more so if they are hot than if they remain cool. Therefore, the electrolyte

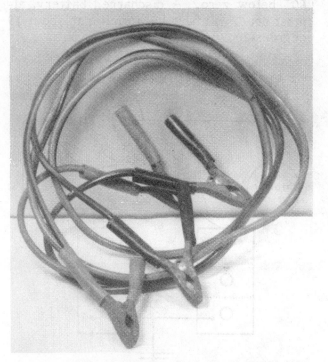

Someone smoking close to this battery during high charge may have ignited the explosive fumes, blowing a hole in the top surface.

A common set of heavy-duty jumper cables. Observe the safety precautions given in the text when using jumper cables.

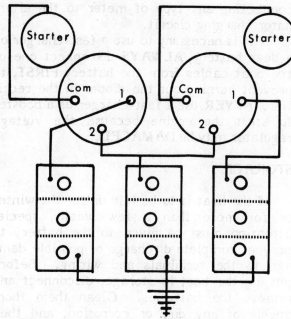

Schematic diagram for a three battery, two engine hookup.

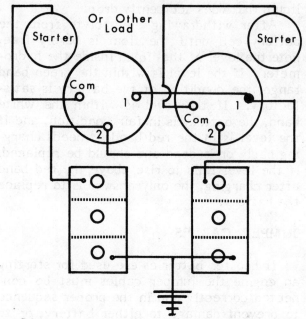

Schematic diagram for a two battery, two engine hookup.

level and the specific gravity should be checked at regular intervals. A drop in the specific gravity reading is cause to charge them back to a full reading.

In cold climates, care should be exercised in selecting the battery storage area. A fully-charged battery will freeze at about 60° below zero. A discharged battery, almost dead, will have ice forming at about 19 degrees above zero.

DUAL BATTERY INSTALLATION

Three methods are available for utilizing a dual battery hookup.

1- A high-capacity switch can be used to connect the two batteries. The accompanying illustration details the connections for installation of such a switch. This type of switch installation has the advantage of being simple, inexpensive, and easy to mount and hookup. However, if the switch is accidently left in the closed position, it will cause the convenience loads to run

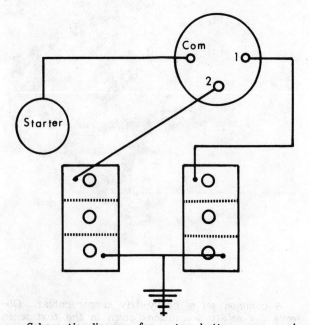

Schematic diagram for a two battery, one engine hookup.

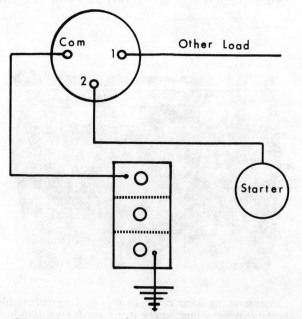

Schematic diagram for a single battery, one engine hookup.

down both batteries and the advantage of the dual installation is lost. The switch may be closed intentionally to take advantage of the extra capacity of the two batteries, or it may be temporarily closed to help start the engine under adverse conditions.

2- A relay can be connected into the ignition circuit to enable both batteries to be automatically put in parallel for charging or to isolate them for ignition use during engine cranking and start. By connecting the relay coil to the ignition terminal of the ignition-starting switch, the relay will close during the start to aid the starting battery. If the second battery is allowed to run down, this arrangement can be a disadvantage since it will draw a load from the starting battery while cranking the engine. One way to avoid such a condition is to connect the relay coil to the ignition switch accessory terminal. When connected in this manner, while the engine is being cranked, the relay is open. But when the engine is running with the ignition switch in the normal position, the relay is closed, and the second battery is being charged at the same time as the starting battery.

3- A heavy duty switch installed as close to the batteries as possible can be connected between them. If such an arrangement is used, it must meet the standards of the American Boat and Yacht Council, Inc. or the Fire Protection Standard for Motor Craft, N.F.P.A. No. 302.

6-3 GAUGES

Gauges or lights are installed to warn the operator of a condition in the cooling and lubrication systems that may need attention. The fuel gauge gives an indication of the amount of fuel in the tank. If the powerhead overheats or the oil pressure drops too low for safety, a gauge or audible device alerts the operator to shut down the powerhead and check the cause of the warning before serious damage is done.

CONSTANT-VOLTAGE SYSTEM

In order for gauges to register properly, they must be supplied with a steady voltage. The voltage variations produced by the powerhead charging system would cause erratic gauge operation, too high when the alternator voltage is high and too low when the alternator is not charging. To remedy this problem, a constant-voltage system is used to reduce the 12-14 volts of the electrical system to an average of 5 volts. This steady 5 volts ensures the gauges will read accurately under varying conditions from the electrical system.

6-4 SERVICE PROCEDURES

Systems utilizing warning lights do not require a constant-voltage system, therefore, this service is not needed.

Service procedures for checking the gauges and their sending units is detailed in the following sections.

TEMPERATURE GAUGES

The body of temperature gauges must be grounded and they must be supplied with 12 volts. Many gauges have a terminal on the mounting bracket for attaching a ground wire. A tang from the mounting bracket makes contact with the gauge. CHECK to be sure the tang does make good contact with the gauge.

Ground the wire to the sending unit and the needle of the gauge should move to the full right position indicating the gauge is in serviceable condition.

See Chapter 3, to test the sender unit.

The gauges and controls on the control panel should be kept clean and protected from water spray, especially when operating in a salt water atmosphere.

WARNING LIGHTS

If a problem arises on a boat equipped with water and temperature lights, the first area to check is the light assembly for loose wires or burned-out bulbs.

When the ignition key is turned on, the light assembly is supplied with 12 volts and grounded through the sending unit mounted on the powerhead. When the sending unit makes contact, because the water temperature is too hot, the circuit to ground is completed and the lamp should light.

Check The Bulb: Turn the ignition switch on. Disconnect the wire at the engine sending unit, and then ground the wire. The lamp on the control panel should light. If it does not light, check for a burned-out bulb or a break in the wiring to the light.

AUDIBLE WARNING DEVICE

A unique audible warning device is used on models with a remote control system. Sounding of the device will alert the operator that a critical operating condition has developed; the powerhead should be shut down; and the cause determined before extensive and expensive damage is done.

Depending on the powerhead being serviced, the warning device may emit:

Continuous Tone indicating a powerhead overheat condition.

Continuous Short Pulse Tone indicating a **NO OIL** condition -- **BAD NEWS** -- very **BAD NEWS**.

Short Pulse Tone -- every 20-seconds indicating a **LOW OIL** condition.

Continuous Tone -- at or near WOT indicating a restriction in the fuel supply.

THERMOMELT STICKS

Thermomelt sticks are an easy method of determining if the powerhead is running at the proper temperature. Thermomelt sticks are not expensive and are available at the local marine dealer.

Start the engine with the propeller in the water and run it for about 5 minutes at roughly 3000 rpm.

CAUTION: Water must circulate through the lower unit to the engine any time the engine is run to prevent damage to the water

pump in the lower unit. Just five seconds without water will damage the water pump.

The 140 degree stick should melt when you touch it to the lower thermostat housing or on the top cylinder. If it does not melt, the thermostat is stuck in the open position and the engine temperature is too low.

Touch the 170 degree stick to the same spot on the lower thermostat housing or on the top cylinder. The stick should not melt. If it does, the thermostat is stuck in the closed position or the water pump is not operating properly because the engine is running too hot. For service procedures on the cooling system, see Chapter 8.

6-5 FUEL SYSTEM

FUEL GAUGE

The fuel gauge is intended to indicate the quantity of fuel in the tank. As the experienced boatman has learned, the gauge reading is seldom an accurate report of the

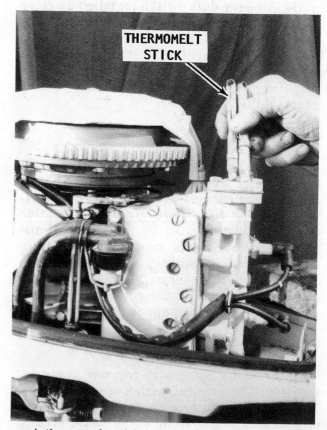

A thermomelt stick is a quick, simple, inexpensive, and fairly accurate method to determine engine running temperature.

fuel available in the tank. The main reason for this false reading is because the boat is rarely on an even keel. A considerable difference in fuel quantity will be indicated by the gauge if the bow or stern is heavy, or if the boat has a list to port or starboard.

Therefore, the reading is usually low. The amount of fuel drawn from the tank is dependent on the location of the fuel pickup tube in the tank. The engine may cutout while cruising because the pickup tube is above the fuel level. Instead of assuming the tank is empty, shift weight in the boat to change the trim and the problem may be solved until you are able to take on more fuel.

FUEL GAUGE HOOKUP

The Boating Industry Association recommends the following color coding be used on all fuel gauge installations:

Black -- for all grounded current-carrying conductors.

Pink -- insulated wire for the fuel gauge sending unit to the gauge.

Red -- insulated wire for a connection from the positive side of the battery to any electrical equipment.

Connect one end of a pink insulated wire to the terminal on the gauge marked **TANK** and the other end to the terminal on top of the tank unit.

Connect one end of a black wire to the terminal on the fuel gauge marked **IGN** and the other end to the ignition switch.

Connect one end of a second black wire to the fuel gauge terminal marked **GRD** and the other end to a good ground. It is important for the fuel gauge case to have a good common ground with the tank unit. Aboard an all-metal boat, this ground wire is not necessary. However, if the control

panel is insulated, or made of wood or plastic, a wire **MUST** be run from the gauge ground terminal to one of the bolts securing the sending unit in the fuel tank, and then from there to the **NEGATIVE** side of the battery.

FUEL GAUGE TROUBLESHOOTING

In order for the fuel gauge to operate properly the sending unit and the receiving unit must be of the same type and preferably of the same make.

The following symptoms and possible corrective actions will be helpful in restoring a faulty fuel gauge circuit to proper operation.

If you suspect the gauge is not operating properly, the first area to check is all electrical connections from one end to the other. Be sure they are clean and tight.

Next, check the common ground wire between the negative side of the battery, the fuel tank, and the gauge on the dash.

If all wires and connections in the circuit are in good condition, remove the sending unit from the tank. Run a wire from the gauge mounting flange on the tank to the flange of the sending unit. Now, move the float up-and-down to determine if the receiving unit operates. If the sending unit does not appear to operate, move the float to the midway point of its travel and see if the receiving unit indicates half full.

If the pointer does not move from the **EMPTY** position one of four faults could be to blame:

1- The dash receiving unit is not properly grounded.

2- No voltage at the dash receiving unit.

3- Negative meter connections are on a positive grounded system.

4- Positive meter connections are on a negative grounded system.

If the pointer fails to move from the **FULL** position, the problem could be one of three faults.

1- The tank sending unit is not properly grounded.

2- Improper connection between the tank sending unit and the receiving unit on the dash.

3- The wire from the gauge to the ignition switch is connected at the wrong terminal.

If the pointer remains at the 3/4 full

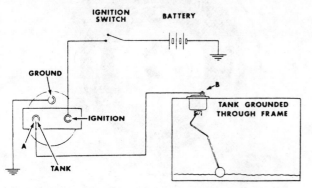

Schematic diagram for a safe fuel tank gauge hookup.

mark, it indicates a 6-volt gauge is installed in a 12-volt system.

If the pointer remains at about 3/8 full, it indicates a 12-volt gauge is installed in a 6-volt system.

Preliminary Inspection

Inspect all of the wiring in the circuit for possible damage to the insulation or conductor. Carefully check:

1- Ground connections at the receiving unit on the dash.

2- Harness connector to the dash unit.

3- Body harness connector to the chassis harness.

4- Ground connection from the fuel tank to the tank floor pan.

5- Feed wire connection at the tank sending unit.

GAUGE ALWAYS READS FULL when the ignition switch is ON:

1- Check the electrical connections at the receiving unit on the dash; the body harness connector to chassis harness connector; and the tank unit connector in the tank.

2- Make a continuity check of the ground wire from the tank to the tank floor pan.

3- Connect a known good tank unit to the tank feed wire and the ground lead. Raise and lower the float and observe the receiving unit on the dash. If the dash unit follows the arm movement, replace the tank sending unit.

GAUGE ALWAYS READS EMPTY when the ignition switch is ON:

Disconnect the tank unit feed wire and do not allow the wire terminal to ground. The gauge on the dash should read FULL.

If Gauge Reads Empty:

1- Connect a spare dash unit into the dash unit harness connector and ground the unit. If the spare unit reads FULL, the original unit is shorted and must be replaced.

2- A reading of EMPTY indicate a short in the harness between the tank sending unit and the gauge on the dash.

If Gauge Reads Full:

1- Connect a known good tank sending unit to the tank feed wire and the ground lead.

2- Raise and lower the float while observing the dash gauge. If the dash gauge follows movement of the float, replace the tank sending unit.

GAUGE NEVER INDICATES FULL

This test requires shop test equipment.

1- Disconnect the feed wire to the tank unit and connect the wire to a good ground through a variable resistor or through a spare tank unit.

2- Observe the dash gauge reading. The reading should be FULL when resistance is increased to about 90 ohms. This resistance would simulate a full tank.

3- If the check indicates the dash gauge is operating properly, the trouble is either in the tank sending unit rheostat being shorted, or the float is binding. The arm could be bent, or the tank may be deformed. Inspect and correct the problem.

6-6 TACHOMETER

An accurate tachometer can be installed on any engine. Such an instrument provides an indication of engine speed in revolutions per minute (rpm). This is accomplished by measuring the number of electrical pulses per minute generated in the primary circuit of the ignition system.

The meter readings range from 0 to 6,000 rpm, in increments of 100. Tachometers have solid-state electronic circuits which eliminate the need for relays or batteries and contribute to their accuracy. The electronic parts of the tachometer, susceptible to moisture, are coated to prolong their life.

Maximum engine performance can only be obtained through proper tuning using a tachometer.

6-7 HORNS

The only reason for servicing a horn is because it fails to operate properly or because it is out of tune. In most cases the problem can be traced to an open circuit in the wiring or to a defective relay.

Cleaning

Crocus cloth and carbon tetrachloride should be used to clean the contact points. **NEVER** force the contacts apart or you will bend the contact spring and change the operating tension.

Check Relay and Wiring

Connect a wire from the battery to the horn terminal. If the horn operates, the problem is in the relay or in the horn wiring. If both of these appear satisfactory, the horn is defective and needs to be replaced.

Before replacing the horn however, connect a second jumper wire from the horn frame to ground to check the ground connection.

Test the winding for an open circuit, faulty insulation, or poor ground. Check the resistor with an ohmmeter, or test the condenser for capacity, ground, and leakage. Inspect the diaphragm for cracks.

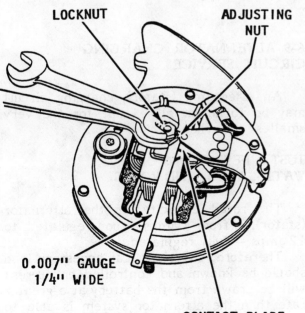

LOCKNUT   ADJUSTING NUT

0.007" GAUGE 1/4" WIDE

CONTACT BLADE INSULATOR

The tone of a horn can be adjusted with a 0.007" feeler gauge, as described in the text. TAKE CARE to prevent the feeler gauge from making contact with the case, or the circuit will be shorted out.

Adjust Horn Tone

Loosen the locknut, and then rotate the adjusting screw until the desired tone is reached. On a dual horn installation, disconnect one horn and adjust each, one-at-a-time. The contact point adjustment is made by inserting a 0.007" feeler gauge blade between the adjusting nut and the contact blade insulator. **TAKE CARE** not to allow the feeler gauge to touch the metallic parts of the contact points because it would short them out. Now, loosen the locknut and turn the adjusting nut down until the horn fails to sound. Loosen the adjusting nut slowly until the horn barely sounds. The locknut **MUST** be tightened after each test. When the feeler gauge is withdrawn the horn will operate properly and the current draw will be satisfactory.

6-8 ELECTRICAL SYSTEM GENERAL INFORMATION

JOHNSON/EVINRUDE WIRING COLOR CODE

At press time, OMC was using the following color code system. We list as "always", but add -- "subject to change".

Red -- always a "hot" lead when outboard cables are connnected to the battery.
Purple -- always a hot lead when key switch is turned to the **ON** or **START** position.
Black -- always a ground lead.
Tan -- always a warning system --low oil or excessive temperature.
Gray -- always a tachometer lead.
Black/Yellow -- always the "Kill" circuit.
Yellow/Red -- always the start circuit.
Purple/White -- always the primer solenoid lead.
Blue/Green, Blue/White, or Green/White -- always the trim/tilt circuits. Lead with Green is **DOWN** circuit; lead with Blue is **UP** circuit.
Yellow, Yellow/Blue, Yellow/Gray -- always the charging circuit.

Probably 75-80% of all Johnson/Evinrude units covered in this manual are started by pulling on a rope. As the manufacturer increased the size and horsepower of the powerheads, it was necessary to replace the rope starter with some form of power cranking system. Today, most small engines are still started by pulling on a rope.

On the larger hp units, an electric cranking motor coupled with a mechanical gear mesh between the cranking motor and the powerhead flywheel, similar to the method used to crank an automobile engine, was added. This system provided an alternate method to the hand cranking rope arrangement. If the electric cranking system is inoperative for any reason, including a dead or weak battery, the powerhead may still be cranked and started by hand. However, on powerheads over 25 hp, considerable strength is required -- perhaps a friend who is a linebacker on an NFL team?

Since the cranking motor requires a large amount of electrical current, it is necessary to have a fully charged battery available for the cranking system. If the boat is equipped with several electrical accessories, such as bait tank with circulating pump, radio, a number of running and accessory lights etc., the charging system must be performing properly to keep the battery charged.

Alternator Charging Circuit

The alternator charging circuit consists of the alternator (stator), rectifier, and the flywheel. This is a direct charge system. A 4 amp non-regulated to 12-amp fully regulated current is constantly produced while the powerhead is operating. A voltage regulator is used with the higher capacity alternator to prevent overcharging the battery.

Small horsepower powerheads -- under say 50hp -- use a hand rewind starter instead of an electric cranking motor. The hand starter may be mounted on top of the powerhead or on the side, depending on the model.

The AC current generated by the alternator (stator), passes through a rectifier where it is changed to DC current to charge the battery.

Choke Circuit

The choke is activated by a solenoid. This solenoid attracts a plunger to close the choke valves. The solenoid is energized when the ignition key is turned to the **START** position and the choke button is depressed, or the key switch is depressed. When using the electric choke, the manual choke **MUST** be in the **NEUTRAL** position.

Only the electric choke is used on the powerheads covered in this manual. On models equipped with a remote control unit, the choke will be activiated if the key is pushed inward while it is being rotated to the **START** position.

Starting Circuit

The starting circuit consists of a cranking motor and a motor-engaging mechanism. A solenoid is used as a heavy-duty switch to carry the heavy current from the battery to the cranking motor. The solenoid is actuated by turning the ignition key to the **START** position. On some models, a pushbutton is used to actuate the solenoid. See Section 6-11 for detailed service procedures on the cranking motor circuit.

6-9 ALTERNATOR CHARGING CIRCUIT SERVICE

An alternator (stator), charging circuit may be found on all except for the very smallest powerheads.

JUST A FEW WORDS ON WATTS, AMPS, AND VOLTS

The rated capacity of the alternator (stator), is from 4 amps -- non-regulated, to 12 amps -- fully regulated.

Therefore, the electrical accessory load should be known and controlled or current will be drawn from the battery at a greater rate than the alternator system is able to produce. Such a negative draw on the battery will result in a run-down condition and failure of the battery to provide the required current to the cranking motor for cranking the powerhead.

To calculate the amperage draw of an accessory, the following simple formula may be used: Amps equals watts divided by volts. $Amps = \dfrac{Watts}{Volts}$

The volts will always be 12. Accessories will usually be given in watts. If the obsolete measurement of candlepower is used, then one candle power is equal to approximately one watt.

Example: A boat has running lights requiring 8 watts; auxiliary lights use 10 watts; and a radio rated at 30 watts.

$Amps = 48\ Watts/12\ volts = 4\ Amps.$

In this case, if all the lights are on and the radio is being used, the total draw on the battery would be 4 amps. If the powerhead is running at 1500 rpm or higher, and the alternator circuit is performing properly by charging the battery with -- say 12-amps, then a net positive gain of 8 amps is being received by the battery.

The circuit consists of the flywheel, generating coils (commonly referred to as the "stator") and the rectifier, including a positive and negative diode. Function of the diodes is to change the alternating current produced to direct current to charge the battery. The diodes permit current to pass in only one direction. Very little service is required for this system. Troubleshooting is confined to work with an ohmmeter. Seldom, if ever, do the magnets in the flywheel lose their magnetism. Problems with the circuit are usually traced to the circuitry, such as a burned coil wire, or defective diodes.

SPECIAL WORDS

A diode in an electrical circuit could be compared with a check valve in a hydraulic system. A check valve will allow fluid to pass in one direction, but close and prevent the fluid from moving in the opposite direction regardless of the pressure buildup. Likewise, a diode will allow electrical current to move in one direction but close the circuit and prevent any current flow in the opposite direction. Therefore, when testing the circuit with an ohmmeter, the meter should indicate current flow in one direction but not in the opposite direction.

TROUBLESHOOTING

Failure in the alternator charging circuit will usually become evident when the battery reaches a run-down condition. To determine why the charging system has failed

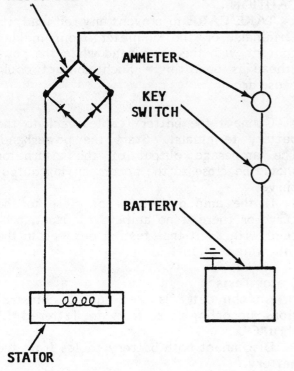

Schematic drawing of the low-output alternator system without a voltage regulator.

Tightening the wingnuts of a cable connection at the battery terminal using a pair of pliers. Tightening the wingnuts by hand is not adequate as explained in the text. The battery terminals MUST be kept clean with an alternator system.

to maintain the battery in satisfactory condition, first, check the condition of the battery and all electrical connections, especially at the battery terminals. Many times, this visual inspection will reveal the problem area.

Verify the battery cables have been **PROPERLY** connected. If the battery has not been connected properly the diodes in the circuit will be blown instantly. Determine the number of accessories connected, their draw, and the capability of the system to maintain the battery in a fully charged condition.

SAFETY WORD

Before conducting any troubleshooting tests or actual service work, the battery cables should be disconnected at the battery terminals. Battery current is **NOT** required for any of the following troubleshooting tests. This safety measure will prevent accidental current from reaching a component resulting in possible damage to the part or personal injury.

ILLUSTRATIONS

The illustrations in this section have been identified by a letter (**"A"** thru **"N"**). The illustration supporting the text will have reference to the same letter designation.

Alternator Output Test

Seldom does the stator cause problems in the charging circuit. However, if other

RECTIFIER Ⓐ

checks have been performed and the stator is suspected as the problem area, first make a careful visual check of the stator for physical damage. If the visual inspection fails to indicate the problem, proceed as follows to test the stator:

NOTE

The alternator output test **MUST** be performed with the outboard mounted in a body of water or in a test tank. **NEVER** attempt to make the test with a flush attachment connected to the lower unit.

The battery cables will be disconnected before making a disconnect in the circuit and then secured back to the battery terminals. The battery should **NOT** be fully charged when starting this test.

Mount the outboard in a test tank or other body of water.

Disconnect both battery cables at the battery.

Remove the red lead from the rectifier at the terminal board.

Connect a 0-40 amp ammeter in series with the rectifier Red lead and the wiring harness Red lead, as depicted in Illustration **"A"**.

CAUTION

TAKE CARE to prevent any red lead, its terminals, or the ammeter from making contact with the powerhead while the powerhead is operating. Such contact could cause arcing.

Connect the battery cables back to the battery terminals. Start the powerhead. The amperage output of the alternator should be close to the accompanying output curve.

If the amp output is not close to the curve or there is no amperage output, proceed with additional testing outlined in the following sections.

Stator Tests

An ohmmeter is used and the illustrations supporting these tests are lettered **"B"** thru **"E"**.

Disconnect both battery cables from the battery.

Disconnect all leads from the stator at the terminal board.

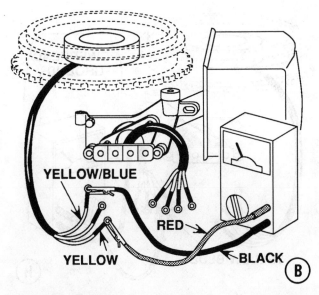

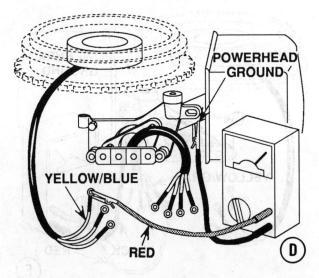

Resistance Test

Set the ohmmeter to the low ohm scale. Connect the meter Black lead to the Yellow/Blue stator lead. Make contact with the Red meter lead to the Yellow stator lead and observe the meter reading, illustration **"B"**.

The meter should indicate 0.50 to 0.60 ohms.

Make contact with the Red meter lead to the Yellow/Gray stator lead, illustration **"C"**.

The meter should indicate 0.50 to 0.60 ohms.

Stator Grounding Test No. 1

Disconnect both meter leads as used in the resistance tests.

Set the ohmmeter to the high ohm scale. Connect the Black meter lead to a good ground on the powerhead. Make contact with the Red meter lead to the Yellow/Blue stator lead, illustration **"D"**.

The ohmmeter should reveal a very high reading. If the ohmmeter indicates a low reading, the stator circuit is grounded.

Stator Grounding Test No. 2

Leave the Black meter lead connected to a good ground on the powerhead.

Make contact with the Red meter lead to the Yellow stator lead and then to the Yellow/Gray stator lead, illustration **"E"**.

In each case, the ohmmeter should reveal a very high reading. If the ohmmeter indicates a low reading, the stator circuit is grounded.

If the stator circuit fails any of the foregoing tests, either the stator lead/s are grounded or the stator is grounded. The lead/s **MUST** be repaired or the stator replaced.

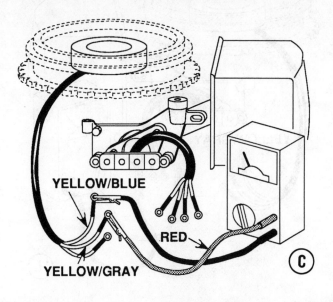

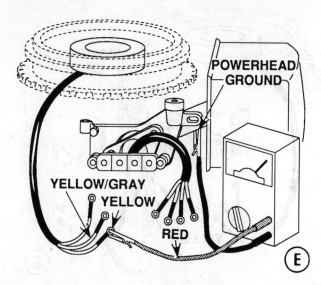

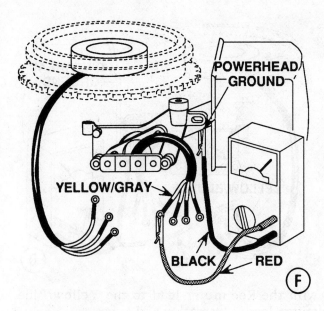

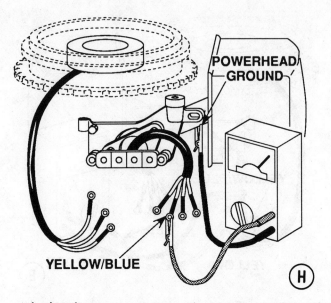

Rectifier Tests

An ohmmeter is used and the illustrations supporting these tests are lettered "F" thru "L".

Disconnect both battery cables from the battery.

Disconnect all leads from the rectifier at the terminal board.

Test No. 1

Set the ohmmeter to the low ohm scale. Connect the Black meter lead to a good ground on the powerhead. Connect the Red meter lead to the Yellow/Gray rectifier lead, illustration "F".

Note the meter reading. Now, if the ohmmeter has a reverse polarity button, depress the button and observe the meter reading. If the meter does not have the

polarity button, reverse the meter leads -- Red to a good ground on the powerhead and the Black to the Yellow/Gray lead and observe the meter reading.

If the meter idicates a high reading with the meter leads connected one way and a low reading when the leads are reverse or the polarity button is depressed, the diode in the rectifier checks out Okay -- allowing current to pass in only one direction. If both readings are high or both readings are low, the diode is defective.

Test No. 2

With the Black meter lead connected to a good ground on the powerhead and the ohmmeter still on the same scale, make contact with the Red meter lead to the

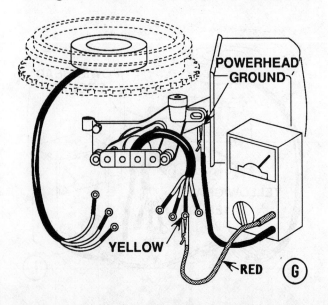

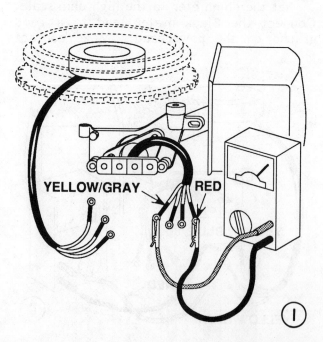

rectifier Yellow lead, illustration "G". Note the meter reading. Depress the polarity button or reverse the meter leads and again note the meter reading.

As with Test No. 1, the meter should indicate a high reading with one connection and a low reading when the connections are reversed.

Test No. 3

With the Black meter lead connected to a good ground on the powerhead, make contact with the Red meter lead to the rectifier Yellow/Blue lead, illustration "H". Depress the polarity button or reverse the meter leads and again observe the meter reading.

As with Tests 1 and 2, a high reading should be indicated in one direction and a low reading when the polarity button is depressed or the leads reversed.

Test No. 4

Connect the Black meter lead to the rectifier Red lead. Make contact with the Red meter lead to the rectifier Yellow/Gray lead, illustration "I". Note the meter reading. Depress the polarity button or reverse the meter leads and again note the meter reading.

A high meter reading with the meter leads connected in one direction and a low meter reading when the polarity button is depressed or the leads are reversed means the diode is satisfactory.

If both meter readings are high or both meter readings are low -- the diode is defective.

Test No. 5

Connect the meter leads between the rectifier Red lead and the rectifier Yellow lead, illustration "J". Note the meter reading. Reverse the leads or depress the polarity button and again note the meter reading.

If the meter indicates a high reading in one direction and a low reading in the opposite direction, the diode is satisfactory. A low reading or high reading in both directions means the diode is defective.

Test No. 6

Connect the meter leads between the rectifier Red lead and the rectifier Yellow/Blue lead, illustration "K". Note the reading. Reverse the leads or depress the

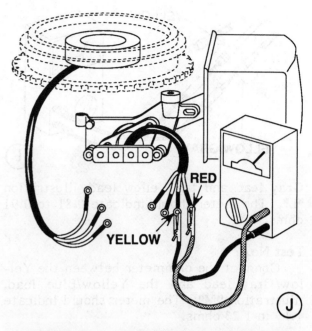

polarity button and again note the reading.

As with the previous test a high reading in one direction and a low reading in the opposite direction indicates the diode is satisfactory. If both readings are low or both readings are high, the diode is defective and must be replaced.

AC Lighting Coil Tests

An ohmmeter is used and the illustrations supporting these tests are lettered "L" thru "N".

Test No. 1

Set the ohmmeter to the low ohm scale. Connect the meter between the Yellow/-

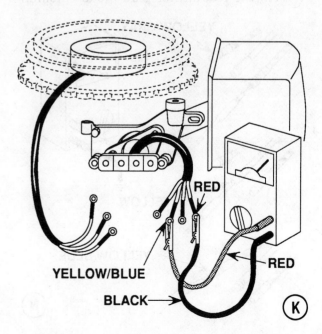

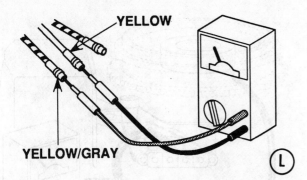

Gray lead and the Yellow lead, illustration **"L"**. The meter should indicate 0.81 to 0.91 ohms.

Test No. 2

Connect the ohmmeter between the Yellow/Gray lead and the Yellow/Blue lead, illustration **"M"**. The meter should indicate 1.19 to 1.23 ohms.

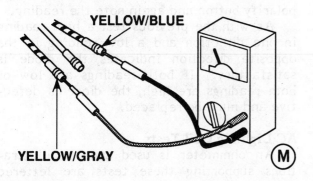

Test No. 3

Set the ohmmeter to the high ohm scale. Connect the black meter lead to a good ground on the powerhead.

With the Red meter lead make momentary contact with each of the connector pins, Illustration "N".

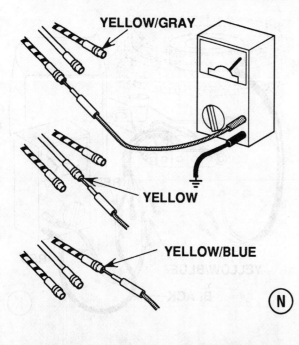

tary contact with each of the connector pins, Illustration **"N"**.

The ohmmeter should indicate a high ohm reading.

If the ohmmeter reading is low, repair the grounded coil lead or replace the stator assembly.

6-10 CHOKE CIRCUIT SERVICE

This short section provides instructions to test the choke circuit. If the system fails the test, the attaching hardware can be removed and the choke assembly replaced.

Functional Test No. 1

Start the engine and allow the powerhead to reach normal operating temperature. Set the powerhead rpm at about 2000 rpm.

Now, while the powerhead is operating, -- electric start models -- push the key inward. Powerhead should run too rich and the rpm drop about 1000 rpm. While the powerhead is operating -- manual start models -- stroke the primer knob. Powerhead should run too rich and speed drop about 1000 rpm.

If solenoid is operating properly --electric choke is in satisfactory condition. If the system fails this test proceed with Test No. 2.

CAUTION

Raw fuel will be discharged from a disconnected fuel line during this test. Therefore, take adequate precautions to catch fuel in a suitable container.

Functional Test No. 2

The solenoid has three fittings. The two small fittings are for discharge lines and the one larger fitting is for the fuel inlet (feed) line. One of the discharge lines is connected to the top of the carburetor behind the throttle plate. With the powerhead **NOT** operating, disconnect the small discharge line between the primer solenoid and the carburetor -- at the carburetor. The line will have enough slack to feed fuel into a container.

Now, squeeze the primer bulb by hand until the bulb is firm. Engage the primer solenoid by pushing in on the key or by manually opening the valve.

Fuel should be discharged from the disconnected fuel line into the container.

If fuel is discharged when the valve is opened manually, the fuel lines are clear. If fuel is discharged when the key is pushed in, both fuel flow and the electrical circuit are verified as in good condition.

If fuel is not discharged when the key is pushed in, proceed with the following electrical circuit test.

Choke Circuit Testing

1- The choke circuit may be quickly tested to determine if it is functioning properly as follows:

a- Obtain an ohmmeter.

b- Connect the black meter lead to an unpainted portion of the engine block for a good ground.

c- Connect the red meter lead to the choke terminal.

d- Test the circuit using the Rx1 scale of the ohmmeter. A satisfactory reading is approximately 4 to 7 ohms.

e- After the test is completed, check to be sure the choke plunger is pulled into the choke solenoid.

If satisfactory results are obtained from the tests -- the solenoid is functioning properly. If the solenoid fails the test, the solenoid **MUST** be replaced -- for efficient powerhead operation.

6-11 CRANKING MOTOR CIRCUIT SERVICE

CIRCUIT DESCRIPTION

As the name implies, the sole purpose of the cranking motor circuit is to control operation of the cranking motor to crank the powerhead until the unit is operating. The circuit includes a solenoid or magnetic switch to connect or disconnect the cranking motor from the battery. The operator controls the switch with a pushbutton or key switch.

A cutout switch is installed in the system to prevent starting the powerhead if the throttle is advanced too far, beyond idle speed. When the throttle is advanced, the cranking motor solenoid is not grounded and the cranking motor will not rotate. On electric shift models, a cutout switch is installed in the circuit to permit operation of the cranking motor **ONLY** if the shift control lever is in **NEUTRAL**. This switch is a safety device to prevent accidental engine start when the outboard is in gear.

CRANKING MOTOR DESCRIPTION

Prestolite and Bosch cranking motors are used on the Johnson/Evinrude powerheads covered in this manual. These two cranking motors are used interchangeably. There-

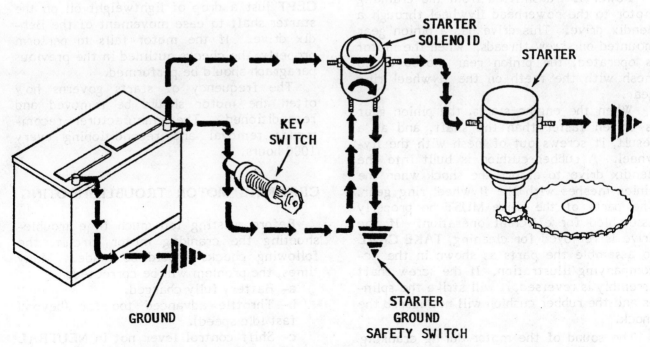

STARTER SOLENOID

STARTER

KEY SWITCH

GROUND

STARTER GROUND SAFETY SWITCH

Functional diagram depicting current flow when the key switch is turned to the ST ART position.

fore, two powerheads of the same horse-power and model year may have either the Prestolite or Bosch unit installed.

Marine cranking motors are very similar in construction and operation to the units used in the automotive industry. All marine cranking motors use the inertia-type drive assembly. This type assembly is mounted on an armature shaft with external spiral splines which mate with the internal splines of the drive assembly.

The cranking motor is a series wound electric motor which draws a heavy current from the battery. It is designed to be used only for short periods of time to crank the powerhead for starting. To prevent over-heating the motor, cranking should not be continued for more than 30 seconds without allowing the motor to cool for at least three minutes. Actually, this time can be spent in making preliminary checks to determine why the powerhead fails to start.

Most cranking motors operate in much the same manner and the service work involved in restoring a defective unit to service is almost identical. Therefore, the information in this chapter is grouped together for the major components of the cranking motor under separate headings. Differences, where they occur, between the manufacturers, are clearly indicated.

Theory of Operation

Power is transmitted from the cranking motor to the powerhead flywheel through a Bendix drive. This drive has a pinion gear mounted on screw threads. When the motor is operated, the pinion gear moves up to mesh with the teeth on the flywheel ring gear.

When the engine starts, the pinion gear is driven faster than the shaft, and as a result, it screws out of mesh with the fly-wheel. A rubber cushion is built into the Bendix drive to absorb the shock when the pinion meshes with the flywheel ring gear. The parts of the drive **MUST** be properly assembled for efficient operation. If the drive is removed for cleaning, **TAKE CARE** to assemble the parts as shown in the ac-companying illustration. If the screw shaft assembly is reversed, it will strike the splines and the rubber cushion will not absorb the shock.

The sound of the motor during cranking is a good indication of whether the cranking

motor is operating properly or not. Natural-ly, temperature conditions will affect the speed at which the cranking motor is able to crank the powerhead. The speed of cranking a cold powerhead will be much slower than when cranking a warm powerhead. An ex-perienced operator will learn to recognize the favorable sounds during cranking under various conditions.

Faulty Symptoms

If the cranking motor spins, but fails to crank the powerhead, the cause is usually a corroded or gummy Bendix drive. The drive should be removed, cleaned, and given an inspection.

If the cranking motor cranks the power-head too slowly, the following are possible causes and the corrective actions that may be taken:

a- Battery charge is low. Charge the battery to full capacity.

b- High resistance connections at the battery, solenoid, or motor. Clean and tighten all connections.

c- Undersize battery cables. Replace cables with sufficient size.

d- Battery cables too long. Relocate the battery to shorten the run to the cranking motor solenoid.

Maintenance

The cranking motor does not require periodic maintenance or lubrication **EX-CEPT** just a drop of lightweight oil on the starter shaft to ease movement of the Ben-dix drive. If the motor fails to perform properly, the checks outlined in the previous paragraph should be performed.

The frequency of starts governs how often the motor should be removed and reconditioned. The manufacturer recom-mends removal and reconditioning every 1000 hours.

CRANKING MOTOR TROUBLESHOOTING

Before wasting too much time trouble-shooting the cranking motor circuit, the following checks should be made. Many times, the problem will be corrected.

a- Battery fully charged.

b- Throttle advanced too far (beyond fast idle speed).

c- Shift control lever not in **NEUTRAL** (electric shift models only).

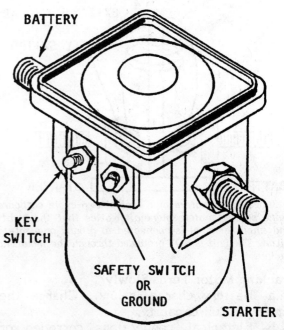

Two handy instruments for use in checking the generating circuit (left) and the starter motor circuit (right). These meters do not require any wire connections. A reading will be obtained by simply placing the meter on the line.

d- All electrical connections clean and tight.

e- Wiring in good condition, insulation not worn or frayed.

f- One of the cutout switches may be defective.

Two more areas may cause the powerhead crankshaft to rotate slowly even though the cranking motor circuit is in excellent condition: A tight or "frozen" powerhead or water in the lower unit causing the bearings to tighten up.

The following troubleshooting procedures are presented in a logical sequence, with the most common and easily corrected areas listed first in each problem area. The connection number refers to the numbered positions in the accompanying illustrations.

Functional diagram of a starter motor solenoid. Notice the separate terminal for a ground wire. This solenoid is NOT grounded through the mounting bracket.

Perform the following quick checks and corrective actions for the problems listed.

TESTING

SAFETY WORD

Before making any test of the cranking system, disconnect the spark plug leads at the spark plugs to prevent the powerhead from possibly starting during the test and causing personal injury.

The following tests are to be performed according to the faulty condition described. The numbers referenced in the steps are correlated with numbers on the accompanying circuit diagram to identify exactly where the connection or test is to be made.

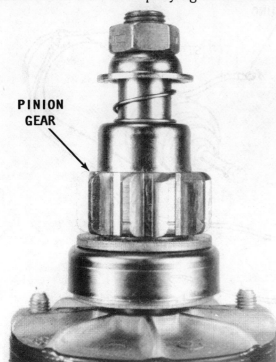

Typical Bendix spring arrangement on a starter motor. A small amount of oil on the shaft in the spring area will prolong satisfactory operation.

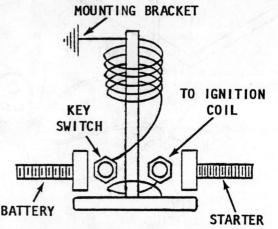

Functional diagram of a "slave-type" starter motor solenoid used on four-cycle engine installations. This solenoid CANNOT be used on a two-cycle engine.

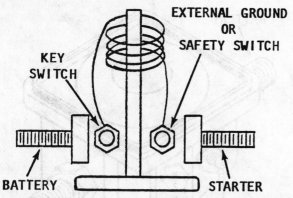

Functional diagram of a typical two-cycle outboard engine starter motor solenoid. Notice that the right-hand small terminal is connected to ground or a safety switch. The unit is NOT grounded through the mounting bracket.

Cranking Motor Turns Slowly

a- Battery charge is low. Charge the battery to full capacity.

b- Electrical connections corroded or loose. Clean and tighten.

c- Defective cranking motor. Perform an amp draw test. Lay an amp draw gauge on the cable leading to the cranking motor No. 5. Turn the key to the START position and attempt to crank the engine. If the gauge indicates an excessive amperage draw, the cranking motor MUST be replaced or rebuilt.

Cranking Motor Drive Fails To Rotate
Voltage Check

a- Check the voltage at No. 2, the battery, and ground.

b- If satisfactory voltage is indicated at the battery, check the voltage at No. 3, the positive side of the cranking motor solenoid. Weak, or no voltage at this point indicates corroded battery terminals, poor connection at the solenoid, or defective wiring between the battery and the solenoid.

c- Test the voltage at No. 4, the key. A full 12-volt reading should be registered at the key. Weak or no voltage at the key indicates a poor connection at the solenoid, or a broken wire between the cranking motor solenoid and the key.

d- If satisfactory voltage is indicated during Steps a, b, and c, connect a volt meter at No. 5 and ground, and then turn the key switch to the START position. If 9-1/2 or more volts is registered at No. 5 and the cranking motor still fails to operate, the motor is defective and requires service. If

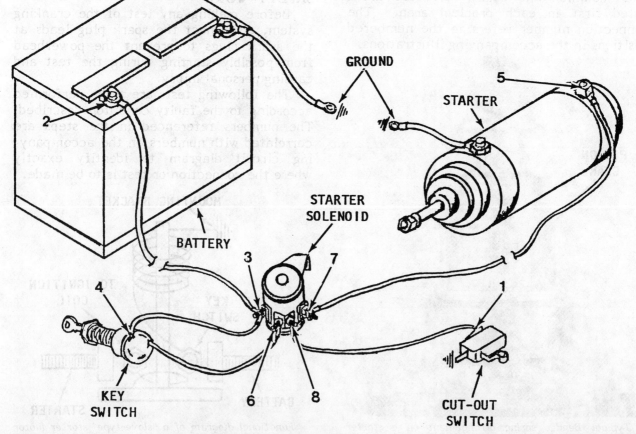

Diagram of hookup for making the various tests outlined in the text. This illustration and the numbers shown are to be used when testing the starter motor components.

voltage is **NOT** present at No. 5, proceed to the next section, Testing Cranking Motor Solenoid.

Testing Cranking Motor Solenoid

FIRST THESE WORDS

The cranking motor solenoid is actually nothing more than a switch between the battery and the cranking motor. Several types of solenoids are used and many appear similar. **NEVER** attempt to use an automotive-type solenoid in a marine installation. Such practice will lead to more problems than can be imagined. An automotive-type solenoid has a completely different internal wiring circuit. If such a solenoid is connected into the cranking motor system, and the system is activated, current will be directed to ground. The wires will be burned and the cutout switch will be burned and rendered useless. Therefore, when installing replacement parts in the cranking motor or other circuits on a marine installation, always take time to obtain parts from a **MARINE** outlet to ensure proper service and to prevent damage to other expensive components.

a- Remove the heavy cranking motor cable at No. 5, at the cranking motor. This cable **MUST** be disconnected prior to performing this test to prevent the cranking motor from turning and cranking the powerhead. Connect a voltmeter to No. 6 (the cranking motor solenoid), and ground. Turn the key to the **START** position. The meter should indicate 12 volts. If voltage is not present at No. 6, the key switch is defective, or the wire is broken between the key switch and the cranking motor solenoid.

b- If voltage is present at No. 6, connect a voltmeter at No. 3 and at No. 7. Connect one end of a jumper wire to No. 2, the positive terminal of the battery and **MOMENTARILY** make contact with the other end at No. 6, the cranking motor solenoid. If voltage is indicated through the cranking motor solenoid, the solenoid is satisfactory and the problem has been corrected while making the tests. Sometimes, when working with electrical circuits, corrective action has been taken almost accidently, a bad connection has been made good, etc. If the solenoid test failed, it does not necessarily mean the solenoid is defective. The solenoid may not be properly grounded through the cutout switch. Therefore, the cutout

switch may be defective and should be checked as outlined later in this section.

c- With the voltmeter still connected at No. 3 and No. 7, connect one end of a jumper wire at No. 8, the cranking motor solenoid, and the other lead to a good ground. Connect a second jumper wire at No. 2, the positive terminal of the battery, to No. 6, the cranking motor solenoid. The voltmeter should indicate voltage is present. If voltage is not present, the cranking motor solenoid is defective and **MUST** be replaced.

Testing Throttle Advance Cutout Switch

Remove the existing wire from No. 1, the switch terminal. Connect one probe lead of an ohmmeter to the terminal. Connect the other test probe lead to a good ground. Depress the switch button and the ohmmeter should indicate continuity. If continuity is not indicated, the switch is defective and **MUST** be replaced. Connect the heavy cable at No. 5, the cranking motor. The cutout switch is located in the shift box and functions in the same manner as the cutout switch mounted on the powerhead.

6-12 CRANKING MOTOR DRIVE GEAR SERVICE

CRANKING MOTOR REMOVAL

Before beginning any work on the cranking motor, disconnect the positive (+) lead from the battery terminal. Remove the hood. Disconnect the red cable at the cranking motor terminal.

Remove the 1/2" bolt securing the motor bracket. This bolt is located on the starboard side of the powerhead just above the carburetor. Remove the three 7/16" bolts (or nuts in some cases) securing the cranking motor bracket to the powerhead. Remove the cranking motor and bracket together.

DRIVE GEAR DISASSEMBLING

Two types of drive gear arrangements are used on cranking motors for powerheads covered in this manual. One has a spring and spring retainer installed above the drive gear, and then the nut securing these parts on the drive shaft. This unit is very simple in construction and therefore, the service procedures, including disassembling and assembling are not difficult or involved. This

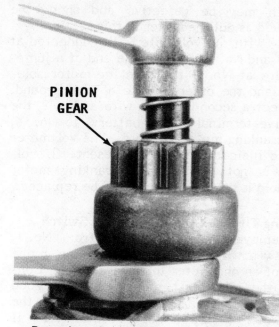

PINION GEAR

Removing the starter drive using an open-end wrench and a box-end wrench to remove the shaft nut.

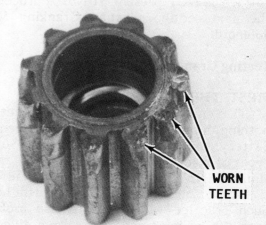

WORN TEETH

Worn teeth on a Bendix drive gear. This gear is no longer fit for service.

drive gear is referred to as Type I. The other is very similar except for the arrangement of parts on the armature shaft. This second unit is referred to as Type II.

To determine which cranking motor drive type is being serviced observe the unit and make a comparison with the two exploded illustrations in this section, especially the pinion gear and the screw shaft.

DISASSEMBLING
Type I Drive Gear

Prevent the armature from turning by holding it with the proper size wrench on the hex nut provided for this purpose on the opposite end from the shaft nut. If the hex

nut is not provided, hold the drive assembly with a pair of water pump pliers. Remove the shaft nut, spring retainer, spring, and then the drive assembly. The shaft nut should be replaced and **NOT** used a second time. The manufacturer **STRONGLY** recommends against using any type of self-locking nut on the shaft.

The exploded drawing accompanying this section will be helpful in assembling the cranking motor in the proper sequence.

CLEANING AND INSPECTING

Inspect the drive gear teeth for chips, cracks, or a broken tooth. Check the spline inside the drive gear for burrs and to be sure the drive gear moves freely on the armature shaft. Check to be sure the return spring is flexible and has not become distorted. Clean the armature shaft with crocus cloth.

PINION GEAR

Removing the drive gear using a pair of pliers to hold the gear and a box-end wrench to remove the shaft nut.

NUT

RETAINER

SPRING

PINION GEAR

Installation sequence of parts when assembling the drive gear.

ASSEMBLING
Type I Drive Gear

Begin by assembling the following parts in the order given. The accompanying illustration, on the previous page, will be most helpful in assembling the parts in the proper sequence. First, slide the pinion gear onto the shaft, then the spring, spring retainer, and finally a **NEW** locking nut. Prevent the armature shaft from turning by holding it with the proper size wrench on the hex nut provided for this purpose on the opposite end from the shaft nut. If the armature hex nut is not provided, hold the drive assembly with a pair of water pump pliers. Tighten the shaft nut securely.

To test the complete cranking motor, proceed directly to Section 6-15.

To install the cranking motor onto the powerhead, if no further work is to be performed, proceed directly to Section 6-16.

DISASSEMBLING

Type II Drive Gear with Rubber Cushion

First, remove the nut from the end of the armature shaft. Scratch a mark on the top of the screw shaft and one on the top of the pinion as an aid during assembling. These marks will identify the top of both parts. Next, remove the following parts in the sequence given: the pinion stop; anti-drift spring; sleeve; screw shaft cap; screw shaft; pinion; thrust washer; cushion cap; cushion; and the cushion retainer. The exploded drawing accompanying this section will be helpful in assembling the starter motor in the proper sequence.

CLEANING AND INSPECTING

Inspect the drive gear teeth for chips, cracks, or a broken tooth. Check the spline inside the drive gear for burrs and to be sure the drive gear moves freely on the armature shaft. Check to be sure the return spring is flexible and has not become distorted. Inspect the rubber cushion for cracks and for signs of oil on the cushion. Clean the armature shaft with crocus cloth.

ASSEMBLING

Type II Drive Gear with Rubber Cushion

Begin by assembling the following parts in the sequence given: The accompanying illustration will be most helpful in assemb-ling the parts in the proper sequence.

First, slide the cushion retainer down onto the drive end cap, with the shoulder facing **UPWARD**. Next, slide the cushion down the armature shaft and seat it over the shoulder of the retainer. Install the cushion cap over the cushion. Slide thrust washer down the armature shaft onto the top of the cap. Rotate the screw shaft clockwise into the pinion, and then slide the pinion and screw shaft down the armature shaft onto the thrust washer. Install the cap over the end of the screw shaft.

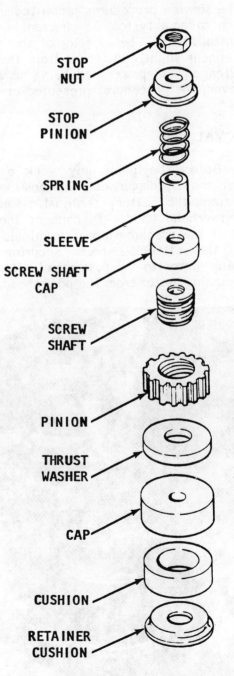

STOP NUT

STOP PINION

SPRING

SLEEVE

SCREW SHAFT CAP

SCREW SHAFT

PINION

THRUST WASHER

CAP

CUSHION

RETAINER CUSHION

Exploded drawing of the Type II drive gear.

Slide the following parts onto the armature shaft in the order given: the sleeve; spring; pinion stop washer; and finally thread the nut onto the end of the shaft. Tighten the nut securely.

To test the complete starter motor, proceed directly to Section 6-15.

To install the starter motor onto the engine, if no further work is to be performed, proceed directly to Section 6-16.

6-13 BOSCH CRANKING MOTOR SERVICE

The service procedures presented in this section cover a typical Bosch cranking motor installation. The exterior of the motor may appear slightly different from the unit installed on the powerhead being serviced. However, the procedures presented are valid.

REMOVAL

1- Before beginning any work on the starter motor, disconnect the positive (+) lead from the battery terminal. Remove the powerhead hood. Disconnect the red cable at the cranking motor terminal. Remove the attaching bolts securing the cranking motor to the powerhead. Remove the cranking motor from the powerhead.

GOOD NEWS

If the only motor repair necessary is replacement of the brushes, the drive gear does not have to be removed. All cranking motors have thru-bolts securing the upper and lower cap to the field frame assembly. In all cases both caps have some type of mark or boss. These marks are used to properly align the caps with the field frame assembly.

ALIGNMENT MARK

BRUSHES

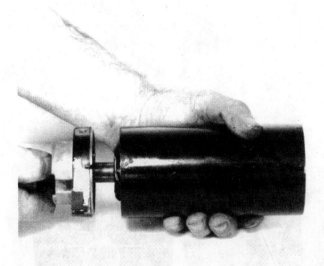

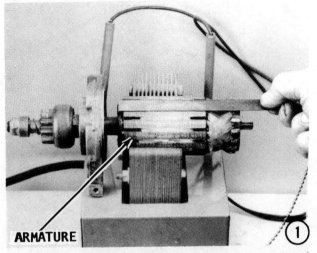

the armature core and the other probe lead on the commutator bar. If the lamp lights, or the meter indicates continuity, the armature is grounded and **MUST** be replaced.

Checking the Commutator Bar

3- Check between or check bar-to-bar as shown in the accompanying illustration. The test light should light, or the meter should indicate continuity. If the commutator fails the test, the armature **MUST** be replaced.

Turning the Commutator

4- True the commutator, if necessary, in a lathe. **NEVER** undercut the mica because the brushes are harder than the insulation. Undercut the insulation between the commutator bars 1/32" to the full width of the insulation and flat at the bottom. A triangular groove is not satisfactory. After the under cutting work is completed, clean out the slots carefully to remove dirt and copper dust. Sand the commutator lightly with No. "00" sandpaper to remove any burrs left

DISASSEMBLING

2- Observe the caps and find the identifying mark or boss on each. If the marks are not visible, make an identifying mark prior to removing the thru-bolts as an essential aid during assembling. Remove the thru-bolts from the bracket and the starter motor.

3- Use a small hammer and **CAREFULLY** tap the lower cap free of the starter motor. On the Bosch starter motor, the brushes are mounted in the end cap. Take care not to lose the four springs and four brushes when the end cap is removed and the brushes pop out.

4- Pull on the armature shaft from the drive gear end and remove it from the field frame assembly.

ARMATURE TESTING

Testing for a Short

1- Position the armature on a growler, then hold a hacksaw blade over the armature core. Turn the growler switch to the **ON** position. Slowly rotate the armature. If the hacksaw blade vibrates, the armature or commutator has a short. Clean the grooves between the commutator bars on the armature. Perform the test again. If the hacksaw blade still vibrates during the test, the armature has a short and **MUST** be replaced.

Testing for a Ground

2- Obtain a test lamp or continuity meter. Make contact with one probe lead on

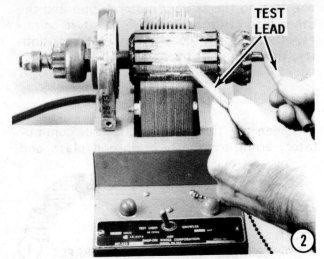

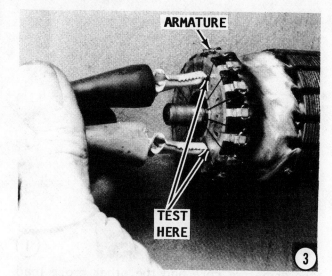

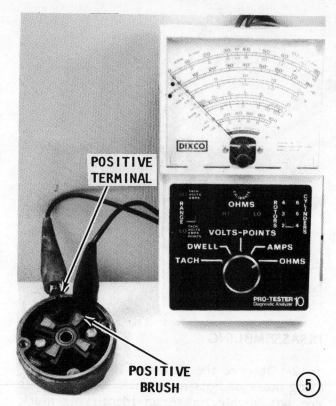

from the undercutting. Check the armature a second time on the growler for possible short circuits.

Positive Brushes

5- The positive brushes can always be identified as the brush with the lead connected to the starter terminal.

Obtain an ohmmeter. Connect one lead of the meter to the positive terminal of the cap and the other lead alternately to the positive brushes. The ohmmeter **MUST** indicate continuity between the brush and the terminal. If the meter indicates any resistance, check the lead to the brush and the lead to the positive terminal solder connection. If the connection cannot be repaired, the brush **MUST** be replaced.

Negative Brush

6- The negative brush can always be identified because the lead is connected to the starter cap.

Obtain an ohmmeter. Make contact with one lead to the starter motor frame and the other lead alternately to the negative brushes. If the meter does not indicate continuity, the field coils open and **MUST** be replaced.

CLEANING AND INSPECTING

Clean the field coils, armature, commutator, armature shaft, brush-end plate and

drive-end housing with a brush or compressed air. Wash all other parts in solvent and blow them dry with compressed air.

Inspect the insulation and the unsoldered connections of the armature windings for breaks or burns.

Perform electrical tests on any suspected defective part, according to the procedures outlined in Section 6-15.

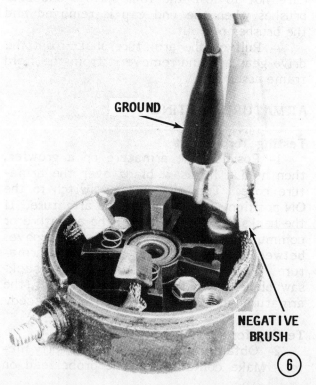

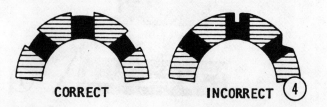

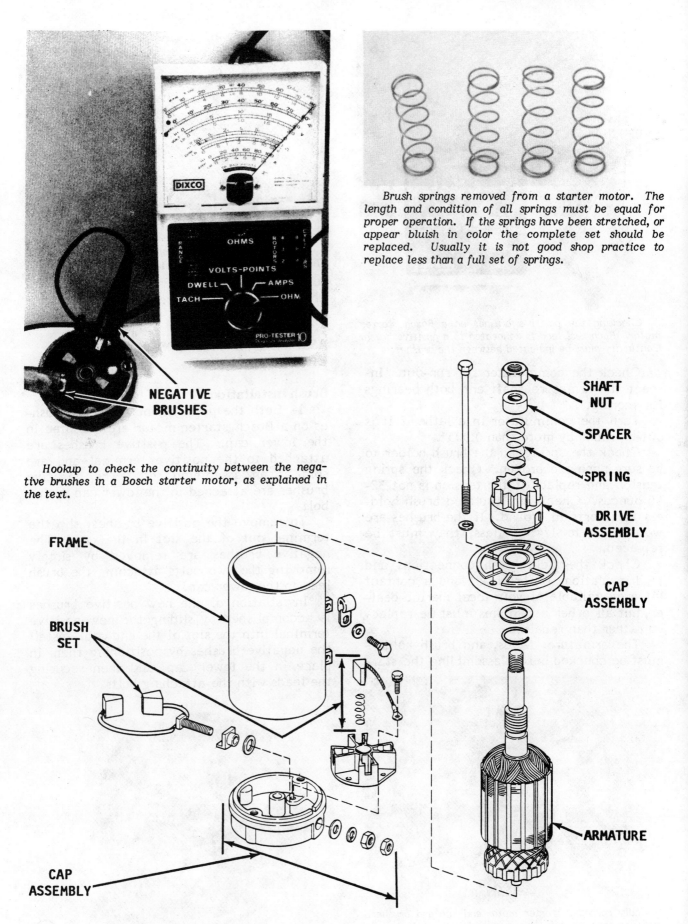

Brush springs removed from a starter motor. The length and condition of all springs must be equal for proper operation. If the springs have been stretched, or appear bluish in color the complete set should be replaced. Usually it is not good shop practice to replace less than a full set of springs.

Hookup to check the continuity between the negative brushes in a Bosch starter motor, as explained in the text.

NEGATIVE BRUSHES

SHAFT NUT

SPACER

SPRING

DRIVE ASSEMBLY

CAP ASSEMBLY

ARMATURE

FRAME

BRUSH SET

CAP ASSEMBLY

Exploded drawing showing arrangement of principle Bosch starter motor parts.

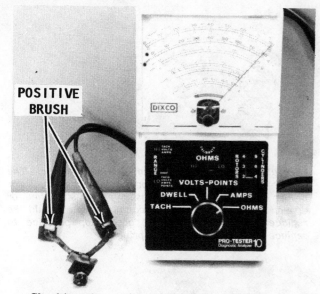

Checking the positive brushes on a Bosch starter motor. Each test lead is connected to a positive brush. Continuity must be indicated between the brushes.

Check the commutator for run-out. Inspect the armature shaft and both bearings for scoring.

Turn the commutator in a lathe if it is out-of-round by more than 0.005".

Check the springs in the brush holder to be sure none are broken. Check the spring tension and replace if the tension is not 32-40 ounces. Check the insulated brush holders for shorts to ground. If the brushes are worn down to 1/4" or less, they must be replaced.

Check the field brush connections and lead insulation. A brush kit and a contact kit are available at your local marine dealer, but all other assemblies must be replaced rather than repaired.

The armature, fields, and brush holders must be checked before assembling the star-

Badly corroded starter motor end cap and brushes. Such damage can only be corrected with replacement parts.

Cracked lower end cap of a Bosch starter motor. This damage was caused by salt water corrosion.

ter motor. See the testing section in this chapter for detailed procedures to test the starter motor.

ASSEMBLING THE BOSCH CRANKING MOTOR

Brush Installation

1- Both the positive and negative brushes on a Bosch starter motor are mounted in the lower cap. The positive brushes are attached to the positive terminal and are sold as an assembled set. The negative brushes are attached to the lower cap with a bolt.

To remove the positive brushes, slip the terminal out of the slot in the cap. The negative brushes are removed by simply removing the two bolts attaching the brush lead to the lower cap.

Installation of the new positive brushes is accomplished by sliding the new positive terminal into the slot of the end cap. Install the negative brushes by positioning them in place in the lower cap, and then securing the leads with the attaching bolts.

Assembling a Bosch Using Special Tool

Make a tool as shown in the accompanying illustration to prevent the brushes from being damaged during installation of the commutator end cap. If a special tool is not possible, see the next section, Assembling a Bosch Without a Special Tool.

1- Slide the brush springs into the brush holders, and then install the positive and negative leads. Position the special tool over the cap and brushes, to hold the brushes in place.

2- Clamp the drive gear in a vise equipped with soft jaws and with the drive gear down. Lower the frame assembly over the armature. Align the marks on the frame assembly with the marks on the upper end cap.

3- Position the lower end cap onto the frame assembly. Lower the cap as far as it will go, and then remove the special tool. Now, align the mark on the cap with the mark on the frame, and then install the thru-bolts and tighten them securely.

4- Place the starter motor on the floor. To test the operation of the motor, first connect one lead from a set of jumper cables to the positive terminal of a battery. Connect the other end of the same lead to the positive terminal of the motor. Connect one end of the second lead of the jumper cables to the negative terminal of the battery.

Now hold the motor firmly on the floor with one foot and at the same time, momentarily make contact with the other end of the second jumper lead to the motor case. The pinion gear should spin rapidly.

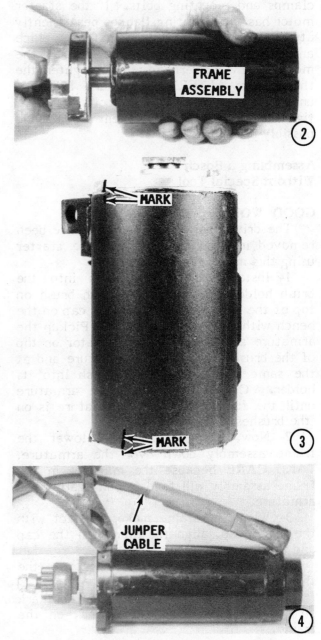

Special tool required to install the end cap on a Bosch starter motor. If this tool is not available, special instructions and illustrations are included later in this section to accomplish the end cap installation work.

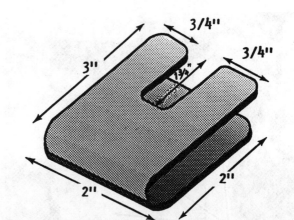

BOLT

⑤

5- Install the starter motor onto the powerhead and secure it in place with the clamps and mounting bolts. If the starter motor has the mounting flanges permanently attached, then position the motor in place and start the three bolts to attach the motor to the powerhead. Now, tighten the three bolts **ALTERNATELY** and **EVENLY** until all bolts are tight. The bolts **MUST** be tightened alternately to prevent binding and possibly bending the bolts.

Assembling a Bosch Without Special Tool

GOOD WORDS

The drive gear assembly must have been removed in order to assemble the starter using this method.

1- Install the brush springs into the brush holder, and then place each brush on top of the springs. Lay the lower cap on the bench with the brush facing up. Pickup the armature and place the commutator on top of the brushes. Lower the armature and at the same time, work each brush into its holder. Continue to lower the armature until the full weight of the armature is on the brushes.

2- Now, very **CAREFULLY** lower the frame assembly down over the armature. **TAKE CARE** because the magnets in the frame assembly will tend to pull against the armature.

3- When the frame makes contact with the lower cap, align the marks on the cap and the frame.

4- Slide the upper cap washer onto the shaft.

5- Install the upper cap with the mark on the cap aligned with the mark on the frame.

ARMATURE

END CAP

①

FRAME ASSEMBLY

②

MARK

③

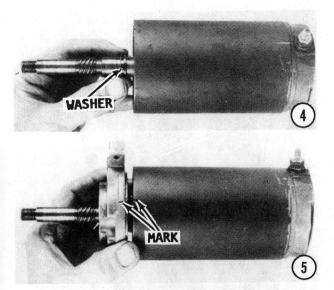

6- Install the thru-bolts and tighten them securely.

7- Place the starter motor on the floor. To test the operation of the motor, first connect one lead from a set of jumper cables to the positive terminal of a battery. Connect the other end of the same lead to the positive terminal of the motor. Connect one end of the second lead of the jumper cables to the negative terminal of the battery.

Now hold the motor firmly on the floor with one foot and at the same time, momentarily make contact with the other end of the second jumper lead to the motor case.

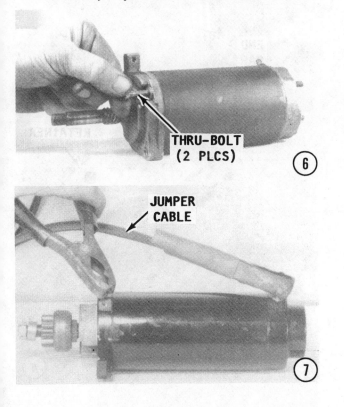

The pinion gear should spin rapidly.

Install the drive gear as described earlier in this section.

8- Install the starter motor onto the engine and secure it in place with the mounting bolts.

To test the complete starter motor, proceed directly to Section 6-15.

To install the starter motor onto the engine, proceed directly to Section 6-16.

6-14 PRESTOLITE SERVICE

REMOVAL

1- Before beginning any work on the cranking motor, disconnect the positive (+) lead from the battery terminal. Remove the powerhead hood. Disconnect the red cable at the cranking motor terminal. Remove the 1/2" bolt securing the cranking motor bracket. This bolt is located on the starboard side of the powerhead just above the carburetor.

Remove the three 7/16" bolts (or nuts in some cases) securing the cranking motor bracket to the powerhead. Some cranking motors have flanges permanently attached to the cranking motor and the attachment to the powerhead is accomplished through these flanges instead of the usual mounting bracket. If the flanges are attached to the motor, remove the three bolts attaching the motor to the powerhead. Remove the cranking motor and bracket together.

GOOD NEWS

If the only motor repair necessary is replacement of the brushes, the drive gear does not have to be removed. All cranking motors have thru-bolts securing the upper and lower cap to the field frame assembly. In all cases both caps have some type of mark or boss. These marks are used to properly align the caps with the field frame assembly.

DISASSEMBLING

2- Observe the caps and find the identifying mark or boss on each. If the marks are not visible, make an identifying mark prior to removing the thru-bolts as an **ESSENTIAL** aid during assembling. Remove the thru-bolts from the bracket and the motor.

3- Use a small hammer and **CAREFULLY** tap the lower cap free of the motor.

4- Pull on the armature shaft from the drive gear end and remove it from the field frame assembly. Remove the brushes from their holders, and then remove the brush springs. Lift the white plastic retainer free from the frame. Observe the location of the notch on the retainer in relation to the frame. The retainer **MUST** be installed in the same position.

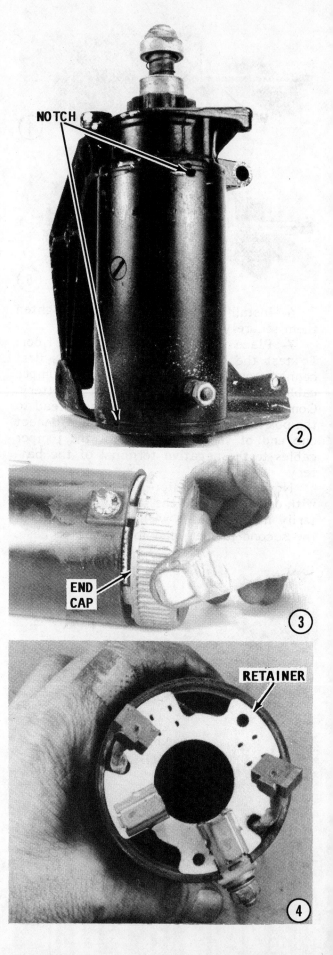

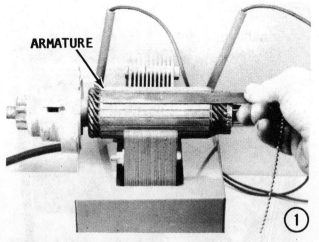

ARMATURE TESTING

Testing for a Short

1- Position the armature on a growler, then hold a hacksaw blade over the armature core. Turn the growler switch to the **ON** position. Slowly rotate the armature. If the hacksaw blade vibrates, the armature or commutator has a short. Clean the grooves between the commutator bars on the armature. Perform the test again. If the hacksaw blade still vibrates during the test, the armature has a short and **MUST** be replaced.

Testing for a Ground

2- Obtain a test lamp or continuity meter. Make contact with one probe lead on the armature core and the other probe lead on the commutator bar. If the lamp lights, or the meter indicates continuity, the armature is grounded and **MUST** be replaced.

Checking the Commutator Bar

3- Check between or check bar-to-bar as shown in the accompanying illustration.

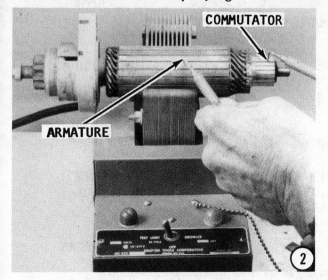

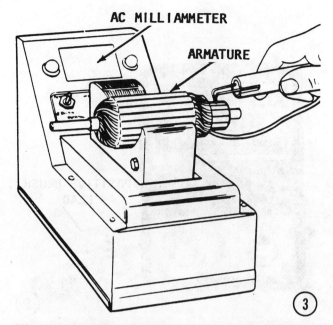

The test light should light, or the meter should indicate continuity. If the commutator fails the test, the armature **MUST** be replaced.

Turning the Commutator

4- True the commutator, if necessary, in a lathe. **NEVER** undercut the mica because the brushes are harder than the insulation. Undercut the insulation between the commutator bars 1/32" to the full width of the insulation and flat at the bottom. A triangular groove is not satisfactory. After the undercutting work is completed, clean out the slots carefully to remove dirt and copper dust. Sand the commutator lightly with No. "00" sandpaper to remove any burrs left from the undercutting. Check the armature a second time on the growler for possible short circuits.

Positive Brushes

5- Notice how the positive brush lead is attached to the terminal on the end of the frame. This is the same terminal to which the heavy battery cable is attached. The terminal may be removed from the frame. Pull the terminal free of the frame.

Obtain an ohmmeter. Connect one test lead of an ohmmeter to the brush and the other test lead to the terminal. Continuity

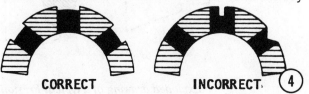

CORRECT   INCORRECT

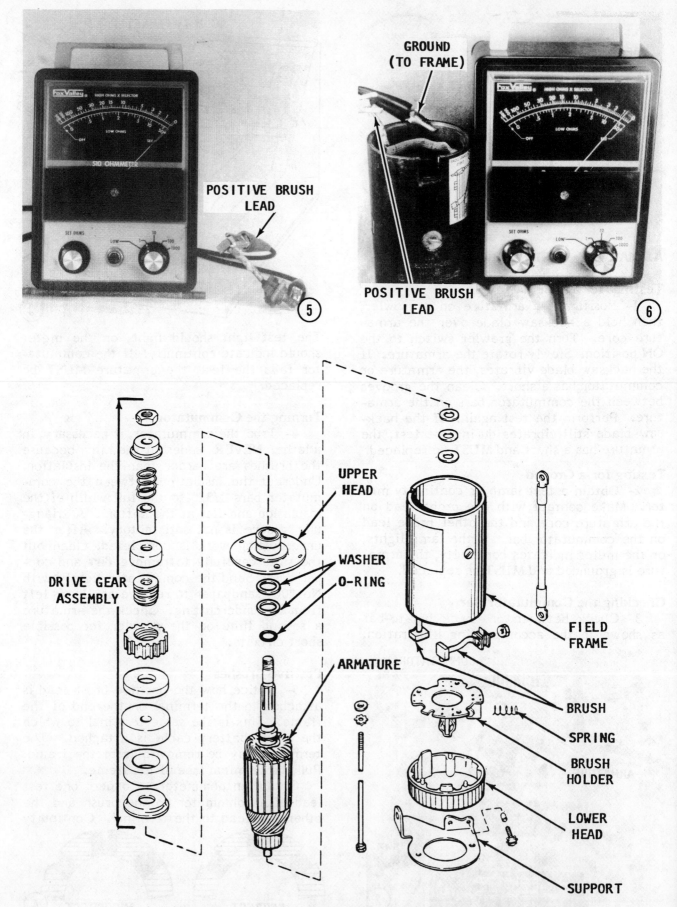

POSITIVE BRUSH
LEAD

(5)

GROUND
(TO FRAME)

POSITIVE BRUSH
LEAD

(6)

UPPER
HEAD

WASHER

O-RING

ARMATURE

DRIVE GEAR
ASSEMBLY

FIELD
FRAME

BRUSH

SPRING

BRUSH
HOLDER

LOWER
HEAD

SUPPORT

Exploded drawing of a typical Prestolite cranking motor, with principle parts identified.

should be indicated on the ohmmeter. If continuity is not indicated, the brush must be replaced. The brush and terminal are sold as an assembly, eliminating the necessity for soldering.

Negative Brushes

6- On Prestolite starters the negative brushes are connected to the field coils inside the starter frame.

Obtain an ohmmeter. Make contact with one test lead to the negative brush and make contact with the other lead to the starter frame. If the meter does not indicate continuity, the field coils are open and **MUST** be replaced.

Check to be sure the soldered connections are **NOT** touching the frame. The fields must not be grounded. If the connections make contact with the frame, the fields would be grounded.

CLEANING AND INSPECTING

Clean the field coils, armature, commutator, armature shaft, brush-end plate and drive-end housing with a brush or compressed air. Wash all other parts in solvent and blow them dry with compressed air.

Inspect the insulation and the unsoldered connections of the armature windings for breaks or burns.

Perform electrical tests on any suspected defective part, according to the procedures outlined earlier in this section.

Check the commutator for runout. Inspect the armature shaft and both bearings for scoring.

Turn the commutator in a lathe if it is out-of-round by more than 0.005".

Check the springs in the brush holder to be sure none are broken. Check the spring tension and replace if the tension is not 32-40 ounces. Check the insulated brush holders for shorts to ground. If the brushes are

worn down to 1/4" or less, they must be replaced.

Check the field brush connections and lead insulation. A brush kit and a contact kit are available at your local marine dealer, but all other assemblies must be replaced rather than repaired.

The armature, fields, and brush holders must be checked before assembling the starter motor. See the testing section in this chapter for detailed procedures to test the starter motor.

ASSEMBLING THE PRESTOLITE

1- Slide the plastic terminal and brush lead retainer into the groove in the frame with the small protrusion on one side facing **DOWNWARD**. Continue pushing the retainer into the groove until it is fully seated.

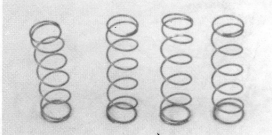

Old brush springs (left) compared with a new set (right). The springs must be in good condition, the same length, and free of any discoloration.

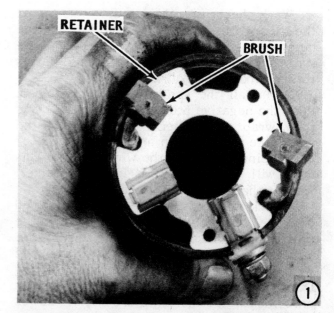

Work the brush retainer down on top of the frame with the positive lead through the cutaway in the retainer plate. Check to be sure the field coil negative brush passes through the **cutaway** in the plate.

2- Install the spring into the retainer. Push the negative brush into its retainer and then, wrap a fine piece of wire around the front side of the brush and the back side of the retainer. Tighten the wire snugly. This wire will hold the brush in the retainer. Repeat the procedure for the positive brush.

Check to be sure the plate is secured onto the frame and the **cutaway** is over the protrusion of the positive plastic terminal.

Clamp the armature in a vise equipped with soft jaws with the drive gear facing **DOWNWARD**. Install the thrust washers onto the end of the armature shaft. Lower the frame assembly down over the armature until the brushes are over the commutator.

3- After the armature is in place, cut and remove the wire wrapped around the brushes to hold them in place. The brushes should then make firm contact with the commutator.

4- Install the end cap onto the end of the starter motor. Observe three small nipples on the inside of the end cap. These nipples **MUST** index with matching dimples in the retaining plate. Align the mark on the side of the end cap with the terminal. Lower the cap onto the frame, and seat it **GENTLY**. **NEVER** tap with a hammer or other tool, because the nipples may not be indexed with the dimples and the tapping may cause damage.

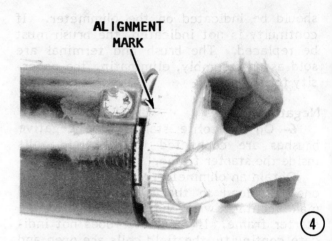

Align the end cap notch or mark with the mark on the frame and the upper cap mark with its mark.

Slide the rubber spacer or collar into the starter bracket, if used. Now, install the starter bracket over the starter motor, if used. Install the thru-bolts through the end cap, the frame, and thread them into the starter bracket. Tighten the thru-bolts securely.

6-15 CRANKING MOTOR TESTING

Place the cranking motor on the floor. Hold the motor firmly with one foot while testing its operation.

CAUTION: The armature will turn rapidly during this test. Therefore, the cranking motor **MUST** be well **SECURED** before making the test to prevent personal **INJURY** or damage to the motor.

Firmly connect one end of a heavy-duty jumper cable to the **POSITIVE** terminal of a

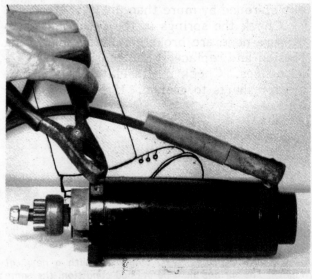

Using foot to hold cranking motor while testing.

battery. Firmly connect the other end of the jumper cable to the cranking motor terminal.

Connect a second heavy-duty jumper cable to the negative terminal of the battery. Now, **MOMENTARILY** make contact with the other end of the second jumper cable anywhere to the frame of the cranking motor.

NEVER make the momentary contact with the positive cable to the terminal, because any arcing at the terminal may damage the terminal threads and the nut may not take to the damaged threads. The motor should turn rapidly. If the cranking motor fails to rotate, the motor must be disassembled again and the service work carefully checked. Sorry about that, but some phase of the rebuild task was not performed properly.

6-16 CRANKING MOTOR INSTALLATION

With Mounting Bracket

Mount the cranking motor and bracket onto the powerhead. Align the top bolt above the carburetor, and then thread it into the block about half-way. Align the other three bolts on the starboard side or start the nuts onto the studs, depending on the model powerhead being serviced. Tighten the three on the side evenly and alternately until they are secure.

With Flanges

If the cranking motor has the mounting flanges permanently attached, then position the motor in place and start the three bolts to attach the motor to the powerhead. Now, tighten the three bolts **ALTERNATELY** and **EVENLY** until all bolts are tight. The bolts **MUST** be tightened alternately to prevent binding and possibly bending the flanges.

Connections

Connect the positive red lead to the cranking motor. Connect the electrical lead to the battery. Connect the fuel line to the fuel pump.

Test the completed work by cranking the powerhead with the cranking motor.

DO NOT, under any circumstances, start the engine unless it is mounted in a test tank or body of water.

CAUTION: Water must circulate through the lower unit to the engine any time the engine is run to prevent damage to the water pump in the lower unit. Just five seconds without water will damage the water pump.

7
REMOTE CONTROLS

7-1 INTRODUCTION

A remote control unit is seldom sold with just an outboard unit. In most cases, the control box is sold separately as an option or it is included with a "package" deal -- boat, outboard, control box, and trailer.

If the control box is included in the "package", the unit will most likely be the latest production model from OMC -- with Johnson or Evinrude colors and decals.

This short chapter covers removal of the unit from the boat, separating the two halves, testing some of the circuits, replacement of the switches and warning horn, and a few words on lubrication.

Non-use is absolutely the greatest enemy of the control unit. The large number of eccentrics, cams, levers, linkages, etc. should be operated at regular intervals -- once a month and the interior parts lubricated -- say every two years.

WOULD YOU BELIEVE

Probably 90% of steering and shifting problems are directly caused by the system not being operated. Without movement, steering and shifting cables and linkages have a tendency to "freeze". **Would you also believe,** service shops report well over 50% of boat cables replaced every year are due to lack of movement.

Therefore, during off-season, when the boat is laid up in a yard, or on a trailer alongside the house, take time to go aboard and operate the steering from hard-over to hard-over and shift the remote control unit through the full range several times to ensure corrosion does not develop causing a fitting or joint to "freeze" preventing proper movement.

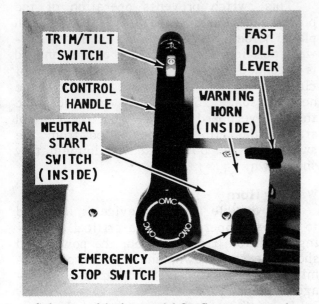

Side view of the late model OMC remote control unit covered in this chapter, with some major parts identified.

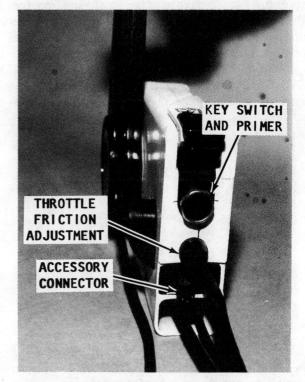

Aft view of the OMC remote control unit covered in this chapter, with some major parts identified.

7-2 DESCRIPTION AND OPERATION

The following components and features are incorporated in the OMC remote control box presented in this chapter and usually associated with the larger OMC outboard units covered in this manual.

Location of the various parts in or on the control box are identified on the accompanying illustration.

The function of the item is almost described in the name.

Control Handle

As the name implies, this handle controls the gear position of the lower unit ---NEUTRAL, FORWARD, and REVERSE and the powerhead rpm. From the vertical position (straight up), movement 32° forward shifts the lower unit into FORWARD gear and movement of the handle 32° aft of vertical shifts the unit into REVERSE gear. Further movement past the 32° position in either FORWARD or REVERSE will increase powerhead rpm.

Key Switch and Primer

Rotating the key switch CLOCKWISE to the first detent energizes all powerhead accessories. Further movement CLOCKWISE to the second detent will energize the solenoid to activate the cranking motor. Pushing the key inward at the first or second detent position, will energize the fuel primer solenoid to choke the carburetor.

When the key is rotated back to the full COUNTERCLOCKWISE position the ignition system and all powerhead controlled accessories.

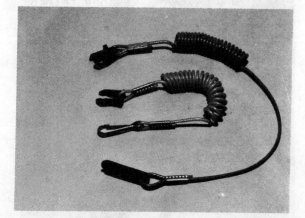

Two different emergency switch lanyards used with the OMC remote control unit. **REMEMBER** -- *it is not possible to start the powerhead if the lanyard is not in place behind the switch knob.*

Neutral Lockout

This safety feature prevents the control handle from moving to the forward or reverse positions when the control handle is in the NEUTRAL position.

With the outboard operating, the helmsperson simply depresses the knob upward against the handle to release the handle for movement -- either forward or aft. The neutral lockout will automatically engage when the control handle is returned to the NEUTRAL position.

Throttle Friction Adjustment

The throttle friction adjustment does exactly what the name implies -- places friction on the control handle to prevent unwanted "creep" of the control handle while the powerhead is operating and/or the boat is underway. Rotating the knob CLOCKWISE will increase friction on the control handle. Logically, rotating the knob COUNTERCLOCKWISE will decrease friction and the handle movement will be more "free".

Neutral Start Switch

Again, the function of this switch is obvious -- to prevent the cranking motor circuit from being energized EXCEPT when the control handle is in the NEUTRAL position. Stated another way, the start switch will only allow the key switch to be rotated to the START position when the control handle is in the NEUTRAL position.

Emergency Stop Switch

This switch prevents operation of the powerhead unless a safety lanyard is in place holding the switch button in an outward position. The other end of the lanyard is intended to be attached to an item of clothing or the life jacket worn by the helmsperson. Should the individual be thrown overboard or away from the control station the clip will be pulled from the switch and the powerhead will immediately shut down.

Warning Horn

This audible warning device is intended to alert the operator that a critical operating condition has developed; the powerhead should be shut down; and the cause determined before extensive and expensive damage is done.

Depending on the powerhead being serviced, the warning device may emit:

Continuous Tone indicating a powerhead overheat condition.

Continuous Short Pulse Tone indicating a **NO OIL** condition -- **BAD NEWS** -- very **BAD NEWS**.

Short Pulse Tone -- every 20-seconds indicating a **LOW OIL** condition.

Continuous Tone -- at or near WOT indicating a restriction in the fuel supply.

Fast Idle Lever

This lever controls powerhead rpm when the control handle is in the **NEUTRAL** position. The lever should be raised to assist during powerhead startup and favorable idle speed. The lever **MUST** be returned to the **RUN** position before moving the control handle out of the **NEUTRAL** position.

Accessory Connector

The accessory connector permits easy access to the powerhead accessory and tachometer circuits. Several OMC tachometers and wiring kits provide a mating connector simplifying accessory installation. **MAXIMUM** draw on the accessory circuit **MUST NOT** exceed 5 amps.

Trim/Tilt Switch

This switch is included on control boxes with trim/tilt installed. The switch permits the helmsperson to raise or lower the outboard through 0-21° while the unit is operating in **FORWARD** gear. When the powerhead is shut down or is operating below 1500 rpm, this same switch allows and controls the outboard to tilt 22-75°.

7-3 FUNCTIONAL TESTS

The following tests can all be performed using an ohmmeter or a continuity test light to test for continuity.

Disconnect the cables from both terminals on the battery. Disconnect the throttle and shift cables at the powerhead.

SPECIAL WORDS

The illustrations supporting the following tests are lettered **"A"** thru **"H"**.

Remove the retainer securing the harness connector together at the powerhead, illustration **"A"**. Disconnect the harness connector. Note the arrow on both halves. During the tests this arrow must **ALWAYS** be on top -- at the 12 o'clock position --for the test probes to be inserted in the correct openings, per the illustrations.

Emergency Stop Switch Circuit

Install the safety clip and lanyard behind the stop switch button.

Obtain an ohmmeter or a continuity light. Set the meter scale for continuity reading. A continuity test light may also be used.

Turn the key to the **ON** position.

With the connector arrow at the 12 o'clock position, insert the ohmmeter probes into the 1 o'clock and 6 o'clock female openings, illustration **"B"**. The meter must show **NO** continuity.

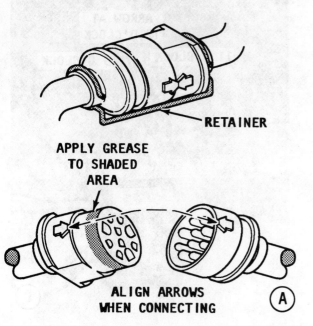

RETAINER

APPLY GREASE TO SHADED AREA

ALIGN ARROWS WHEN CONNECTING (A)

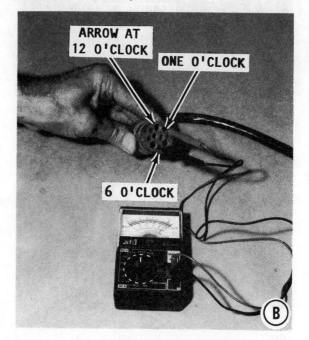

ARROW AT 12 O'CLOCK   ONE O'CLOCK

6 O'CLOCK   (B)

Keep the test probes in the same openings and remove the emergency stop switch clip from behind the knob. The ohmmeter **MUST** indicate continuity.

If the tests are not satisfactory, check the wiring carefully. Final solution -- replace the emergency stop switch.

Key Switch Circuit

An ohmmeter or a continuity light may be used to make the following tests.

Install the safety switch clip behind the knob.

With the connector arrow at the 12 o'clock position, leave the meter test probes in the 1 o'clock and 6 o'clock openings, as in the previous tests, illustration **"B"**.

Turn the key switch to the **OFF** position. The meter should indicate continuity.

Now, turn the key switch to the **ON** position. The meter should indicate **NO** continuity.

Keep the connector arrow at the 12 o'clock position, and move the meter probes to the 5 o'clock and 9 o'clock positions, illustration **"C"**. With probes in place, move the key switch to the **ON** and **START** positions and at the same time push the key inward. The meter should indicate continuity. Allow the key to move outward and the meter should indicate **NO** continuity.

With the control handle in the **NEUTRAL** position, insert the meter probes into the 5 o'clock and 11 o'clock positions, illustration **"D"**. After the probes are in place, turn the key switch to the **START** position. The meter should show continuity. Release the key and the meter should indicate **NO** continuity. If the test results are not satisfactory, check the wiring and test the switch.

Neutral Start Switch Circuit

An ohmmeter or a continuity light may be used to make the following tests.

Check to be sure the control handle is in the **NEUTRAL** position. Insert the meter probes in the 5 o'clock and 11 o'clock positions, illustration **"D"**. Turn the key to the **START** position and the meter should indicate continuity.

Move the control handle to the **FORWARD** position and again turn the key to the **START** position. The meter should indicate **NO** continuity.

Move the control handle to the **REVERSE** position and turn the key to the **START** position. The meter should indicate **NO** continuity.

If the test results are not satisfactory, check the wiring and test the neutral start switch.

Trim/Tilt Switch Circuit

An ohmmeter or a continuity light may be used to make the following tests.

At the outboard, disconnect the control handle trim/tilt harness from the trim/tilt

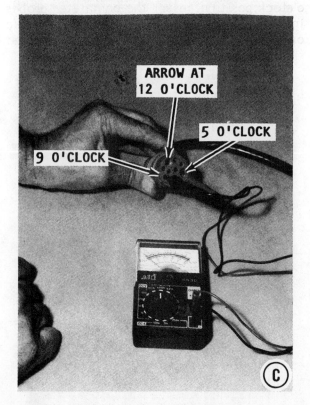

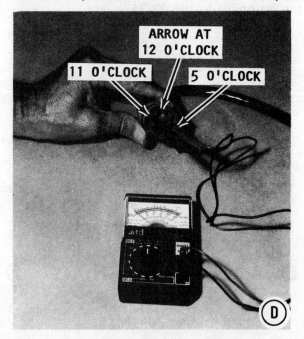

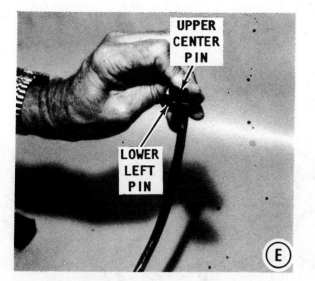

motor harness at the connector.

Make contact with the meter test probes to the upper center pin and the lower left pin, as indicated in illustration **"E"**. Depress the trim/tilt switch for the **UP** direction. The meter should indicate continuity. Release the switch and the meter should indicate **NO** continuity.

Make contact with the meter probes to the upper center pin and the lower right pin, illustration **"F"**.

Depress the trim/tilt switch for the **DOWN** direction. The meter should indicate continuity. Release the switch and the meter should indicate **NO** continuity.

If the test results are not satisfactory check the wire location or replace the switch.

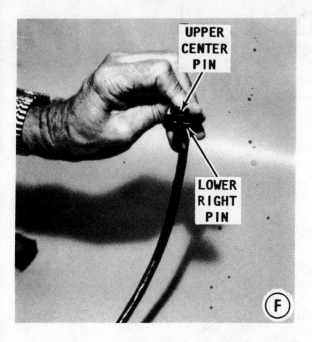

7-4 REMOTE CONTROL BOX SERVICING

The remote control box should give years of satisfactory service. As mentioned earlier, at the beginning of this chapter, non-use -- lack of movement -- is the biggest enemy of the control box. From a practical standpoint, the control handle should be operated every month and the box opened every couple years. The cams, joints, eccentrics, etc. can then be easily lubricated.

Procedures in this section for the remote control box include removal from the bulkhead in the boat, separating the two halves, replacement of the key switch, warning horn, neutral safety switch, and the trim/-tilt switch in the control handle. Once the box is opened the throttle cable and the shift cable can be disconnected.

Removal and Opening

1- Remove the two screws securing the control box to the bulkhead in the boat.

2- Lay the control box on a couple pieces of wood -- 2x4's will do fine -- with the back side of the box facing up. Back out the hex head screw securing the control handle **THREE** full turns. Splines in the handle index with splines in the hub in the box. Using a center punch on the head of the hex head screw and a soft head mallet, disengage the control handle splines from the splines in the hub. Once the handle is

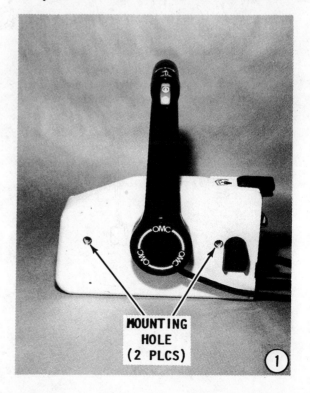

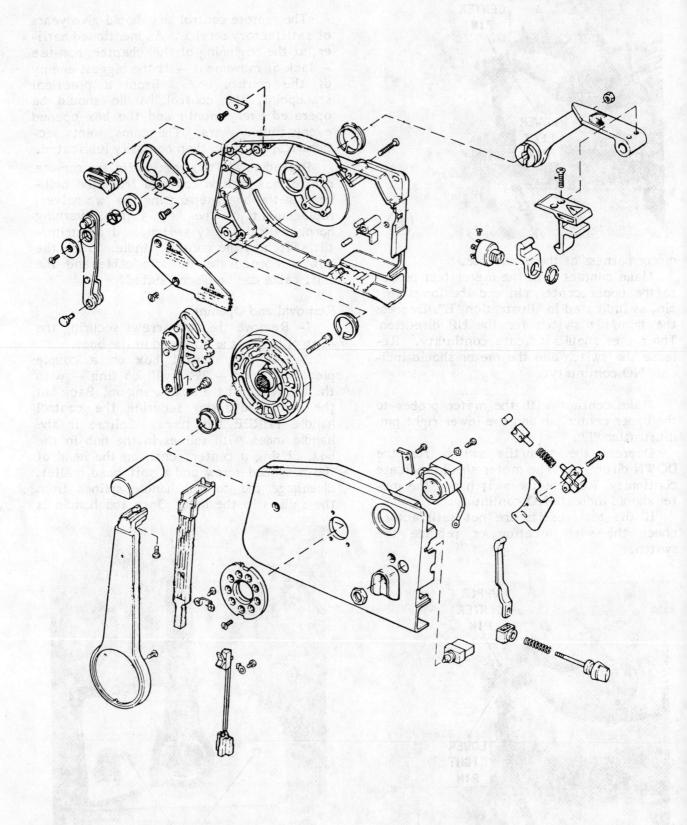

Exploded drawing of the remote control unit covered in this manual.

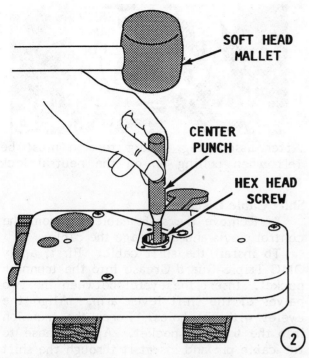

SOFT HEAD MALLET

CENTER PUNCH

HEX HEAD SCREW

②

COVER HALF

HOUSING HALF

④

free, remove the hex head screw and the control handle. Because the wires into the handle are still intact, the handle will be left in its approximate position.

3- Remove the three screws securing the two halves of the box together.

4- Begin to separate the box along the top seam. Once the box is opened, everything inside is exposed and individual items can be replaced without difficulty.

Closing Control Box

Apply a coating of RTV Sealer, or equivalent, to the mating surfaces of the control box. This sealer will help protect everything inside the box.

Lay the cover half on a flat surface. The flat surface will keep the cable pins from falling free. Move the housing half

over the cover half with the electrical cable grommet properly indexed into the housing.

Secure the two halves together with the three screws. Tighten the screws to a torque value of 40-50 in lbs (4.6-5.5Nm).

REPLACING INTERIOR COMPONENTS

Neutral Start Switch

5- Remove the two screws securing the neutral start switch, disconnect the electrical leads and remove the switch.

During installation, apply OMC Black Neoprene Dip to the switch terminals.

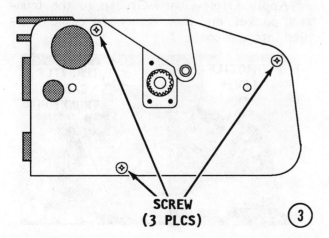

SCREW (3 PLCS)

③

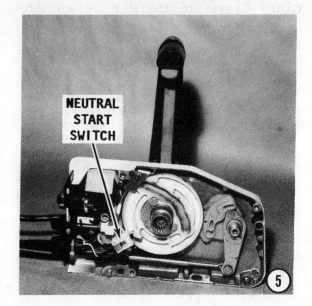

NEUTRAL START SWITCH

⑤

Key Switch

6— Back off the securing nut, and disconnect the electrical leads.

During installation, apply OMC Black Neoprene Dip to the switch terminals.

Warning Horn

One screw and two push-on connectors secure the horn in the control box. Remove the screw and separate the connectors, illustration "6".

During installation of the horn, apply OMC Black Neoprene Dip to the switch terminals.

Trim/Tilt Switch

7— After the control handle has been removed, as outlined in Step 2 -- from the back side of the control handle, remove the two screws securing the neutral lock plate. Lift the neutral lock slide and the return spring out of the handle. Remove the control handle knob screw, and then the four screws securing the cover to the control handle. Remove the knob and the cover. The trim/tilt switch is now exposed. Remove the screw and the switch.

During installation, apply OMC Triple-Guard Grease to the slide return spring.

After assembling, spring tension must be felt when pulling up on the neutral lock slide.

Shift Cable Disconnect

8— Remove the shift cable pin from the control clevis and disengage the cable.

To install the shift cable: First, apply OMC Triple-Guard Grease into the trunnion pocket. Insert the eyelet between the two halves of the shift lever arm. Align the eyelet with the holes. Force the trunnion into the trunnion pocket. Apply grease to the cable pin and insert it through the shift lever halves to secure the cable end in place.

Throttle Cable Disconnect

9— First, lift the throttle cable trunnion from the pocket, and then move the throttle lever rearward until the cable pin is exposed. Remove the pin, and then the throttle cable.

To install the throttle cable: First, move the throttle lever rearward out from under the plastic plate until the hole in the lever arm is exposed. Place the eylet between the halves of the lever arm and align the eyelet with the holes. Apply Triple-Guard Grease to the cable pin. Align the hole in the cable end with the holes in the lever arm, and then insert the pin through lever arm, eyelet and the other part of the lever arm.

Apply Triple-Guard Grease to the trunnion pocket, and then insert the cable trunnion into the pocket.

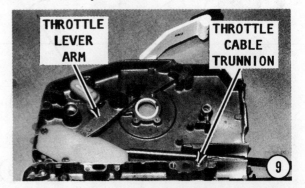

8
LOWER UNIT

8-1 DESCRIPTION

The lower unit is considered as that part of the outboard below the exhaust housing. The unit contains the propeller shaft, the driven and pinion gears, the driveshaft from the powerhead and the water pump. On models equipped with shifting capabilities, the forward and reverse gears, together with the clutch, shift assembly, and related linkage, are all housed within the lower unit.

Chapter Coverage

Eight lower units have been used by the manufacturer on the 1- and 2-cylinder outboards covered in this manual. Each lower unit has differences requiring a separate section of procedures for service work. Instructions for the type of lower unit and the model years used will be presented as follows:

Type "A" — 2.3hp and 3.3hp -- 1991 & On -- Section 8-4, beginning on Page 8-8.

Type "B" -- 3hp and 4hp -- 1990 & On -- Section 8-5, beginning on Page 8-19.

Type "C" — 4hp Delux, 6hp and 8hp -- 1990 & On -- Section 8-6, beginning on Page 8-30.

Type "D" — 9.9hp and 15hp -- 1990 & On -- Section 8-7, beginning on Page 8-47.

Type "E" — 20hp, 25hp and 30hp -- 1990 & On -- Section 8-8, beginning on Page 8-66.

Type "F" — 40hp and 50hp -- 1990 & On -- Section 8-9, beginning on Page 8-91.

Type "G" -- Jet Drive -- 1990 & On -- Section 8-10, beginning on Page 8-117.

Type "H" -- Colt, Junior -- 1990 only -- Section 8-11, beginning on Page 8-134.

Water Pump

Water pump service work is by far the most common reason for removal of the lower unit. Each lower unit service section contains complete detailed procedures to rebuild the water pump. On all units covered in this manual, **EXCEPT** 2.3hp and 3.3hp, the water pump is installed on top of the lower unit. Therefore, the lower unit must be separated from the intermediate housing in order to perform service work on the pump. On the 2.3hp and the 3.3hp, the water pump is installed on the propeller shaft directly behind the propeller. On these models, only the propeller and the water pump cover need be removed to service the water pump.

A neglected lower unit cannot be expected to perform to maximum efficiency, compared with a unit receiving TLC (tender loving care).

The instructions given to prepare for the water pump work must be performed as listed. However, once the pump is ready for installation, if no other work is to be performed on the lower unit, the reader may jump to the pump assembling procedures and continue with installation of the water pump.

Each section is presented with complete detailed instructions to remove, disassemble, clean and inspect, assemble and install virtually all parts in the lower unit. If a part is found to be in serviceable condition, simply skip the steps involved and proceed with the necessary work to restore the lower unit to "like new" condition.

ILLUSTRATIONS

The illustrations included with the procedural steps in this chapter were taken of the most popular lower units. In some rare cases, the unit being serviced may not appear to be absolutely identical with the unit illustrated. However, the step-by-step work sequence will be valid in all cases. If there is a special procedure for a unique lower unit, the differences will be clearly indicated in the step.

Typical water pump installation in place on the lower unit — ready for mating to the intermediate housing.

SPECIAL WORDS

All threaded parts are right-hand unless otherwise indicated.

If there is any water in the lower unit or metal particles are discovered in the gear lubricant, the lower unit should be completely disassembled, cleaned, and inspected.

Actually, problems in the lower unit can be classified into three broad areas:

1- Lack of proper lubrication in the lower unit. Most often this is caused by failure of the operator to check the gear oil level frequently and to add lubricant when required.

2- A faulty seal allowing water to enter the lower unit. Water allowed to remain in the lower unit over a period of non-use time will separate from the oil and can be destructive.

3- Excessive clutch dog and clutch ear wear on the forward and reverse gears. This condition is caused by excessive wear in the bellcrank under the powerhead. A worn bellcrank will result in sloppy shifting of the lower unit and cause the clutch components to wear and develop shifting problems. Improper shifting techniques at the shift box will also result in excessive wear to the clutch dog and clutch ears of the forward and reverse gears.

Time will also take its toll. Continued service over a long period of time will cause parts to wear and require replacement.

8-2 PROPELLER SERVICE

PROPELLER WITH SHEAR PIN REMOVAL

If the unit being serviced has the shear pin located between the propeller nut and the propeller, the propeller nut should be removed and the shear pin checked.

To remove the propeller, first pull the cotter key, and then remove the propeller nut, shear pin, and washer. Because the shear pin is not a tight fit, the propeller is able to move on the pin and cause burrs on the hole. The propeller may be difficult to remove because of these burrs. To overcome this problem, the propeller hub has two grooves running the full length of the hub. Hold the shaft from turning, and then rotate the propeller 1/4 turn to position the grooves over the drive pin holes. The pro-

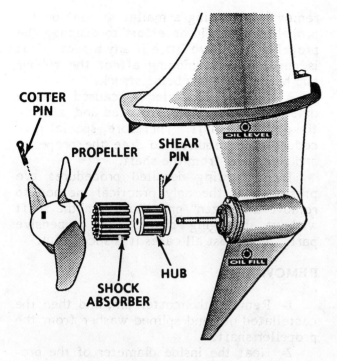

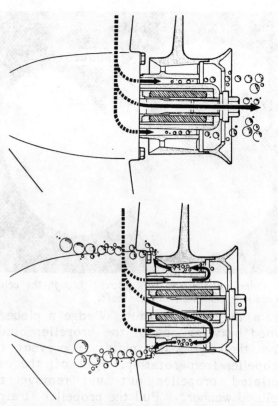

Arrangement of parts on the propeller shaft of small units, without propeller exhaust.

Cross-section drawing of the lower unit showing route of the exhaust gases with the unit in forward gear (top), and in reverse gear (bottom).

peller can then be pulled straight off the shaft. After the propeller has been removed, file the drive pin holes on both sides of the shaft to remove the burrs.

If the propeller is the type with the shear pin installed next to the gearcase head, first remove the cotter key, then the propeller nut. Next, slide the propeller free of the shaft.

EXHAUST PROPELLER

Propellers with the exhaust passing through the hub **MUST** be removed more frequently than the standard propeller. Re-

Propeller with exhaust hub. The defuser ring is clearly visible.

moval after each weekend use or outing is not considered excessive. These propellers do not have a shear pin. The shaft and propeller have splines which **MUST** be coated with an anti-corrosion lubricant prior to installation as an aid to removal the next time the propeller is pulled. Even with the lubricant applied to the shaft splines, the propeller may be difficult to remove.

The propeller with the exhaust hub is more expensive than the standard propeller and therefore, the cost of rebuilding the unit, if the hub is damaged, is justified.

A replaceable diffuser ring on the backside of the propeller disperses the exhaust away from the propeller blades as the boat moves through the water. If the ring becomes broken or damaged "ventilation" would be created pulling the exhaust gases back into the negative pressure area behind the propeller. This condition would create considerable air bubbles and reduce the effectiveness of the propeller.

PROPELLER WITH EXHAUST -- REMOVAL

First, disconnect the high tension leads to the spark plugs to prevent accidental powerhead start. Next, pull the cotter pin

Propeller with the two grooves through the center to assist in removal from the shaft.

from the propeller nut. Wedge a piece of wood between one of the propeller blades and the cavitation plate to prevent the propeller from rotating. Back off the castellated propeller nut and remove the splined washer. Pull the propeller straight off the shaft. It may be necessary to carefully tap on the front side of the propeller with a soft headed mallet to jar it loose. Remove the thrust washer from the propeller shaft.

"FROZEN" PROPELLER

DESCRIPTION

If an exhaust propeller is "frozen" to the propeller shaft and the usual methods of

Propeller exhaust arrangement showing the thrust washer, propeller, splined washer, and propeller nut.

removal fail, using a mallet to beat on the propeller blades in an effort to dislodge the propeller will have little if any affect. This is due to the cushioning affect the rubber hub has on the blow being struck.

A "frozen" propeller is caused by the inner sleeve becoming corroded and stuck to the propeller shaft. Therefore, special procedures are required to free the propeller and remove it from the shaft.

The following detailed procedures are presented as the only practical method to remove a "frozen" propeller from the shaft without damaging other more expensive parts. In almost all cases it is successful.

REMOVAL

1- Remove the cotter pin, and then the castellated nut and splined washer from the propeller shaft.

2- Heat the inside diameter of the propeller with a torch. Do not apply the heat to the outside surface of the propeller. Concentrate the heat in the area of the hub and shaft where the nut was removed and as far into the hub as possible. Continue applying heat, and at the same time have an assistant use a piece of 2" x 4" wooden block wedged between one of the blades and the lower unit housing. Use a prying force on the propeller while the heat is being applied. As the heat melts the inner rubber hub, the propeller will come free.

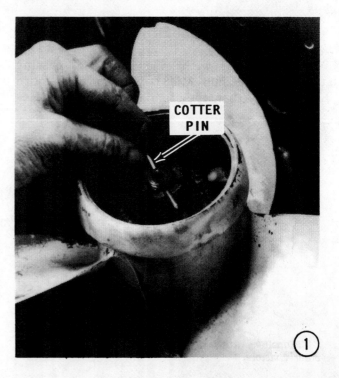

WARNING

As the force and heat are applied, the propeller may "pop" loose suddenly and without warning. Therefore, stand to one side while applying the heat as a precaution against personal injury.

3- After the propeller has been removed, the sleeve and what's left of the rubber hub will still be stuck to the propeller shaft. Attach a puller to the thrust washer. Apply more heat to the sleeve and rubber hub, and at the same time take up on the puller. When the sleeve reaches the proper temperature, it will be released and come free.

4- If a puller is not available, as described in Step 3, use a sharp knife and cut the rubber hub from the sleeve. An alternate method is to use canned heat, or the equivalent, and set fire to the rubber hub. Allow the hub to burn away from the sleeve. When the fire burns out, only a small amount may be left on the sleeve. Heat the sleeve again, and then while it is still hot, use a chisel, punch, or similar tool with a hammer, and drive the sleeve free of the shaft. Allow the propeller shaft to cool, and then clean the splines thoroughly. Take time to remove any corrosion. Install the propeller with a **NEW** hub and sleeve according to the instructions outlined in following paragraphs.

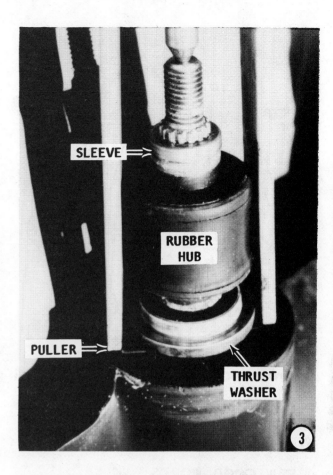

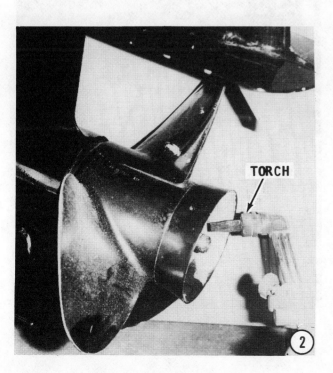

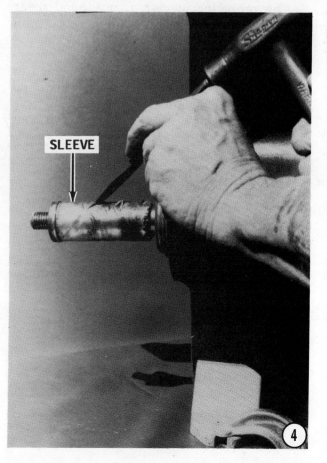

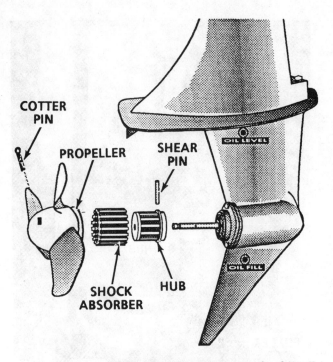

Arrangement of parts on the propeller shaft of small units, without propeller exhaust.

PROPELLER INSTALLATION WITH SHEAR PIN

A FEW GOOD WORDS

The propeller washer, if used, and shear pin, play an extremely important role. When shifting gears during normal operation, or if the propeller should hit an underwater obstacle, the propeller is subjected to considerable shock. A washer is installed between the propeller and drive pin. This washer **MUST** always be in place for proper operation. If the hub should slip, the propel-

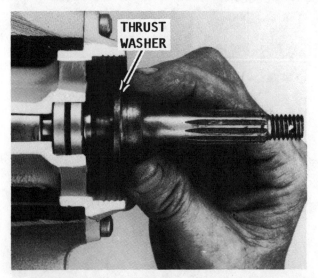

Installing the thrust washer onto the propeller shaft of a propeller exhaust unit.

ler will move back towards the propeller nut and lock against the drive pin. The washer is designed to stop propeller movement so the drive pin can be easily removed for service. Now, on with the installation.

Install the propeller. Coat the propeller shaft with an anti-corrosion grease. Install the propeller with the drive pin holes aligned. Install the washer and drive pin. Slide the propeller cap into place and secure it with the cotter pin.

If the unit being serviced uses the shear pin between the propeller and the bearing carrier, proceed as follows: Install the shear pin and then coat the propeller shaft with anti-corrosion grease; install the propeller; propeller nut; and then the cotter pin.

EXHAUST PROPELLER INSTALLATION

Slide the thrust washer onto the propeller shaft. Coat the propeller shaft with Perfect Seal No. 4, Triple Guard Lubricant,

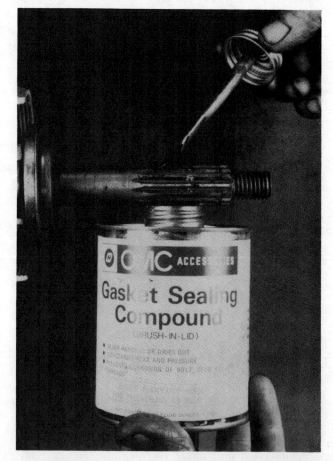

Applying gasket sealer to the propeller shaft splines to prevent the propeller from becoming "frozen" to the shaft.

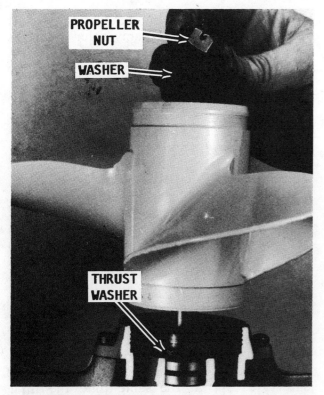

Installation of a propeller exhaust propeller with principle parts identified.

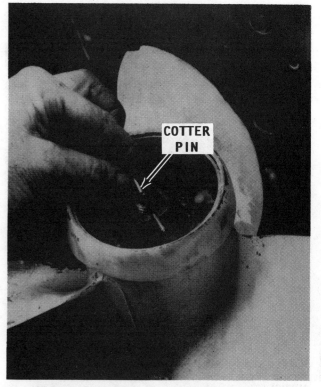

Installing the cotter pin through the castellated nut on a propeller exhaust unit.

OMC Gasket Sealing Compound, or similar good grade of lubricant to prevent the propeller from becoming "frozen" to the shaft. Slide the propeller onto the shaft with the splines in the propeller indexing with the splines on the shaft. Slide the splined washer onto the shaft. Thread the castellated nut onto the shaft. Jam a piece of wood between one of the propeller blades and the cavitation plate to prevent the

propeller from turning. Tighten the propeller nut securely and then a bit more to align the hole through the nut with the hole through the propeller shaft. Install the cotter pin through the nut and propeller shaft.

8-3 LOWER UNIT LUBRICATION

DRAINING LOWER UNIT

Position a suitable container under the lower unit, and then remove the **FILL** screw and the **VENT** screw.

CRITICAL WORD

On many lower units, the Phillips screw securing the shift fork in place is located very close to the fill screw, as shown in the accompanying illustration. On some units the Phillips screw is located on the other side. If the wrong screw is removed, **BAD NEWS, VERY BAD NEWS.** The lower unit will have to be disassembled in order to return the shift fork to its proper location.

Allow the gear lubricant to drain into the container. As the lubricant drains, catch some with your fingers, from time-to-time, and rub it between your thumb and

Tightening the nut on a propeller exhaust unit. The cotter pin is installed after the nut is secure.

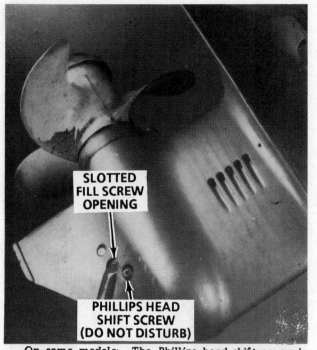

On some models: The Phillips head shift screw is located very close to the SLOTTED fill screw. If the Phillips head screw is removed by mistake -- BAD NEWS! The lower unit must then be disassembled in order to return the shift mechanism to its proper location.

finger to determine if any metal particles are present. If metal is detected in the lubricant, the unit must be completely disassembled, inspected, and the damaged parts replaced.

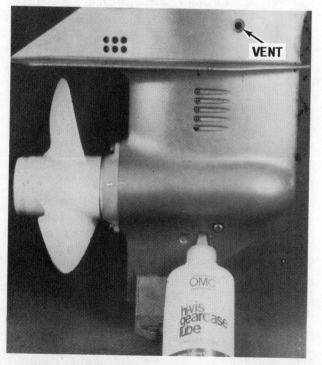

Filling a lower unit with OMC Gearcase Lubricant. Notice the vent plug has been removed to allow air to escape, as the unit fills with lubricant.

Check the color of the lubricant as it drains. A whitish or creamy color indicates the presence of water in the lubricant. Check the drain pan for signs of water separation from the lubricant. The presence of any water in the gear lubricant is **BAD NEWS.** The unit must be completely disassembled, inspected, the cause of the problem determined, and then corrected.

FILLING LOWER UNIT

Fill the lower unit with lubricant. Insert the lubricant tube into the bottom opening, and then fill the unit until lubricant is visible at the vent hole. Install the vent plug. Remove the gear lubricant tube and install the drain/fill plug.

After the lower plug has been installed, remove the vent plug again and using a squirt-type oil can, add lubricant through this vent hole. A squirt-type oil can must be used to allow the trapped air in the lower unit to escape at the same time the final lubricant is added. Once the unit is completely full, install and tighten the vent plug. See capacities in the Appendix.

8-4 TYPE "A" LOWER UNIT
2.3HP AND 3.3HP
NO SHIFT

DESCRIPTION

The Type **"A"** lower unit is a direct drive unit -- the pinion gear on the lower end of the driveshaft is in constant mesh with the forward gear. Reverse action of the propeller is accomplished by the boat's operator swinging the outboard with the tiller handle 180° and holding it in this position while the boat moves sternward. When the operator is ready to move the boat forward again, he/she simply swings the tiller handle back to the normal forward position.

LOWER UNIT REMOVAL

Disconnect the high tension spark plug lead to prevent accidental firing of the cylinder as the work progresses.

The following procedures present complete instuctions to remove, disassemble and assemble virtually all parts of the 2.3hp and 3.3hp lower unit. Perform only those procedures necessary to restore the unit to serviceable condition.

1- Position a suitable container under the lower unit, and then remove the oil level screw and the oil fuel screw. Allow the gear lubricant to drain into the container. As the lubricant drains, catch some with your fingers from time to time, and rub it between a thumb and finger to determine if any metal particles are present.

If metal particles are detected in the lubricant, the unit must be completely disassembled, inspected, and the damaged parts replaced.

Check the color of the lubricant as it drains. A whitish or creamy color indicates the presence of water in the lubricant. Check the drain pan for signs of water separation from the lubricant. The presence of any water in the gear lubricant is **BAD NEWS**. The unit must be completely disassembled, inspected, the cause of the problem determined, and corrected.

After the lubricant has drained, temporarily install the oil level screw and the oil fill screw.

Remove the two attaching bolts securing the lower unit to the intermediate housing. Separate the lower unit from the intermediate housing. The water tube will come free

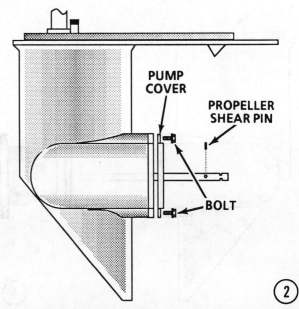

of the grommet on top of the lower unit and remain with the intermediate housing. The driveshaft will remain with the lower unit.

Lower Unit Disassembling

Straighten the cotter pin, and then pull it free of the propeller with a pair of pliers or cotter pin removal tool. Remove the propeller, and then push the shear pin out of the propeller shaft.

2- Remove the propeller shear pin. Remove the two bolts securing the water pump cover to the lower unit. Using two small screwdrivers -- one working on each side -- pry the water pump cover free of the lower unit. **TAKE CARE** not to mar the sealing surfaces of the pump cover or the mating lower unit.

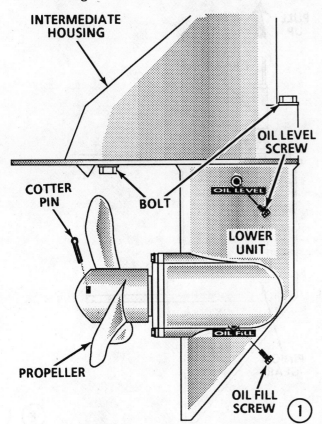

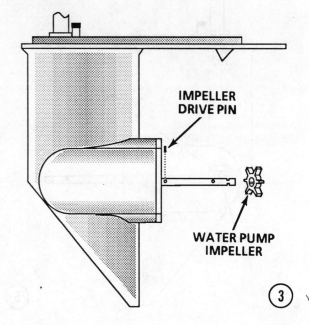

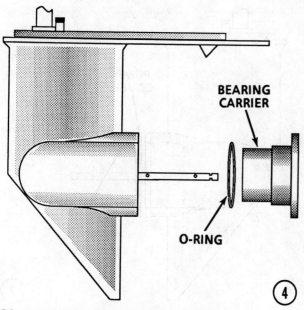

BEARING CARRIER

O-RING

④

Alternate Pump Cover Removal

Using a soft head mallet, rotate the cover through 90°, and then gently tap the ears of the cover to jar it free.

3- Slide the water pump impeller free of the propeller shaft. Pull the impeller drive pin out of the propeller shaft.

Bearing Carrier Removal

4- Remove the two bolts securing the bearing carrier in the lower unit. Now, using two screwdrivers -- one working on the opposite side of the carrier -- pry the carrier clear of the lower unit housing, and

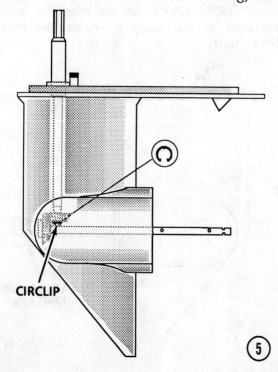

CIRCLIP

⑤

then move it off the propeller shaft. Remove and discard the **O**-ring.

Alternate Carrier Removal

Using a soft head mallet, rotate the carrier through 90°, and then gently tap the ears of the carrier to jar it free. Remove and discard the **O**-ring.

5- Using a thin screwdriver, pry the circlip free from the end of the driveshaft. This clip holds the pinion gear onto the driveshaft. The clip may not come free on the first try, but have patience and it can be worked free.

6- With one hand, pull the driveshaft up and out of the lower unit housing and at the same time be prepared to catch the pinion gear with the other hand when the gear comes free of the shaft.

Propeller Shaft Removal

7- Remove the propeller shaft from the lower unit housing. The forward gear will come with the shaft. The thrust washer may come with the forward gear **OR** it may remain in the lower unit against the forward bearing. If the thrust washer remained in the housing, reach in and pull it out. The forward bearing is a press-fit and will be

PULL UP

DRIVESHAFT

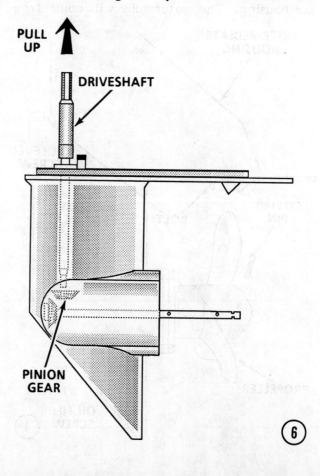

PINION GEAR

⑥

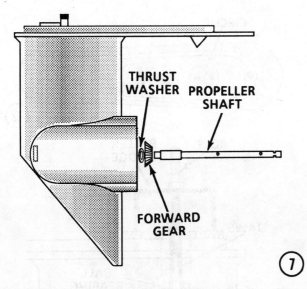

THRUST WASHER

PROPELLER SHAFT

FORWARD GEAR

(7)

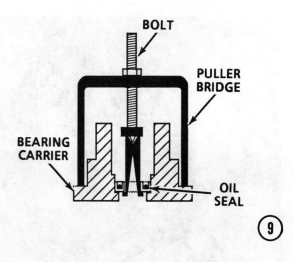

BOLT

PULLER BRIDGE

BEARING CARRIER

OIL SEAL

(9)

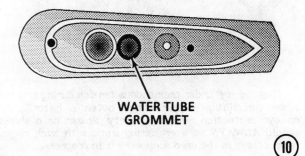

WATER TUBE GROMMET

(10)

"pulled" free later, as described in the Step 16.

Bearing Carrier Disassembling

8- Obtain the following special tools: OMC Puller Bridge P/N 432127, OMC Two Jaw Puller P/N 432130, and OMC Backing Plate P/N 115312. Make a setup with the tools to pull the ball bearing from the forward end of the bearing carrier. Hook the ends of Jaw Puller behind the inner race of the bearing. Tighten the bolt to pull the bearing free of the carrier.

9- Use the same setup as described in Step 8 to pull the oil seal from the forward end of the bearing carrier.

Evacuating Driveshaft Bore

10- Remove and discard the water tube grommet.

11- Using the same tools listed in Step 8, make a setup on top of the lower unit housing and pull the oil seal free.

SAFETY WORDS

Circlips are made of spring steel and may pop out of the groove, or slip free of the pliers, with considerable force. Therefore, warn others in the area and **WEAR** eye protection glasses or a shield while removing this type of ring.

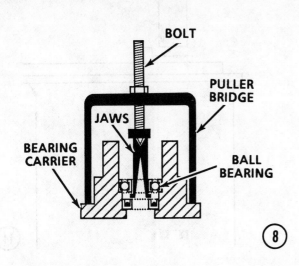

BOLT

PULLER BRIDGE

JAWS

BEARING CARRIER

BALL BEARING

(8)

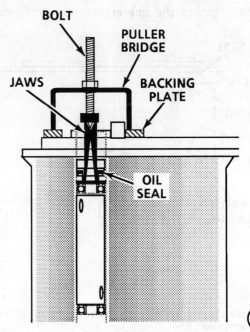

BOLT

PULLER BRIDGE

JAWS

BACKING PLATE

OIL SEAL

(11)

Circlips are under tremendous tension during removal and installation — presenting a potential hazard. As an eye protection measure, safety glasses or a shield should ALWAYS be worn during work with such rings. Warn others in the area such work is in progress.

12- Obtain a pair of internal circlip pliers and remove the circlip from the groove in the driveshaft bore.

13- Using the same tools and setup as described in Step 11, pull the upper ball bearing from the driveshaft bore.

14- Lift the driveshaft sleeve free of the bore.

15- Obtain OMC Bearing Puller Kit P/N 115307. Insert the driver into the bottom of

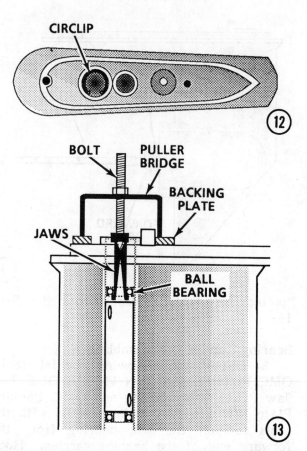

the lower ball bearing. Pass the threaded rod through the plate, down into the driveshaft bore and thread it into the driver. Tighten the bolt on the upper end of the threaded rod. This action will draw the lower ball bearing upward and free of its installation position. Once free, it may be easily removed from the bore.

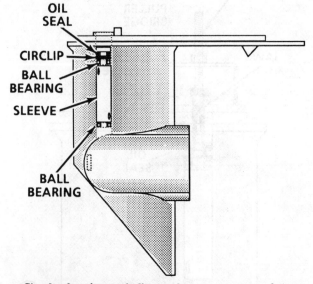

Simple drawing to indicate the arrangement of parts in the driveshaft bore.

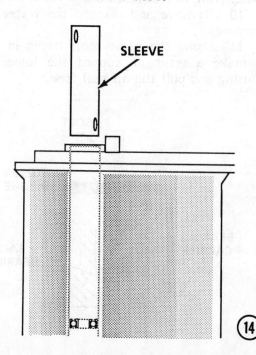

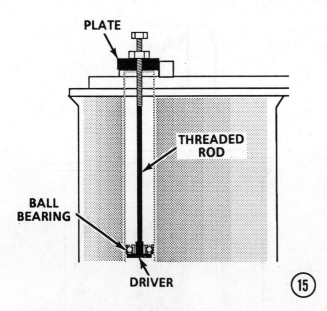

PLATE

THREADED
ROD

BALL
BEARING

DRIVER   ⑮

Forward Bearing Removal

16- Using the same setup as outlined in Steps 8, 9, 11, and 13, pull the forward bearing free of the lower unit housing.

CLEANING AND INSPECTING

Clean all water pump parts with solvent, and then dry them with compressed air. Inspect the water pump cover and base for cracks and distortion. If possible, **ALWAYS** install a new water pump impeller while the lower unit is disassembled. A new impeller will ensure extended satisfactory service and give "peace of mind" to the owner. If the old impeller must be returned to service, **NEVER** install it in reverse to the original direction of rotation. Installation in reverse will cause premature impeller failure.

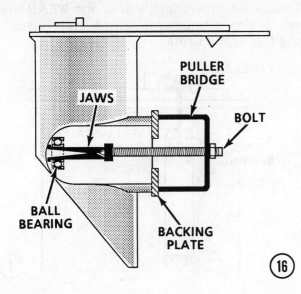

PULLER
BRIDGE

JAWS

BOLT

BALL
BEARING

BACKING
PLATE

⑯

Inspect the ends of the impeller blades for cracks, tears, and wear. Check for a glazed or melted appearance, caused from operating without sufficient water. If any question exists, as previously stated, install a **NEW** impeller if at all possible.

GOOD WORDS

If an old impeller is installed be **SURE** the impeller is installed in the same manner from which it was removed -- the blades will rotate in the same direction. **NEVER** turn the impeller over thinking it will extend its life. On the contrary, the blades would crack and break after just a short time of operation.

Inspect the bearing surface of the propeller shaft. Check the shaft surface for pitting, scoring, grooving, imbedded particles, uneven wear and discoloration.

Check the straightness of the propeller shaft with a set of V-blocks. Rotate the propeller on the blocks.

Good shop practice dictates installation of new O-rings and oil seals **REGARDLESS** of their appearance.

Clean the pinion gear and the propeller shaft with solvent. Dry the cleaned parts with compressed air.

Check the pinion gear and the drive gear for abnormal wear.

ASSEMBLING

The following procedures provide detailed instructions to assemble and install virtually every part of the lower unit. If a component was not removed or disassembled, simply skip the steps involved with that part and continue with the necessary work to return the lower unit to service.

GOOD WORDS

Lubricate all bearings, gears, and shafts, with OMC Hi-Vis Gearcase lubricant before assembling the part.

Forward Bearing Installation

1- Obtain OMC Collar P/N 115313 and OMC Bearing and Seal Installer P/N 115310. Position the lower unit on a block of wood on the deck, as indicated in the accompanying illustration. If necessary, instruct an assistant to hold the lower unit steady while components are being installed into the lower unit housing.

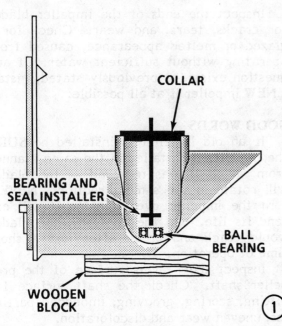

COLLAR

BEARING AND SEAL INSTALLER

BALL BEARING

WOODEN BLOCK

①

Insert the new forward bearing into the propeller shaft bore with the lettered side of the bearing facing **UP**. Fit the Collar over the Bearing and Seal Installer. The purpose of this collar is to ensure the bearing is driven in squarely.

Position the Bearing and Seal Installer, with the collar in place, over the forward bearing. Use a soft head mallet and drive the forward bearing into place until it is seated.

Lower Unit Housing Buildup

2- Obtain OMC Collar P/N 115314 and Bearing and Seal Installer P/N 115310.

Insert the new lower driveshaft bearing into the driveshaft bore with the lettered side of the bearing facing **UP**. Fit the Collar over the Bearing and Seal Installer. The purpose of this collar is to ensure the bearing is driven in squarely.

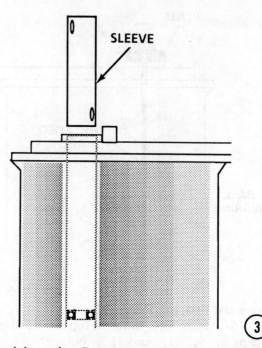

COLLAR

BEARING AND SEAL INSTALLER

BALL BEARING

②

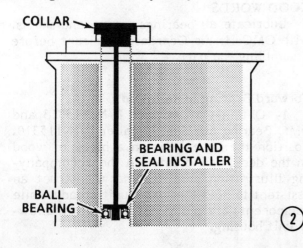

SLEEVE

③

Position the Bearing and Seal Installer, with the collar in place, over the forward bearing. Use a soft head mallet and drive the forward bearing into place until it is seated.

3- Insert the sleeve into the driveshaft bore. Either end may go in first.

4- Using the same tool setup as used in Step 2, install the upper bearing into the driveshaft bore, with the lettered side facing **UP**. Continue to drive the bearing into place until the tool makes contact with the shoulder in the bore.

SAFETY WORDS

Circlips are made of spring steel and may pop out of the groove, or slip free of the pliers, with considerable force. Therefore, warn others in the area and **WEAR** eye protection glasses or a shield while installing this type of ring.

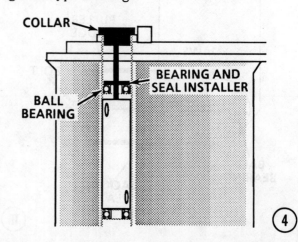

COLLAR

BEARING AND SEAL INSTALLER

BALL BEARING

④

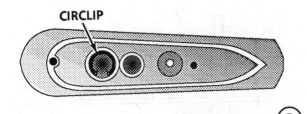

CIRCLIP

(5)

5- Wearing an eye protection device, and using a pair of internal circlip pliers, install the circlip into the driveshaft bore groove, with the sharp side of the circlip facing **UP**.

6- Apply OMC Gasket Sealing Compound to the exterior surface of the oil seal. Obtain OMC Bearing and Seal Installer P/N 115310. Insert the oil seal into the driveshaft bore with the spring facing **DOWN**. Drive the seal with the installer tool and a soft head mallet until the seal seats.

7- Apply Scotch-Grip Rubber Adhesive 1300 to the exterior surface of the water tube grommet. Install the grommet over the water tube.

Bearing Carrier Assembling

8- Pack the lip of the oil seal with OMC Triple-Guard Lubricant. Apply OMC Gasket Sealing Compound, or equivalent, to the

*Circlips are under tremendous tension during removal and installation — presenting a potential hazard. As an eye protection measure, safety glasses or a shield should **ALWAYS** be worn during work with such rings. Warn others in the area such work is in progress.*

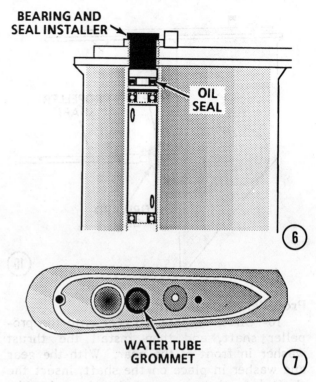

BEARING AND SEAL INSTALLER

OIL SEAL

(6)

WATER TUBE GROMMET

(7)

exterior surface of the oil seal. Position the oil seal into the forward end of the bearing carrier, with the lettered side of the seal facing **UP**. The tool will then bear against this lettered surface. Drive the seal into place using OMC Bearing and Seal Installer P/N 115311 and a soft head mallet.

9- Install a new ball bearing into the bearing carrier, in the same manner as the seal in the previous step was installed. The lettered side of the bearing must face UP. The bearing is correctly installed when the tool makes contact with the shoulder in the bearing carrier.

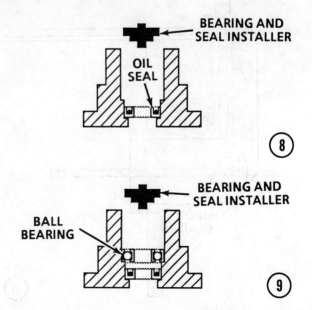

BEARING AND SEAL INSTALLER

OIL SEAL

(8)

BEARING AND SEAL INSTALLER

BALL BEARING

(9)

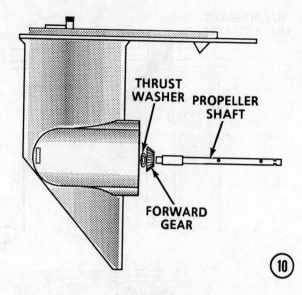

Propeller Shaft Installation

10- Slide the forward gear onto the propeller shaft, and then install the thrust
washer in front of the gear. With the gear
and washer in place on the shaft, insert the
shaft into the lower unit housing with the
forward end of the shaft indexed into the
forward bearing.

11- Lower the driveshaft down into the
bore. Slide the pinion gear up onto the
lower end of the driveshaft. The internal

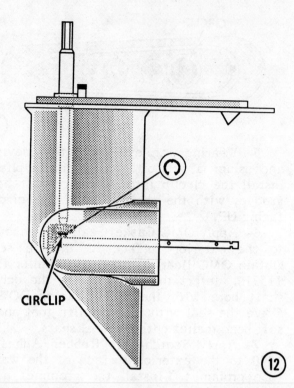

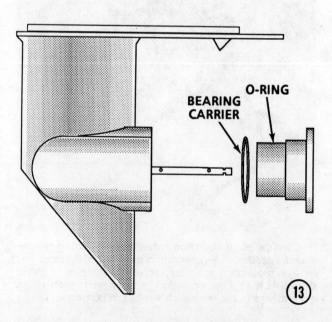

splines on the pinion gear will index with the
splines on the driveshaft and the pinion gear
teeth will mesh with the forward gear teeth.
Rotating the pinion gear slightly will permit
the splines to index and the gears to mesh.

12- Snap the circlip into the groove on
the end of the driveshaft. The circlip is
installed with the sharp side of the clip
facing **OUTWARD.**

Bearing Carrier Installation

13- Apply a light coating of OMC
Triple-Guard Lubricant to a new bearing

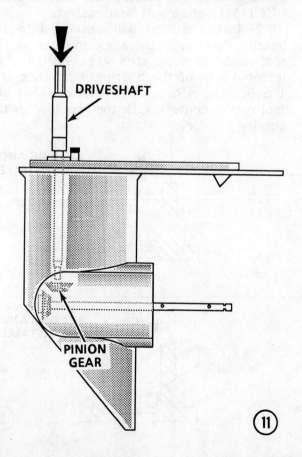

carrier O-ring. Install the O-ring into the groove in the carrier. Now, cover the sharp end of the propeller shaft with several layers of available tape to protect the seals already installed in the bearing carrier as the carrier is installed. After the end of the propeller shaft has been wrapped with the tape, carefully slide the bearing carrier onto the propeller shaft and forward into the lower unit housing.

Water Pump Installation

14- Push the water pump impeller drive pin into the hole in the propeller shaft. Cover the impeller vanes with a light coating of engine oil. Install the water pump impeller onto the propeller shaft and slide it forward on the shaft until the groove in the impeller begins to index over the pin. Rotate the water pump impeller and shaft **CLOCKWISE** and at the same time, push the impeller into the bearing carrier cavity. The impeller is properly installed when the pin is indexed to the full depth of the groove.

15- Position the water pump cover over the impeller. Apply a coating of OMC Gasket Sealing Compound to the threads of the two water pump cover bolts. Secure the cover in place with the two bolts. Tighten the bolts to a torque value of 60-84 in lbs (7-9Nm).

Lower Unit Installation

16- Apply a coating of OMC Nut Lock to the threads of the two lower unit mounting

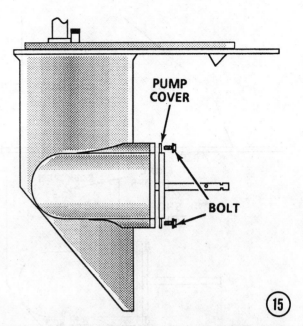

PUMP COVER

BOLT

(15)

bolts. Apply a light coating of OMC Moly Lube or Multi-purpose Water Resistant Lubricant to the powerhead driveshaft flats and some lubricant into the intermediate tube. Move the lower unit housing up to mate with the intermediate housing. Bring the two

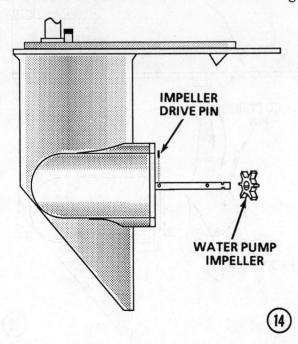

IMPELLER DRIVE PIN

WATER PUMP IMPELLER

(14)

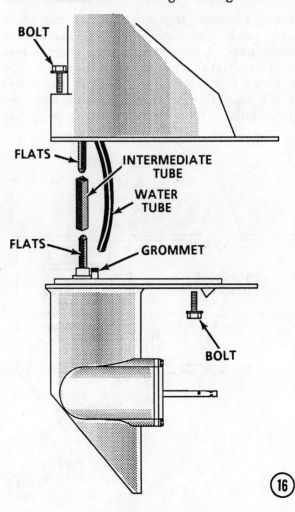

BOLT

FLATS

INTERMEDIATE TUBE

WATER TUBE

FLATS

GROMMET

BOLT

(16)

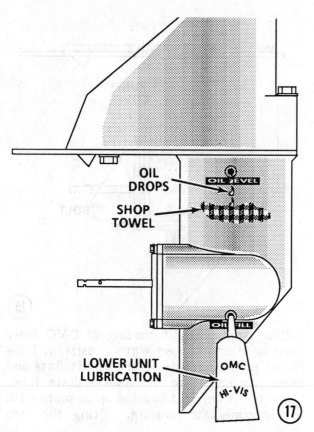

OIL DROPS

SHOP TOWEL

OIL LEVEL

OIL FILL

LOWER UNIT LUBRICATION

OMC HI-VIS

⑰

shaft may be slowly rotated **CLOCKWISE** and only **CLOCKWISE** -- to prevent damage to the water pump vanes.

Secure the two housings with the two bolts. Tighten the bolts to a torque value of 89-115 in lbs (10-13Nm).

Filling Lower Unit With Lubricant

17- Remove oil level opening plug and the oil fill opening plug. Obtain a tube of OMC Hi-Vis Lubricant. Insert the tube into the lower oil fill opening. Fill the lower unit with lubricant until the lubricant begins to escape from the oil level opening.

18- Install the oil level plug **FIRST,** and then the oil fill plug.

The lower unit oil capacity is approximately three fluid ounces (90ml). Tighten both plugs to a torque value of 30-40 in lb (3.5-4.5Nm).

Apply OMC Triple Guard Lubricant to a **NEW** propeller shear pin and insert the pin into the hole in the propeller shaft.

Propeller Installation

19- Apply OMC Triple Guard to the end of the propeller shaft. Install the propeller onto the shaft, indexing the propeller over the shear pin. Secure the propeller to the shaft with a **NEW** cotter pin. Separate the ends of the pin and spread them apart around the propeller to properly secure the propeller to the shaft.

housings closer together and at the same time guide the lower unit driveshaft to index into the lower end of the short intermediate driveshaft tube and the water tube to index through the grommet in the lower unit housing. As an aid to indexing the intermediate tube and shaft, the propeller

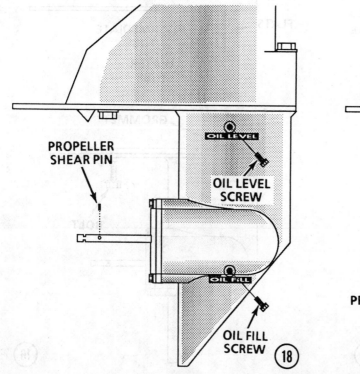

PROPELLER SHEAR PIN

OIL LEVEL

OIL LEVEL SCREW

OIL FILL

OIL FILL SCREW

⑱

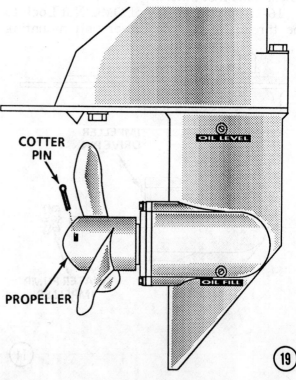

COTTER PIN

OIL LEVEL

PROPELLER

OIL FILL

⑲

8-5 TYPE "B" LOWER UNIT
3HP AND 4HP
NO REVERSE

DESCRIPTION

The **Type "B"** lower unit does not have a reverse gear, but a shift arrangement allows the forward drive gear to be disengaged from the pinion gear, thus providing a "neutral" condition. Reverse action of the propeller is accomplished by the boat's operator swinging the outboard with the tiller handle 180°, with the unit in **FORWARD** gear, and holding it in this position while the boat moves sternward. When the operator is ready to move the boat forward again, he/she simply swings the tiller handle back to the normal forward position.

LOWER UNIT REMOVAL

Disconnect the high tension spark plug lead to prevent accidental firing of the cylinder as the work progresses.

The following procedures present complete instructions to remove, disassemble, assemble virtually all parts of the 3hp and 4hp lower unit.

If a particular part is believed to be in satisfactory condition and does not need replacement, simply skip the step/s involved in removing the part and continue with the work required to restore the lower unit to service.

1- Position a suitable container under the lower unit, and then remove the oil level screw and the oil fuel screw. Allow the gear lubricant to drain into the container. As the lubricant drains, catch some with your fingers from time to time, and rub it between a thumb and finger to determine if any metal particles are present.

If metal particles are detected in the lubricant, the unit must be completely disassembled, inspected, and the damaged parts replaced.

Check the color of the lubricant as it drains. A whitish or creamy color indicates the presence of water in the lubricant. Check the drain pan for signs of water separation from the lubricant. The presence of any water in the gear lubricant is **BAD NEWS**. The unit must be completely disassembled, inspected, the cause of the problem determined, and corrected.

After the lubricant has drained, temporarily install the oil level screw and the oil fill screw.

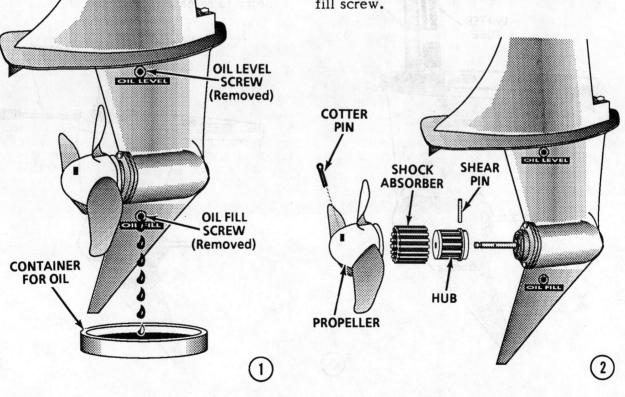

Propeller Removal

2- Straighten the cotter pin, and then pull it free of the propeller with a pair of pliers or cotter pin removal tool. Remove the propeller, the shock absorber, the shear pin, and the hub.

LOWER UNIT DISASSEMBLING

3- Remove the two attaching bolts securing the lower unit to the intermediate housing. Separate the lower unit from the intermediate housing. The water tube will come free of the grommet on top of the water pump cover and remain with the intermediate housing. The driveshaft will remain with the lower unit. The shift rod will slide out of the shift rod bushing in the lower unit and remain with the intermediate housing.

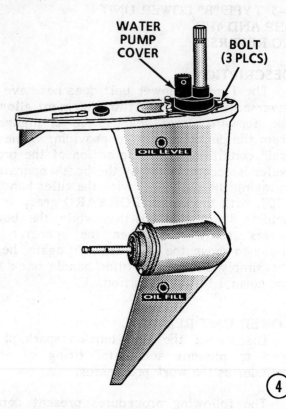

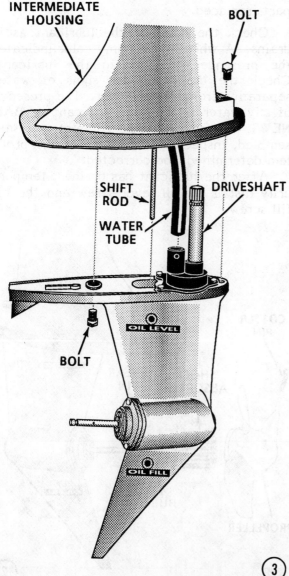

Water Pump Removal

4- Remove the three bolts securing the water pump cover to the lower unit. Slide the cover and its contents up and off the driveshaft.

5- Using a thin flat blade screwdriver, pry up evenly around the pump cover perimeter to release it from the plate.

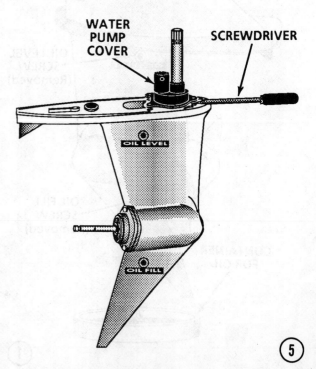

6- Remove and discard the O-ring from the cover. Using a pair of needlenose pliers, pull the pump impeller free of the cover. If the insert is damaged, pull the insert out of the cover with the needlenose pliers.

7- Remove the impeller drive pin from the flat on the driveshaft.

8- Using a thin flat blade screwdriver pry evenly around the perimeter of the pump plate to release it from the lower unit surface. Lift off the plate, gasket, and the water intake screen. Discard the gasket and clean the intake screen. A used tooth brush is a very useful tool to clean the screen.

Driveshaft Removal

9- Pull the driveshaft up and out of the lower unit. The pinion gear on the lower end of the driveshaft will slide free because it is not held in place with a circlip or other restraining device.

Propeller Shaft Removal and Disassembling

10- Remove the two bolts securing the bearing carrier in the lower unit. Now, using two screwdrivers -- one working on the opposite side of the carrier -- pry the carrier clear of the lower unit housing. Pull the propeller shaft assembly out of the lower unit.

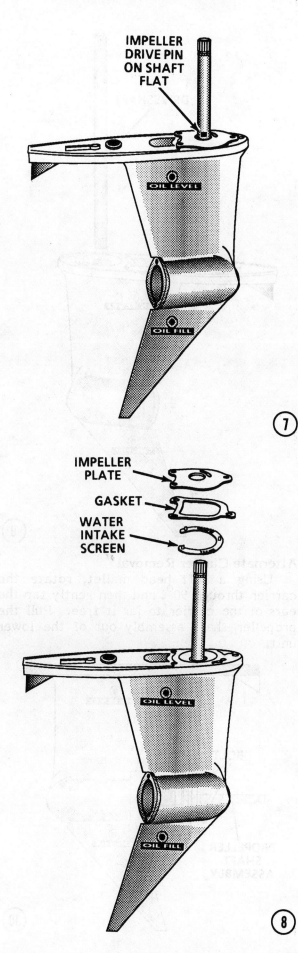

IMPELLER DRIVE PIN ON SHAFT FLAT

OIL LEVEL

OIL FILL

⑦

IMPELLER PLATE

GASKET

WATER INTAKE SCREEN

OIL LEVEL

OIL FILL

⑧

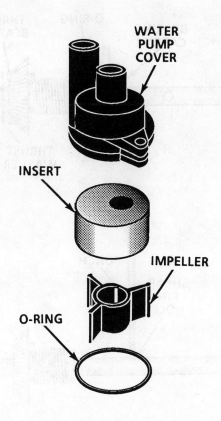

WATER PUMP COVER

INSERT

IMPELLER

O-RING

⑥

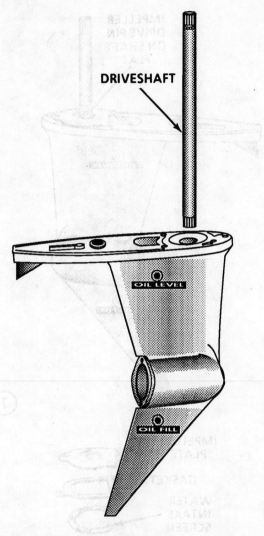

DRIVESHAFT

⑨

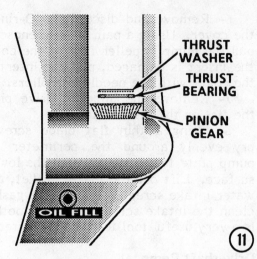

THRUST
WASHER

THRUST
BEARING

PINION
GEAR

OIL FILL

⑪

11- Reach inside the lower unit cavity and pull out the pinion gear, the thrust bearing, and the thrust washer.

12- Pull the bearing carrier free of the propeller shaft. Remove and discard the O-ring around the carrier. Remove the thrust washer and bearing from the forward end of the shaft.

13- Insert a punch into the shear pin hole in the propeller shaft. Hold the forward gear to prevent it from rotating, and at the same time rotate the propeller shaft **COUNTERCLOCKWISE** with the punch. Continue rotating and pulling on the shaft to work the gear free of the shaft.

Alternate Carrier Removal

Using a soft head mallet, rotate the carrier through 90°, and then gently tap the ears of the carrier to jar it free. Pull the propeller shaft assembly out of the lower unit.

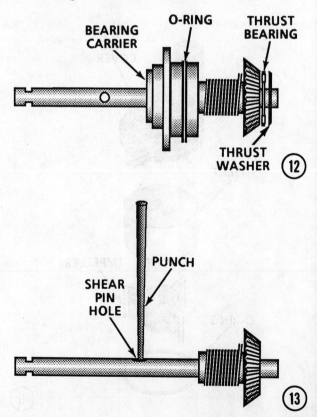

BEARING
CARRIER

O-RING

THRUST
BEARING

THRUST
WASHER

⑫

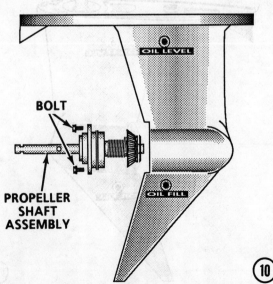

BOLT

PROPELLER
SHAFT
ASSEMBLY

OIL LEVEL

OIL FILL

⑩

PUNCH

SHEAR
PIN
HOLE

⑬

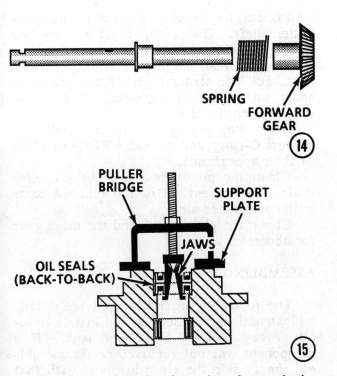

SPRING

FORWARD GEAR

(14)

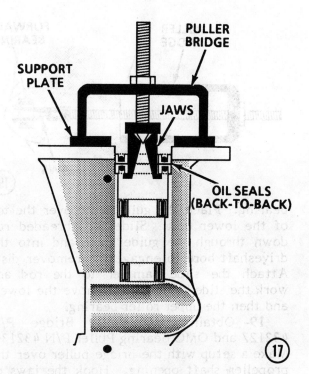

PULLER BRIDGE

SUPPORT PLATE

JAWS

OIL SEALS (BACK-TO-BACK)

(15)

(17)

14- Slide the clutch spring free of the forward gear hub.

Bearing Carrier Disassembling

15- Obtain OMC Puller Bridge P/N 432127, OMC Small Puller Jaws P/N 432131 and a suitable support plate. Make a setup with the puller bridge on the aft end of the bearing carrier. Insert the small puller jaws behind the two oil seals and pull both seals free of the carrier.

16- Obtain OMC Bearing Remover and Installer P/N 392091. Insert the short end of the special tool into the rear opening of the bearing carrier. Using a soft head mallet, drive the roller bearing free of the carrier.

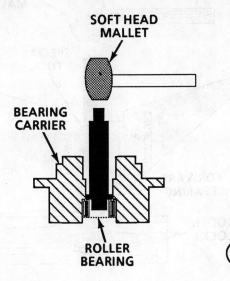

SOFT HEAD MALLET

BEARING CARRIER

ROLLER BEARING

(16)

Lower Unit Housing Disassembling

17- Obtain OMC Puller Bridge P/N 432127 and OMC Puller Jaws P/N 432130. Make a setup with the puller bridge on top of the lower unit. Insert the puller jaws behind the two oil seals and pull them free of the housing.

18- Obtain OMC Driveshaft Bearing Remover and Installer Kit P/N 392092 (consisting of a guide plate, a remover disc, and threaded rod) and OMC Slide Hammer P/N 391008. Position the remover disc inside the lower unit cavity under the lower roller

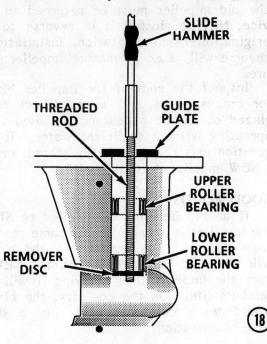

SLIDE HAMMER

THREADED ROD

GUIDE PLATE

UPPER ROLLER BEARING

LOWER ROLLER BEARING

REMOVER DISC

(18)

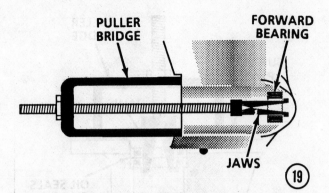

PULLER BRIDGE   FORWARD BEARING   JAWS

⑲

bearing. Place the guide plate over the top of the lower unit. Slide the threaded rod down through the guide plate and into the driveshaft bore to engage the remover disc. Attach the slide hammer to the rod and work the slide hammer to remove the lower, and then the upper roller bearing.

19- Obtain OMC Puller Bridge P/N 432127 and OMC Bearing Puller P/N 432130. Make a setup with the bridge puller over the propeller shaft opening. Hook the jaws of the bearing puller behind the forward bearing and pull the bearing free.

CLEANING AND INSPECTING

Clean all water pump parts with solvent, and then dry them with compressed air. Inspect the water pump cover and base for cracks and distortion. If possible, **ALWAYS** install a new water pump impeller while the lower unit is disassembled. A new impeller will ensure extended satisfactory service and give "peace of mind" to the owner. If the old impeller must be returned to service, **NEVER** install it in reverse to the original direction of rotation. Installation in reverse will cause premature impeller failure.

Inspect the ends of the impeller blades for cracks, tears, and wear. Check for a glazed or melted appearance, caused from operating without sufficient water. If any question exists, as previously stated, install a **NEW** impeller if at all possible.

GOOD WORDS

If an old impeller is installed be **SURE** the impeller is installed in the same manner from which it was removed -- the blades will rotate in the same direction. **NEVER** turn the impeller over thinking it will extend its life. On the contrary, the blades would crack and break after just a short time of operation.

Inspect the bearing surface of the propeller shaft. Check the shaft surface for pitting, scoring, grooving, imbedded particles, uneven wear and discoloration.

Check the straightness of the propeller shaft with a set of V-blocks. Rotate the propeller on the blocks.

Good shop practice dictates installation of new O-rings and oil seals **REGARDLESS** of their appearance.

Clean the pinion gear and the propeller shaft with solvent. Dry the cleaned parts with compressed air.

Check the pinion gear and the drive gear for abnormal wear.

ASSEMBLING

The following procedures provide detailed instructions to assemble and install virtually every part of the lower unit. If a component was not removed or disassembled, simply skip the steps involved with that part and continue with the necessary work to return the lower unit to service.

GOOD WORDS

Lubricate all bearings, gears, and shafts, with OMC Hi-Vis Gearcase lubricant before assembling the part.

Forward Bearing Installation

1- Obtain OMC Bearing and Seal Installer P/N 392091. Position the lower unit on a block of wood on the deck, as indicated in the accompanying illustration. If necessary, instruct an assistant to hold the lower unit

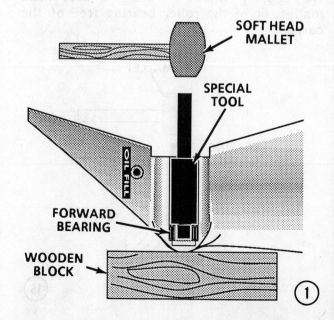

SOFT HEAD MALLET

SPECIAL TOOL

OIL FILL

FORWARD BEARING

WOODEN BLOCK

①

steady while components are being installed into the lower unit housing.

Insert the new forward bearing into the propeller shaft bore with the lettered side of the bearing facing **UP**.

Insert the **SHORT** end of the installer tool over the forward bearing. Use a soft head mallet and drive the forward bearing into place until it seats against the shoulder in the lower unit housing.

Lower Unit Buildup

2- Obtain OMC Driveshaft Bearing Remover and Installer Kit P/N 393092, consisting of a long threaded rod, a guide plate, a pin, and an installer disc. Insert the lower roller bearing into the driveshaft bore with the lettered side of the bearing facing **UP**. Make a setup with the kit parts over the driveshaft bore, with the installer disc threaded onto the rod with the concave side of the disc facing **DOWN** to bear against the bearing. Drive the bearing into the bore using a soft head mallet striking the end of the rod until the pin makes contact with the drive plate. The bearing is then seated properly.

3- Insert the upper driveshaft bearing into the driveshaft bore, with the lettered side of the bearing facing **UP**. Now, using the same special tool kit and set up as described in the previous step for the lower

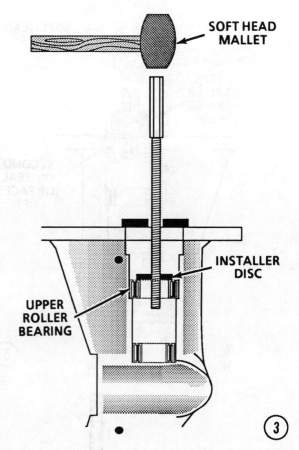

bearing, drive the upper bearing into the bore until it seats against a shoulder in the bore.

4- Pack the lips of both driveshaft oil seals with OMC Triple Guard Lubricant. Apply a coating of OMC Gasket Sealing Compound, or equivalent, to the exterior surfaces of both seals. Position the first seal in the driveshaft bore with the lip of the

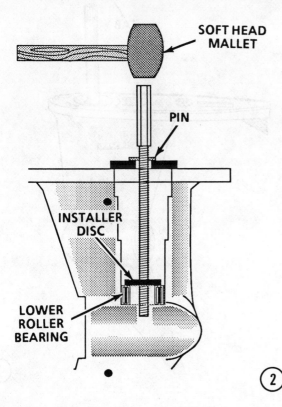

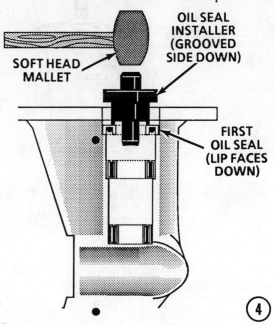

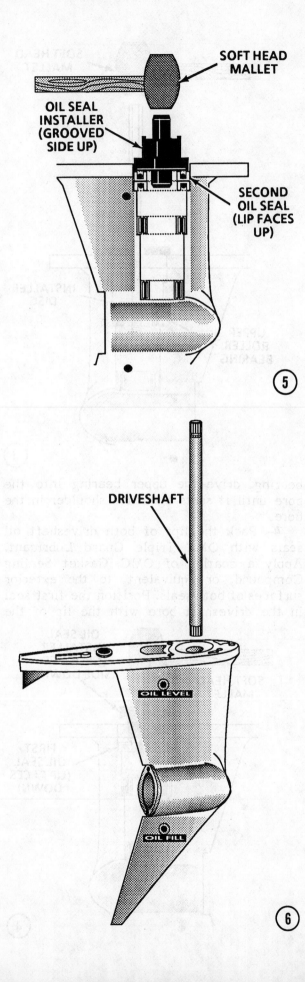

SOFT HEAD MALLET

OIL SEAL INSTALLER (GROOVED SIDE UP)

SECOND OIL SEAL (LIP FACES UP)

⑤

DRIVESHAFT

OIL LEVEL

OIL FILL

⑥

seal facing **DOWN**. Obtain OMC Seal Installer P/N 327572. Insert the tool into the bore with the grooved side facing **DOWN** to bear against the seal. Now, drive the seal into place using a soft head mallet, until the seal seats against a shoulder in the bore.

5- Install the second seal in the same manner as the first seal with **TWO** important changes. First, the lip of the seal **MUST** face **UP** and second, the grooved side of the tool faces **UP**. Drive the seal into place until it makes contact with the first seal. These seals are now installed back-to-back. One prevents lubricant from escaping and the other prevents water from entering the housing.

6- Lower the driveshaft into the lower unit bore through the seals and bearings.

Water Pump Assembling

7- Slide the water intake screen down over the driveshaft, with the three prongs on the screen facing **UP**. Install a new gasket and the impeller plate down over the screen.

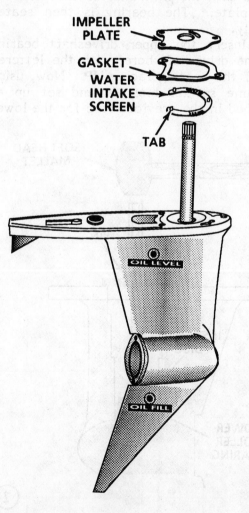

IMPELLER PLATE

GASKET

WATER INTAKE SCREEN

TAB

OIL LEVEL

OIL FILL

⑦

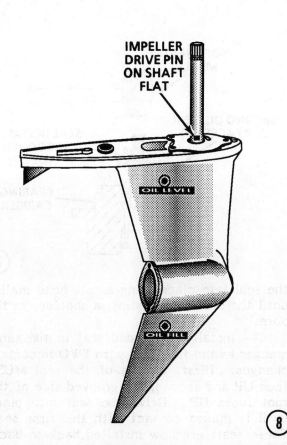

IMPELLER
DRIVE PIN
ON SHAFT
FLAT

OIL LEVEL

OIL FILL

8

8- Apply a dab of lubricant onto the pump impeller drive pin, and then place it against the flat on the driveshaft.

9- Push a new insert into the water pump cover. Install a new water pump impeller inside the insert, with each vane of the impeller folded back slightly to ride in

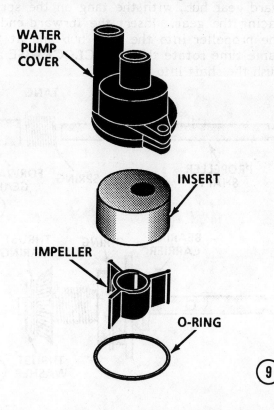

WATER
PUMP
COVER

INSERT

IMPELLER

O-RING

9

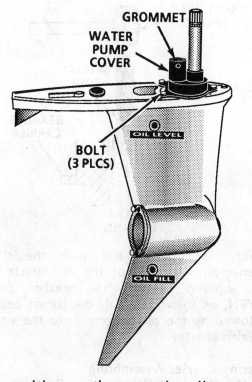

GROMMET
WATER
PUMP
COVER

OIL LEVEL

BOLT
(3 PLCS)

OIL FILL

10

this position as the pump impeller rotates **CLOCKWISE.** Install a new O-ring into the groove in the water pump cover.

10- Slide the assembled water pump down onto the driveshaft and over the plate. Secure the pump in place with the two attaching bolts. Tighten the bolts to a torque value of 25-35 in lb (2.8-4Nm). Apply Scotch-Grip Rubber Adhesive 1300 to the exterior surface of the water tube grommet. Install the grommet into the water pump cover.

Pinion Gear Installation

11- Invert the lower unit (turn it upside down), and clamp the driveshaft in a vise equipped with soft jaws. Reach into the

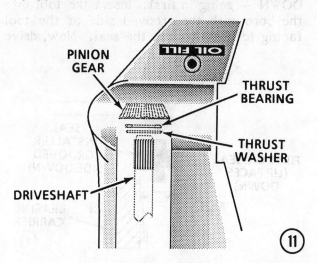

OIL FILL

PINION
GEAR

THRUST
BEARING

THRUST
WASHER

DRIVESHAFT

11

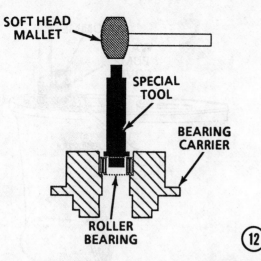

SOFT HEAD MALLET / SPECIAL TOOL / BEARING CARRIER / ROLLER BEARING ⑫

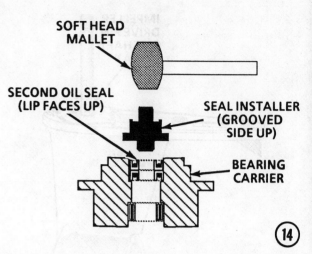

SOFT HEAD MALLET / SECOND OIL SEAL (LIP FACES UP) / SEAL INSTALLER (GROOVED SIDE UP) / BEARING CARRIER ⑭

lower unit cavity and slide the thrust washer onto the end of the driveshaft with the tapered side of the washer facing **DOWN,** as shown. Install the thrust bearing followed by the pinion gear onto the end of the driveshaft.

Bearing Carrier Assembling

12- Obtain OMC Bearing Remover and Installer P/N 392091. Position a new roller bearing into the forward end of the bearing carrier, with the lettered side of the bearing facing **UP** -- outward. Now, drive the bearing into place in the carrier with the special tool and a soft head mallet until the bearing seats against a shoulder in the carrier.

13- Obtain OMC Seal Installer P/N 327572. Pack the lips of both propeller shaft oil seals with OMC Triple Guard Lubricant. Apply a coating of OMC Gasket Sealing Compound, or equivalent, to the exterior surfaces of both seals. Position the first seal into the aft end of the bearing carrier, with the lip of the seal facing **DOWN** -- going in first. Insert the tool into the bore with the grooved side of the tool facing to bear against the seal. Now, drive

the seal into place using a soft head mallet until the seal seats against a shoulder in the bore.

14- Install the second seal in the same manner as the first seal with **TWO** important changes. First, the lip of the seal **MUST** face **UP** and second, the grooved side of the tool faces **UP**. Drive the seal into place until it makes contact with the first seal. These seals are now installed back-to-back. One prevents lubricant from escaping and the other prevents water from entering the housing.

Propeller Shaft Assembling and Installation

15- Slide the clutch spring onto the forward gear hub, with the tang on the spring facing the gear. Insert the forward end of the propeller into the gear hub and at the same time rotate the shaft **CLOCKWISE** and push the shaft into place.

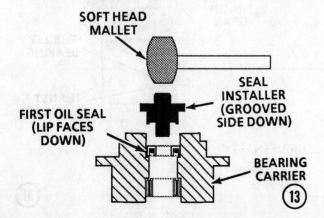

SOFT HEAD MALLET / SEAL INSTALLER (GROOVED SIDE DOWN) / FIRST OIL SEAL (LIP FACES DOWN) / BEARING CARRIER ⑬

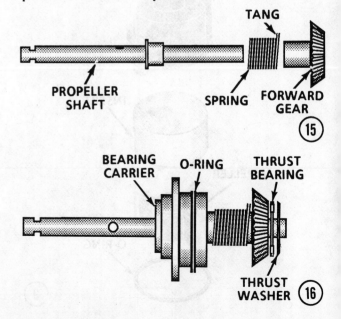

TANG / PROPELLER SHAFT / SPRING / FORWARD GEAR ⑮

BEARING CARRIER / O-RING / THRUST BEARING / THRUST WASHER ⑯

16- Apply a light coating of OMC Triple-Guard Lubricant to a new bearing carrier O-ring. Install the O-ring into the groove in the carrier. Now, cover the sharp end of the propeller shaft with several layers of available tape to protect the seals already installed in the bearing carrier as the carrier is installed. After the end of the propeller shaft has been wrapped with the tape, carefully slide the bearing carrier onto the propeller shaft. Install the thrust bearing followed by the thrust washer onto the forward end of the propeller shaft. The tapered end of the thrust washer **MUST** face toward the **END** of the shaft, after it is installed.

17- With the lower unit still upside down and the driveshaft still clamped in the vise, install the assembled propeller shaft into the lower unit. It will probably be necessary to rotate the propeller shaft **CLOCKWISE** slightly as the shaft moves into place to permit the teeth of the forward gear to mesh with the teeth of the pinion gear. Secure the bearing carrier and propeller shaft with the two bolts. Tighten the bolts to a torque value of 60-84 in lb (7-9Nm).

Lower Unit Installation

18- Apply a coating of OMC Nut Lock to the threads of the two lower unit mounting bolts. Apply a light coating of OMC Moly Lube or Multi-purpose Water Resistant Lubricant to the powerhead internal splines. Move the lower unit housing up to mate with the intermediate housing. Bring the two housings closer together and at the same time three things must be done almost si-

multaneously: Guide the lower unit driveshaft to permit the splines to index with the splines in the lower end of the crankshaft; guide the water tube to index through the grommet in the lower unit housing; and guide the shift rod into the shift rod bushing in the lower unit. As an aid to indexing the driveshaft and crankshaft splines, the propeller shaft may be slowly rotated **CLOCKWISE** and only **CLOCKWISE** --to prevent damage to the water pump vanes.

Secure the two housings with the two bolts. Tighten the bolts to a torque value of 60-84 in lbs (7-9Nm).

Filling Lower Unit With Lubricant

19- Remove the oil level opening plug and the oil fill opening plug. Obtain a tube of OMC Hi-Vis Lubricant. Insert the tube into the lower oil fill opening. Fill the

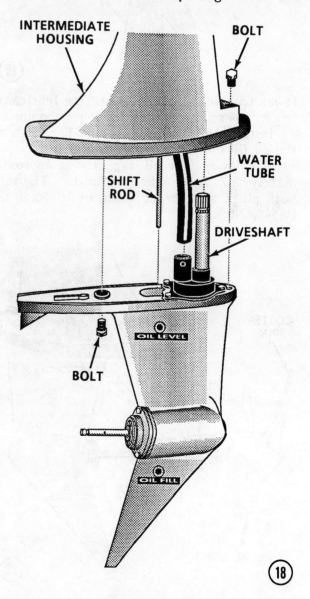

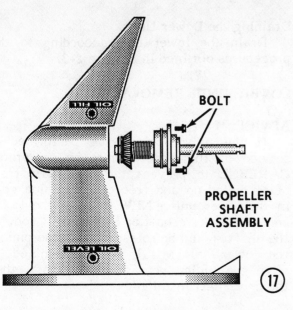

BOLT

PROPELLER
SHAFT
ASSEMBLY

⑰

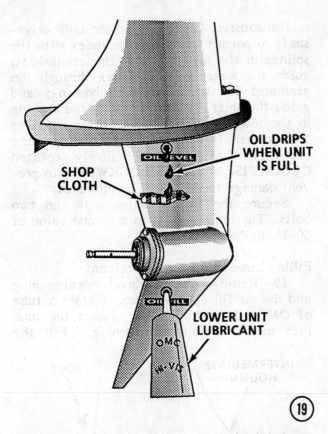

SHOP CLOTH

OIL DRIPS WHEN UNIT IS FULL

LOWER UNIT LUBRICANT

⑲

lower unit with lubricant until the lubricant begins to escape from the oil level opening.

Install the oil level plug **FIRST**, and then the oil fill plug.

The lower unit oil capacity is approximately 2.7 fluid ounces (80ml). Tighten both plugs to a torque value of 40-50 in lb (4.5-5.6Nm).

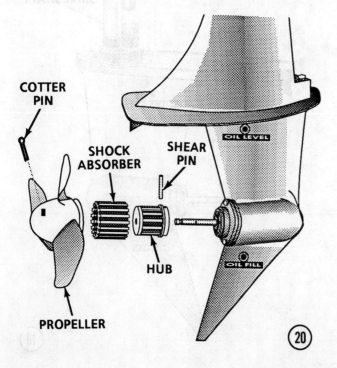

COTTER PIN

SHOCK ABSORBER

SHEAR PIN

HUB

PROPELLER

⑳

Propeller Installation

20- Apply OMC Triple-Guard Lubricant to a **NEW** propeller shear pin and to the end of the propeller shaft. Install the hub onto the propeller shaft and secure it in place with the shear pin. Install the shock absorber followed by the propeller. Secure the propeller on the shaft with a **NEW** cotter pin. Separate the ends of the pin and spread them apart around the propeller to properly secure the propeller to the shaft.

8-6 TYPE "C" LOWER UNIT 4 DELUX, 6HP, AND 8HP SHIFT WITH NO DISCONNECT

Description

This unit has a shift mechanism controlled through a shift handle on the starboard side of the exhaust housing. A shaft extends through the housing and is connected to a gear on the inside. Another gear transfers the motion downward by means of a shaft. The lower end of this shaft indexes into shift cam in the lower unit. Therefore, the shift rod is not disconnected in order to remove the lower unit.

The actual shift system consists of a cam and plunger arrangement, four detent balls, and a spring. Shifting is accomplished through action of the shift rod moving the cam which pushes the plunger to engage the detent balls, placing the unit in either forward or reverse gear operation.

Propeller Removal

Remove the propeller according to the detailed procedures outlined in Section 8-2.

Draining the Lower Unit

Drain the lower unit according to the procedures outlined in Section 8-3.

LOWER UNIT REMOVAL

ADVICE

If the only work to be performed is service of the water pump, be extremely **CAREFUL** to prevent the driveshaft from being pulled up and free of the pinion gear in the lower unit. **NEVER** carry the lower unit by the driveshaft or by the shift rod. If the shaft should be released from the pinion, the lower unit **MUST** be disassembled to align the pinion gear and driveshaft, then the driveshaft installed.

1- Disconnect the spark plug wire from the plug. Turn the propeller shaft **CLOCK-WISE,** and at the same time shift the unit into **FORWARD** gear using the shift handle. Remove the retaining bolts securing the lower unit to the exhaust housing. Two bolts are located under the cavitation plate on each side of the zinc. A third bolt is located just above the cavitation plate at the forward leading edge of the exhaust housing. **CAREFULLY** pull directly downward on the lower unit, to prevent damage to the water tube, driveshaft, or shift rod, and remove the lower unit. The shift rod and the driveshaft will come free of the powerhead with the lower unit.

GOOD WORDS

If this shift rod is frozen in the gear at the shift handle, the rod will not come free with the lower unit. In this case, the shift handle mechanism must be disassembled in order to remove the gear on the end of the shift rod. The powerhead would have to be removed to gain access to the gears.

MORE GOOD WORDS

In **MOST** cases, if any unit being serviced has the 6-inch extension, it is **NOT** neces-sary to remove the extension in order to "drop" the lower unit.

2- Position the lower unit in a vertical position on the edge of the work bench resting on the cavitation plate. Secure the lower unit in this position with a C-clamp. The lower unit will then be held firmly in a favorable position for further service work. An alternate method is to cut a groove in a short piece of 2" x 6" wood to accommodate the lower unit with the cavitation plate resting on top of the wood. Clamp the wood in a vise and service work may then be performed with the lower unit erect (in its normal position), or inverted (upside down). In both positions, the cavitation plate is the supporting surface.

WATER PUMP DISASSEMBLING

TAKE TIME

Take time to notice the four bolts passing through the water pump. If a water pump replacement is the only work to be performed, **DO NOT** remove the three bolts in the gearcase cover. If these bolts should mistakenly be removed, the shift rod will be dislodged -- **BAD NEWS.** It is then necessary to rebuild the lower unit.

3- With the lower unit in position as described in Step 2 above, remove the O-ring from the top of the driveshaft. Remove the four bolts through the water pump. Work the water pump upward and free of the driveshaft.

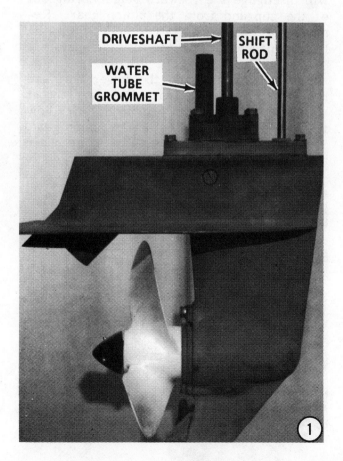

Photo 3 — WATER PUMP BOLT

4- Work the impeller up the driveshaft. Remove the impeller key.

5- Remove the water pump plate and gasket. Notice the plate and two gaskets under the water pump impeller. Notice how the plate is installed with the beveled, or rounded side of the plate facing **UP**. Both gaskets are identical and either one may be installed on top of the plate.

If the only work to be performed is service of the water pump, proceed directly to Page 8-45, Water Pump Installation.

LOWER UNIT DISASSEMBLING

6- Remove the two screws securing the bearing carrier in the lower unit. Tap the bearing carrier in either direction with a soft-headed mallet, to turn the bearing carrier in the housing. The ears of the bearing carrier will then protrude out the side of the lower unit. After the ears are exposed, tap

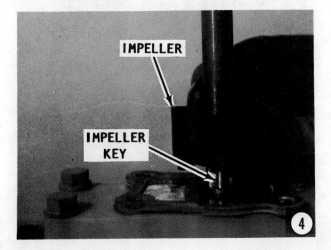

Photo 4 — IMPELLER / IMPELLER KEY

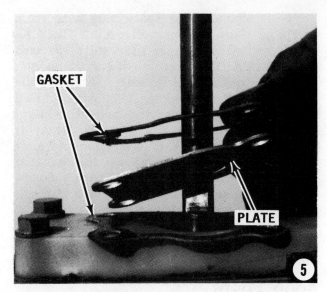

Photo 5 — GASKET / PLATE

on the bottom side of the ears and remove the bearing carrier.

7- Pull out the propeller shaft assembly. Slide the reverse gear and thrust washer from the aft end of the shaft. Remove the plunger, spring, and four detent balls from the forward end of the shaft.

8- Insert the propeller shaft down into the forward gear. Use a clenched fist and rap sharply on the end of the propeller shaft towards the cavitation plate. This action will disengage the forward gear bearing out of the cup and from the pinion gear. Remove the propeller shaft. Reach into the lower unit and remove the forward gear.

9- Pull upward on the driveshaft, and at the same reach inside the lower unit and remove the pinion gear, the two thrust washers, and the thrust bearing. Notice how the washers are installed on the pinion gear. They **MUST** be installed in the same order and position from which they are removed.

Photo 6 — BEARING CARRIER

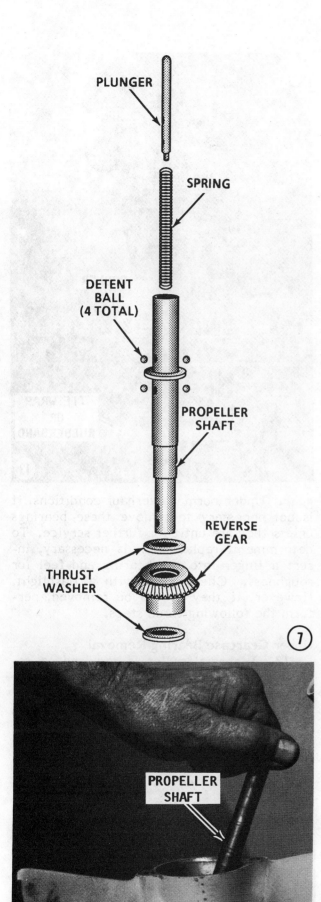

PLUNGER

SPRING

DETENT BALL (4 TOTAL)

PROPELLER SHAFT

REVERSE GEAR

THRUST WASHER

⑦

DRIVESHAFT

PINION GEAR

⑨

10 Clean the shift rod and then coat it with oil, as an aid to removing the gearcase cover. Remove the three retaining bolts, and then slide the gearcase cover up and free of the shift rod.

11- Lift the shift rod free of the gearcase, and at the same time reach into lower unit cavity and remove the shift cam.

PROPELLER SHAFT

⑧

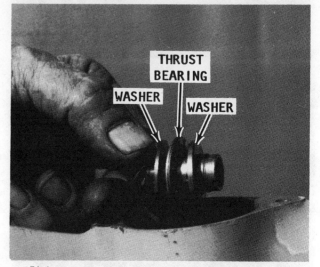

THRUST BEARING

WASHER   WASHER

Pinion gear with the beveled washers and thrust washer in the correct order on the shank.

Removing the Forward Bearing Race

12- Use a slide hammer with fingers to remove the forward bearing race from the lower unit housing.

GOOD WORDS

A driveshaft bearing is installed at the top of the lower unit. A pinion gear bearing is installed at the lower end of the driveshaft. A bearing is attached to the forward

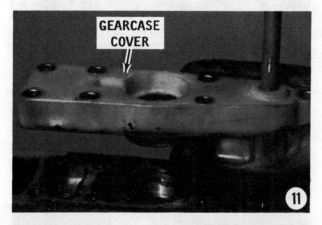

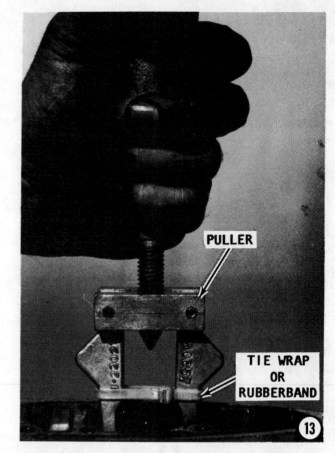

gear. Under normal overhaul conditions, it is not necessary to remove these bearings unless they are unfit for further service. To determine if replacement is necessary, insert a finger into the bearing and feel for roughness. Check them with a flashlight. However, if they are to be removed, perform the following three steps.

Upper Gearcase Bearing Removal

13- A special puller is required to remove the upper bearing installed under the gearcase cover. The complete puller consists of a slide hammer, OMC P/N 380658,

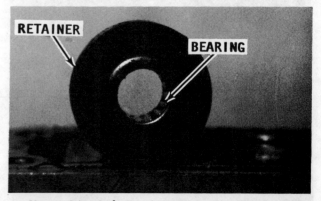

Upper driveshaft bearing and bearing housing. The housing is removed first, and then the bearing, but only if the bearing is unfit for further service.

and a bearing remover, OMC P/N 380657. If the special puller is not available, a puller similar to the one shown in the accompanying illustration may be used effectively. Fit the fingers of the puller underneath the bearing and hold them in place with a heavy rubberband. Use the slide hammer to pull the bearing free of the recess. The slide hammer in this case, is used in the opposite application for which it was designed.

Pinion Gear Bearing Removal

14- If the bearing has been damaged to the point where it must be replaced, it may be driven out with a punch, or other suitable tool, because further damage is no loss. Drive the bearing downward into the bore of the lower unit.

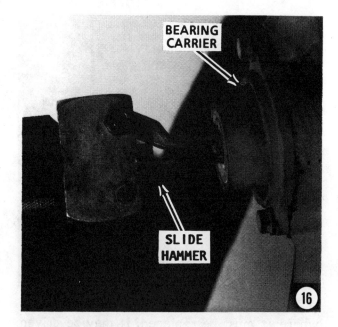

Forward Gear Bearing Removal

This bearing should only be removed if the bearing or gear is damaged and is to be replaced.

15- A beveled clamp-type arrangement, as shown in the accompanying illustration, is required. Such a tool will exert a force between the bearing and the gear in order to remove the bearing without harm. Install the clamp and tighten it around the bearing. Now, push in the center of the bearing. The gear and bearing will separate.

Seal Removal from the Bearing Carrier

16- Remove the O-ring and then reinstall the bearing carrier back onto the lower unit. Tighten the two retaining bolts. Use a slide hammer to remove the two seals from the bearing carrier. Remove the two retaining bolts and again remove the carrier from the lower unit.

Seal Removal from the Gearcase Cover

17- Use special tool OMC P/N 319880 or any type of punch to remove the seals. If

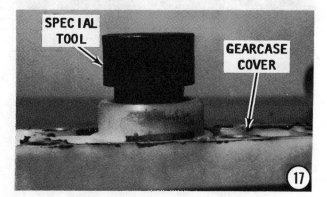

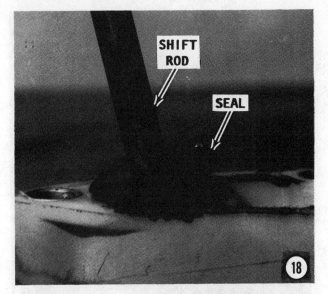

the seal has been damaged to the point where it must be replaced, it may be driven out with a punch, or other suitable tool, because further damage is no loss.

Shift Rod Seal Removal

18- Use the shift rod to remove the seal. Work the shift rod in the seal, and then push outward and the seal will pop out of the gearcase cover. After the seal has been removed, remove the O-ring.

CLEANING AND INSPECTING

Clean all water pump parts with solvent, and then dry them with compressed air.

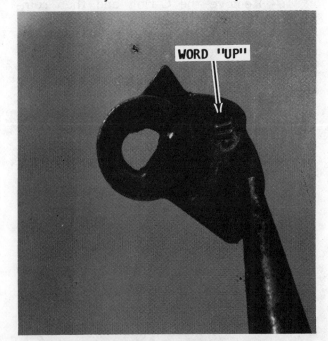

Shift cam with the word UP embossed on the arm to ensure proper installation.

Inspect the water pump cover and base for cracks and distortion, possibly caused from overheating. Inspect the face plate and water pump insert for grooves and/or rough surfaces.

If possible, **ALWAYS** install a complete new water pump while the lower unit is disassembled. A new impeller will ensure extended satisfactory service and give "peace of mind" to the owner. If the old impeller must be returned to service, **NEVER** install it in reverse to the original direction of rotation. Installation in reverse will cause premature impeller failure.

Inspect the impeller side seal surfaces and the ends of the impeller blades for cracks, tears, and wear. Check for a glazed or melted appearance, caused from operating without sufficient water. If any question exists, and as previously stated, install a new impeller if at all possible.

Clean all parts with solvent and dry them with compressed air. **DISCARD** all O-rings and gaskets. Inspect and replace the driveshaft if the splines are worn. Inspect the gearcase and exhaust housing for dam-

Pinion gear with the first thrust washer to be installed. The inside diameter of the washer is beveled.

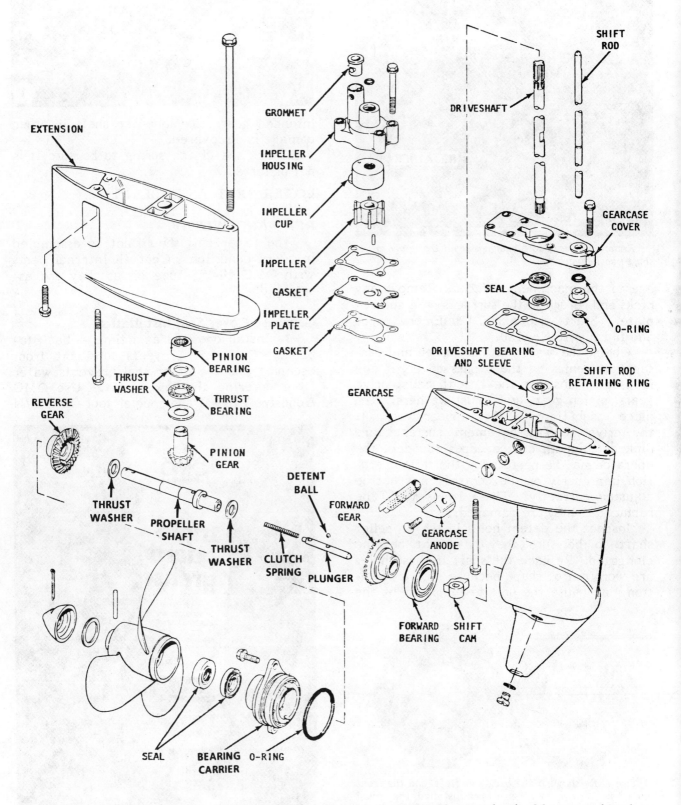

EXTENSION

GROMMET

IMPELLER
HOUSING

IMPELLER
CUP

IMPELLER

GASKET

IMPELLER
PLATE

GASKET

PINION
BEARING

THRUST
WASHER

THRUST
BEARING

PINION
GEAR

REVERSE
GEAR

THRUST
WASHER

PROPELLER
SHAFT

THRUST
WASHER

CLUTCH
SPRING

DETENT
BALL

PLUNGER

FORWARD
GEAR

DRIVESHAFT

SHIFT
ROD

GEARCASE
COVER

SEAL

O-RING

DRIVESHAFT BEARING
AND SLEEVE

SHIFT ROD
RETAINING RING

GEARCASE

GEARCASE
ANODE

FORWARD
BEARING

SHIFT
CAM

SEAL

BEARING
CARRIER

O-RING

Exploded drawing of the Type "C" lower unit matched with the 4D thru 8hp powerheads. Major parts have been identified.

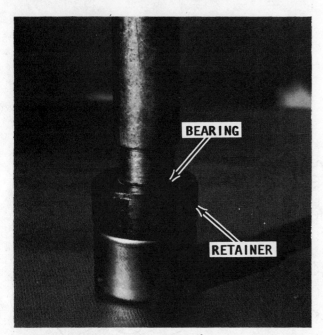

Removing the needle bearings from the upper drive-shaft bearing retainer.

age to the machined surfaces. Remove any nicks and refurbish the surfaces on a surface plate. Start with a No. 120 Emery paper and finish with No. 180.

Check the water intake screen and passages by removing the bypass cover, if one is used. Inspect the detent balls, drive gears, pinion gear, and thrust washers. Replace these items if they appear worn. If the drive gear arrangement surfaces are nicked, chipped, or the edges rounded, the operator may be performing the shift operation improperly or the controls may not be adjusted correctly. These items **MUST** be replaced if they are damaged.

Inspect the detent holes in the propeller shaft to be sure they have not become elongated. Replace the shaft if these holes are worn out of shape because such a condition may cause the unit to jump out of the

Two seals showing the back side (left) and the front side (right). When double seals are installed, they must always be installed back-to-back with Triple Guard Grease between the flat surfaces.

intended gear. Replace the shaft if other damage is discovered.

Check the clutch spring to be sure it is not distorted.

LOWER UNIT ASSEMBLING

READ AND BELIEVE

The lower unit should not be assembled in a dry condition. Coat all internal parts with OMC HI-VIS lube oil as they are assembled.

Gearcase Cover Seal Installation

1- Install one seal at a time -- back-to-back. One seal prevents lubricant from escaping and the other seal prevents water from entering the lower unit. Use OMC Adhesive **Type-M** and special tool OMC P/N

SEAL INSTALLER

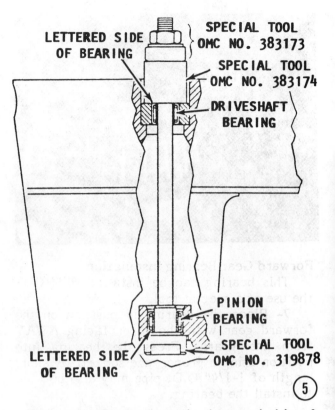

LETTERED SIDE OF BEARING — SPECIAL TOOL OMC NO. 383173
SPECIAL TOOL OMC NO. 383174
DRIVESHAFT BEARING
PINION BEARING
SPECIAL TOOL OMC NO. 319878
LETTERED SIDE OF BEARING

326547 or a socket the same size as the seal. Press the first seal into place from the bottom side of the cover.

2- Coat the seal with Triple Guard Lubricant. This type lubricant will not dissipate, but remains to perform its lubricating job between the seals.

3- Press the second seal into the gear case cover from the bottom side using the special tool or the proper size socket.

Shift Rod O-ring and Bushing Installation

4- Slide the O-ring into place in the gearcase cover. Glue the shift rod bushing into the cover using OMC Adhesive Type-M.

Pinion and Driveshaft Bearings — Installation

If the pinion and driveshaft bearings were removed, they are installed simultaneously using special tools. There is no other way to install these bearings properly without the special OMC tools. Sorry about that!

5- Obtain special tool OMC P/N 383173 and Spacer No. 383174. The pinion bearing is installed with the lettered side facing DOWN. The driveshaft bearing is installed with the lettered side facing UP. The tool

will be pressing against the lettered side of each bearing. Work the bearings into place by tightening on the tool nut.

Forward Gear Bearing Race Installation

If the forward gear bearing race was removed, a new bearing can only be properly installed using a special OMC tool. As with the pinion and driveshaft bearings, there is no other way. Again, sorry about that!

6- Obtain special tool OMC P/N 326025. Set the bearing race in the housing, and then use the tool and tap the race into position.

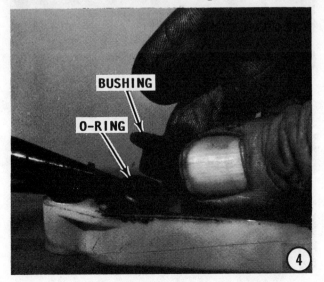

BUSHING
O-RING

SPECIAL TOOL

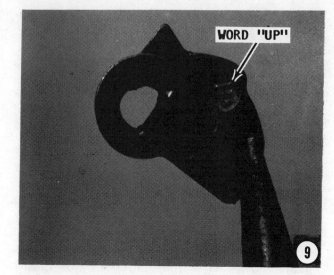

Forward Gear Bearing Installation

This bearing can be installed **WITHOUT** the use of a special tool.

7- Place the bearing in position on the forward gear with the taper facing **AWAY** from the gear. Press the bearing into position using a proper size socket. A short length of 1-1/4" O.D. pipe may also be used to install the bearing.

Bearing Carrier Seal Installation

These two seals are of different sizes -- one small -- the other larger. The smaller

(narrow) seal is installed on the inside of the bearing carrier with the lip facing **INWARD**. The larger (wide) seal is installed on the outside of carrier with the lip facing **OUTWARD**.

8- Install each seal from the proper side of the bearing carrier. Use Triple Guard Lubricant between the seals. This type of lubricant will not dissipate, but remain to perform its lubricating job. Install the O-ring into the groove of the bearing carrier.

Set the assembled bearing carrier aside for installation later.

Shift Rod and Shift Cam Installation

9- Pick up the cam with a pair of needle-lenose pliers and observe the word **UP** embossed on the arm of the cam. Take special notice of the irregular shape of the cam arm. The long flat surface must butt up against the back side of the lower unit

housing. Also observe the inside diameter of the cam with the flat area on the side next to the cam arm.

10- Using the needlenose pliers, lower the shift cam down into the lower unit with the flat edge of the cam arm facing the **PORT** side and towards the **BACK** of the housing, as shown. Push the cam all the way in with the flat side of the cam against the back of the housing. In this position, when the shift rod is installed, the correct half-moon area on the inside diameter of the cam will index in the cam and the shift rod will be in the proper position.

11- Snap the E-clip into the recess of the shift rod, as shown. Insert the shift rod down through the lower unit housing and into the shift cam with the retaining ring going in **FIRST.** Check the shift rod to be sure the flat area on the end of the rod is in the same plane (fore-and-aft) as shown in the accompanying illustration, **"A"**, when viewed from the **PORT** side.

SPECIAL NOTE

When the shift rod is rotated to the position shown in illustration **"B"**, as viewed from the **PORT** side, the unit will be in **NEUTRAL.** When the shift rod is rotated further to the position shown in illustration **"C"**, still viewed from the **PORT** side, the unit is in **REVERSE.**

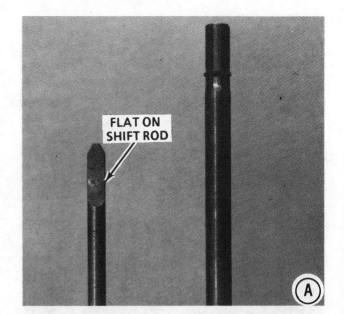

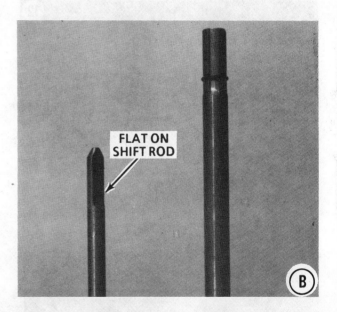

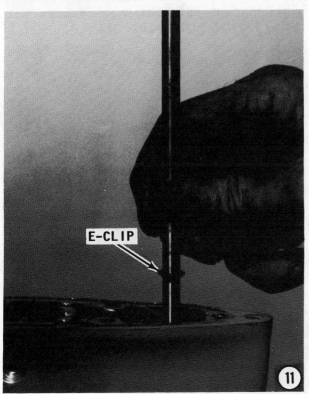

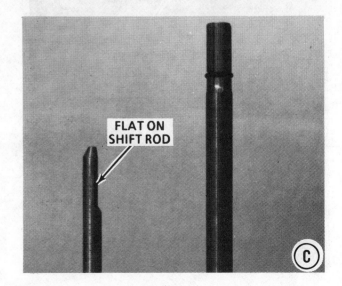

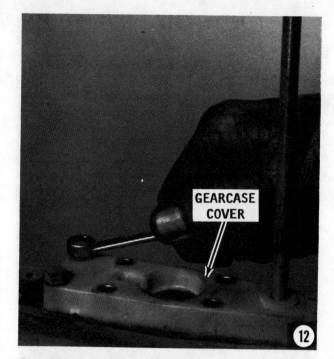

Gear Case Cover Installation

12- Place a **NEW** gasket on the bottom of the gearcase cover. Coat the bottom side of the gearcase cover and the mating surface of the lower unit with OMC Adhesive **Type-M.** Coat the shift rod with oil as an aid to installing the gearcase cover.

Slide the gearcase cover down over the shift rod. Secure the gearcase cover to the lower unit with the retaining bolts. Tighten the bolts evenly and alternately to the torque value given in the Appendix.

Pinion Gear Bearings
Installation Onto Pinion Gear

A bearing, two thrust washers, and a thrust bearing, are used on the pinion gear. The method and order of installation is most important. Notice how the inner edge of one thrust washer is beveled. This washer is installed first with the beveled edge side facing **DOWN.** The thrust bearing is instal-

led next, and finally the second thrust washer, with the outer beveled edge side facing **UP**.

13- Coat the thrust washers, the thrust bearing, and the shank of the pinion gear with OMC HI-VIS Gearcase Lubricant. Slide the thrust washer with the beveled side facing the pinion gear shank.

14- Slide the thrust bearing onto the shank with the flat side facing **DOWN**.

15- Slide the second thrust washer onto the pinion gear shank with the outside beveled side facing **UP**.

Lower Unit Assembling

16- Work the pinion gear assembly up into the lower unit housing through the pinion gear bearing previously installed into the housing.

17- Secure the lower unit in a horizontal position (90° from the normal position). Lower the forward gear into the lower unit housing with the gear tilted toward the cavitation plate and the gear teeth facing backward toward the opening.

18- Insert the propeller shaft through the opening and through the forward gear. Forcefully pull the propeller shaft toward the skeg to snap the teeth of the forward into mesh with the teeth of the pinion gear. When the two gears have meshed, the propeller shaft will be in a vertical position. Withdraw the propeller shaft.

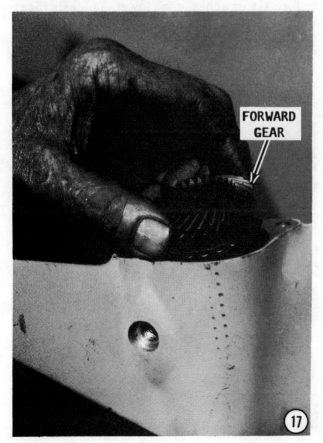

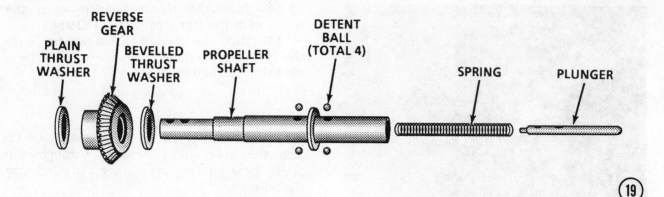

PLAIN THRUST WASHER — REVERSE GEAR — BEVELLED THRUST WASHER — PROPELLER SHAFT — DETENT BALL (TOTAL 4) — SPRING — PLUNGER

⑲

Propeller Shaft Assembling and Installation

19- Slide the bevelled thrust washer onto the aft end of the propeller shaft, with the tapered side going on **FIRST**, to make contact with the shoulder on the shaft. Next, install the reverse gear on the shaft up against the thrust washer. Install the plain thrust washer.

Insert the spring and plunger into the forward end of the propeller shaft.

Coat two of the detent balls with OMC Needle Bearing Assembly Lubricant. Push the plunger inward against the spring until the reverse gear detent notches align with the holes in the propeller shaft. Now, insert the two detent balls into the notches. Continue to hold the plunger inward and at the same time slide the reverse gear into position over the balls. This action will trap the detent balls and hold the plunger in position.

Fill the forward gear notches with the same lubricant, and then install the two remaining balls. The lubricant will hold the balls in place while other parts are being assembled. Push the reverse gear hub forward to hold the assembled propeller shaft in the reverse gear condition, and at the same time insert the propeller shaft into the forward gear. Check to be sure the detent balls are positioned 90° from the driving lugs on the forward gear.

20- Coat the exposed end of the propeller shaft with lubricant. Place the O-ring in position on the bearing carrier. Slide the carrier down over the propeller shaft. Just before it reaches the lower unit surface, coat the O-ring with gasket sealer compound, and then secure the carrier to the lower unit with the attaching bolts.

21- Tighten the bolts **ALTERNATELY** and **EVENLY** to a torque value of 48-60 in lb (5-7Nm).

O-RING

⑳

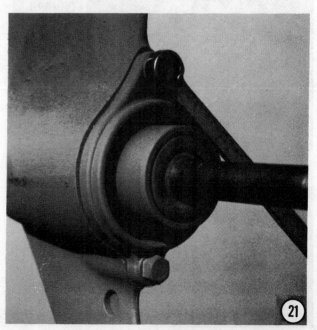

㉑

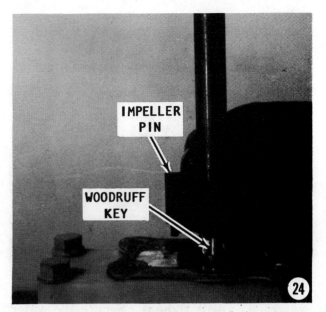

WATER PUMP INSTALLATION

22- Secure the lower unit in the vertical (normal) position. Slide the driveshaft down through the lower unit until it indexes with the pinion gear. If necessary, rotate the shaft slightly to allow the splines on the shaft to index with the splines in the pinion gear.

23- Apply OMC Sealer to both surfaces of both water pump gaskets. Place one gasket in position on the surface of the gearcase cover. Place the water pump plate into position on the gasket with the rounded edge facing **UPWARD**. Place the second gasket in position on the top of the water pump plate.

24- Slide the impeller down the driveshaft. Before it reaches the keyway in the shaft, insert the key in the keyway, and then slide the impeller down onto the surface of the water pump plate.

25- Coat the inside surface of the water pump housing with light-weight oil, and then slide the housing down the driveshaft and over the impeller. Rotate the driveshaft as the water pump housing is moved into place to allow the impeller blades to enter and lay back properly inside the housing.

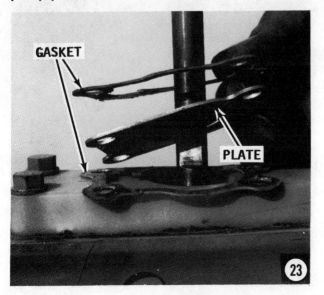

26- Apply a coating of Loctite to the water pump bolt threads. Secure the water pump to the lower unit with the bolts. Tighten the bolts **EVENLY** and **ALTERNATELY** to a torque value of 60-84 in lb (7-9Nm). Slide the O-ring down the driveshaft and into place on top of the water pump.

LOWER UNIT INSTALLATION

SPECIAL INSTRUCTIONS

Fill the lower unit with lubricant. Follow the procedures listed in Section 8-3.

Install the propeller, see Section 8-2.

Check to be sure the spark plug wires are disconnected from the plugs. Verify the shift handle is in the **FORWARD** gear position. Check to be sure the lower unit shift is in the **FORWARD** gear position. This check can be accomplished by verifying the shift rod is turned tightly **CLOCKWISE** and that the unit is in forward gear. When the

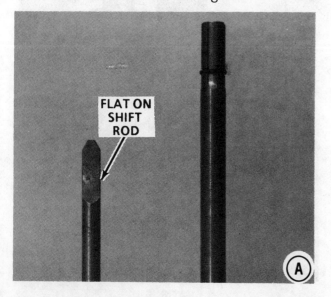

FLAT ON SHIFT ROD

unit is in **FORWARD** gear, the flat portion of the shift rod will be in the same plane (fore-and-aft) as shown in the accompanying illustration, **"A"**, when viewed from the **PORT** side.

Check to be sure the water pump tubes are clean, and then coat them with lubricant. Coat the O-rings with oil. Coat the splines of the driveshaft with OMC Moly-lube.

27- Bring the lower unit together with the exhaust housing, and at the same time guide the water tubes and the driveshaft through the openings in the exhaust housing. Bring the assembled lower unit and the exhaust housing together with the powerhead, and at the same time guide the water tubes and the driveshaft into the place. Have an assistant slowly turn the propeller **CLOCKWISE** as the units are brought together to allow the splines on the end of the driveshaft to index with the splines in the crankshaft. Check to be sure the end of the shift rod indexes into the shift handle gear.

28- Install the retaining bolts securing the lower unit to the exhaust housing. Two bolts are installed up through the lower unit cavitation plate on each side of the zinc. A third bolt is installed down through the forward leading edge of the exhaust housing.

Tighten the bolts **ALTERNATELY** and **EVENLY** to a torque value of 120-144 in lb (14-16Nm). Install a new zinc and secure it in place with the Phillips screw.

FUNCTIONAL CHECK

Perform a functional check of the completed work by mounting the outboard in a test tank, in a body of water, or with a flush attachment connected to the lower unit. If the flush attachment is used, **NEVER** operate the powerhead above an idle speed, because the no-load condition on the propeller would allow the powerhead to **RUNAWAY** resulting in serious damage or destruction of the engine.

CAUTION: Water must circulate through the lower unit to the engine any time the engine is run to prevent damage to the water pump in the lower unit. Just five seconds without water will damage the water pump.

Start the powerhead and observe the tattle-tale flow of water from idle relief in the exhaust housing. The water pump

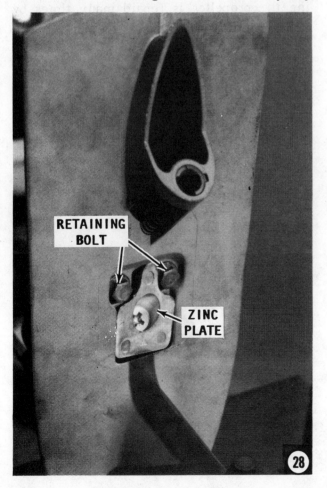

RETAINING BOLT

ZINC PLATE

28

installation work is verified. If a "Flushette" is connected to the lower unit, **VERY LITTLE** water will be visible from the idle relief port. Shift the outboard into the three gears and check for smoothness of operation and satisfactory performance.

8-7 TYPE "D" LOWER UNIT
9.9HP AND 15HP
PROPELLER EXHAUST
MECHANICAL SHIFT

DESCRIPTION

This lower unit has forward, neutral, and reverse shifting capabilities, with exhaust gases routed through the propeller.

The lower unit must be separated slightly from the exhaust housing in order to disconnect the shift rod.

GOOD WORDS

Propellers with the exhaust passing through the hub **MUST** be removed more frequently than the standard propeller. Removal after each weekend use or outing is not considered excessive. These propellers do not have a shear pin. The shaft and propeller have splines which **MUST** be coated with an anti-corrosion lubricant prior to installation as an aid to removal the next time the propeller is pulled. Even with the lubricant applied to the shaft splines, the propeller may be difficult to remove.

The propeller with the exhaust hub is more expensive than the standard propeller, therefore, the cost of rebuilding the unit, if the hub is damaged, is justified, illustration **"A"**.

A replaceable diffuser ring on the aft side of the propeller disperses the exhaust gases away from the propeller blades. If the

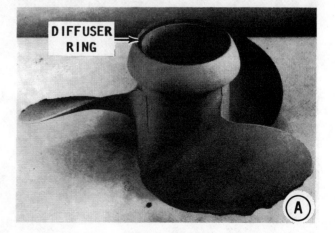

DIFFUSER RING

A

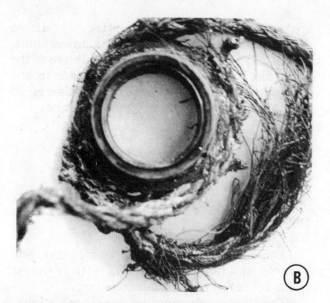

ring becomes broken or damaged "ventilation" would be created pulling the exhaust gases back into the negative pressure area behind the propeller. This condition would create considerable air bubbles and reduce the effectiveness of the propeller.

If the propeller is "frozen" onto the propeller shaft, see the special instructions for removal outlined in Section 8-2 Frozen Propeller Removal.

TROUBLESHOOTING

Troubleshooting **MUST** be done **BEFORE** the unit is removed from the powerhead to

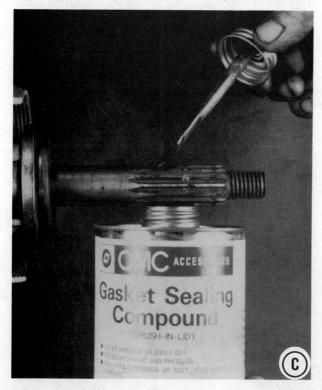

permit isolating the problem to one area. Always attempt to proceed with troubleshooting in an orderly manner. The shotgun approach will only result in wasted time, incorrect diagnosis, replacement of unnecessary parts, and frustration.

The following procedures are presented in a logical sequence with the most prevalent, easiest, and less costly items to be checked, listed first.

Water in the Lower Unit

Water in the lower unit is usually caused by fish line becoming entangled around the propeller shaft behind the propeller and damaging the propeller seal. If the line is not removed, it will cut the propeller shaft seal and allow water to enter the lower unit. Fish line has also been known to cut a groove in the propeller shaft, illustration "B".

The propeller should be removed each time the boat is hauled from the water at the end of an outing and any material entangled behind the propeller removed before it can cause extensive damage. The small amount of time and effort involved in pulling the propeller is repaid many times by reduced maintenance and service work, including the replacement of expensive parts, illustration "C".

Slippage in the Lower Unit

If the shift seems to be slipping as the boat moves through the water: Check the propeller and the rubber hub. If the propel-

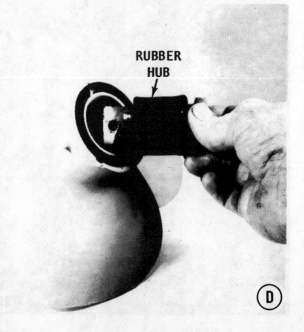

RUBBER HUB

ler has been subjected to many strikes against underwater objects, it could slip on its hub. If the hub is damaged or excessively worn on the small propellers, it is not economical to have the hub or propeller rebuilt. A new propeller may be purchased for considerably less than meeting the expense of rebuilding an old worn propeller, illustration **"D"**.

Jumping Out of Gear

If a loud thumping sound is heard at the transom while the boat is underway, the unit is jumping out of gear, resulting in a no-load condition on the propeller. When this happens, the rushing water under the hull forces the lower unit in a backward direction. The unit jumps back into gear; the propeller catches hold; the lower unit is forced forward again; and the result is the thumping sound as the action is repeated. Normally this type of action occurs perhaps once a day, then more frequently each time the clutch is operated, until finally the unit will not stay in gear for even a short time.

The following areas must be checked to locate the cause:

1- Check the upper shift rod at the connection underneath the carburetor.

2- Attempt to correct the problem with better shift habits by the operator.

3- Check for water in the lower unit as described earlier in this section.

4- Normal wear at the upper and lower shift rod connection may cause the unit to jump out of gear.

Frozen Powerhead

This condition is suggested when the operator unsuccessfully attempts to crank the powerhead, either with a hand starter or with a cranking motor. The flywheel will

Bellcrank located under the carburetor. The shift linkage is disconnected at this point, as described in the text.

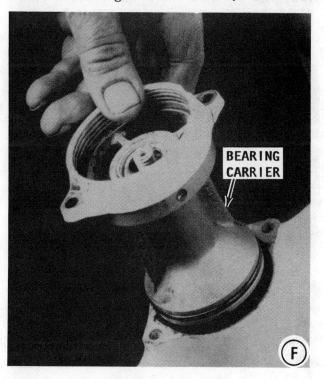

not rotate. Do not assume the powerhead is "frozen" until the lower unit has been removed and thoroughly checked. If the lower unit is "locked" (the driveshaft or propeller shaft will not rotate), the powerhead will have the indication of being "frozen" (failure to rotate the flywheel), illustration **"E"**.

The first step to perform under these conditions is to "pull" the lower unit, and then again attempt to crank the powerhead. If the attempt is successful with the lower unit disconnected, the problem is in the lower unit. If the attempt to crank the powerhead is still unsuccessful, the problem is in the powerhead.

LOWER UNIT REMOVAL

SPECIAL WORDS

The lower unit must be separated slightly from the exhaust housing in order to disconnect the shift rod.

The bearing carrier in the lower unit may be removed and the seals replaced without removing the lower unit. **HOWEVER**, this is not considered good practice because there is no way to check the bearings for damage if water has been allowed to enter the lower unit. Therefore, if the bearing carrier seals must be replaced, the lower unit should be removed and a complete inspection made of all other bearings and seals for water damage, illustration **"F"**.

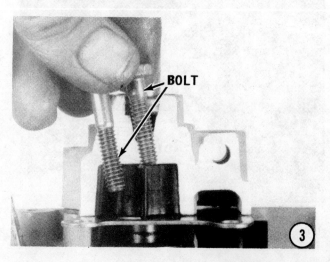

Preliminary Tasks

Drain the lower unit according to the procedures outlined in Section 8-3.

Remove the propeller, see Section 8-2. If the propeller is "frozen" to the shaft, see special instructions outlined in the same section.

1- Remove the powerhead cowling. Disconnect the spark plug leads at the spark plugs. Remove the bolts securing the lower unit to the exhaust housing or to the 6-inch extension. In **MOST** cases, it is not necessary to remove the 6-inch extension in order to "drop" the lower unit. Separate the lower unit slightly from the exhaust housing or the 6-inch extension. Reach in and remove the bottom bolt through the shift rod connector. The shift rod is now free to separate from the connector. **CAREFULLY** remove the lower unit straight away from the exhaust housing or the 6-inch extension. **TAKE CARE** not to pull backward, sideways, etc., when removing the lower unit, because the driveshaft may be bent.

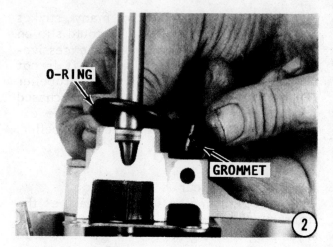

WATER PUMP REMOVAL

2- Remove the O-ring from the top of the driveshaft. Remove the O-ring and grommet from the top of the water pump.

3- Remove the bolts securing the water pump housing to the lower unit. Notice how the bolts from the aft holes are longer than the two bolts forward. Slide the impeller housing up and free of the driveshaft.

4- Remove the impeller and driveshaft key. Remove the water pump plate and gaskets from the lower unit.

If the only work to be performed is service of the water pump, proceed directly to Page 8-63, Water Pump Installation.

Bearing Carrier Removal
FIRST, THESE WORDS

A hole is drilled through the propeller shaft just forward of the reverse gear. A detent spring is installed in the hole along with a detent ball on each side of the propeller shaft. The clutch dog is installed over the top of the two detent balls. This arrangement assists in holding the unit in the specific gear desired. When the bearing carrier, propeller shaft, and associated parts are removed, the clutch dog will remain in the lower unit. As soon as the clutch dog slides free of the propeller shaft, the detent balls and spring will fly free of the shaft, but be contained within the lower unit housing. Therefore, after the propeller shaft, bearing carrier, etc., are removed, take time to retrieve the two detent balls and the spring from inside the lower unit, illustration "G".

5- Remove the two screws securing the bearing carrier in the lower unit.

DETENT BALL

G

GOOD WORDS

Three methods are available to free the carrier from the lower unit.

The first method involves the use of a special tool, OMC P/N 386631. This tool is installed over the propeller shaft; the propeller nut is threaded onto the shaft behind the tool; and then the tool extended with a wrench on both sides of the shaft; "pulling" the propeller shaft, bearing carrier, and reverse gear from the lower unit.

On some bearing carriers, two threaded holes are provided in the carrier. The second method involves the use of a flywheel puller. Two long bolts are installed into the threaded holes of the bearing carrier, while the center bolt is tightened against the end of the propeller shaft to pull the carrier free, illustration #H.

The third method of removing the bearing carrier also utilizes the two threaded

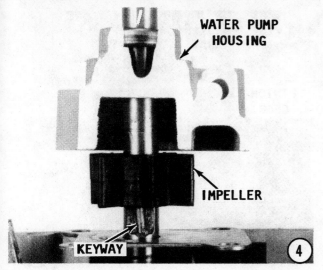

WATER PUMP HOUSING

IMPELLER

KEYWAY

4

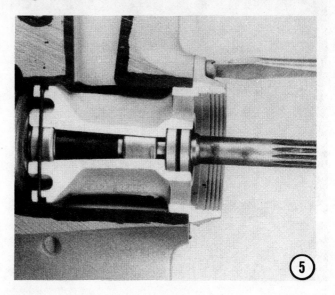

5

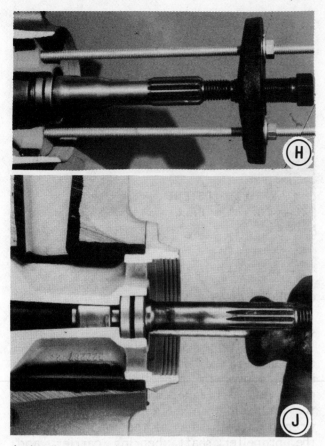

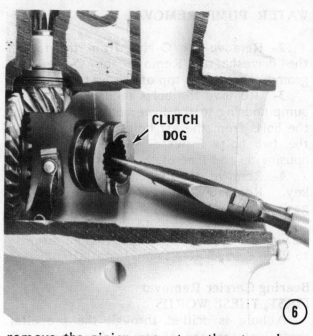

holes in the carrier. A slide hammer with a long rod is attached to the carrier and the carrier removed in that manner, illustration **"J"**. The propeller shaft and reverse gear are then removed, illustration **"K"**.

6- Reach in with a pair of needle-nose pliers and remove the clutch dog from the cradle, as shown.

7- Remove the driveshaft from the lower unit and at the same time, reach in and

remove the pinion gear, two thrust washers, and the thrust bearing. Pay particular attention to the washers and how they are positioned on top of the pinion gear. One washer is noticeably thicker than the other. Also, one washer is beveled on the inside diameter and the other washer is beveled on the outside diameter. Take time to identify one of the washers with a dab of paint or other mark to ensure they will be installed in the same location from which they were removed.

8- Back out the shift rod from the yoke by turning it **COUNTERCLOCKWISE** until it is free. Withdraw the shift rod from the lower unit.

9- Remove the Phillips screw from outside of the lower unit. This is the screw very close to the lubricant drain plug. The screw secures the shift lever and yoke as-

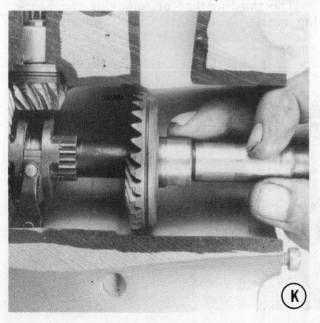

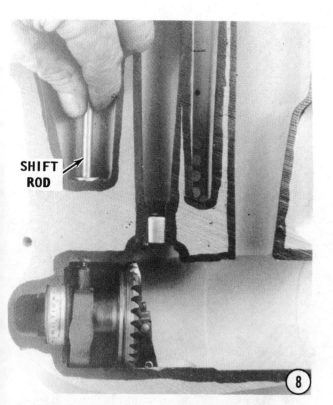

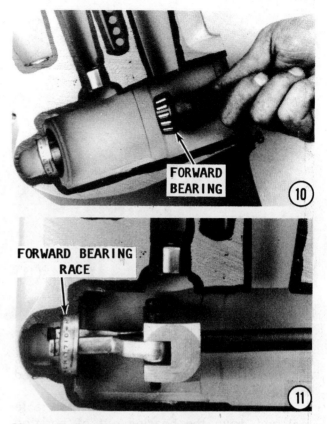

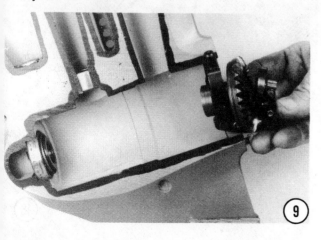

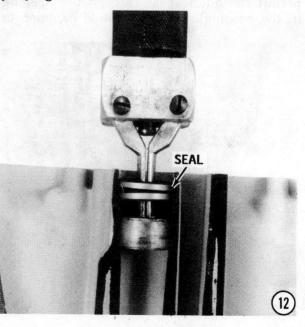

sembly in position. After the screw is removed, the shift lever, yoke assembly, and forward gear may be removed from the lower unit.

10- Remove the forward gear tapered bearing from the lower unit.

11- Remove the forward gear bearing race. This race need **NOT** be removed unless the forward gear tapered bearing is to be replaced. The bearing and the race are sold as a matched pair.

Upper Driveshaft Seals — Removal

12- Use a slide hammer with finger-type pullers and remove the two seals. Notice how the two seals are installed back-to-back. It is most important that they be installed in the same position from which they were removed.

Upper Driveshaft Bearing Removal

This bearing need **NOT** be removed, unless it is unfit for further service. Check the condition of the bearing by inserting a finger and rotating the bearing while checking for rough spots or evidence of binding. Use a flashlight and check to be sure there is no evidence of corrosion or other damage.

13- Use a slide hammer with fingers to remove the bearing, as shown in the accompanying illustration.

BEARING

(13)

Pinion Gear Bearing — Removal

As with the upper driveshaft bearing, the lower bearing need **NOT** be removed unless it is unfit for further service. Check its condition in the same manner as described for the upper bearing.

14- Use a drift punch or other suitable tool and drive the bearing out of position into the lower unit. A special tool is not required because if the bearing is to be removed it is unfit for further service and additional damage will be of no consequence.

Shift Rod O-ring and Bushing Removal

15- Insert a length of 1/4" rod down through the O-ring and bushing. Thread a nut onto the lower end of the rod. (The edges of the nut must be first rounded to permit the nut to pass through the opening in the housing). Attach a slide hammer to

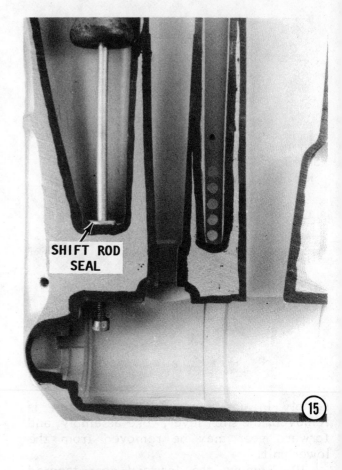

SHIFT ROD SEAL

(15)

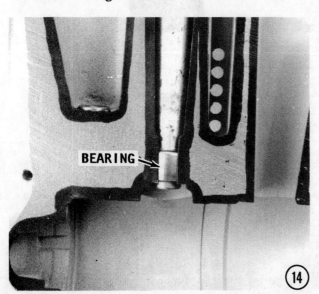

BEARING

(14)

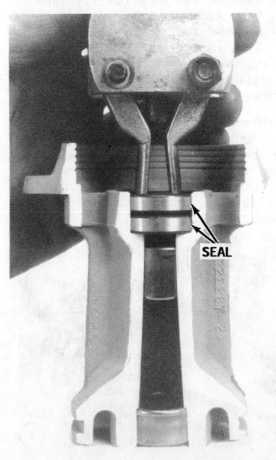

SEAL

(16)

the rod and "pull" the **O**-ring, bushing and washer.

Bearing Carrier Seal Removal

16- Use a puller with two fingers and work the puller down into the seals. Take up on the puller and remove the seals. Notice how the seals are installed back-to-back. They must be installed in this manner to provide a proper seal to prevent the lubricant in the lower unit from escaping and water from entering.

Bearing Carrier Bearing Removal

The bearings in the bearing carrier need **NOT** be removed unless they are unfit for further service. Check their condition for damage and corrosion. Insert a finger inside and rotate the bearing while checking for roughness or any sign of binding. Use a flashlight and check for evidence of corrosion.

17- Use a punch and drive the bearing out. The bearing is being removed because it is unfit for service, therefore further damage is of no consequence.

CLEANING AND INSPECTING

Clean all water pump parts with solvent, and then dry them with compressed air.

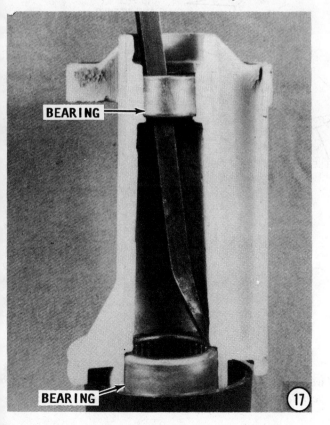

Inspect the water pump cover and base for cracks and distortion, possibly caused from overheating. Inspect the face plate and water pump insert for grooves and/or rough surfaces. If possible, **ALWAYS** install a complete new water pump while the lower unit is disassembled. A new impeller will ensure extended satisfactory service and give "peace of mind" to the owner. If the old impeller must be returned to service, **NEVER** install it in reverse to the original direction of rotation. Installation in reverse will cause premature impeller failure.

Inspect the impeller side seal surfaces and the ends of the impeller blades for cracks, tears, and wear. Check for a glazed or melted appearance, caused from operating without sufficient water. If any question exists, and as previously stated, install a new impeller if at all possible.

Clean all parts with solvent and dry them with compressed air. **DISCARD** all O-rings and gaskets. Inspect and replace the driveshaft if the splines are worn. Inspect the gearcase and exhaust housing for damage to the machined surfaces. Remove any nicks and refurbish the surfaces on a surface plate. Start with a No. 120 Emery paper and finish with No. 180.

Check the water intake screen and passages by removing the bypass cover, if one is used. Inspect the clutch dog, drive gears, pinion gear, and thrust washers. Replace these items if they appear worn. If the clutch dog and drive gear arrangement surfaces are nicked, chipped, or the edges

Badly worn pinion gear from a lower unit. The teeth of the gears must be carefully inspected for wear and damage.

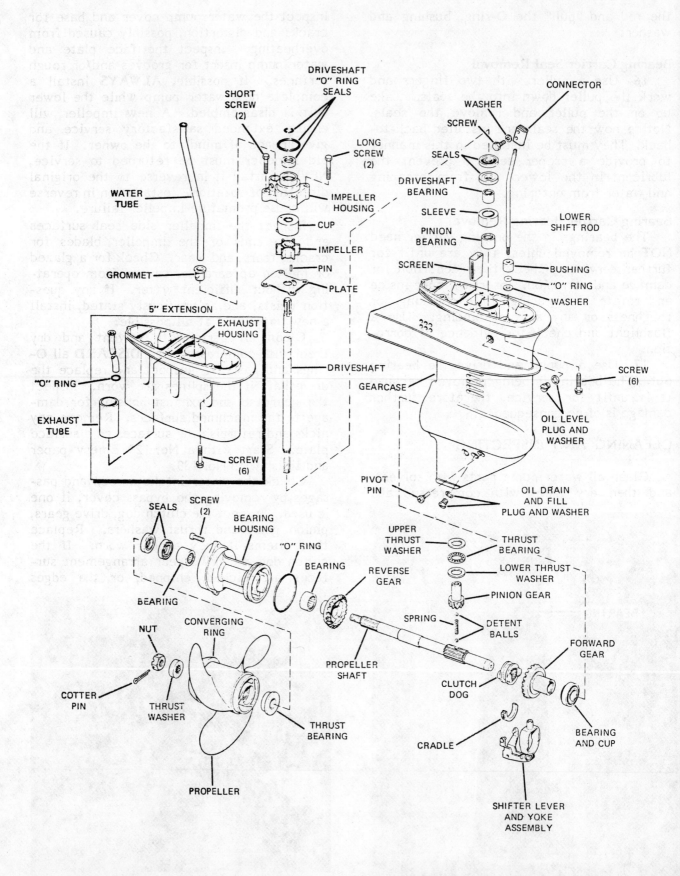

Exploded drawing of the Type "D" lower unit matched with the 9.9 thru 15hp powerheads. A breakdown of an extension unit is also included in the inset. Major parts have been identified. Notice the detent spring and two detent balls in the propeller shaft.

The ears on this clutch dog and the teeth on the gear are badly worn. Both items are unfit for further service.

rounded, the operator may be performing the shift operation improperly or the controls may not be adjusted correctly. These items **MUST** be replaced if they are damaged.

Inspect the dog ears on the inside of the forward and reverse gears. The gears must be replaced if they are damaged.

Check the cradle that rides on the inside diameter of the clutch dog. The sides of the cradle must be in good condition, free of any damage or signs of wear. If damage or wear has occurred, the cradle must be replaced.

Check the shift lever and the two prongs that fit inside the cradle. Check to be sure the prongs are not worn or rounded. Damage or wear to the prongs indicates the lever must be replaced.

New clutch dog and gear. Compare these two parts with those shown at the top of this column.

A rusted and corroded gear. Water was allowed to enter the lower unit through a badly worn seal and cause this damage to the gear and other expensive parts.

LOWER UNIT ASSEMBLING

READ AND BELIEVE

The lower unit should not be assembled in a dry condition. Coat all internal parts

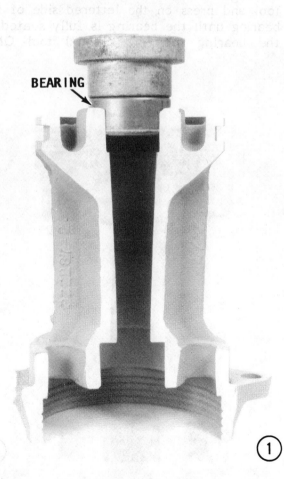

BEARING

①

A set of double seals showing the back side (left) and the front side (right). These seals are installed back-to-back (flat side-to-flat side) with Triple Guard Grease between the surfaces. This arrangement prevents fluid from passing in either direction.

with OMC HI-VIS lube oil as they are assembled. All seals should be coated with OMC Gasket Seal Compound. When two seals are installed back-to-back, use Triple Guard Grease between the seal surfaces.

Forward and Rear Bearing Installation Into the Bearing Carrier

1- If these two bearings were removed during disassembling because they were unfit for service, special tools are required to install the new bearings. Special tool OMC P/N 319876 is required to install the rear bearing into the bearing carrier. Obtain the tool and press on the lettered side of the bearing until the bearing is fully seated in the bearing carrier. Special tool OMC

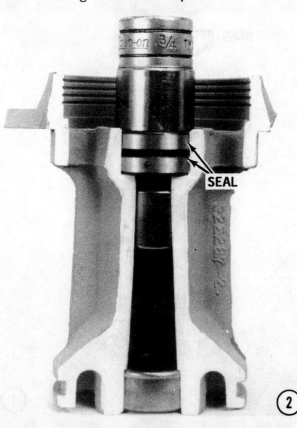

SEAL

②

P/N 319875 is required to install the front bearing into the bearing carrier. Obtain the special tool and press against the lettered side of the bearing until the bearing is in place in the bearing carrier.

Bearing Carrier Seals -- Installation

These seals are installed one at-a-time, back-to-back.

2- Special tool OMC P/N 319877, or a socket the same size as the seal may be used to install the seals. Coat the outside surface of the seal with OMC Lubricant and press the first seal into place in the bearing carrier bore, with the back side of the seal facing **OUTWARD**. After the seal is in place, apply a coating of Triple Guard Lubricant to the seal surface. Install the second seal into the bearing carrier, with the back side of the seal facing **INWARD**. Install a **NEW** O-ring onto the outside surface of the bearing carrier. Set the assembled bearing carrier aside for later installation.

Pinion Gear Bearings -- Installation
SPECIAL WORDS

The lower pinion gear bearing **MUST** be installed from the bottom cavity of the lower unit. The upper pinion gear bearing is pressed into place from the top.

3- If the lower pinion gear bearing was removed during disassembling because it was unfit for further service, obtain special tools OMC P/N 319878 and No. 383173. Using these special tools is the only way the

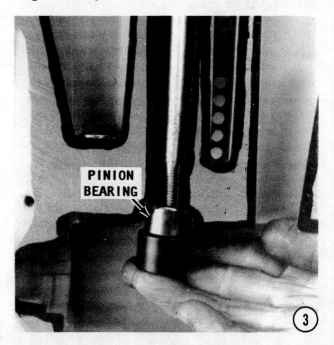

PINION BEARING

③

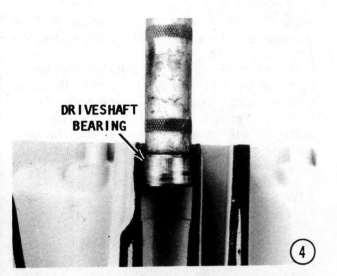

DRIVESHAFT BEARING

④

bearing may be installed properly. Actually, the bearing is "pulled" up into place with the special tools. Place the bearing in position for installation with the lettered side facing **DOWN**. Insert the special tool through the opening in the lower unit and through the bearing. Thread the bolt into the special tool and "pull" the bearing up into place in the lower unit.

4- To install the upper pinion gear bearing, obtain special tool OMC P/N 319931 or P/N 326566. Place the bearing in position with the lettered side facing **UP**. Use the special tool and press the bearing into place.

5- Install the upper driveshaft seals back-to-back. Coat the surfaces between the two seals with Triple Guard Grease.

Shift Rod O-ring and Bushing — Installation

6- Lower the washer, O-ring, and bushing for the shift rod into place in the lower unit housing. Obtain special tool OMC P/N 304515. Tap the bushing into place with a hammer and the special tool until the bushing is fully seated in the housing. If the special tool is not available, a socket or

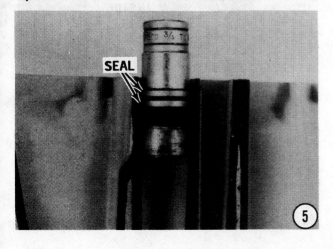

SEAL

⑤

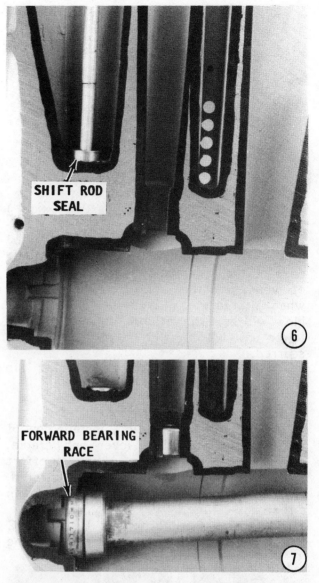

SHIFT ROD SEAL

⑥

FORWARD BEARING RACE

⑦

other similar tool may be used to install the bushing **PROVIDED** the tool will not damage the bushing during the installation process.

Forward Gear Bearing Race — Installation

7- Obtain special tool OMC P/N 319929 and drive handle OMC No. 311880. Seat the

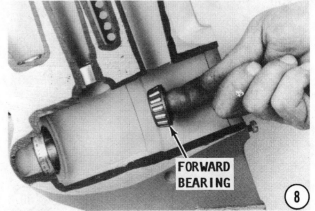

FORWARD BEARING

⑧

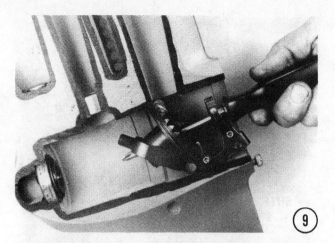

(9)

tool into the bearing race. Drive the race into place in the lower unit. If the leading edge of the lower unit is placed on a block of wood, or other solid non-mar surface, when the race is driven into place, each blow will be firm without "bounce" and the lower unit housing will not be damaged or scarred.

8- Insert the forward gear bearing into the race installed in the previous step.

WORDS OF ADVICE

The following operation, installation of the forward gear shift lever and cradle, requires time and patience. Make an effort to stay calm without becoming frustrated, and the installation will be accomplished in a reasonable time.

A special flexible tool is listed, OMC P/N 319991, to install these parts. The

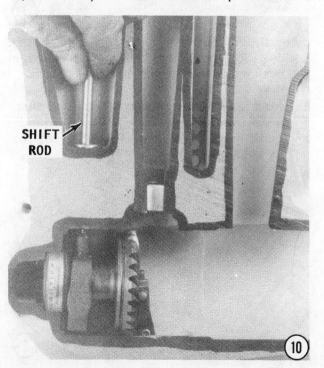

SHIFT ROD

(10)

cable is threaded into the yoke and fed up through the lower unit. When the assembled propeller shaft is inserted into the lower unit, the cable is pulled through the lower unit and thus the short extension of the yoke will be guided into the shift rod hole. However, if the tool is not available, time and patience will result in victory, and the unit will function properly. Without the special tool, proceed as outlined in the next step.

ADVICE

Perform Steps 9 and 10 together.

9- Slide the forward gear into the shift lever and yoke assembly. Slip the cradle into the fingers of the shift lever. Coat the shift rod with oil. Insert the assembly into the lower unit and work the top part of the yoke up into the recess where the shift rod comes through.

10- At the same time, insert the shift rod down through the lower unit and shift rod bushing. Thread the shift rod into the shift lever yoke assembly four or five complete turns. An adjustment for the shift rod will be made later in Step 21.

Pinion Gear Installation

11- Observe the two pinion gear wash-

INSIDE BEVELED EDGE

(11)

THRUST BEARING

12

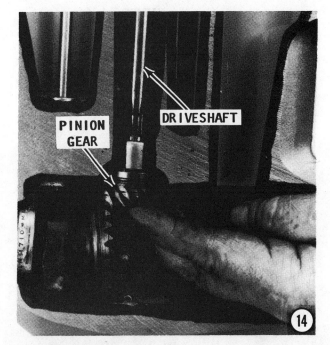

PINION GEAR

DRIVESHAFT

14

ers. One washer is beveled on an inner edge and the other washer is beveled on an outer egde. Slide the washer with the inner edge bevel onto the shank of the pinion gear with the bevel facing **DOWN**.

12- Slide the thrust bearing onto the shank of the pinion gear.

13- Slide the remaining beveled washer onto the shank with the bevel on the outer egde facing **UP**.

14- Slide the pinion gear shank up into the pinion gear bearing, and at the same time, lower the driveshaft down through the lower unit and into the pinion gear. Rotate the driveshaft very slightly after it makes contact with the pinion gear to allow the splines on the shaft to index with the splines of the gear.

Clutch Dog Installation

15- Pick up the clutch dog and notice the grooves on one side of the outside surface. When the clutch dog is installed, these grooves **MUST** face the forward gear, illustration **"A"**. Grasp the clutch dog with a pair of needle-nose pliers and insert it into the cradle of the yoke assembly with the

OUTSIDE BEVELED EDGE

13

GROOVE

A

CLUTCH DOG

15

grooved side going in **FIRST**. Life is not a bowl of cherries, and this is not the easiest task, but if the cradle is tilted back slightly it will help with the installation. Work slowly and with patience, and the clutch dog will be properly seated in the yoke assembly.

Propeller Shaft Installation

16- Coat the propeller shaft, the detent spring, and two detent balls, with needle bearing grease or similar lubricant to hold them in place during installation of the shaft. The detent balls do not seat all the way into the shaft holes. Therefore, if grease is not used, they will not remain in place while the propeller shaft is installed into the lower unit.

Slide the spring into the shaft and place the two balls in position. Place the lower unit in a horizontal position with the opening facing up. Insert the propeller shaft into the lower unit with the shaft in the position that places the balls on each side, as shown. In this position the balls should be aligned with the ramps in the clutch dog. The detent balls must ride in a groove inside the clutch dog.

Continue moving the propeller shaft into the lower unit housing until the shaft indexes into the clutch dog. Align the detent balls with the ramps on the clutch dog by turning the shaft very slowly. If the detent balls will not align in the center of the ramp, withdraw the shaft slightly to clear the splines in the clutch dog, and then rotate the shaft 180° and the balls should align with the ramp when the splines index again. The spring will allow the detent balls to be depressed slightly and then to be held in place in the clutch dog groove. As the shaft moves into its proper place, a definite "click" sound will be heard, indicating the detent balls have "popped" into the clutch dog groove and the unit is in **NEUTRAL**.

17- Coat the threads, and then install the Phillips screw through the outside of the lower unit housing to secure the shift lever and yoke assembly in place. This is accomplished by reaching inside the lower unit and moving the shift lever and yoke assembly to align the hole in the housing with the hole in the assembly. When the holes are aligned, install the screw and tighten it securely.

Reverse Gear Installation

18- Insert the reverse gear into the lower unit over the propeller shaft.

DETENT BALL

CLUTCH DOG

16

PHILLIPS SCREW

17

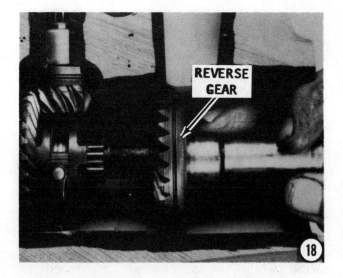

REVERSE GEAR

18

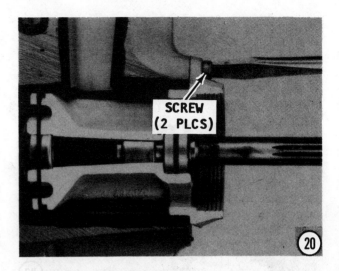

SCREW (2 PLCS)

20

Bearing Carrier Installation

19- Thoroughly lubricate the bearings in the bearing carrier with HI-VIS Lubricant. Check to be sure the O-ring is in place. Insert the bearing carrier into the lower unit housing.

20- Coat the threads of the attaching screws with OMC Sealant. Secure the bearing carrier in place with the screws. Tighten the screws **EVENLY** and **ALTERNATELY** to the torque value for the size bolt used as given in the Appendix.

Shift Rod Adjustment

21- After installation, the bend in the shift rod must be toward the forward (leading) edge of the lower unit. Lay a straightedge over the top of the lower unit housing. Measure the distance from the straightedge to the top corner edge of the shift connector. This dimension must be between 1/16"-13/32" (1.6-10.3mm). To adjust, thread the shift rod into, or out of, the yoke assembly until the required dimension is obtained.

WATER PUMP INSTALLATION

22- Apply a thin bead of OMC Sealant Type-**M** onto the bottom side of the water pump plate where it will contact the lower unit. Slide the plate down over the driveshaft and into place on the lower unit surface. Apply a coating of needle bearing grease in the keyway of the driveshaft.

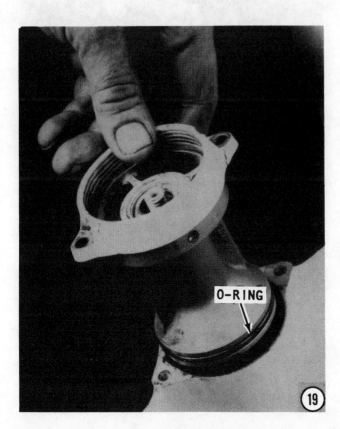

O-RING

19

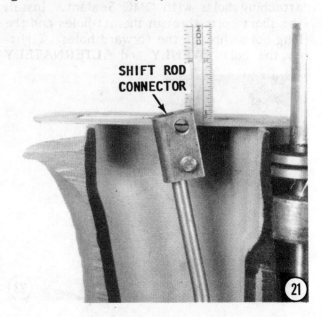

SHIFT ROD CONNECTOR

21

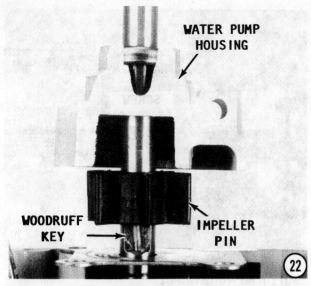

WATER PUMP HOUSING

WOODRUFF KEY

IMPELLER PIN

22

Slide the impeller down the driveshaft. Just before it covers the driveshaft keyway, insert the key into the keyway. Slide the impeller the remaining way down the driveshaft until it seats on the water pump plate with the key indexed into the slot of the impeller.

Coat the inside surfaces of the water pump housing with light-weight oil. Lower the water pump housing down the driveshaft over the impeller and onto the water pump plate. Rotate the driveshaft **CLOCKWISE** while lowering the water pump housing to allow the impeller blades to enter and to assume their natural and proper position inside the housing. Continue to rotate the driveshaft and work the water pump housing downward until it is seated on the water pump plate.

23- Coat the threads of the water pump attaching bolts with OMC Sealant. Install the short bolts through the aft holes and the long bolts through the forward holes. Tighten the bolts **EVENLY** and **ALTERNATELY**

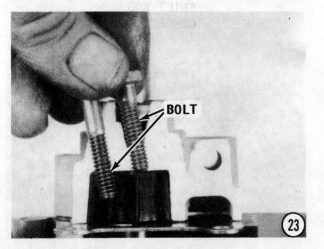

BOLT

23

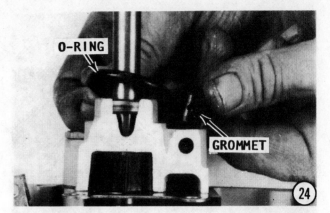

O-RING

GROMMET

24

to the torque value for the size bolt used as given in the Appendix.

ALWAYS rotate the driveshaft **CLOCKWISE** while the screws are tightened to prevent damaging the impeller vanes. If the impeller is not rotated, the housing could damage or cut the end of the vanes as the screws are brought up tight. The rotation allows them to spring back in a natural position.

24- Place **NEW** grommets into the water pump housing for the water pickup. If a new water pump was installed, these grommets will already be in place. Check to be sure the seal is in place on top of the water pump

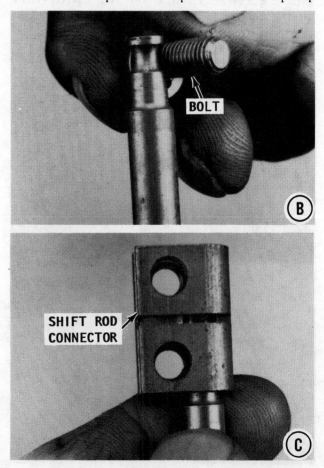

BOLT

B

SHIFT ROD CONNECTOR

C

housing. Slip the O-ring onto the end of the driveshaft.

Filling the Lower Unit

Fill the lower unit with lubricant according to the procedures in Section 8-3.

Propeller Installation

Install the propeller, see Section 8-2.

LOWER UNIT INSTALLATION

GOOD WORDS

Connecting the shift rod with the connector is not an easy task but can be accomplished as follows: First, notice the cutout area on the end of the shift rod. This area permits the bolt to pass through the connector, past the shift rod, and into the other side of the connector. It is this bolt that holds the shift rod in the connector. Now, in order for the bolt to be properly installed, the cutout area on the shift rod **MUST** be aligned in such a manner to allow the bolt to be properly installed. Therefore, as the lower unit is mated with the exhaust housing, exercise patience as the two units come together, to enable the bolt to be installed at the proper time. If the rod is allowed to move too far into the connector before the bolt is installed, it may be possible to force the bolt into place, past the shift rod. The threads on the bolt will be stripped, and the shift rod will eventually come out of the connector, illustration **"B"** and **"C"**.

25- Install the connector onto the lower unit shift rod, with the **NO THREAD** section facing towards the starboard side of the lower unit. With the connector in this position, the bolt may be inserted through the connector and "catch" the threads on the far side. Install the connector bolt in the manner described in the previous paragraph.

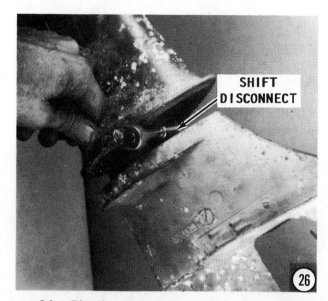

26- Check to be sure the water pickup tubes are clean, smooth, and free of any corrosion. Coat the water pickup tubes and grommets with lubricant as an aid to installation. Guide the lower unit up into the exhaust housing with the water tube sliding into the rubber grommet of the water pump. Continue to work the lower unit towards the exhaust housing, and at the same time rotate the propeller shaft as an aid to indexing the driveshaft splines with the crankshaft. Insert the bolt into the connector. **TAKE TIME** to read and understand the "Good Words" just before Step 25, before making this connection.

Start the bolts securing the lower unit to the exhaust housing. Tighten the bolts **EVENLY** and **ALTERNATELY** to the torque value for the size bolt used as given in the Appendix.

FUNCTIONAL CHECK

27- Perform a functional check of the completed work by mounting the outboard in a test tank, in a body of water, or with a flush attachment connected to the lower unit. If the flush attachment is used, **NEVER** operate the powerhead above an idle speed, because the no-load condition on the propeller would allow the powerhead to **RUNAWAY** resulting in serious damage or destruction of the unit.

CAUTION: Water must circulate through the lower unit to the engine any time the engine is run to prevent damage to the water pump in the lower unit. Just five seconds without water will damage the water pump.

Start the powerhead and observe the tattle-tale flow of water from idle relief in the exhaust housing. The water pump installation work is verified. If a "Flushette" is connected to the lower unit, **VERY LITTLE** water will be visible from the idle relief port. Shift the outboard into the three gears and check for smoothness of operation and satisfactory performance.

8-8 TYPE "E" LOWER UNIT PROPELLER EXHAUST-- MECHANICAL SHIFT 20HP, 25HP AND 30HP

DESCRIPTION

As the name implies, the unit covered in this section is a mechanical shift, propeller exhaust lower unit. Forward, neutral, and reverse shift capabilities are incorporated. A pinion gear is splined onto the lower end of the driveshaft. This pinion gear rotates constantly while the powerhead is operating and drives the forward and reverse gears. A clutch dog splined to the propeller shaft is centered between the two gears when the unit is in neutral gear. A shift lever causes the clutch dog to engage either the forward or reverse gear. Power is then transferred from the direction gear through the clutch dog to the propeller shaft and propeller.

TROUBLESHOOTING

Troubleshooting **MUST** be done **BEFORE** the unit is removed from the exhaust housing, to permit isolating the problem to one area. Always attempt to proceed with troubleshooting in an orderly manner. The shotgun approach will only result in wasted time, incorrect diagnosis, replacement of unnecessary parts, and frustration.

The following procedures are presented in a logical sequence with the most prevalent, easiest, and less costly items to be checked, listed first.

Unable to Shift into Forward or Reverse

Remove the propeller according to the procedures outlined in Section 8-2. Make a careful check of the rubber hub to determine if it has been slipping in the propeller. If there is any evidence the rubber has melted, or if pieces of rubber have been torn from the hub, it is a clear indication the hub has been slipping.

If the check reveals the hub has been slipping, the propeller must be sent to a propeller shop with the proper equipment and trained personnel to perform the necessary service work.

Water in the Lower Unit

Water in the lower unit is usually caused by fish line becoming entangled around the propeller shaft ahead of the propeller and damaging the propeller seal. If the line is not removed, it will cut the propeller shaft seal and allow water to enter the lower unit. Fish line has also been known to cut a groove in the propeller shaft.

The shift rod seal may be damaged and require replacement. The seal under the water pump may be damaged and allowing water to enter the lower unit.

The propeller should be removed each time the boat is hauled from the water at the end of an outing and any material entangled behind the propeller removed before it can cause extensive damage. The small amount of time and effort involved in pulling the propeller is repaid many times by reduced maintenance and service work, including the replacement of expensive parts.

Slippage in the Lower Unit

If the shift seems to be slipping as the boat moves through the water: First, check the propeller and the rubber hub. If the propeller has been subjected to many strikes against underwater objects, it could slip on its hub. If the hub is damaged or excessively worn on the small propellers, it is questionable whether it economical to have the hub or propeller rebuilt. Sometimes a new propeller may be purchased for less than meeting the expense of rebuilding an old worn propeller. It will pay to check it out.

Difficult Shifting

Verify that the ignition switch is **OFF**, or better still, disconnect the spark plug wires from the plugs, to prevent possible personal injury, should the powerhead start. Shift the unit into **REVERSE** gear at the shift control box, and at the same time have an assistant turn the propeller shaft to ensure the clutch is fully engaged. If the shift handle is hard to move, the trouble may be in the lower unit, the shift cable, the handle passing through the exhaust housing, or in the shift box if one is used.

To Isolate the Problem:

Disconnect the shift cable, if used, at the powerhead. Operate the shift lever at the shift box. If shifting is still hard, the problem is in the shift cable or control box, see Chapter 7. If the shifting feels normal with the shift cable disconnected, the prob-

lem must be in the lower unit or in the area where the shift lever passes through the cowling to the bellcrank. Lack of lubrication is usually the cause of problems with the shift lever and bellcrank. To verify the problem is in the lower unit, have an assistant turn the propeller and at the same time move the shift cable back-and-forth. Determine if the clutch engages properly.

Jumping Out of Gear

If a loud thumping sound is heard at the transom while the boat is underway, the unit is jumping out of gear, resulting in a no-load condition on the propeller. When this happens, the rushing water under the hull forces the lower unit in a backward direction. The unit jumps back into gear; the propeller catches hold; the lower unit is forced forward again; and the result is the thumping sound as the action is repeated. Normally this type of action occurs perhaps once a day, then more frequently each time the clutch is operated, until finally the unit will not stay in gear for even a short time.

The following areas must be checked to locate the cause:

1- Check the bellcrank under the powerhead. Remove the window on the port and starboard side of the lower unit. Hold the shift rod with a pair of pliers and at the

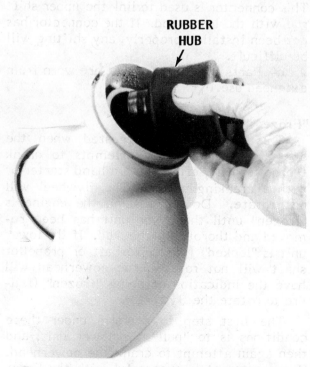

The rubber hub on the propeller exhaust unit was found to be slipping. A new hub is being installed.

Window in a lower unit to permit access to the shift rod disconnect. Notice how the ramps face forward to allow water to enter the cooling system. If the window is not installed properly, the engine will quickly overheat from lack of cooling water, causing damage to the pump impeller and the powerhead.

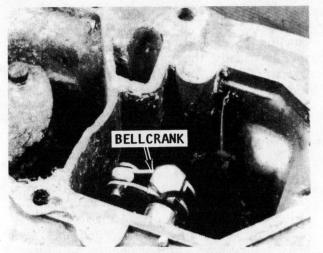

View into the exhaust housing after the power head has been removed. The shift rod bellcrank is visible.

same time attempt to move the shift lever on the starboard side of the powerhead. If it is possible to move the shift lever, the bellcrank is damaged.

2- Disconnect the shift cable at the engine. Attempt to shift the unit into forward gear with the shift lever on the starboard side of the powerhead and at the same time rotate the propeller in an effort to shift into gear. Shift the control lever at the control box into forward gear. Move the shift cable at the powerhead up to the shift handle and determine if the cable is properly aligned. If the inner cable should slip on the end cable guide, the adjustment would be lost.

3- Move the shift lever at the powerhead into the neutral position and the shift lever at the control box to the neutral position. Now, move the shift cable up to the shift lever and see if it is aligned. Shift the unit into reverse at the powerhead and shift the control lever at the control box into reverse. Move the cable up and see if it is aligned. If the cable is properly aligned, but the unit still jumps out of gear when the cable is connected, one of three conditions may exist.

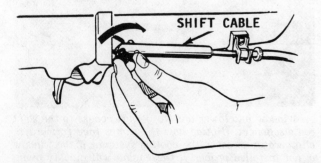

Detailed drawing to depict removal of the cable from the shift handle.

Damaged pistons in a "frozen" powerhead. If the shift problem is isolated to the powerhead, the lower unit need not be disassembled.

a- The bellcrank is worn excessively or damaged.

b- The shift rod connector is misaligned. This connector is used to link the upper shift rod with the lower rod. If the connector has not been installed properly, any shifting will be difficult.

c- Parts in the lower unit are worn from extended use.

"Frozen" Powerhead

This condition is suggested when the operator unsuccessfully attempts to crank the powerhead, either with a hand starter or with a cranking motor. The flywheel will not rotate. Do not assume the engine is "frozen" until the lower unit has been removed and thoroughly checked. If the lower unit is "locked" (the driveshaft or propeller shaft will not rotate), the powerhead will have the indication of being "frozen" (failure to rotate the flywheel).

The first step to perform under these conditions is to "pull" the lower unit, and then again attempt to crank the powerhead. If the attempt is successful with the lower unit disconnected, the problem is in the lower unit. If the attempt to crank the powerhead is still unsuccessful, the problem is in the powerhead.

LOWER UNIT SERVICE

Access to the shift connector for all units covered in this section, unit is gained by removing the window in the lower unit.

Propeller Removal

Remove the propeller according to the procedures outlined in Section 8-2.

Draining Lower Unit

Drain the lubricant in the lower unit, see Section 8-3.

GOOD WORDS

If water is discovered in the lower unit and the determination is made the propeller shaft seal is damaged and requires replacement, the lower unit does **NOT** have to be removed in order to accomplish the work.

The bearing carrier can be removed and the seal replaced without disassembling the lower unit. **HOWEVER**, and this is a big **HOWEVER**, such a procedure is not considered good shop practice, but merely a quick-fix. If water has entered the lower unit, the unit should be disassembled and a detailed check made to determine if any other seals, bearings, bearing races, O-rings or other parts have been rendered unfit for further service by the water.

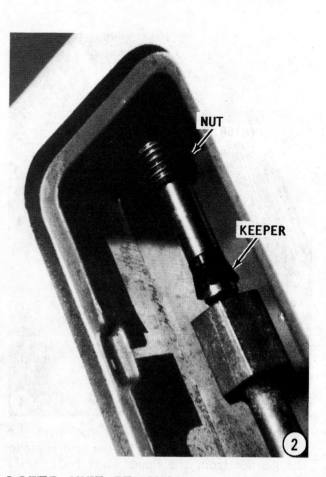

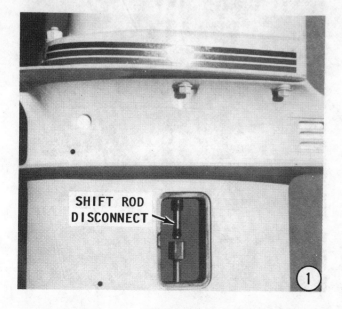

LOWER UNIT REMOVAL

1- Disconnect the spark plug leads from the spark plugs. Remove the port and starboard water inlet screens, below the anti-cavitation plate. After the screens have been removed, the shift rod connection will be visible inside the lower unit. Notice the two nuts on the shift rod. Use two wrenches and loosen the top nut. Back off the upper nut onto the upper portion of the shift rod.

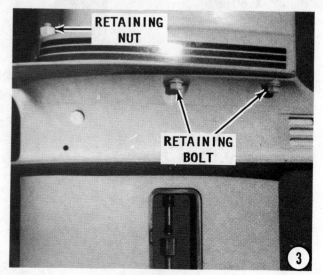

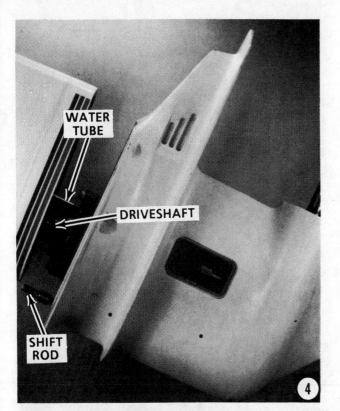

WATER
TUBE

DRIVESHAFT

SHIFT
ROD

4

2- Work a knife or similar tool into the split of the black plastic keeper and remove the keeper from the upper shift rod. Shift the unit into forward gear, and then remove the nut from the upper shift rod.

3- Remove the forward nut and two bolts, one on each side of the lower unit, securing the unit to the exhaust housing.

4- Separate the lower unit from the exhaust housing. **TAKE CARE** not to twist or pull backward, sideways, etc., when removing the lower unit, because the driveshaft

may be bent. If there is restricted clearance between the bottom of the lower unit and the floor, tilt the complete unit forward in order to gain the necessary distance for the lower unit to clear.

5- Position the lower unit in a vertical position on the edge of the work bench resting on the cavitation plate. Secure the lower unit in this position with a C-clamp. The lower unit will then be held firmly in a favorable position for further service work. An alternate method is to cut a groove in a short piece of 2" x 6" wood to accommodate the lower unit with the cavitation plate resting on top of the wood. Clamp the wood in a vise and service work may then be performed with the lower unit erect (in its normal position), or inverted (upside down). In both positions, the cavitation plate is the supporting surface.

WATER PUMP REMOVAL

ADVICE

If the only work to be performed is service of the water pump, be extremely **CAREFUL** to prevent the driveshaft from being pulled up and free of the pinion gear in the lower unit. **NEVER** carry the lower unit by the driveshaft or by the shift rod. If the shaft should be released from the pinion gear, the lower unit **MUST** be disassembled to align the pinion gear and driveshaft, then the driveshaft installed.

2" X 6"
WOODEN
BLOCK

5

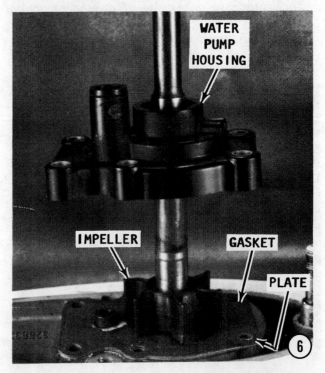

WATER
PUMP
HOUSING

IMPELLER

GASKET

PLATE

6

6- Remove the bolts securing the water pump housing to the lower unit. Slide the water pump housing up and free of the driveshaft. Remove the water pump impeller and key from the driveshaft. Remove the long spacer and bushing from the pump housing. Slide the gasket, pump plate, and second gasket, up and free of the driveshaft. Discard the gaskets. New gaskets are included in a water pump repair kit.

GOOD WORDS

If the only work to be performed is service of the water pump, proceed directly to Page 8-88, Water Pump Installation.

LOWER UNIT DISASSEMBLING

7- Grasp the driveshaft firmly and withdraw it from the lower unit.

Bearing Carrier Removal

8- Use a thin-wall socket and remove the bolts securing the bearing carrier in the lower unit.

SPECIAL WORDS

Two methods are available to pull the bearing carrier from the lower unit. One method involves the use of a flywheel puller and a couple of bolts. The second method requires the use of a special OMC puller and a couple bolts. Both methods are described

in the following step. If the bearing carrier is stubborn and refuses to budge, apply heat to the outside surface of the lower unit while taking up on either type of puller. **TAKE CARE** not to overheat the lower unit. Aluminum will start to bubble at a relatively low temperature.

9- Obtain an OMC Flywheel Puller and two 1/4" x 20 bolts. Thread the bolts into the bearing carrier opposite one another. Take up on the center nut of the puller and pull the bearing carrier free of the lower unit.

The second method involves the use of a special puller tool, OMC P/N 378103 and two long 1/4" x 20 bolts. Thread the bolts into the bearing carrier. Use the special puller to remove the bearing carrier, as shown in the accompanying illustration **"A"**.

WARNING

The next step involves a dangerous procedure and should be executed with care while wearing **SAFETY GLASSES**. The retaining ring is under tremendous tension in the groove and while it is being removed. If it should slip off the Truarc pliers, it will travel with incredible speed causing personal injury if it should strike a person. Therefore, continue to hold the ring and pliers firm after the ring is out of the groove and clear of the lower unit. Place the ring on the floor and hold it securely with one foot before releasing the grip on the pliers. An

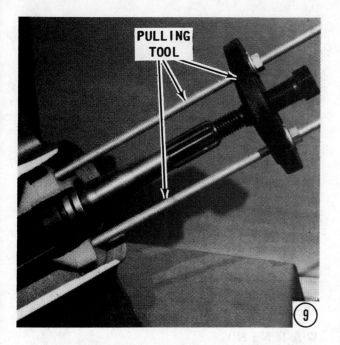

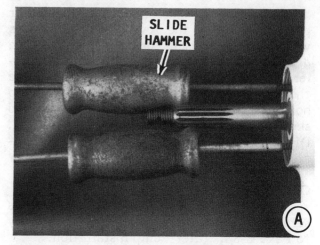

alternate method is to hold the ring inside a trash barrel, or other suitable container, before releasing the pliers.

10- Obtain a pair of Truarc pliers. Insert the tips of the pliers into the holes of the retaining ring. Now, **CAREFULLY** remove the retaining ring from the groove and gear case without allowing the pliers to slip. Release the grip on the pliers in the manner

Retaining rings are under tremendous tension during removal and installation — presenting a potential hazard. As an eye protection measure, safety glasses or a shield should ALWAYS be worn during work with such rings. Warn others in the area such work is in progress.

described in the above **WARNING**. Remove the retainer plate. As the plate is removed, notice which surface is facing into the housing, as an aid during installation.

11- Use the proper size wrench extended through the window of the lower unit, and back the shift rod out of the shift yoke by rotating it **COUNTERCLOCKWISE** until it is free.

12- Using a pair of needlenose pliers, reach in, grasp the shift yoke, and slide it back off the propeller shaft.

NOW THESE WORDS

A hole is drilled through the propeller shaft just forward of the reverse gear. A detent spring is installed in the hole along with a detent ball on each side of the propeller shaft. The clutch dog is installed over the top of the two detent balls. This arrangement assists in holding the unit in the specific gear desired.

When the bearing carrier, propeller shaft, and associated parts are removed, the clutch dog will remain in the lower unit. As soon as the clutch dog slides free of the

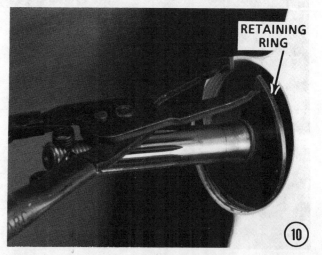

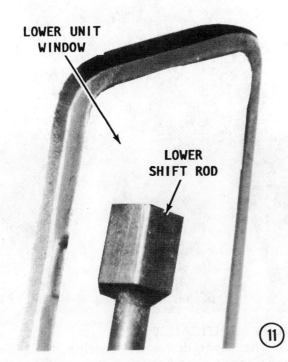

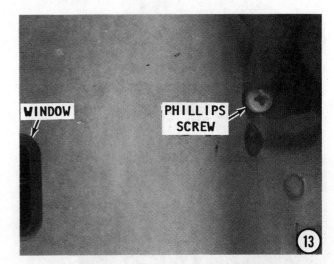

propeller shaft, the detent balls and spring will fly free of the shaft, but will be contained within the lower unit housing. Therefore, after the propeller shaft, bearing carrier, etc., are removed (next step) take time to retrieve the two detent balls and the spring from inside the lower unit.

Propeller Shaft — Removal

13- Remove the Phillips screw from the outside of the lower unit. This is the screw securing the shift lever in place and is located close to the lubricant drain plug.

14- After the screw has been removed, grasp the propeller shaft firmly and withdraw it from the lower unit. The reverse gear, clutch dog, cradle, and associated parts will come out with the shaft. If these parts fail to come out with the shaft, reach in and remove them one at-a-time. Remember the two detent balls and the detent spring.

15- Remove the pinion gear, two thrust washers, and thrust bearing. Take time to notice the arrangement of the two thrust washers and the bearing. Notice how one thrust washer is beveled on the inside diameter and the other is beveled on the outside diameter. It is extremely important that these washers and the thrust bearing are installed properly.

16- Reach in and remove the forward gear. **TAKE CARE** not to lose the thrust washer installed in the inside diameter of the forward gear.

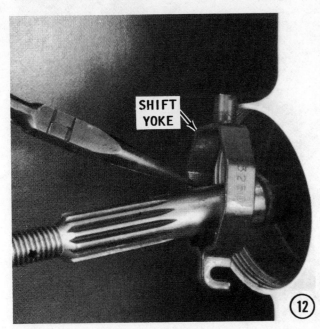

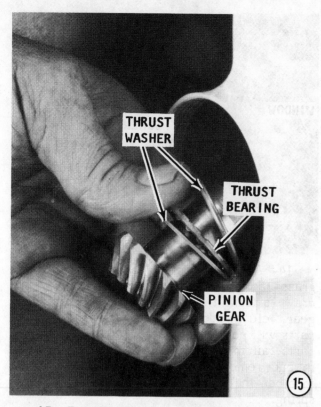

THRUST WASHER

THRUST BEARING

PINION GEAR

(15)

17- Remove the forward gear tapered bearing from the lower unit.

Upper Driveshaft Seals -- Removal

18- Use a slide hammer with external jaws. Fit the jaws down inside the seals, and then operate the slide hammer to remove the seals. Notice how the seals were installed back-to-back (flat side against flat side).

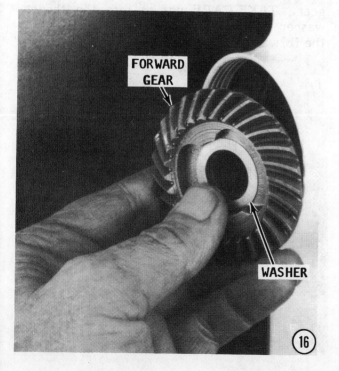

FORWARD GEAR

WASHER

(16)

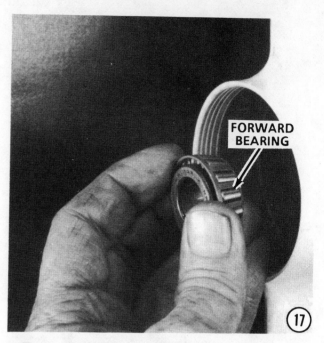

FORWARD BEARING

(17)

VERY CRITICAL WORDS

The upper driveshaft bearing, the pinion gear bearing, and the forward gear bearing race **NEED NOT** be removed unless they are unfit for further service. These bearings, especially the pinion gear bearing, can only be installed using special OMC tools. Even with the tools, the task is not an easy one.

SLIDE HAMMER

(18)

Therefore, determine their condition by first checking with a flashlight for signs of corrosion or damage, and then by inserting a finger into the bearing and rotating it while checking for "rough" spots or binding. If they appear to be in satisfactory condition, "let a sleeping dog lie." Continue with the other work.

Upper Driveshaft Bearing — Removal

19- The upper driveshaft bearing is housed in a sleeve. If the bearing is to be removed, first punch out the bearing downward into the housing. Use special tool OMC P/N 391010, and remove the bearing sleeve from the housing. Turn the lower unit upside down and the upper driveshaft bearing will fall free.

Pinion Gear Bearing — Removal

20- Use any type of punch to drive the bearing free. Further damage to the bearing is of no concern because it is being removed due to its unfitness for further service.

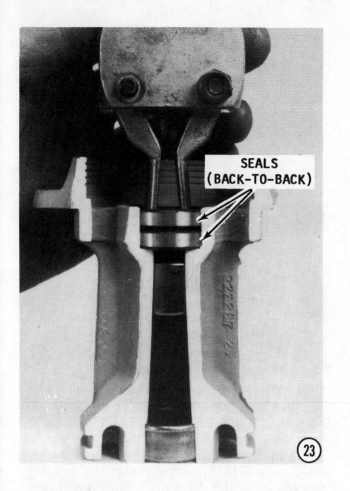

SEALS
(BACK-TO-BACK)

(23)

Forward Gear Bearing Race -- Removal

21- Use a slide hammer equipped with fingers and "pull" the race from the housing.

Shift Rod 0-ring and Bushing — Removal

22- Insert a length of 1/4" rod, with threads on both ends, down through the O-ring and bushing. Thread a nut onto the lower end of the rod. (The edges of the nut must be first rounded to permit the nut to pass through the opening in the housing.) Attach a slide hammer to the rod and "pull" the O-ring, bushing and washer.

Bearing Carrier Seals — Removal

23- Use a puller with two fingers and work the puller down into the seals. Take up on the puller and remove the seals. Notice how the seals are installed back-to-back. They must be installed in this manner to provide a proper seal to prevent the lubricant in the lower unit from escaping and water from entering.

An alternate method to remove the seal is to wedge a heavy duty screwdriver underneath the seal, and then to pull back on the bearing carrier, as shown, illustration "B". The seal will pop out. Repeat the procedure for the second seal.

Bearing Carrier Bearing — Removal

The bearings in the bearing carrier need

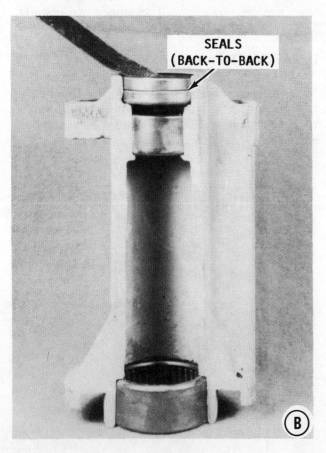

SEALS
(BACK-TO-BACK)

(B)

SLIDE
HAMMER

(24)

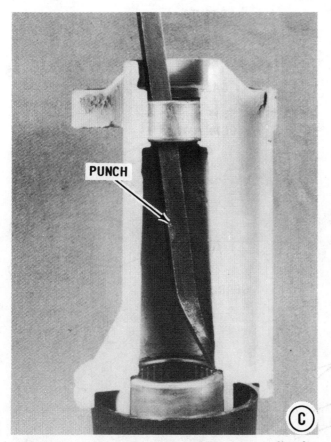

New clutch dog and gear. Compare these two parts with those shown at the bottom of the previous column.

CLEANING AND INSPECTING

Clean all water pump parts with solvent, and then dry them with compressed air. Inspect the water pump cover and base for cracks and distortion, possibly caused from overheating. Inspect the face plate and water pump insert for grooves and/or rough surfaces. If possible, **ALWAYS** install a complete new water pump while the lower unit is disassembled. A new impeller will ensure extended satisfactory service and give "peace of mind" to the owner. If the old impeller must be returned to service, **NEVER** install it in reverse to the original direction of rotation. Installation in reverse will cause premature impeller failure.

NOT be removed unless they are unfit for further service. Check their condition for damage and corrosion. Insert a finger inside and rotate the bearing while checking for roughness or any sign of binding. Check for evidence of corrosion.

24- Use a slide hammer with fingers and remove the bearings outward from the carrier. An alternate method is to use a punch and drive the bearing free of the carrier, illustration "C". If the bearing is damaged and no longer fit for service, further damage will be of no consequence.

The ears on this clutch dog and the teeth on the gear are badly worn. Both items are unfit for further service.

A rusted and corroded gear. Water was allowed to enter the lower unit through a badly worn seal and cause this damage to the gear and other expensive parts.

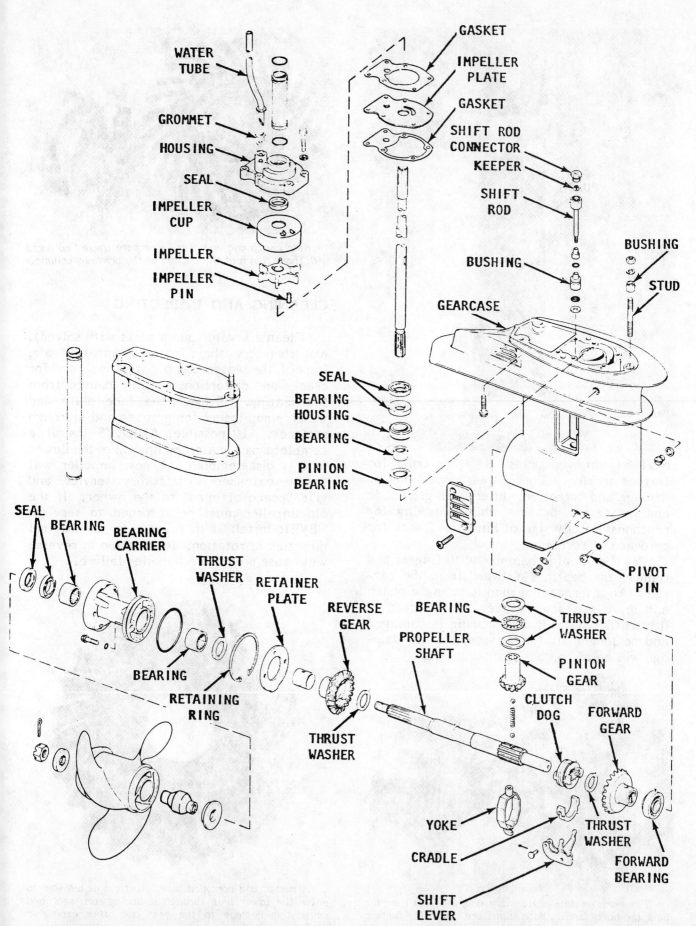

WATER TUBE

GROMMET

HOUSING

SEAL

IMPELLER CUP

IMPELLER

IMPELLER PIN

GASKET

IMPELLER PLATE

GASKET

SHIFT ROD CONNECTOR

KEEPER

SHIFT ROD

BUSHING

GEARCASE

BUSHING

STUD

SEAL

BEARING HOUSING

BEARING

PINION BEARING

PIVOT PIN

SEAL

BEARING

BEARING CARRIER

THRUST WASHER

RETAINER PLATE

REVERSE GEAR

BEARING

PROPELLER SHAFT

THRUST WASHER

PINION GEAR

BEARING

RETAINING RING

CLUTCH DOG

FORWARD GEAR

THRUST WASHER

THRUST WASHER

FORWARD BEARING

YOKE

CRADLE

SHIFT LEVER

Exploded drawing of the Type "E" lower unit matched with the 20hp, 25hp, and the 30hp powerheads. Major parts have been identified.

Inspect the impeller side seal surfaces and the ends of the impeller blades for cracks, tears, and wear. Check for a glazed or melted appearance, caused from operating without sufficient water. If any question exists, and as previously stated, install a new impeller if at all possible.

Clean all parts with solvent and dry them with compressed air. **DISCARD** all O-rings and gaskets. Inspect and replace the driveshaft if the splines are worn. Inspect the gearcase and exhaust housing for damage to the machined surfaces. Remove any nicks and refurbish the surfaces on a surface plate. Start with a No. 120 Emery paper and finish with No. 180.

Check the water intake screen and passages by removing the bypass cover, if one is used. Inspect the clutch dog, drive gears, pinion gear, and thrust washers. Replace these items if they appear worn. If the clutch dog and drive gear arrangement surfaces are nicked, chipped, or the edges rounded, the operator may be performing the shift operation improperly or the controls may not be adjusted correctly. These items **MUST** be replaced if they are damaged.

Inspect the dog ears on the inside of the forward and reverse gears. The gears must be replaced if they are damaged.

Check the cradle that rides on the inside diameter of the clutch dog. The sides of the

cradle must be in good condition, free of any damage or signs of wear. If damage or wear has occurred, the cradle must be replaced.

Check the shift lever and the two prongs that fit inside the cradle. Check to be sure the prongs are not worn or rounded. Damage or wear to the prongs indicates the lever must be replaced.

LOWER UNIT ASSEMBLING

READ AND BELIEVE

The lower unit should **NOT** be assembled in a dry condition. Coat all internal parts with OMC HI-VIS lube oil as they are assembled. All seals should be coated with OMC Gasket Seal Compound. When two seals are installed back-to-back, use Triple Guard Lubricant between the seal surfaces.

Bearing Carrier Bearings — Installation

1- Obtain special tool OMC P/N 321429. Place the bearing in position with the lettered side of the bearing facing **UP**. Press against the lettered side of the forward bearing, using the special tool and an arbor press. Turn the bearing carrier end-for-end and press the other bearing into place in the same manner, using the arbor press and special tool OMC P/N 321428. **ALWAYS** press against the lettered side of the bearing.

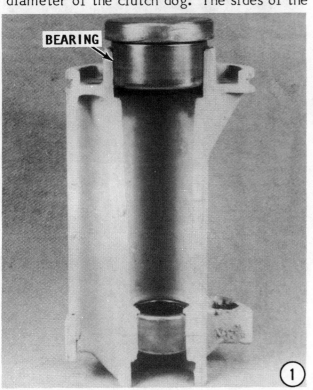

A set of double seals showing the back side (left) and the front side (right). These seals are installed back-to-back (flat side-to-flat side) with Triple Guard Grease between the surfaces. This arrangement prevents fluid from passing in either direction.

2- The two bearing carrier seals are installed back-to-back. The inner seal prevents the lower unit lubricant from escaping, and the outer seal prevents water from entering the lower unit. Coat the outside surfaces of the seals with seal compound and press the first seal into place with the flat side facing **UP**. Coat the flat side of the installed seal and the flat side of the second seal with Triple Guard Grease. Install the second seal with the flat side facing **DOWN**. After installation, the lip of the second seal should be flush with the surface of the bearing carrier.

3- Check to be sure the O-ring groove of the bearing carrier is clean. Coat the O-ring with OMC HI-VIS lube oil and then

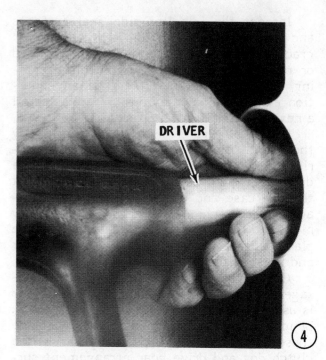

install it into the groove. Set the bearing carrier aside for later installation.

Forward Gear Bearing Race — Installation

4- If the forward gear bearing race was removed, install a new race by first coating the race with OMC HI-VIS lube oil. Obtain special tool, OMC P/N 319929 and driver handle, OMC P/N 311880. Drive the race into place squarely.

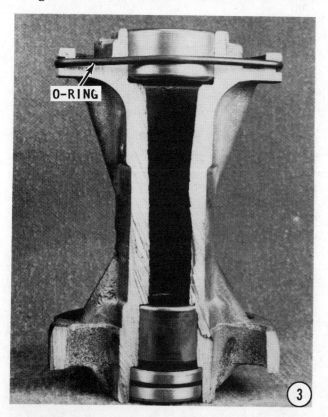

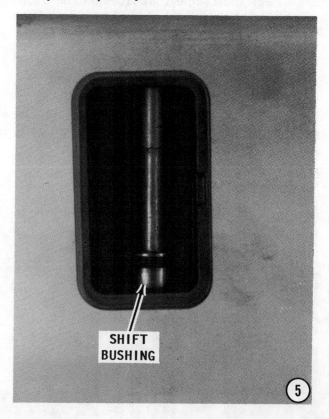

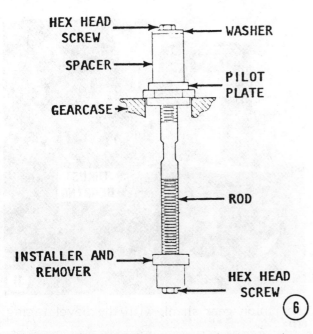

Shift Rod Bushing and O-ring — Installation

5- Obtain special tool, OMC P/N 304515. Install the washer and then the O-ring. Install the bushing using the special tool. If the special tool is not available, a tool the same size as the outside diameter as the bushing may be used to drive the bushing into place. The bushing does not go in hard. **TAKE CARE** not to distort the inside diameter of the bushing.

Pinion Gear Bearing — Installation

6- Special tool OMC P/N 391257 is required to install this bearing. Assemble the tool, as shown in the accompanying illustration. Drive the bearing into place from the top until the plate of the tool makes contact with the lower unit. The bearing will then be installed to the proper depth in the lower unit.

Upper Driveshaft Bearing — Installation

7- Obtain special tool, OMC P/N 322923. Press the bearing into the bearing retainer with the lettered side of the bearing facing the tool. Press the assembled bearing and retainer into the lower unit housing until it seats. A socket the same size as the outside diameter of the retainer may be used. **TAKE CARE** not to damage the bearing during installation.

Driveshaft Seals — Installation

8- These seals are installed back-to-back, flat side-to-flat side, illustration **"A"**. The inner seal prevents the lower unit lubricant from escaping, and the outer seal prevents water from entering the lower unit. Coat the outside surface of the first seal with Seal Compound. Install the seal with the flat side facing **UP**. Coat the flat side of the installed seal and the flat side of the second seal with Triple Guard Grease. Install the second seal with the flat side facing **DOWN**.

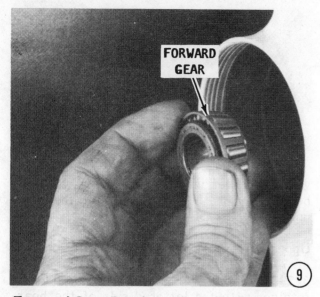

FORWARD GEAR

⑨

Forward Gear Bearing — Installation

9- Insert the bearing into the race with the tapered side of the bearing going in **FIRST**.

Pinion Gear — Installation

10- Install the thin thrust washer onto the pinion gear shank with the bevel on the inside diameter facing **DOWN**.

11- Slide the thrust bearing onto the shank.

12- Slide the thick second washer onto

BEVEL

PINION GEAR SHANK

⑩

THRUST BEARING

⑪

the pinion gear shank with the bevel facing **UP**.

13- Insert the assembled pinion gear into the opening in the lower unit housing.

Forward Gear — Installation

14- Insert the thrust washer into the forward gear. Lower the forward gear down into the lower unit housing past the pinion gear. By tilting the gear slightly it will pass the pinion gear. The teeth of the forward gear must index with the teeth of the pinion gear.

Shift Mechanism — Installation

15- Lower the shift lever into the lower unit housing and into place in the recess.

BEVEL

⑫

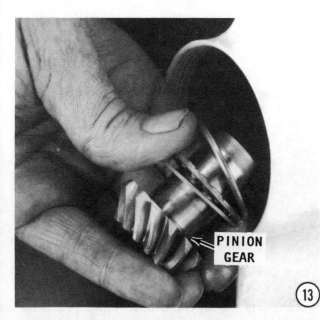

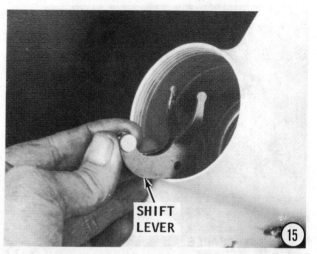

with the two detent balls. If the cutouts do not align with the detent balls, withdraw the clutch dog, rotate it 180°, and then slide it back onto the shaft, reference illustration "B".

Notice the small groove on the outside diameter on one side of the clutch dog. This groove **MUST** face the forward gear. Carefully slide the clutch dog over the two detent balls and into the neutral position on the shaft.

Propeller Shaft — Installation

VERY GOOD WORDS

Two illustrations accompany most of the next 10 steps. One set is numbered and is the same as the step number. The other set has a letter identification. **NOW**, the numbered illustrations show the work being performed in the normal manner with the lower unit. The lettered illustrations show the work being performed on a work bench in order to give you a clear understanding of what is happening and the relationship of the parts inside the lower unit.

16- Insert the detent spring through the hole in the propeller shaft. Apply a small amount of needle bearing grease to the hole on both sides of the propeller shaft. Stick a detent ball into place in the grease on each side of the shaft. Slide the clutch dog down the propeller shaft with the cutouts aligned

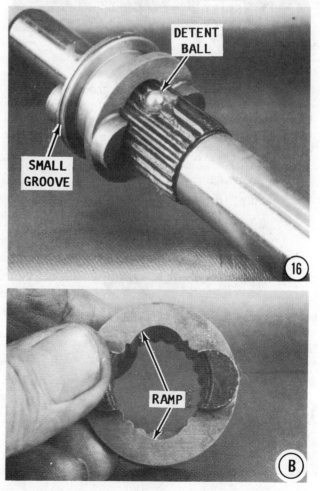

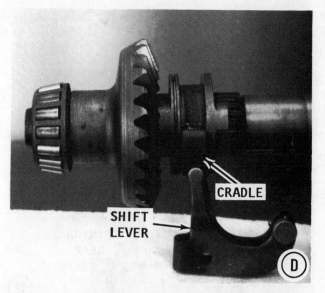

17- Lubricate the large groove in the clutch dog with needle bearing grease, reference illustration "C". This illustration will be helpful in understanding the relationship of the forward gear, forward gear bearing, and the shift lever. Notice how the shaft passes over the top of the shift lever. Slide the cradle into the large groove of the

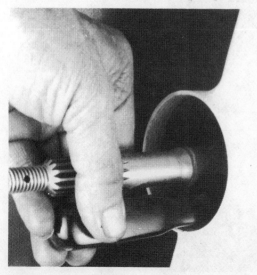

clutch dog. Insert the assembled propeller shaft into the lower unit, with the forward end of the shaft indexed into the forward gear and forward gear bearing. At the same time, the shaft is worked over the top of the shift lever.

18- Slide a screwdriver blade under the shift lever and work the fork fingers up into the cradle, reference illustration "D".

19- After the fingers are in place in the cradle, move the shift lever until the hole in the lever is aligned with the hole in the lower unit housing, reference illustration "E". When the holes align, start the Phillips screw through the housing and into the lever. Apply a coating of OMC 1000 sealer onto the threads of the screw. Tighten the screw securely.

PHILLIPS SCREW

E

Reverse Gear — Installation

20- Insert the thrust washer into the reverse gear. Slide the reverse gear down the propeller shaft into the lower unit housing.

21- Slide the shift yoke down the propeller shaft and hook it into the shift lever, reference illustration **"F"**.

22- Coat the shift rod with light-weight oil and then insert the shift rod down through the **O**-ring and thread it into the shift yoke. Tighten the shift rod securely in the yoke, with a wrench, reference illustration **"G"**.

Bearing Carrier — Installation

23- Slide the retainer plate onto the propeller shaft and against the lower unit housing, with the lip of the plate indexed into the short slot on the bottom side of the lower unit, reference illustration **"H"**.

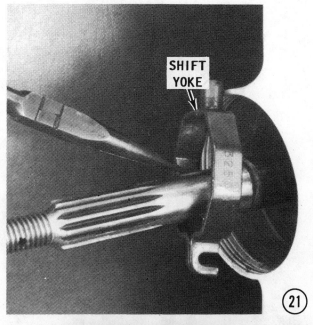

SHIFT YOKE

21

SHIFT YOKE

SHIFT LEVER

F

REVERSE GEAR

20

LOWER SHIFT ROD

G

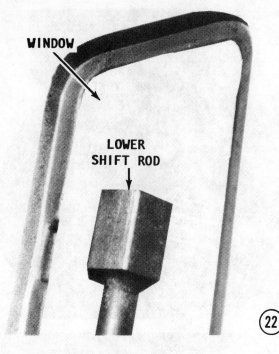

WINDOW

LOWER
SHIFT ROD

(22)

RETAINER
PLATE

(23)

WARNING

This next step can be dangerous. The snap ring is placed under tremendous tension with the Truarc pliers while it is being placed into the groove. Therefore, wear **SAFETY GLASSES** and exercise care to prevent the snap ring from slipping out of the

pliers. If the snap ring should slip out, it would travel with incredible speed and cause personal injury if it struck a person.

24- Use a pair of Truarc pliers and install the Truarc snap ring into the groove in the lower unit, next to the retainer plate. Check to be sure the ring is properly seated all the way around in the groove.

25- Obtain two 1/4" rods about 12" long with 1/4 x 28 threads on one end. Thread these two rods into the retainer plate to act as guides for installation of the bearing carrier. Observe the word **UP** embossed into the metal of the bearing carrier rim, reference illustration **"J"**. Slide the bearing carrier onto the propeller shaft over the guide rods and into the lower unit, with the word **UP** facing **UPWARD** in relation to the lower unit housing. Check to be sure the thrust washer is seated in the recess of the bearing carrier towards the reverse gear.

Retaining rings are under tremendous tension during removal and installation — presenting a potential hazard. As an eye protection measure, safety glasses or a shield should ALWAYS be worn during work with such rings. Warn others in the area such work is in progress.

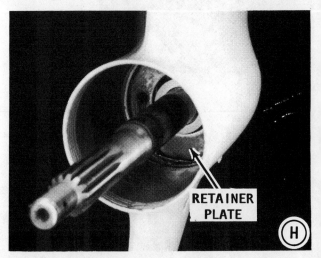

RETAINER
PLATE

(H)

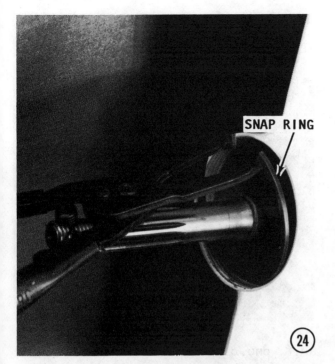

Figure 24 — SNAP RING

26- Slide new O-rings onto the bearing carrier bolts. Coat the threads of the bolts with OMC Sealer. Install two bolts through the bearing carrier and into the retainer. Back out the two guide rods used to install the bearing carrier. Install and tighten the bearing carrier bolts to the torque value given in the Appendix, for the size bolt being used.

Driveshaft — Installation

27- Install the guide bushing down over the stud on top of the housing, as shown.

28- Apply sealer to the upper housing surface.

Figure 25 — GUIDE ROD

Figure J — UP

29- Coat the driveshaft with light-weight oil as an aid to installation. Slide the driveshaft down into the lower unit. As the driveshaft is lowered, rotate the driveshaft slightly to permit the splines on the shaft to index with the splines of the pinion gear.

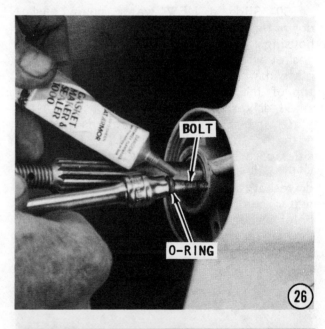

Figure 26 — BOLT, O-RING

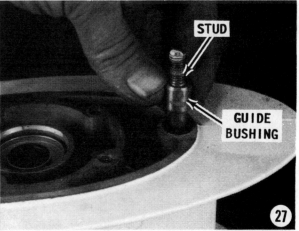

Figure 27 — STUD, GUIDE BUSHING

SEALER

(28)

WATER PUMP INSTALLATION

30- Slide a **NEW** water pump gasket down the driveshaft and into place on the housing. Coat the upper surface of the gasket with sealer. Slide the water pump plate down the driveshaft and into place on top of the gasket. Check to be sure the small driveshaft grommet that seats in the plate and through the gaskets is installed with the small side facing **UP**. Coat both sides of a **NEW** second water pump gasket with sealer, and then slide it down the driveshaft and into place on top of the water pump plate. Check to be sure the holes in both gaskets, the plate, and in the housing are aligned. If the holes do not

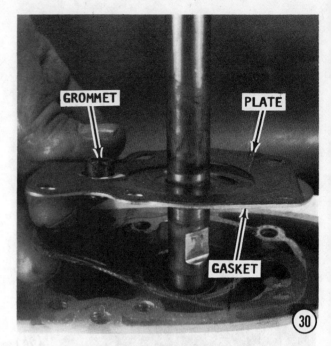

GROMMET   **PLATE**

GASKET

(30)

align, one of the gaskets or the plate is upside down. Correct the error.

31- Slide the water pump impeller down the driveshaft. Just before the impeller covers the cutout for the impeller pin, install the pin. Align the slot in the impeller with the impeller pin, and then continue to work the impeller down the driveshaft until it is firmly in place on the surface of the upper water pump gasket.

32- Check to be sure **NEW** seals and O-rings have been installed in the water pump. Lubricate the inside surface of the water pump with light-weight oil. Lower the water pump housing down the driveshaft and over the impeller. **ALWAYS** rotate the driveshaft slowly **CLOCKWISE** as the hous-

DRIVESHAFT

(29)

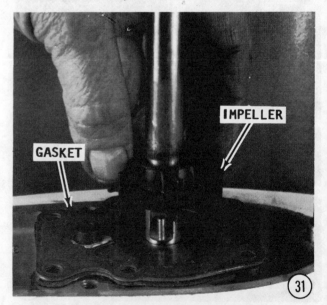

IMPELLER

GASKET

(31)

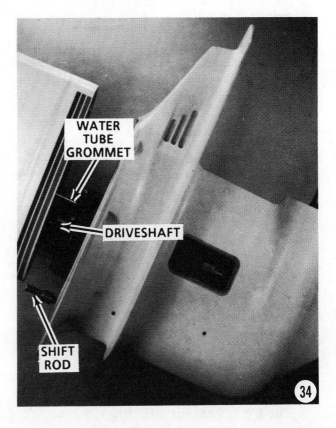

ing is lowered over the impeller to allow the impeller blades to assume their natural and proper position inside the housing. Continue to rotate the driveshaft and work the water pump housing downward until it is seated on the gasket and plate.

33- Coat the threads of the water pump attaching bolts with sealer, and then secure the pump in place with the bolts. Tighten the bolts **ALTERNATELY** and **EVENLY**. Check to be sure a **NEW** grommet has been installed in the top of the water pump. Install a **NEW O**-ring onto the top of the driveshaft.

LOWER UNIT INSTALLATION

34- Check to be sure the water tubes are clean, smooth, and free of any corrosion. Coat the water pickup tubes and grommets with lubricant as an aid to installation. Check to be sure the spark plug wires are disconnected from the spark plugs. Bring the lower unit housing together with the exhaust housing, and at the same time, guide the water tube into the rubber grommet of the water pump. As the two units come together, rotate the flywheel slowly

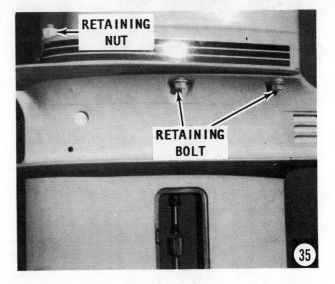

to permit the splines of the driveshaft to index with the splines of the crankshaft.

35- After the surfaces of the lower unit and exhaust housing make contact, start the nut on the stud on the leading edge of the lower unit. Start the four bolts on the bottom side of the lower unit. **DO NOT** tighten this hardware at this time.

36- At the powerhead, move the shift lever to the **FORWARD** gear position. The shift connection is made through the water pickup openings in the lower unit. Slip the upper shift lever nut upward, and then snap the keeper onto the end of the shift rod. Move the shift lever to the **REVERSE** gear position. Lower the shift rod into the lower unit shift rod section. Tighten the nut to secure the lower portion of the shift rod to the upper portion.

37- Install and secure the two water pickup windows in place, with the cutout slots facing **FORWARD**.

Filling the Lower Unit

Fill the lower unit with lubricant according to the procedures in Section 8-3.

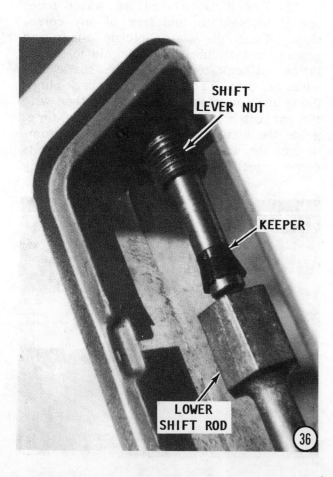

Propeller Installation

Install the propeller, see Section 8-2.

FUNCTIONAL CHECK

Perform a functional check of the completed work by mounting the outboard in a test tank, in a body of water, or with a flush attachment connected to the lower unit. If the flush attachment is used, **NEVER** operate the powerhead above an idle speed, because the no-load condition on the propeller would allow the powerhead to **RUNAWAY** resulting in serious damage or destruction of the unit.

CAUTION: Water must circulate through the lower unit to the engine any time the engine is run to prevent damage to the water pump in the lower unit. Just five seconds without water will damage the water pump.

Start the powerhead and observe the tattle-tale flow of water from idle relief in the exhaust housing. The water pump installation work is verified. If a "Flushette" is connected to the lower unit, **VERY LITTLE** water will be visible from the idle relief port. Shift the outboard into the three gears and check for smoothness of operation and satisfactory performance.

8-9 TYPE "F" LOWER UNIT
40HP AND 50HP
MECHANICAL SHIFT
SHIFT DISCONNECT UNDER
LOWER CARBURETOR

DESCRIPTION

The lower unit covered in this section is a complete mechanical shift unit. A shift cable connects the shift box to the shift linkage at the powerhead. A shift rod extends from the powerhead down through the exhaust housing to the lower unit.

Shifting into forward, neutral, and reverse, is accomplished directly through mechanical means from the shift control handle through the cable and linkage to the clutch dog in the lower unit.

The lower unit houses the driveshaft and pinion gear, the forward and reverse driven gears, the propeller shaft, shift lever, cradle, shift shaft, clutch dog, shift rod, and the necessary shims, bearings, and associated parts to make it all work properly. A detent ball and spring is installed on some models.

The water pump is considered a part of the lower unit.

Two different lower units are covered in this section. One unit is used with the electric start model and the other with manual start. The upper driveshaft bearing and the shift mechanism differ between the two units. These differences are clearly indicated in the procedural steps and illustrations.

TROUBLESHOOTING

Preliminary Checks

At rest, and without the powerhead running, the lower unit is in forward gear. Mount the outboard in a test tank, in a body of water, or with a flush attachment connected to the lower unit. If the flush attachment is used, **NEVER** operate the engine above an idle speed, because the no-load condition on the propeller would allow the powerhead to **RUNAWAY**, resulting in serious damage or destruction of the powerhead.

CAUTION: Water must circulate through the lower unit to the engine any time the engine is run to prevent damage to the water pump in the lower unit. Just five seconds without water will damage the water pump.

Attempt to shift the unit into **NEUTRAL** and **REVERSE**. It is possible the propeller may turn very slowly while the unit is in neutral, due to "drag" through the various gears and bearings. If difficult shifting is encountered, the problem is in the shift linkage or in the lower unit.

The second area to check is the quantity and quality of the lubricant in the lower unit. If the lubricant level is low, contaminated with water, or is broken down because of overuse, the shift mechanism may be affected. Water in the lower unit is **VERY BAD NEWS** for a number of reasons.

Remember, when the powerhead is not running, the unit should be in **FORWARD** gear.

BEFORE making any tests, remove the propeller, see Section 8-2. Check the propeller carefully to determine if the hub has been slipping and giving a false indication the unit is not in gear, reference illustration **"A"**. If there is any doubt, the propeller should be taken to a shop properly equipped for testing, before the time and expense of disassembling the lower unit is undertaken. The expense of the propeller testing and possible rebuild is justified.

The following troubleshooting procedures are presented on the assumption the lower unit lubricant, and the propeller have been checked and found to be satisfactory.

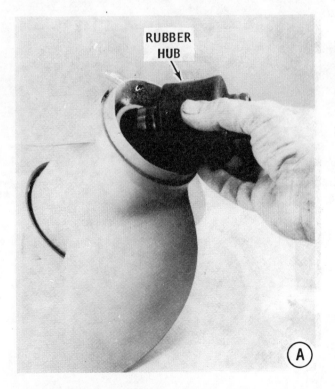

RUBBER HUB

Ⓐ

Lower Unit Locked

Determine if the problem is in the powerhead or in the lower unit. Attempt to rotate the flywheel. If the flywheel can be moved even slightly in either direction, the problem is most likely in the lower unit. If it is not possible to rotate the flywheel, the problem is a "frozen" powerhead. To absolutely verify the powerhead is "frozen", separate the lower unit from the exhaust housing and then again attempt to rotate the flywheel. If the attempt is successful, the problem is definitely in the lower unit. If the attempt to rotate the flywheel, with the lower unit removed, still fails, a "frozen" powerhead is verified, reference illustration **"B"**.

Unit Fails to Shift
Neutral, Forward, or Reverse

With the outboard mounted on a boat, in a test tank, or with a flush attachment connected, disconnect the shift lever at the powerhead. Attempt to manually shift the unit into **NEUTRAL, REVERSE, FORWARD.** At the same time move the shift handle at the shift box and determine that the linkage and shift lever are properly aligned for the shift positions. If the alignment is not correct, adjust the shift cable at the trunnion. It is also possible the inner wire may have slipped in the connector at the shift box. This condition would result in a lack of

inner cable to make a complete "throw" on the shift handle. Check to be sure the shift handle is moved to the full shift position and the linkage to the lower unit is moved to the full shift position.

If an adjustment is required at the shift box, see Chapter 7.

If it is not possible to shift the unit into gear by manually operating the shift rod while the powerhead is running, the lower unit requires service as described in this section.

LOWER UNIT SERVICE

Propeller Removal

If the propeller was not removed, as directed for the troubleshooting, remove it now, according to the procedures outlined in Section 8-2.

Draining the Lower Unit

Drain the lower unit of lubricant, see Section 8-3.

GOOD WORDS

If water is discovered in the lower unit and the propeller shaft seal is damaged and requires replacement, the lower unit does **NOT** have to be removed in order to accomplish the work.

The bearing carrier can be removed and the seal replaced without disassembling the

Helm-type shift box and steering wheel.

lower unit. **HOWEVER**, such a procedure is not considered good shop practice, but merely a quick-fix. If water has entered the lower unit, the unit should be disassembled and a detailed check made to determine if any other seals, bearings, bearing races, O-rings or other parts have been rendered unfit for further service by the water.

LOWER UNIT REMOVAL

1- Disconnect and ground the spark plug leads. Remove the cranking motor. Remove the bolt or bolts from the shift connector under the lower carburetor.

2- Scribe a mark on the trim tab and a matching mark on the lower unit to ensure the trim tab will be installed in the same position from which it is removed. Use an Allen wrench and remove the trim tab.

3- On some units, an attaching bolt is installed inside the trim tab cavity. Use a 1/2" socket with a short extension and remove this bolt. Failure to remove this bolt from inside the trim tab cavity may result in an expensive part being broken in an attempt to separate the lower unit from the exhaust housing. Remove the 5/8" countersunk bolt located just ahead of the trim tab position.

4- Remove the four 9/16" bolts, two on each side, securing the lower unit to the exhaust housing. Work the lower unit free of the exhaust housing. If the unit is still

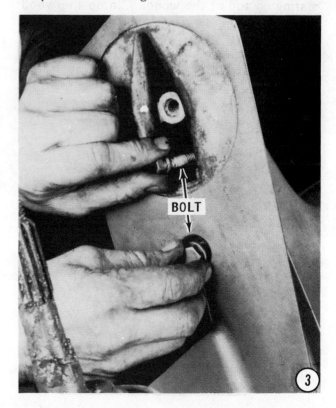

mounted on a boat, tilt the outboard forward to gain clearance between the lower unit and the deck (floor, ground, whatever). **EXERCISE CARE** to withdraw the lower unit straight away from the exhaust housing to prevent bending the driveshaft.

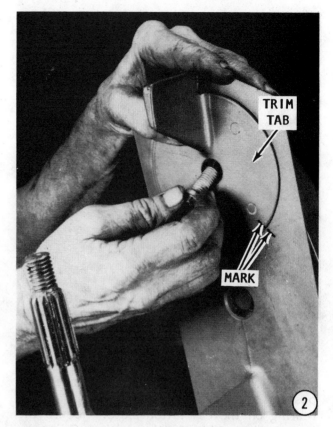

WATER PUMP REMOVAL

5- Position the lower unit in the vertical position on the edge of the work bench resting on the cavitation plate. Secure the lower unit in this position with a C-clamp. The lower unit will then be held firmly in a favorable position during the service work. An alternate method is to cut a groove in a short piece of 2" x 6" wood to accommodate the lower unit with the cavitation plate resting on top of the wood. Clamp the wood in a vise and service work may then be performed with the lower unit erect (in its normal position), or inverted (upside down). In both positions, the cavitation plate is the supporting surface.

6- Remove the O-ring from the top of the driveshaft. Remove the bolts securing the water pump to the lower unit housing. Pull the water pump housing up and free of the driveshaft. Slide the water pump impeller up and free of the driveshaft.

7- Pop the impeller Woodruff key out of the driveshaft keyway. Slide the water pump base plate up and off of the driveshaft.

GOOD WORDS

If the only work to be performed is service of the water pump, proceed directly to Page 8-114, Water Pump Installation.

Bearing Carrier — Removal

8- Remove the four 5/16" bolts from inside the bearing carrier. Notice how each bolt has an O-ring seal. These O-rings should be replaced each time the bolts are

removed. In most cases, the word **UP** is embossed into the metal of the bearing carrier rim. This word must face **UP** in relation to the lower unit during installation.

9- Remove the bearing carrier using one of the methods described in the following paragraphs, under Special Words.

Model 40hp, 48hp & 50hp -- 1990 & On
Remove the two anode retaining screws and pull the anode free.

SPECIAL WORDS
Several models of bearing carriers are used on the lower units covered in this section.

The bearing carriers fit very tightly into the lower unit opening. Therefore, it is not uncommon to apply heat to the outside surface of the lower unit with a torch, at the same time the puller is being worked to remove the carrier. **TAKE CARE** not to overheat the lower unit.

One model carrier has two threaded holes on the end of the carrier. These threads permit the installation of two long bolts. These bolts will then allow the use of a flywheel puller to remove the bearing carrier, illustration **"9"**.

Another model does not have the threaded screw holes. To remove this type bearing carrier, a special puller with arms must be used. The arms are hooked onto the carrier web area, and then the carrier removed, reference illustration **"C"**.

WARNING
The next step involves a dangerous procedure and should be executed with care while wearing **SAFETY GLASSES**. The retaining rings are under tremendous tension in the groove and while they are being

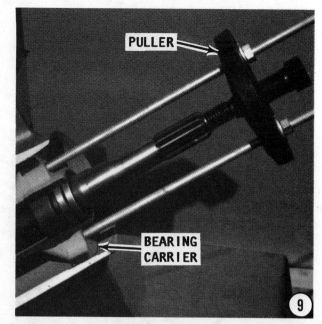

removed. If a ring should slip off the Truarc pliers, it will travel with incredible speed causing personal injury if it should strike a person.

Therefore, continue to hold the ring and pliers firm after the ring is out of the groove and clear of the lower unit. Place the ring on the floor and hold it securely with one foot before releasing the grip on the pliers. An alternate method is to hold the ring inside a trash barrel, or other suitable container, before releasing the pliers.

10- Obtain a pair of Truarc pliers. Insert the tips of the pliers into the holes of

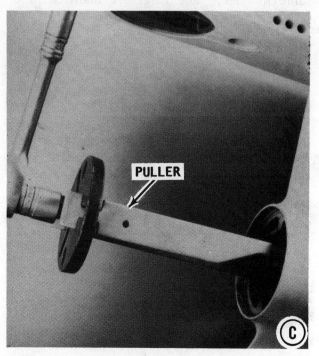

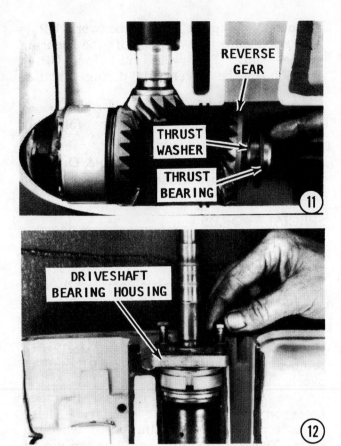

Retaining rings are under tremendous tension during removal and installation — presenting a potential hazard. Safety glasses or a shield should ALWAYS be worn during such work for eye and face protection. Warn others in the area such work is in progress.

the first retaining ring. Now, **CAREFULLY** remove the retaining ring from the groove and gear case without allowing the pliers to slip. Release the grip on the pliers in the manner described in the above **WARNING**. Remove the second retaining ring in the same manner. The rings are identical and either one may be installed first.

11- Remove the retainer plate, thrust washer, thrust bearing, and reverse gear from the propeller shaft.

12- Remove the four bolts from the driveshaft bearing housing. Pull up on the shift rod. This action will place shift dog into the forward gear position. Rotate the propeller shaft a bit as a check to be sure the unit is in forward gear.

Pinion Gear — Removal

Special tool, OMC P/N 316612 is required to turn the driveshaft in order to remove the pinion gear nut.

13- Obtain the special tool and slip it over the end of the driveshaft with the splines of the tool indexed with the splines

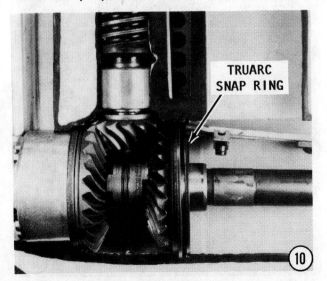

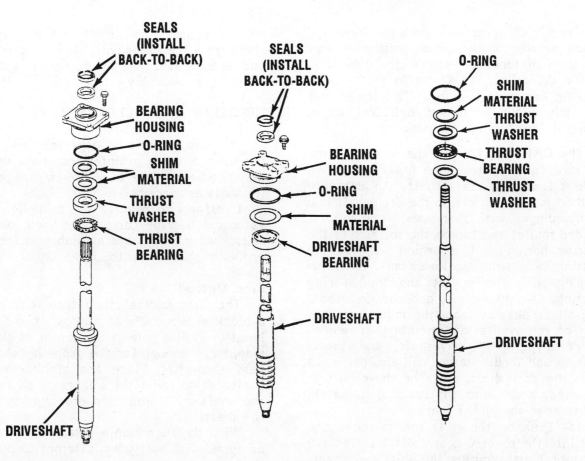

SEALS
(INSTALL
BACK-TO-BACK)

BEARING
HOUSING

O-RING

SHIM
MATERIAL

THRUST
WASHER

THRUST
BEARING

DRIVESHAFT

SEALS
(INSTALL
BACK-TO-BACK)

BEARING
HOUSING

O-RING

SHIM
MATERIAL

DRIVESHAFT
BEARING

DRIVESHAFT

O-RING

SHIM
MATERIAL

THRUST
WASHER

THRUST
BEARING

THRUST
WASHER

DRIVESHAFT

Exploded drawing the of the Type "F1" driveshaft (left), the Type "F2" (center), and the Model 40h-, 48hp and 50hp -- 1990 and on (right). Full exploded lower unit drawings are presented on Pages 8-104 and 8-105.

on the driveshaft. Hold the pinion gear nut with the proper size wrench, and at the same time rotate the driveshaft, with the special tool and wrench **COUNTERCLOCKWISE** until the nut is free. If the special tool is not available, clamp the driveshaft in a vise equipped with soft jaws, in an area below the splines but not in the water pump impeller area. Now, with the proper size wrench on the pinion gear nut, rotate the complete lower unit **COUNTERCLOCKWISE** until the nut is free.

This procedure will require the lower unit to be rotated, then the wrench released, the lower unit turned back, the wrench again attached to the nut, and the unit rotated again. Continue with this little maneuver until the nut is free. After the nut is free, proceed with the next step. The driveshaft will be withdrawn from the pinion gear.

Driveshaft — Removal

FIRST, THESE CRITICAL WORDS

Two different driveshaft arrangements and two different shift mechanism are used in the lower units covered in this manual. These units will be identified here as **"F1"** and **"F2"**.

On Type **"F1"** unit, a thrust bearing, thrust washer and shim material are used on the upper end of the driveshaft, as shown in the accompanying exploded drawing. This unit has two detent balls and two springs in the shift mechanism.

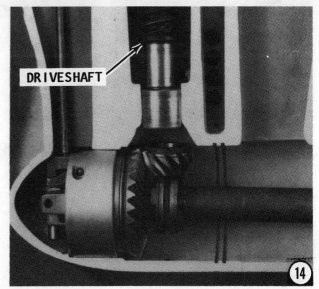

DRIVESHAFT

14

Type "F2" lower unit does not have the thrust bearing or the thrust washer -- only the shim material is used on the upper end of the driveshaft, as shown in the accompanying exploded drawing. This lower unit has only a single detent ball and single spring in the shift mechanism.

14- **CAREFULLY** pry the bearing housing upward away from the lower unit, then slide it free of the driveshaft. An alternate method is to again clamp the driveshaft in a vise equipped with soft jaws. Use a soft-headed mallet and tap on the top side of the bearing housing. This action will jar the housing loose from the lower unit. Continue tapping with the mallet and the bearing housing, O-rings, shims, and the driveshaft will all breakaway from the lower unit and may be removed as an assembly. If removing a Type "F1" driveshaft, the thrust bearing and thrust washer will also come out with the driveshaft. As the driveshaft is removed, reach into the lower unit, catch, and remove the pinion gear.

15- Push on the shift rod to move the unit into the reverse gear position. Remove the four bolts securing the shift rod cover. Rotate the shift rod **COUNTERCLOCKWISE** until it is free of the shifter detent in the lower unit. Remove the shift rod and cover as an assembly. As the plate is removed, notice which surface is facing into the housing, as an aid during installation.

Propeller Shaft -- Removal

16- Grasp the propeller shaft firmly with one hand and remove the propeller shaft from the lower unit. The forward gear and bearing housing will come out with the shaft as an assembly.

"FROZEN" PROPELLER SHAFT

On rare occasions, especially if water has been allowed to enter the lower unit, it may not be possible to withdraw the propeller shaft as described in Step 16.

If efforts to remove the propeller shaft after the bearing carrier has been removed fail: Two methods are available to free the propeller shaft from the lower unit.

First Method

The first and safest method is to obtain a block of wood 2"x 4" approx. one foot in length. Drill a hole in the center of the flat side, large enough for the propeller shaft to pass through. Place the block over the shaft. Slide some thick large washers over the shaft and thread the propeller nut onto the shaft.

With the skeg clamped securely in a vise equipped with soft jaws, attempt to pull the shaft free. If necessary hammer on the wood, rotating the block at intervals to prevent wedging the shaft in any one direction.

Second Method

The second and less desirable method is as follows: Clamp the propeller shaft horizontally in a vise equipped with soft jaws. Two blocks of wood may be substituted for the soft jaws, but **NEVER** clamp the shaft in the vise without protection, to prevent dam-

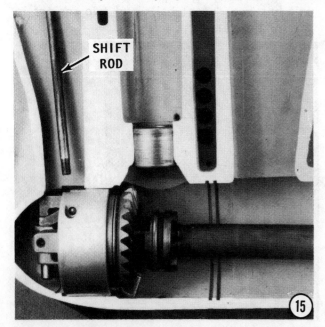

SHIFT ROD

15

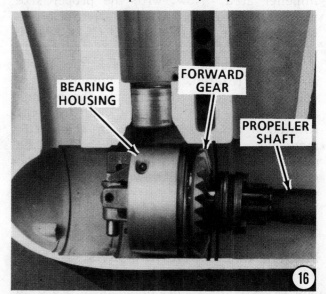

BEARING HOUSING

FORWARD GEAR

PROPELLER SHAFT

16

age to the threads or splines. Use a soft head mallet and strike the lower unit with quick sharp blows midway between the anti-cavitation plate and the propeller shaft. This action will drive the lower unit from the propeller shaft. **TAKE CARE** to hold the lower unit to keep it in line with the propeller shaft and prevent wedging the shaft in any one direction. Holding the lower unit will also prevent the housing from falling to the floor when the housing comes free of the shaft.

Driveshaft Disassembling
Lower Unit Type "F2" Only

17- The driveshaft upper bearing and cone are replaced as an assembly. If replacement is required, obtain special tool OMC P/N 387131. Clamp the tool below the bearing, and then place the unit in an arbor press equipped with a deep throat pedestal. Press the shaft from the bearing.

An alternate method is to clamp special tool OMC P/N 387206 to the driveshaft just above the shoulder. Now, invert the driveshaft and place it in an arbor press with special tool OMC P/N 387206 seated on the press. Place a piece of pipe over the driveshaft, and then press against tool OMC P/N 387131 to remove the bearing.

SEALS
BACK-TO-BACK

(18)

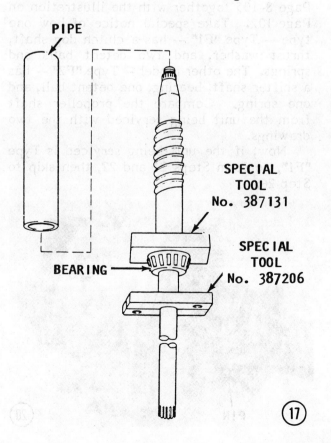

PIPE

SPECIAL
TOOL
No. 387131

SPECIAL
TOOL
No. 387206

BEARING

(17)

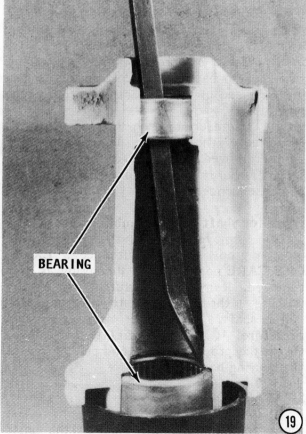

BEARING

(19)

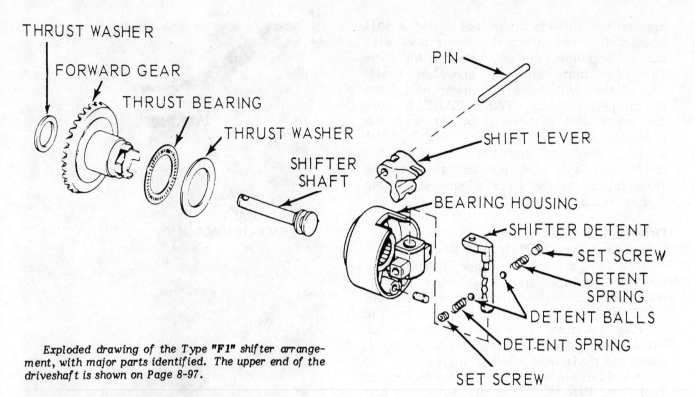

THRUST WASHER

FORWARD GEAR

THRUST BEARING

THRUST WASHER

SHIFTER SHAFT

PIN

SHIFT LEVER

BEARING HOUSING

SHIFTER DETENT

SET SCREW

DETENT SPRING

DETENT BALLS

DETENT SPRING

SET SCREW

Exploded drawing of the Type "F1" shifter arrangement, with major parts identified. The upper end of the driveshaft is shown on Page 8-97.

Bearing Carrier — Disassembling
INSPECTION FIRST

The bearings in the carrier need **NOT** be removed unless they are unfit for further service. Insert a finger and rotate the bearing. Check for "rough" spots or binding. Inspect the bearing for signs of corrosion or other types of damage. If the bearings must be replaced, proceed with the next step.

18- Use a seal remover to remove the two back-to-back seals or clamp the carrier in a vise and use a pry bar to pop each seal out.

19- Use a drift punch to drive the bearings free of the carrier. The bearings are being removed because they are unfit for service, therefore, additional damage is of no consequence.

Propeller Shaft — Disassembling

20- Remove the coil spring from the outside groove of the clutch dog. **EXERCISE CARE** not to distort the spring as it is removed. Remove the clutch dog pin by pushing it through the clutch dog and propeller shaft. This pin is not a tight fit, therefore, it is not difficult to remove. Grasp the propeller shaft and pull it free of the bearing housing.

SPECIAL WORDS

Two different shifter arrangements on the propeller shaft are used on the units

covered in this section. An exploded drawing of one type -- Type **"F1"** -- is shown on Page 8-104 together with the illustration on this Page. An exploded drawing of the other lower unit -- Type **"F2"** -- will be found on Page 8-105 together with the illustration on Page 102. Take special notice of how one type -- Type **"F1"** -- has a clutch dog shaft, thrust washer, and two detent balls and springs. The other model -- Type **"F2"** -- has a shifter shaft, bearing, one detent ball, and one spring. Compare the propeller shaft from the unit being serviced with the two drawings.

Now, if the unit being serviced is Type **"F1"**, perform Steps 21 and 22, then skip to Step 26.

PIN

20

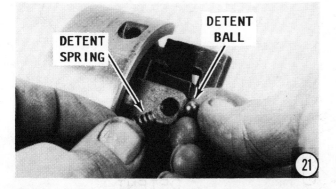

If the unit being serviced is Type **"F2"**, skip to Step 23 and perform the work thru Step 25, then continue with Step 26.

NOTE: The following two steps are to be performed if servicing a Type **"F1"** unit.

21- Remove the two Allen screws from the back side of the bearing housing and at the same time be prepared to catch the two detent balls and springs.

22- Remove the shift lever pin. Remove the shift lever and shifter detent, from the bearing housing. Remove the shifter shaft.

NOTE: The following three steps are to be performed if servicing a Type **"F2"** unit.

23- Remove the shift lever pin and disengage the shift lever from the cradle in the shift shaft.

24- Remove the shifter shaft and cradle. Remove the shift lever.

SAFETY WORD

The next step is dangerous. The detent balls are under tension from the springs. The balls are released with pressure. Therefore, **SAFETY GLASSES** should be worn as personal protection for the eyes.

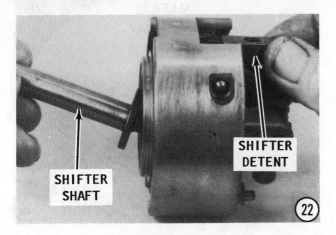

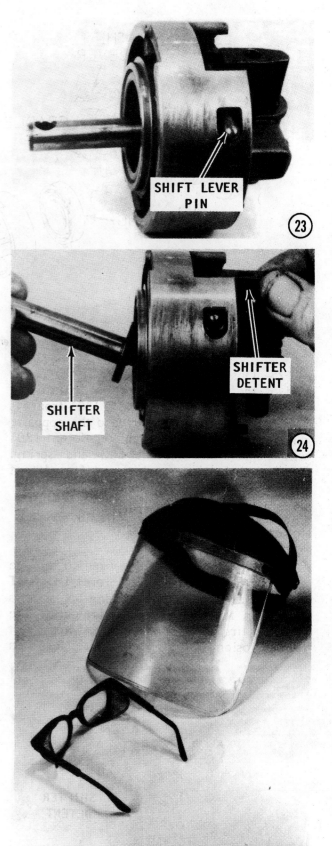

The detent balls are under tremendous tension during removal and installation — presenting a potential hazard. Safety glasses or a shield should **ALWAYS** be worn during such work for eye and face protection. Warn others in the area such work is in progress.

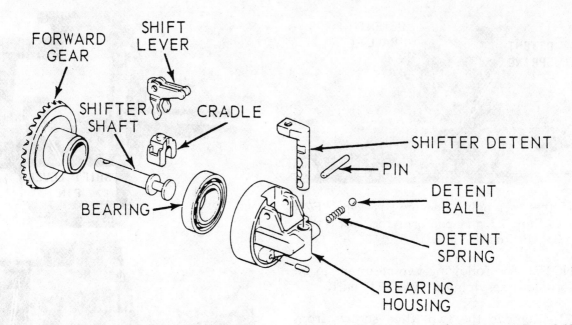

FORWARD
GEAR

SHIFT
LEVER

SHIFTER
SHAFT

CRADLE

BEARING

SHIFTER DETENT

PIN

DETENT
BALL

DETENT
SPRING

BEARING
HOUSING

Exploded drawing of the Type "F2" shifter arrangement. The upper end of the driveshaft is shown on Page 8-97.

25- Observe the short arm at the upper end of the shifter detent. Now, rotate the shifter detent until this arm is 180° from its original position (facing toward the opposite direction). Rotating the shifter detent 180° will depress the detent ball and spring. From this position, work the shifter detent up out of the housing.

Lower Driveshaft Bearing — Removal
SPECIAL WORDS

A special tool **MUST** be used to remove either type bearing. Therefore, **DO NOT** attempt to remove this bearing unless it is unfit for further service. To check the bearing, first use a flashlight and inspect it for corrosion or other damage. Insert a finger into the bearing, and then check for "rough" spots or binding while rotating it.

26- Remove the Allen screw, from the water pickup slots in the starboard side of the lower unit housing. This screw secures the bearing in place and **MUST** be removed before an attempt is made to remove the bearing. The bearing must be actually

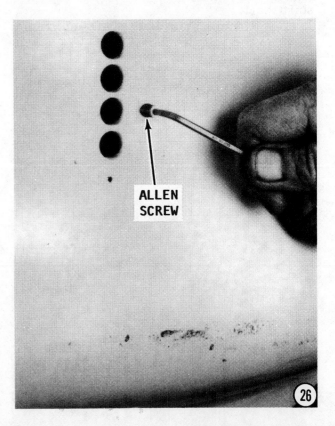

ALLEN
SCREW

SHIFTER
DETENT

25

26

"PULLED" upward to come free. **NEVER** make an attempt to "drive" it down and out.

27- If servicing a Type **"F1"** unit, obtain special tool, OMC P/N 385546. If servicing a Type **"F2"** unit, obtain special tool, OMC P/N 391257. Use the special tool and "pull" the bearing from the lower unit.

CLEANING AND INSPECTING

Wash all parts in solvent and dry them with compressed air. Discard all O-rings and seals that have been removed. A new seal kit for this lower unit is available from the local dealer. The kit will contain the necessary seals and O-rings to restore the lower unit to service.

Inspect all splines on shafts and in gears for wear, rounded edges, corrosion, and damage.

Carefully check the driveshaft and the propeller shaft to verify they are straight and true without any sign of damage. A complete check must be performed by turning the shaft in a lathe. This is only necessary if there is evidence to suspect the shaft is not true.

Check the water pump housing for corrosion on the inside and verify the impeller

and base plate are in good condition. Actually, good shop practice dictates to rebuild or replace the water pump each time the lower unit is disassembled. The small cost is rewarded with "peace of mind" and satisfactory service.

Inspect the lower unit housing for nicks, dents, corrosion, or other signs of damage. Nicks may be removed with No. 120 and No. 180 emery cloth. Make a special effort to ensure all old gasket material has been removed and mating surfaces are clean and smooth.

Inspect the water passages in the lower unit to be sure they are clean. The screen may be removed and cleaned.

Check the gears and clutch dog to be sure the ears are not rounded. If doubt exists as to the part performing satisfactorily, it should be replaced.

Inspect the bearings for "rough" spots, binding, and signs of corrosion or damage.

ASSEMBLING

READ AND BELIEVE

The lower unit should **NOT** be assembled in a dry condition. Coat all internal parts with OMC HI-VIS lube oil as they are assembled. All seals should be coated with OMC Gasket Seal Compound. When two seals are installed back-to-back, use Triple Guard Grease between the seal surfaces.

Propeller Shaft Assembling

SPECIAL WORDS

If servicing a Type **"F1"** lower unit, perform Steps 1 thru 4, then skip to Step 10.

If servicing a Type **"F2"** lower unit, skip to Step 5 and perform the work thru Step 9, then continue with Step 10.

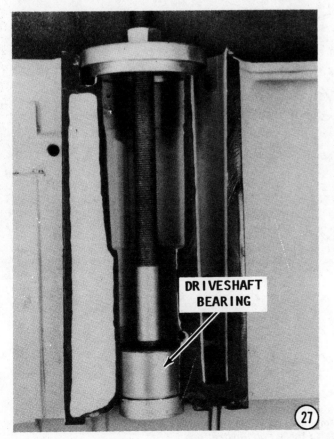

DRIVESHAFT BEARING

27

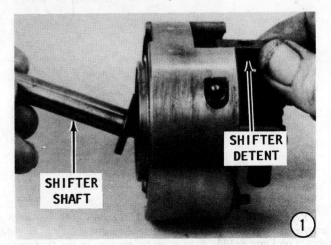

SHIFTER SHAFT

SHIFTER DETENT

1

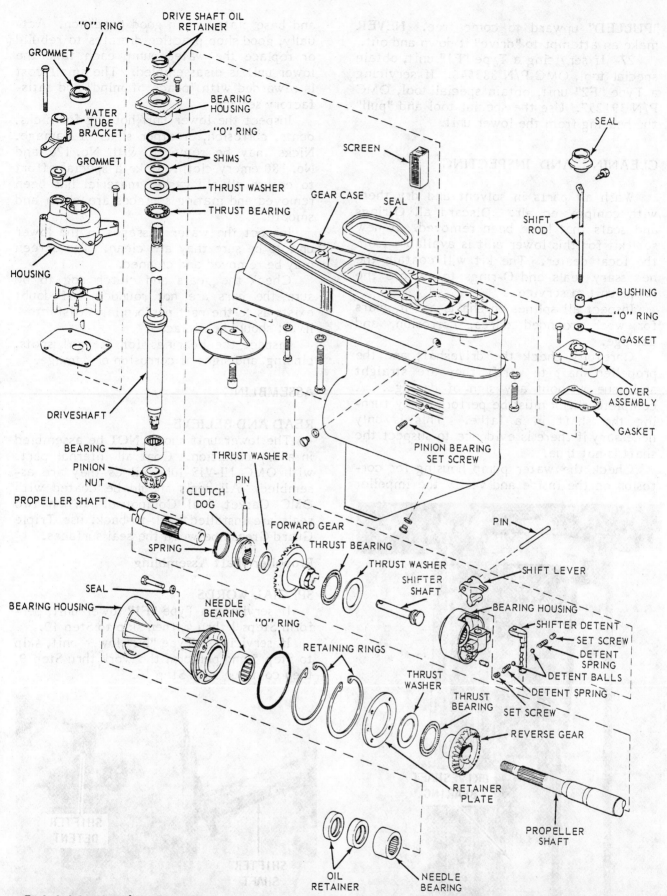

"O" RING

GROMMET

DRIVE SHAFT OIL RETAINER

WATER TUBE BRACKET

GROMMET

BEARING HOUSING

"O" RING

SHIMS

THRUST WASHER

THRUST BEARING

HOUSING

SCREEN

SEAL

GEAR CASE

SEAL

SHIFT ROD

BUSHING

"O" RING

GASKET

COVER ASSEMBLY

GASKET

DRIVESHAFT

BEARING

PINION

NUT

PROPELLER SHAFT

CLUTCH DOG

PIN

SPRING

THRUST WASHER

FORWARD GEAR

THRUST BEARING

THRUST WASHER

SHIFTER SHAFT

PINION BEARING SET SCREW

PIN

SHIFT LEVER

BEARING HOUSING

SHIFTER DETENT

SET SCREW

DETENT SPRING

DETENT BALLS

DETENT SPRING

SET SCREW

REVERSE GEAR

SEAL

BEARING HOUSING

NEEDLE BEARING

"O" RING

RETAINING RINGS

THRUST WASHER

THRUST BEARING

RETAINER PLATE

PROPELLER SHAFT

OIL RETAINER

NEEDLE BEARING

Exploded drawing of the Type "F1" lower unit covered in this section. This unit has a thrust bearing and thrust washer on the driveshaft and two detent balls and springs for the shift mechanism. Enlarged details of the shift mechanism are shown on Page 8-100. Numerous references in the text are made to these two illustrations.

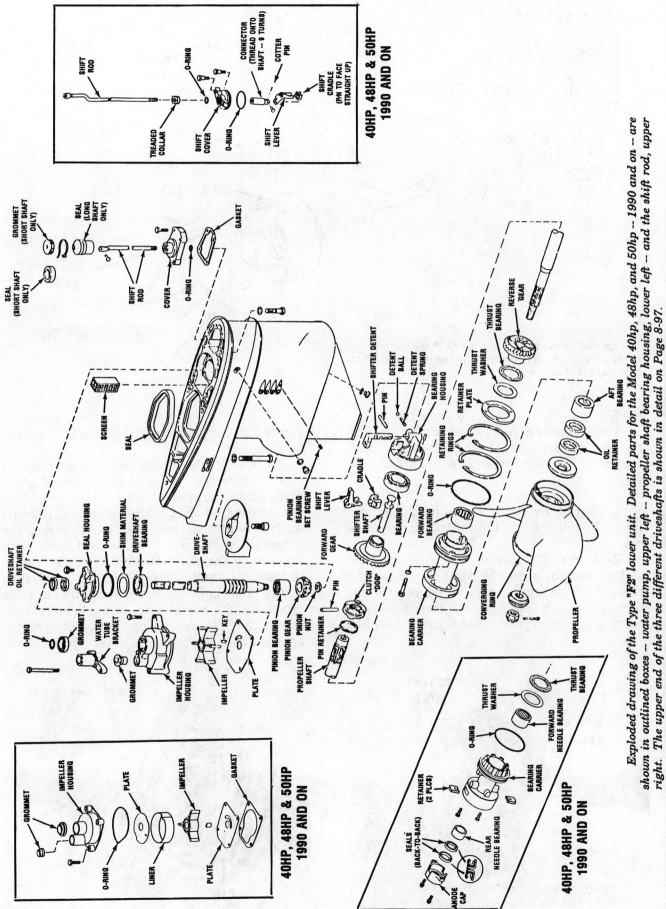

40HP, 48HP & 50HP 1990 AND ON

SHIFT ROD

TREADED COLLAR

O-RING

SHIFT COVER

O-RING

CONNECTOR (THREAD ONTO SHAFT — 9 TURNS)

COTTER PIN

SHIFT LEVER

SHIFT CRADLE (PIN TO FACE STRAIGHT UP)

GROMMET (SHORT SHAFT ONLY)

SEAL (LONG SHAFT ONLY)

GASKET

SEAL (SHORT SHAFT ONLY)

SHIFT ROD

COVER

O-RING

SCREEN

SEAL

DRIVESHAFT OIL RETAINER

SEAL HOUSING

O-RING

SHIM MATERIAL

DRIVESHAFT BEARING

DRIVE-SHAFT

O-RING

GROMMET

WATER TUBE BRACKET

GROMMET

IMPELLER HOUSING

IMPELLER

KEY

PLATE

PINION BEARING

PINION GEAR

PINION NUT

PIN RETAINER

PROPELLER SHAFT

CLUTCH "DOG"

PIN

FORWARD GEAR

PINION BEARING SET SCREW

CRADLE

SHIFTER SHAFT

SHIFT LEVER

SHIFTER DETENT

PIN

DETENT BALL

DETENT SPRING

BEARING HOUSING

BEARING

FORWARD BEARING

O-RING

RETAINING RINGS

RETAINER PLATE

THRUST WASHER

THRUST BEARING

REVERSE GEAR

OIL RETAINER

AFT BEARING

BEARING CARRIER

CONVERGING RING

PROPELLER

Exploded drawing of the Type "F2" lower unit. Detailed parts for the Model 40hp, 48hp, and 50hp — 1990 and on — are shown in outlined boxes — water pump, upper left — propeller shaft bearing housing, lower left — and the shift rod, upper right. The upper end of the three different driveshafts is shown in detail on Page 8-97.

40HP, 48HP & 50HP 1990 AND ON

GROMMET

IMPELLER HOUSING

PLATE

IMPELLER

GASKET

O-RING

LINER

PLATE

40HP, 48HP & 50HP 1990 AND ON

THRUST WASHER

THRUST BEARING

O-RING

RETAINER (2 PLCS)

BEARING CARRIER

FORWARD NEEDLE BEARING

SEALS (BACK-TO-BACK)

REAR NEEDLE BEARING

ANODE CAP

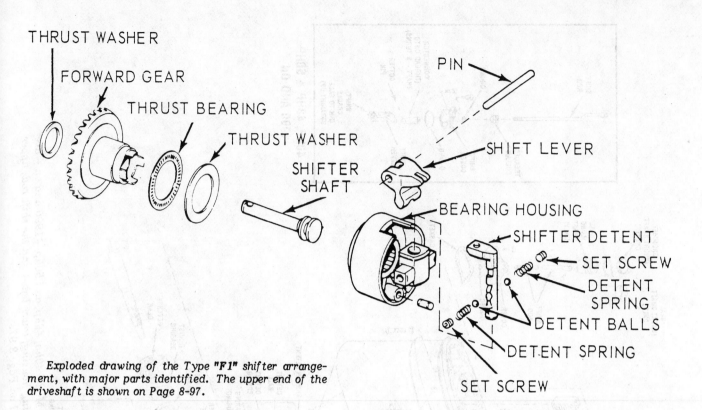

Exploded drawing of the Type "F1" shifter arrangement, with major parts identified. The upper end of the driveshaft is shown on Page 8-97.

NOTE

The following four steps are to be performed if servicing a Type **"F1"** lower unit, shown in the accompanying illustration on this page.

1- Install the shift lever and shifter detent. Install the shifter shaft and then the pin.

2- Insert the two detent balls into the shifter detent, then the springs. Coat the threads of the set screws with Loctite, and then secure the springs and detent balls in place with the screws. The screws should be

tightened until each head is flush with the housing.

3- Install the thrust bearing and thrust washer onto the shank of the forward gear. Now, slide the assembled forward gear into the bearing housing. Slide the small thrust washer onto the propeller shaft. Insert the propeller shaft into the forward gear with the slot in the shaft aligned with the hole in the shifter shaft. Insert the pin, and then install the coil spring around the outside groove of the clutch dog. Check to be sure one coil of the spring does not overlap another coil. Press the shifter detent down

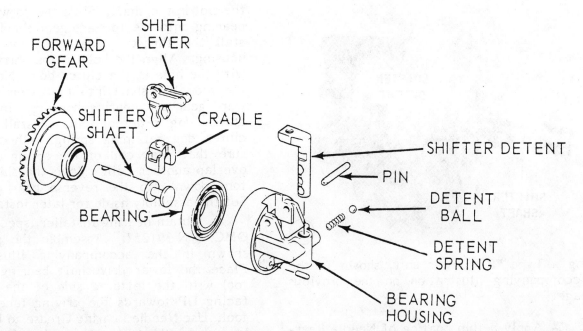

FORWARD GEAR

SHIFT LEVER

SHIFTER SHAFT

CRADLE

BEARING

SHIFTER DETENT

PIN

DETENT BALL

DETENT SPRING

BEARING HOUSING

Exploded drawing of the Type "F2" shifter arrangement. The upper end of the driveshaft is shown on Page 8-97.

to move the clutch dog into the reverse gear position. Set the unit aside for later installation.

4- Obtain bearing installer, special tool, OMC P/N 385546. Assemble the washer, plate, guide sleeve, and installer portion of

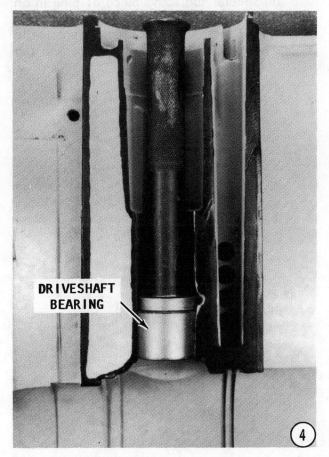

DRIVESHAFT BEARING

④

the tool to the screw in the order given, and with the shoulder of the tool facing down. Place the bearing onto the tool with the lettered side of the bearing **TOWARD** the shoulder. Place the lower driveshaft bearing and tool in position in the lower unit. Now, drive the bearing into the lower unit until the plate makes contact with the lower unit surface. Coat the threads of the setscrew with Loctite and then secure the bearing in place with the setscrew.

NOTE

Perform the following 5 steps if servic-

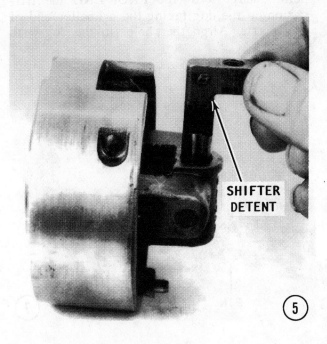

SHIFTER DETENT

⑤

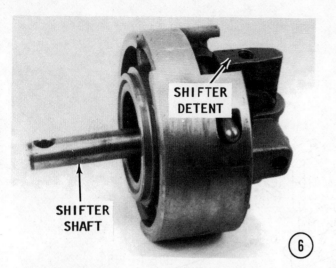

SHIFTER
DETENT

SHIFTER
SHAFT

6

ing a Type **"F2"** lower unit, shown in the accompanying illustration on the previous page.

5- Apply a thin coating of Needle Bearing Grease onto the detent ball and spring. Insert the detent ball and spring into the bearing housing. Observe the short arm on the shifter detent. Position the detent with this short arm facing forward. Now, use a punch, or similar tool, and depress the ball and spring, and **AT THE SAME TIME** press the shift detent down into the bearing housing. Rotate the shifter detent 180°.

6- Place the cradle onto the shifter shaft and in position in the bearing housing. Install shifter lever with the cradle engaged with the shifter detent. Work the pin through the shifter detent and cradle.

7- Install the clutch dog onto the propeller shaft with the **PROP END** identification on the dog facing the propeller end of the shaft. Align the holes in the dog with

the slot in the shaft. Slide the forward gear bearing onto the forward gear shoulder. Install the thrust washer into the bearing housing. Align the hole in the shifter shaft with the hole in the clutch dog. Now, slide the propeller shaft into the assembled forward gear and bearing housing. Insert the clutch dog retaining pin. Install a **NEW** clutch dog retaining spring. Check to be sure that one coil of the spring does not overlap another coil. Move the shifter detent down into the reverse gear position. Set the assembly aside for later installation.

8- Obtain bearing installer, special tool, OMC P/N 391257. Assemble the parts as shown in the accompanying illustration. Place the lower driveshaft bearing on the tool with the lettered side of the bearing facing **UP** towards the driving face of the tool. Use Needle Bearing Grease to hold the bearing on the tool. Now, drive the bearing into place until the washer makes contact with the spacer. Coat the threads of the setscrew with Loctite. Secure the bearing in place with the screw.

9- Obtain special tool, OMC P/N 387131, OMC P/N 387206, and a short piece of pipe to fit over the driveshaft, as shown. Use the special tools and piece of pipe in an arbor press to install the upper driveshaft bearing. Press with the tool against the inner race until the bearing is seated on the driveshaft shoulder.

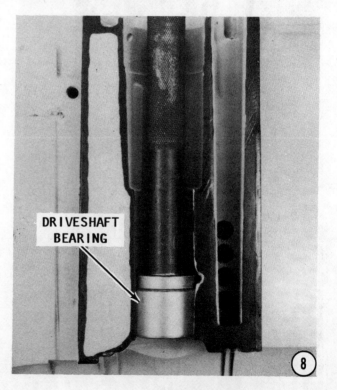

DRIVESHAFT
BEARING

8

PIN

7

10- Clamp the driveshaft in a vise equipped with soft jaws in such a manner that the splines, water pump area, or other critical portions of the shaft cannot be damaged. Slide the pinion gear onto the driveshaft with bevel of the gear teeth facing toward the lower end of the shaft. Install the pinion gear nut and tighten it to a torque value of 40 to 45 ft lb (54 to 61Nm). No parts should be installed on the upper end of the driveshaft at this point.

Slide the shim material removed during disassembling onto the driveshaft and seat it against the driveshaft shoulder. If servicing a Type "F1" lower unit, obtain special shimming tool, OMC P/N 315767. If servicing a Type "F2" lower unit, obtain special shimming tool, OMC P/N 320739.

Slip the special tool down over the driveshaft and onto the upper surface of the top piece of shim material. Measure the distance between the top of the pinion gear and the bottom of the tool. The tool should just barely make contact with the pinion gear surface for **ZERO** clearance. If servicing a Type "F2" unit, remove 0.007" (0.178mm), of shim material **AFTER** the zero clearance has been obtained. Remove the tool and set the shim material aside for installation. Back off the pinion gear nut and remove the pinion gear. Remove the driveshaft from the vise.

Model 40hp, 48hp and 50hp
1990 and On

Shim procedures are only required if a **NEW** thrust bearing, thrust washer, pinion gear, shaft or bearing housing is being installed on the driveshaft.

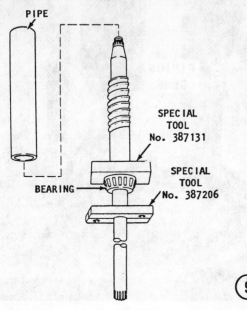

PIPE

SPECIAL TOOL No. 387131

SPECIAL TOOL No. 387206

BEARING

9

FEELER GAUGE

SPECIAL SHIM GAUGE

10

Detailed illustrated procedures are presented as an **Addendum** on Page A-21 in the Appendix.

11- Insert the complete assembled propeller shaft into the lower unit housing with the pin on the back side of the forward gear housing indexed into the hole in the lower unit.

12- Thread the shift rod into the shift detent assembly. Coat both sides of the shift rod gasket with sealer. Place the gasket in position, and then install the shift rod cover onto the housing. The shift rod is adjusted in the next step. Slide the shift rod boot down over the shift rod cover.

11

SHIFT ROD

12

Shift Rod Adjustment

13- Measure the distance from the top of the lower unit housing to the center of the hole in the shift rod. This distance should be 15-29/32" (40.4 cm for a standard short shaft unit, and 16-29/32" for a 40hp, 48hp or 50hp unit -- 1990 and on. Add exactly 5" (12.7 cm), for a unit with a long shaft.

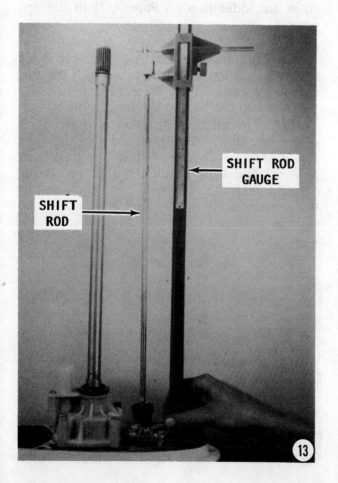

SHIFT ROD GAUGE

SHIFT ROD

13

Rotate the shift rod clockwise or counterclockwise until the required dimension is obtained.

14- Insert the pinion gear into the cavity in the lower unit with the flat side of the bearing facing **UPWARD**. Hold the pinion gear in place and at the same time lower the driveshaft down into the lower unit. As the driveshaft begins to make contact with the pinion gear, rotate the shaft slightly to permit the splines on the shaft to index with the splines of the pinion gear. After the shaft has indexed with the pinion gear, thread the pinion gear nut onto the end of the shaft. Obtain special tool, OMC P/N 316612. Slide the special tool over the upper end of the driveshaft with the splines of the tool indexed with the splines on the shaft. Attach a torque wrench to the special tool. Now, hold the pinion gear nut with the proper size wrench and rotate the driveshaft **CLOCKWISE** with the special tool until the pinion gear nut is tightened to a torque value of 40-45 ft lbs (54-68Nm). Remove the special tool.

15- If servicing a Type **"F1"** lower unit, slide the thrust bearing, thrust washer, and the shims, set aside after the shim gauge procedure in Step 10, onto the driveshaft. If servicing a Type **"F2"** lower unit, slide the

PINION GEAR

14

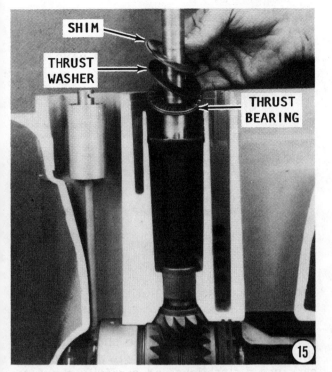

SHIM

THRUST WASHER

THRUST BEARING

15

BEARING HOUSING

17

ing with the flat side of the seal facing **INWARD**. Insert a **NEW** O-ring into the bottom opening of the bearing housing.

17- Wrap friction tape around the splines of the driveshaft to protect the seals in the bearing housing as the housing is installed. Now, slide the assembled bearing housing down the driveshaft and seat it in the lower unit housing. Secure the housing in place with the four bolts. Tighten the bolts **ALTERNATELY** and **EVENLY**.

Reverse Gear — Installation

18- Apply a light coating of grease to the inside surface of the reverse gear to hold the thrust washer in place. Place the thrust washer onto the front side of the reverse gear. Install the thrust bearing and thrust washer onto the shank on the back side of the reverse gear. Slide the reverse gear down the propeller shaft and index the teeth of the reverse gear with the teeth of the pinion gear.

19- Insert the retainer plate into the lower unit against the reverse gear.

WARNING

This next step can be dangerous. Each snap ring is placed under tremendous tension with the Truarc pliers while it is being placed into the groove. Therefore, wear **SAFETY GLASSES** and exercise care to pre-

shims only onto the driveshaft. (This unit does not have the other items.)

16- Install the two seals back-to-back into the opening on top of the bearing housing. One seal prevents lubricant from escaping and the other seal prevents water from entering. Coat the outside surface of the **NEW** seals with OMC Lubricant. Press the first seal into the housing with the flat side of the seal facing **OUTWARD**. After the seal is in place, apply a coating of Triple Guard Grease to the flat side of the installed seal and the flat side of the second seal. Press the second seal into the bearing hous-

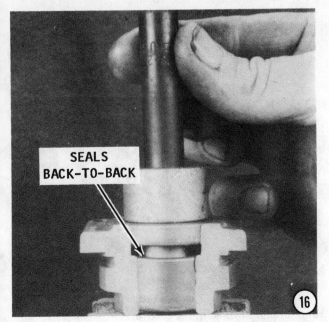

SEALS BACK-TO-BACK

16

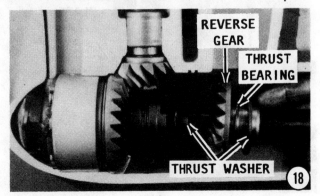

REVERSE GEAR

THRUST BEARING

THRUST WASHER

18

Snap rings are under tremendous tension during removal and installation — presenting a potential hazard. Safety glasses or a shield should ALWAYS be worn during such work for eye and face protection. Warn others in the area such work is in progress.

vent the snap ring from slipping out of the pliers. If the snap ring should slip out, it would travel with incredible speed and cause personal injury if it struck a person.

20- Install the Truarc snap rings one at-a-time following the precautions given in the **WARNING** and the **ADVICE** given in the following paragraph.

WORDS OF ADVICE

The two snap rings index into separate grooves in the lower unit housing. As the first ring is being installed, depth perception may play a trick on your eyes. It may

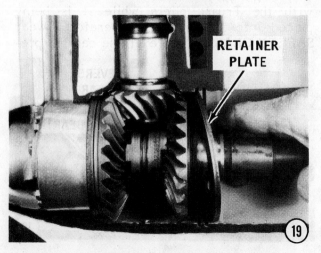

RETAINER PLATE

⑲

TRUARC SNAP RING

⑳

appear that the first ring is properly indexed all the way around in the proper groove, when in reality, a portion may be in one groove and the remainder in the other groove. Should this happen, and the Truarc pliers be released from the ring, it is extremely difficult to get the pliers back into the ring to correct the condition. If necessary use a flashlight and carefully check to be sure the first ring is properly seated all the way around **BEFORE** releasing the grip on the pliers. Installation of the second ring is not so difficult because the one groove is filled with the first ring.

Bearing Carrier Bearings & Seals Installation

21- Install the reverse gear bearing into the bearing carrier by pressing against the **LETTERED** side of the bearing with the proper size socket. Press the forward gear bearing into the bearing carrier in the same

TRUARC SNAP RING

View into the lower unit showing the Truarc snap rings properly locked in the grooves.

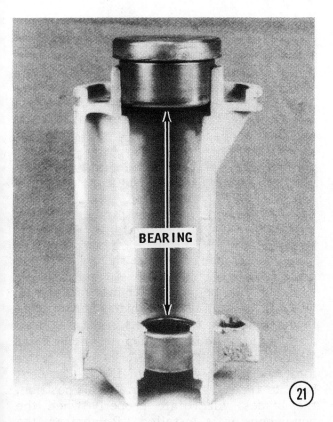

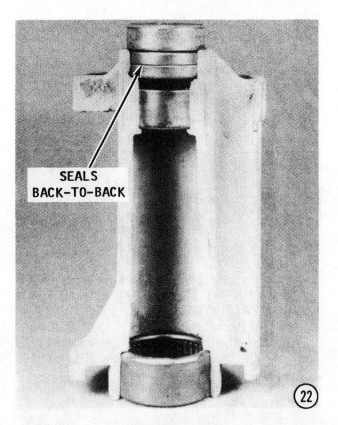

manner. Press against the **LETTERED** side of the bearing.

22- Coat the outside surfaces of the seals with HI-VIS oil. Install the first seal with the flat side facing **OUT.** Coat the flat surface of both seals with Triple Guard Lubricant and then install the second seal with the flat side going in **FIRST.** The seals are then back-to-back with the lubricant between the two surfaces. The outside seal prevents water from entering the lower unit and the inside seal prevents the lubricant in the lower unit from escaping.

Bearing Carrier Installation

23- Obtain two long 1/4" rods with threads on one end. Thread the rods into

the retainer plate opposite each other to act as guides for the bearing carrier. Check the bearing carrier to be sure a **NEW O**-ring has been installed. Position the carrier over the guide pins with the embossed word **UP** on the rim of the carrier facing **UP** in relation to the lower unit housing, reference illustration **"D".** Now, lower the bearing carrier down over the guide pins and into place in the lower unit housing.

24- Slide **NEW** little O-rings onto each bolt, and apply some OMC Sealer onto the threads. Install the bolts through the carrier and into the retaining plate. After a couple bolts are in place, remove the guide pins and install the remaining bolts. Tighten the bolts **EVENLY** and **ALTERNATELY** to

A set of double seals showing the back side (left) and the front side (right). These seals are installed back-to-back (flat side-to-flat side) with Triple Guard Grease between the surfaces. This arrangement prevents fluid from passing in either direction.

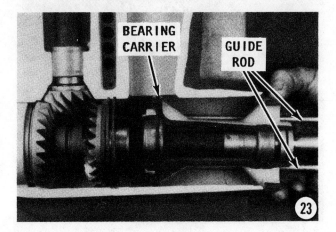

D

the torque value for the size bolt given in the Appendix.

WATER PUMP ASSEMBLING AND INSTALLATION

25- Insert the plate into the housing, as shown. The tang on the bottom side of the

BOLT

24

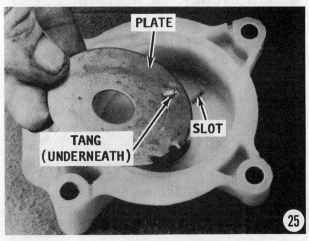

PLATE

TANG (UNDERNEATH)   SLOT

25

LINER

26

plate **MUST** index into the short slot in the pump housing.

26- Slide the pump liner into the housing with the two small tabs on the bottom side indexed into the two cutouts in the plate.

27- Coat the inside diameter of the liner with light-weight oil. Work the impeller into the housing with all of the blades bent back to the right, as shown. In this position, the blades will rotate properly when the pump housing is installed. Remember, the pump and the blades will be rotating **CLOCKWISE** when the housing is turned over and installed in place on the lower unit.

28- Coat the mating surface of the lower unit with 1000 Sealer. Slide the water

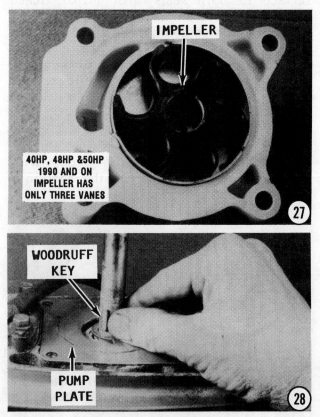

IMPELLER

40HP, 48HP &50HP 1990 AND ON IMPELLER HAS ONLY THREE VANES

27

WOODRUFF KEY

PUMP PLATE

28

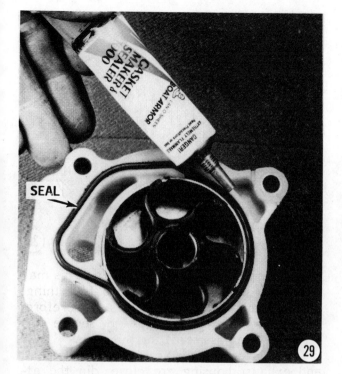

SEAL

29

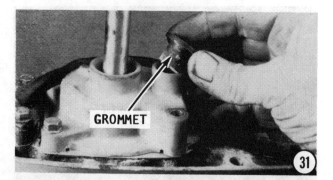

GROMMET

31

pump base plate down the driveshaft and into place on the lower unit. Insert the Woodruff key into the key slot in the drive-shaft.

29- Lay down a very thin bead of 1000 Sealer into the irregular shaped groove in the housing. Insert the seal into the groove, and then coat the seal with the 1000 Sealer.

30- Begin to slide the water pump down the driveshaft, and at the same time observe the position of the slot in the impeller. Continue to work the pump down the driveshaft, with the slot in the impeller indexed over the Woodruff key. The pump must be fairly well aligned before the key is covered because the slot in the impeller is not visible as the pump begins to come close to the base plate.

31- Install the short forward bolt through the pump and into the lower unit. **DO NOT** tighten this bolt at this time. Insert the grommet into the pump housing.

32- Install the grommet retainer and water tube guide onto the pump housing. Install the remaining pump attaching bolts. Tighten the bolts **ALTERNATELY** and **EVENLY,** and at the same time rotate the driveshaft **CLOCKWISE.** If the driveshaft is not rotated while the attaching bolts are being tightened, it is possible to pinch one of the impeller blades underneath the housing.

33- Slide the large grommet down the driveshaft and seat it over the pump collar. This grommet does not require a sealer substance. Its function is to prevent exhaust gases from entering the water pump.

WATER PUMP HOUSING

30

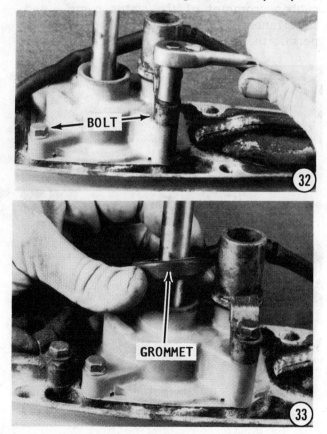

BOLT

32

GROMMET

33

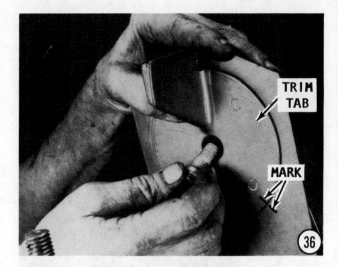

LOWER UNIT INSTALLATION

34- Check to be sure the water tubes are clean, smooth, and free of any corrosion. Coat the water pickup tubes and grommets with lubricant as an aid to installation. Check to be sure the spark plug leads are disconnected from the spark plugs. Bring the lower unit housing together with the exhaust housing, and at the same time, guide the water tube into the rubber grommet of the water pump. Continue to bring the two units together, at the same time guide the water pickup tubes into the rubber grommet of the water pump, and, simultaneously rotate the flywheel slowly to permit the splines of the driveshaft to index

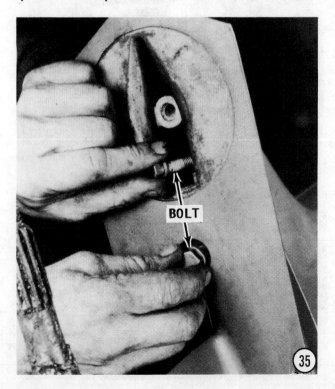

with the splines of the crankshaft. This may sound like it is necessary to do four things at the same time, and so it is. Therefore, make an earnest attempt to secure the services of an assistant for this task.

35- After the surfaces of the lower unit and exhaust housing are close, dip the attaching bolts in OMC Sealer and then start them in place. Install the two bolts on each side of the two housings. Install the bolt in the recess of the trim tab. Install the bolt just forward of the trim tab that extends up through the cavitation plate. Tighten the bolts **ALTERNATELY** and **EVENLY** to the torque value for this size bolt given in the Appendix.

36- Install the trim tab with the mark made on the tab during disassembling aligned with the mark made on the lower unit housing.

37- The shift rod is visible beneath and to the rear of the lower carburetor. Install

the shift connector with a bolt or pin, depending on the model being serviced.

Filling the Lower Unit

Fill the lower unit with lubricant according to the procedures in Section 8-3.

Propeller Installation

Install the propeller, see Section 8-2.

FUNCTIONAL CHECK

Perform a functional check of the completed work by mounting the outboard in a test tank, in a body of water, or with a flush attachment connected to the lower unit. If the flush attachment is used, **NEVER** operate the powerhead above an idle speed, because the no-load condition on the propeller would allow the powerhead to **RUNAWAY** resulting in serious damage or destruction of the powerhead.

CAUTION: Water must circulate through the lower unit to the engine any time the engine is run to prevent damage to the water pump in the lower unit. Just five seconds without water will damage the water pump.

Start the engine and observe the tattletale flow of water from the idle relief in the exhaust housing. The water pump installation work is verified. If a "Flushette" is connected to the lower unit, **VERY LITTLE** water will be visible from the idle relief port. Shift the outboard into the three gears and check for smoothness of operation and satisfactory performance.

8-10 TYPE "G" LOWER UNIT OUTBOARD JET DRIVE

INTRODUCTION

Outboard Jet Drive units are manufactured by Specialty Manufacturing Co., San Leandro, California. The units have been designed to permit boating in areas prohibited to a boat equipped with a conventional propeller outboard drive system. The housing of the jet drive barely extends below the hull of the boat allowing passage in ankle deep water, white water rapids, and over sand bars or in shoal water which would foul a propeller drive.

Description and Operation

The Outboard Jet Drive provides reliable propulsion with a minimum of moving parts. The units operate under the same laws and principles employed for the other jet drive units covered in this manual. Simply stated, water is drawn into the unit through an intake grille by an impeller driven by a driveshaft off the crankshaft of the powerhead. The water is immediately expelled under pressure through an outlet nozzle directed away from the stern of the boat.

As the speed of the boat increases and reaches planing speed, the jet drive discharges water freely into the air, and only the intake grille makes contact with the water.

The jet drive is provided with a gate arrangement and linkage to permit the boat to be operated in reverse. When the gate is moved downward over the exhaust nozzle, the pressure stream is reversed by the gate and the boat moves sternward.

Conventional controls are used for powerhead speed, movement of the boat, shifting and power trim and tilt.

Model Identification and Serial Numbers

A model letter identification indicating "size" is stamped on the rear, port side of the jet drive housing. A serial number for the unit is stamped on the starboard side of the jet drive housing, as indicated in the accompanying illustration.

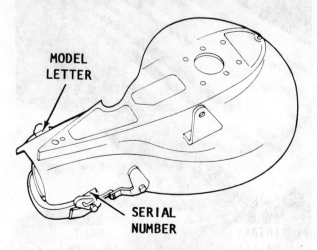

MODEL LETTER

SERIAL NUMBER

The model letter designation and the serial numbers are embossed on the jet drive housing.

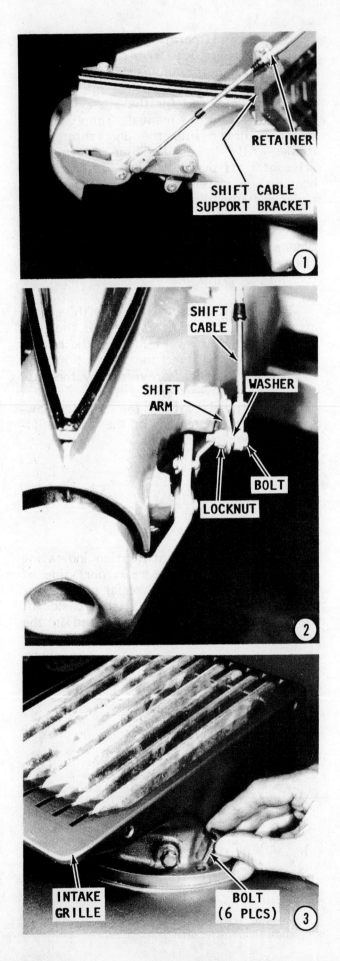

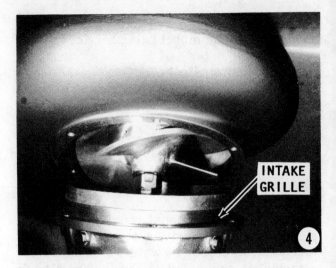

REMOVAL & DISASSEMBLING

1- Remove the two bolts and retainer securing the shift cable to the shift cable support bracket.

2- Remove the locknut, bolt, and washer securing the shift cable to the shift arm. Try not to disturb the length of the cable.

3- Remove the six bolts securing the intake grille to the jet casing.

4- Ease the intake grille from the jet drive housing.

5- Pry the tab or tabs of tabbed washer away from the nut to allow the nut to be removed.

6- Loosen, and then remove the nut.

7- Remove the tabbed washer and spacers. Make a careful count of the spacers behind the washer. If the unit is relatively new, there could be as many as eight or nine

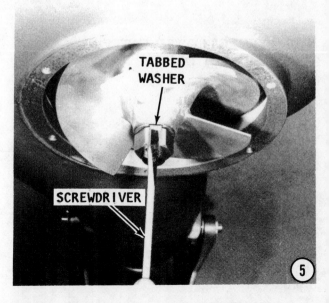

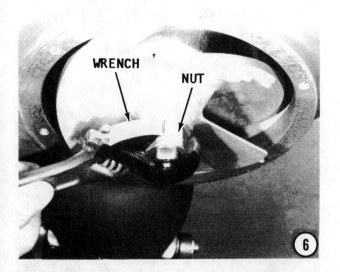

WRENCH   NUT

6

JET
IMPELLER

8

spacers behind the washer. If less than
eight or nine spacers, depending on the
model being serviced, are removed from
behind the washer, the others will be found
behind the jet impeller, which is removed in
the following step. A **TOTAL** of either
EIGHT or **NINE** spacers will be found.

8- Remove the jet impeller from the
shaft. If the impeller is "frozen" to the
shaft, obtain a block of wood and a hammer.
Tap the impeller in a **CLOCKWISE** direction
to release the shear key.

9- Slide the nylon sleeve and shear key
free of the driveshaft and any spacers found
behind the impeller. Make a note of the
number of spacers at both locations --behind
the impeller **AND** on top of the impeller,
under the nut and tabbed washer.

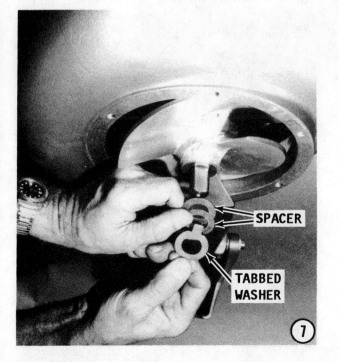

SPACER

TABBED
WASHER

7

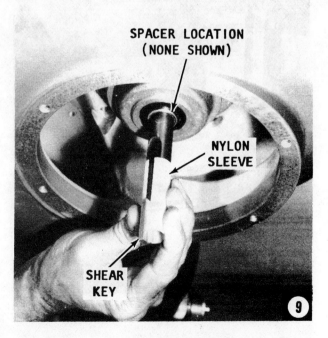

SPACER LOCATION
(NONE SHOWN)

NYLON
SLEEVE

SHEAR
KEY

9

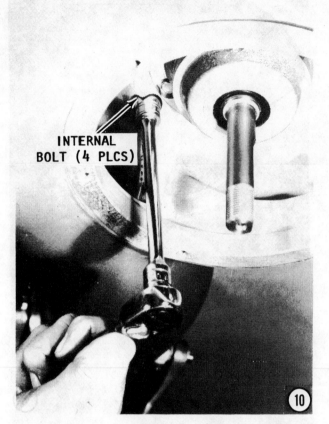

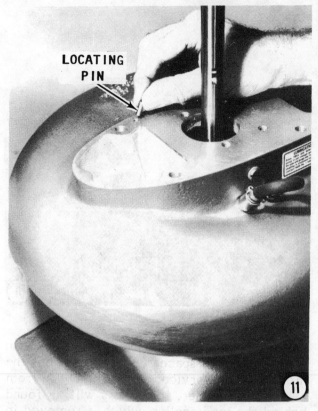

10- One external bolt and four internal bolts are used to secure the jet drive to the intermediate housing. The external bolt is located at the aft end of the anti-cavitation plate. The four internal bolts are located inside the jet drive housing, as indicated in the accompanying illustration. Remove the five attaching bolts.

11- Lower the jet drive from the intermediate housing. Remove the locating pin from the forward starboard side (or center forward, depending on the model being serviced) of the upper jet housing.

SPECIAL WORDS

There will be a total of **SIX** locating pins to be removed in the following steps. Make careful note of the size and location of each

Location of the one exterior bolt securing the jet drive to the outboard intermediate housing.

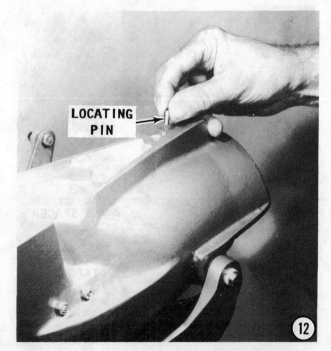

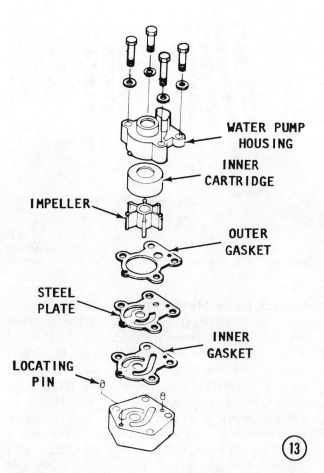

WATER PUMP HOUSING

INNER CARTRIDGE

IMPELLER

OUTER GASKET

STEEL PLATE

INNER GASKET

LOCATING PIN

⑬

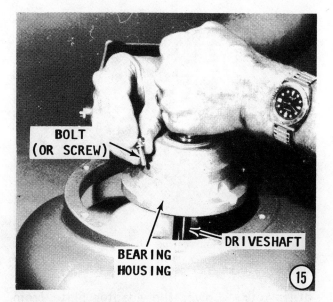

BOLT (OR SCREW)

BEARING HOUSING

DRIVESHAFT

⑮

SPECIAL WORDS

Two different length bolts are used at this location on some models, and one of these models uses special "D" shaped washers.

Pull the water pump housing, the inner cartridge and the water pump impeller, up and free of the driveshaft. Remove the Woodruff key from its recess in the driveshaft. Next, remove the outer gasket, the steel plate and the inner gasket.

14- Remove the two small locating pins and lift the aluminum spacer up and free of the driveshaft.

15- Remove the driveshaft and bearing assembly from the housing. One model has two #10-24x5/8" screws and lockwashers securing the driveshaft assembly to the housing. All other models have four ¼-20x7/8" bolts and lockwashers securing the driveshaft assembly to the housing.

16- Remove the large thick adaptor plate from the intermediate housing. This plate is secured with five bolts and lockwashers. The aft bolt is smaller than the

when they are removed, as an assist during assembling.

12- Remove the locating pin from the aft end of the housing. This pin and the one removed in the previous step should be of identical size.

13- Remove the four bolts and washers from the water pump housing.

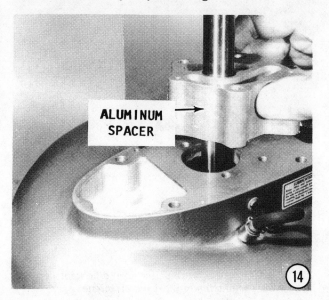

ALUMINUM SPACER

⑭

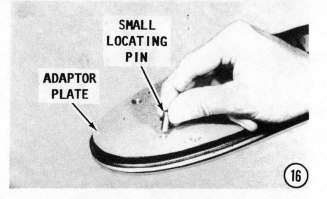

SMALL LOCATING PIN

ADAPTOR PLATE

⑯

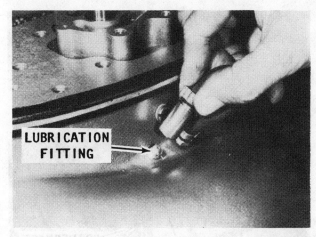

Cleaning and lubricating the bearing assembly is best accomplished by completely replacing the old lubricant with new, as described in the text.

other four. Lower the adaptor plate from the intermediate housing and remove the two small locating pins, one on the forward port side and another from the last aft hole in the adaptor plate. Both pins are identical size.

CLEANING AND INSPECTING

Wash all parts, except the driveshaft assembly, in solvent and blow them dry with compressed air. Rotate the bearing assembly on the driveshaft to inspect the bearings for "rough" spots, binding, and signs of corrosion or damage.

Saturate a shop towel with solvent and wipe both extensions of the driveshaft.

*Take extra precautions to **PREVENT** cleaning solution from entering the three lubrication passages.*

The slats of the grille must be carefully inspected and any bent slats straightened, for maximum performance of the jet drive.

Bearing Assembly

Lightly wipe the exterior of the bearing assembly with the same shop towel. Do not allow solvent to enter the three lubricant passages of the bearing assembly. The best way to clean these passages is **NOT** with solvent - because any solvent remaining in the assembly after installation will continue to dissolve good useful lubricant and leave bearings and seals dry. This condition will cause bearings to fail through friction and seals to dry up and shrink - losing their sealing qualities.

The only way to clean and lubricate the bearing assembly is after installation to the jet drive - via the exterior lubrication fitting. This procedure is described at the end of this chapter, on Page 8-133. The text explains how the old lubricant may be completely replaced with new.

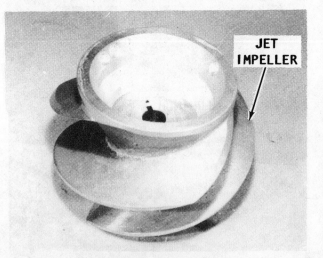

The edges of the jet impeller should be kept as sharp as possible for maximum jet drive efficiency.

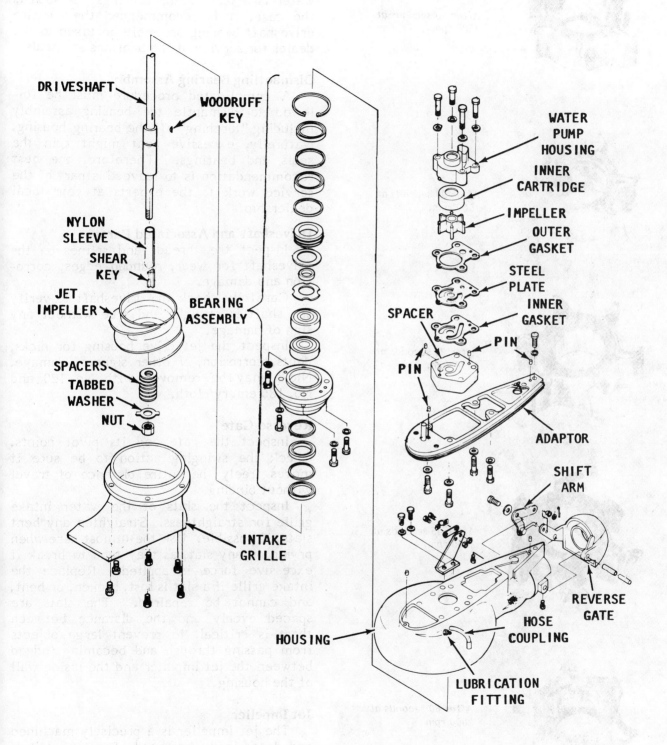

DRIVESHAFT

WOODRUFF KEY

NYLON SLEEVE

SHEAR KEY

JET IMPELLER

BEARING ASSEMBLY

SPACERS

TABBED WASHER

NUT

INTAKE GRILLE

HOUSING

WATER PUMP HOUSING

INNER CARTRIDGE

IMPELLER

OUTER GASKET

STEEL PLATE

INNER GASKET

SPACER

PIN

PIN

ADAPTOR

SHIFT ARM

REVERSE GATE

HOSE COUPLING

LUBRICATION FITTING

Exploded drawing of a jet drive lower unit, with major parts identified.

After 60 seconds at 1500 rpm.

After 90 seconds at 1500 rpm.

After 30 seconds at 2000 rpm.

After 45 seconds at 2000 rpm.

After 60 seconds at 2000 rpm

Cautions throughout this manual point out the danger of operating the engine without water passing through the water pump. The above photographs are self evident.

If the old lubricant emerging from the hose coupling is a dark, dirty, grey color, the seals have already broken down and water is attacking the bearings. If such is the case, it is recommended the entire driveshaft bearing assembly be taken to the dealer for service of the bearings and seals.

Dismantling Bearing Assembly

A complicated procedure must be followed to dismantle the bearing assembly including "torching" off the bearing housing. Naturally, excessive heat might ruin the seals and bearings. Therefore, the best recommendation is to leave this part of the service work to the experts at your local dealership.

Driveshaft and Associated Parts

Inspect the threads and splines on the driveshaft for wear, rounded edges, corrosion and damage.

Carefully check the driveshaft to verify the shaft is straight and true without any sign of damage.

Inspect the jet drive housing for nicks, dents, corrosion, or other signs of damage. Nicks may be removed with No. 120 and No. 180 emery cloth.

Reverse Gate

Inspect the gate and its pivot points. Check the swinging action to be sure it moves freely the entire distance of travel without binding.

Inspect the slats of the water intake grille for straightness. Straighten any bent slats, if possible. Use the utmost care when prying on any slat, as they tend to break if excessive force is applied. Replace the intake grille if a slat is lost, broken, or bent, and cannot be repaired. The slats are spaced evenly and the distance between them is critical, to prevent large objects from passing through and becoming lodged between the jet impeller and the inside wall of the housing.

Jet Impeller

The jet impeller is a precisely machined and dynamically balanced aluminum spiral. Observe the drilled recesses at exact locations to achieve this delicate balancing. Some of these drilled recesses are clearly shown in the accompanying illustration.

Excessive vibration of the jet drive may be attributed to an out-of-balance condition caused by the jet impeller being struck excessively by rocks, gravel or cavitation "burn".

The term cavitation "burn" is a common expression used throughout the world among people working with pumps, impeller blades, and forceful water movement.

"Burns" on the jet impeller blades are caused by cavitation air bubbles exploding with considerable force against the impeller blades. The edges of the blades may develop small "dime size" areas resembling a porous sponge, as the aluminum is actually "eaten" by the condition just described.

Excessive rounding of the jet impeller edges will reduce efficiency and performance. Therefore, the impeller should be inspected at regular intervals.

If rounding is detected, the impeller should be placed on a work bench and the edges restored to as sharp a condition as possible, using a file. Draw the file in only one direction. A back-and-forth motion will not produce a smooth edge. **TAKE CARE** not to nick the smooth surface of the jet impeller. Excessive nicking or pitting will create water turbulence and slow the flow of water through the pump.

Inspect the shear key. A slightly distorted key may be reused although some difficulty may be encountered in assembling the jet drive. A cracked shear key should be discarded and replaced with a new key.

Water Pump

Clean all water pump parts with solvent, and then blow them dry with compressed air. Inspect the water pump housing for cracks and distortion, possibly caused from overheating. Inspect the steel plate, the thick aluminum spacer and the water pump cartridge for grooves and/or rough spots. If possible **ALWAYS** install a new water pump impeller while the jet drive is disassembled. A new water pump impeller will ensure extended satisfactory service and give "peace of mind" to the owner. If the old water pump impeller must be returned to service, **NEVER** install it in reverse of the original direction of rotation. Installation in reverse will cause premature impeller failure.

If installation of a new water pump impeller is not possible, check the sealing surfaces and be satisfied they are in good condition. Check the upper, lower, and ends

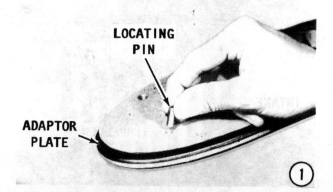

of the impeller vanes for grooves, cracking, and wear. Check to be sure the indexing notch of the impeller hub is intact and will not allow the impeller to slip.

ASSEMBLING

1- Identify the two small locating pins used to index the large thick adaptor plate to the intermediate housing. Insert one pin into the last hole aft on the topside of the plate. Insert the other pin into the hole forward toward the port side, as shown.

Lift the plate into place against the intermediate housing with the locating pins indexing with the holes in the intermediate housing.

Secure the plate with the five bolts. One of the five bolts is shorter than the other four. Install the short bolt in the most aft location.

Tighten the long bolts to a torque value of 22 ft lbs (30Nm). Tighten the short bolt to a torque value of 11 ft lbs (15Nm).

2- Place the driveshaft bearing assembly into the jet drive housing. Rotate the bear-

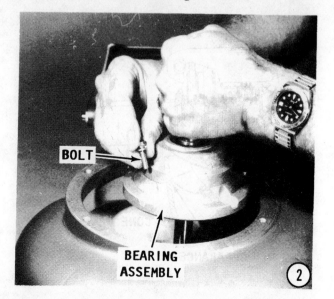

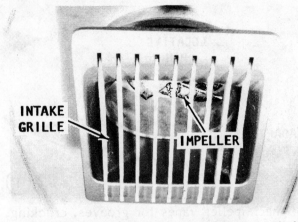

The clearance between the jet impeller and the casing cone can be fairly well estimated by shining a flashlight up through the grille and visually checking the distance between the impeller and the cone.

ing assembly until all bolt holes align. There is only **ONE** correct position.

Tighten the four bolts and lockwashers to a torque value of 5 ft lbs (7Nm).

SPECIAL WORDS

If installing a new jet impeller, place all the spacers (eight or nine, depending on the model being serviced) at the lower or "nut" end of the impeller, and skip the following step.

Shimming Jet Impeller

The clearance between the outer edge of the jet drive impeller and the water intake housing cone wall should be maintained at approximately 0.020-0.030" (0.5-0.8mm). This distance can be visually checked by

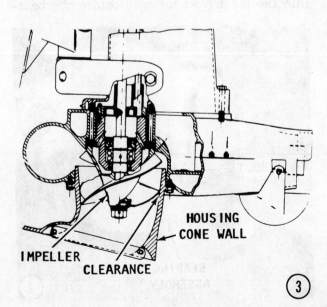

shining a flashlight up through the intake grille and estimating the distance between the impeller and the casing cone, as indicated in the accompanying illustrations. It is not humanly possible to accurately measure this clearance, but by observing closely and estimating the clearance, the results should be fairly accurate.

After continued use, the clearance will increase. The spacers removed in Steps 7 & 8 are used to position the impeller along the driveshaft with the desired clearance between the jet impeller and the housing wall.

3- A total of either **EIGHT** or **NINE** spacers are used depending on the model being serviced. When new, all spacers are located at the tapered (or "nut") end of the impeller. As the clearance increases, the spacers are transferred from the tapered ("nut") end and placed at the wide ("intermediate housing") end of the jet impeller.

This procedure is best accomplished while the jet drive is removed from the intermediate housing.

Secure the driveshaft with the attaching hardware. Installation of the shear key and nylon sleeve is not vital to this procedure. Place the unit on a convenient work bench. Shine a flashlight through the intake grille into the housing cone and "eyeball" the clearance between the jet impeller and the cone wall, as indicated in the accompanying line drawing. Move spacers one-at-a-time from the tapered end to the wide end to

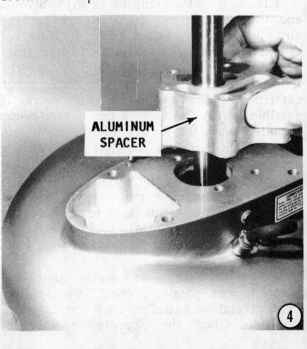

obtain a satisfactory clearance. Dismantle the driveshaft and note the exact count of spacers at both ends of the bearing assembly. This count will be recalled in Steps 9 & 11 of Assembly to properly install the jet impeller.

Water Pump Assembling

4- Place the aluminum spacer over the driveshaft with the two holes for the indexing pins facing **UPWARD.** Fit the two locating pins into the holes of the spacer.

GOOD WORDS

The manufacturer recommends **NO** sealant be used on either side of the water pump gaskets.

5- Slide the inner water pump gasket (the gasket with two curved openings) over the driveshaft. Position the gasket over the two locating pins. Slide the steel plate down over the driveshaft with the tangs on the plate facing **DOWNWARD** and with the holes in the plate indexed over the two locating pins.

Check to be sure the tangs on the plate fit into the two curved openings of the gasket beneath the plate. Now, slide the outer gasket (the gasket with the large center hole) over the driveshaft. Position the gasket over the two locating pins.

Fit the Woodruff key into the driveshaft. Just a dab of grease on the key will help to hold the key in place. Slide the water pump impeller over the driveshaft with the rubber membrane on the top side and the keyway in the impeller indexed over the Woodruff key. **TAKE CARE** not to damage the membrane. Coat the impeller blades with water resistant lubricant.

Install the insert cartridge, the inner plate, and finally the water pump housing over the driveshaft. Rotate the insert cartridge **COUNTERCLOCKWISE** over the impeller to tuck in the impeller vanes. Seat all parts over the two locating pins.

On some smaller models two different length bolts are used at this location, and one of these models uses special **"D"** shaped washers. All other models use plain washers.

All models: Tighten the four bolts to a torque value of 11 ft lbs (15Nm).

6- Install one of the small locating pins into the aft end of the jet drive housing.

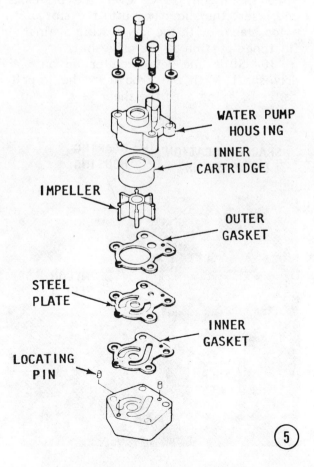

WATER PUMP HOUSING

INNER CARTRIDGE

IMPELLER

OUTER GASKET

STEEL PLATE

INNER GASKET

LOCATING PIN

(5)

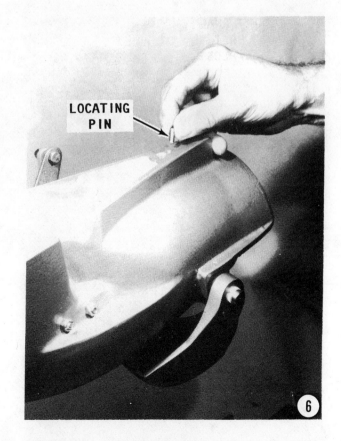

LOCATING PIN

(6)

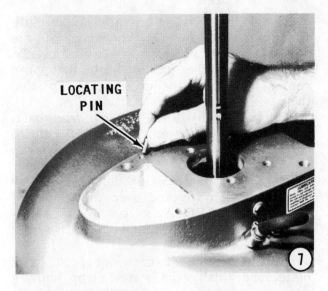

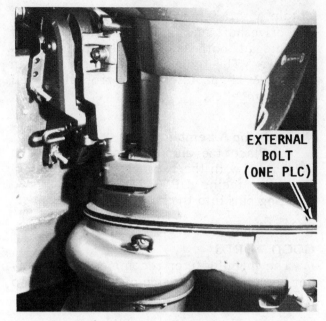

Location of the one exterior bolt securing the jet drive to the outboard intermediate housing.

Jet Drive Installation

7- Install the other small locating pin into the forward starboard side (or center forward end, depending on the model being serviced).

8- Raise the jet drive unit up and align it with the intermediate housing, with the small pins indexed into matching holes in the adaptor plate. Install the four internal bolts -- two shorter bolts in the two forward holes and the two longer bolts in the two aft holes. Install the one external bolt. Location of the external bolt is at the aft end of the anticavitation plate. Tighten the five bolts to a torque value of 11 ft lbs (15Nm).

9- Place the required number of spacers (if any) as determined in Step 3, **OR** in the paragraphs following "Shimming the Jet Impeller", up against the bearing housing. Slide the nylon sleeve over the driveshaft and insert the shear key into the slot of the nylon sleeve with the key resting against the flattened portion of the driveshaft.

10- Slide the jet impeller up onto the driveshaft, with the groove in the impeller collar indexing over the shear key.

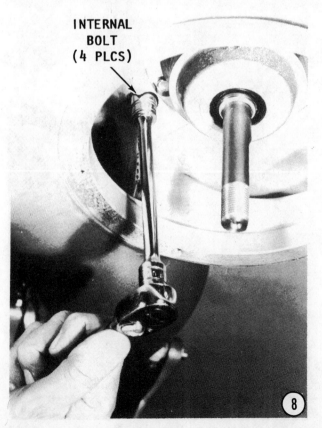

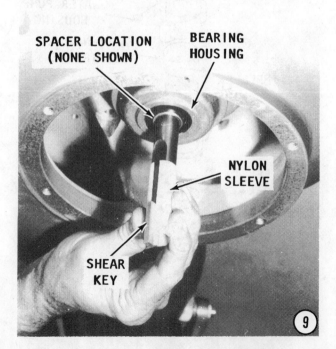

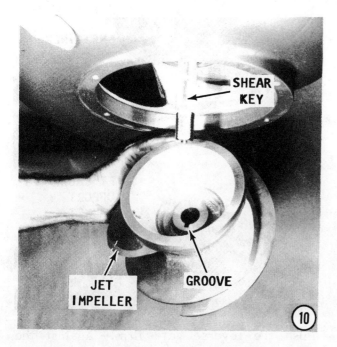

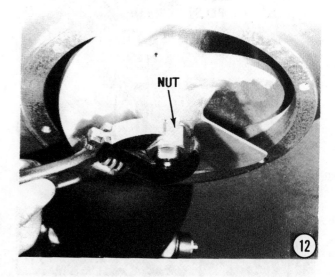

11- Place the remaining spacers over the driveshaft. The number of spacers will be eight or nine less the number used in Step 9, depending on the model being serviced.

12- Tighten the nut to a torque value of 17 ft lbs (23Nm). If neither of the two tabs on the tabbed washer aligns with the sides of the nut, remove the nut and washer. Invert the tabbed washer. Turning the washer over will change the tabs by approximately 15°. Install and tighten the nut to

the required torque value. The tabbed washer is designed to align with the nut in one of the two positions described.

13- Bend the tabs up against the nut to prevent the nut from backing off and becoming loose.

14- Install the intake grille onto the jet drive housing with the slots facing aft. Install and tighten the six securing bolts. Tighten the bolts to a torque value of 5 ft lbs (7Nm).

15- Slide the bolt through the end of the shift cable, washer, and into the shift arm. Install the locknut onto the bolt and tighten the bolt securely.

16- Install the shift cable against the shift cable support bracket and secure it in place with the two bolts.

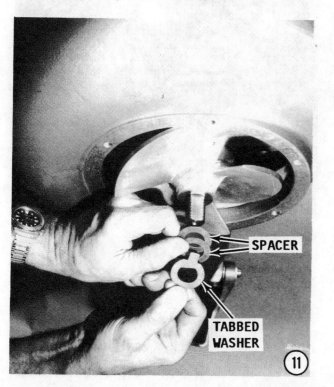

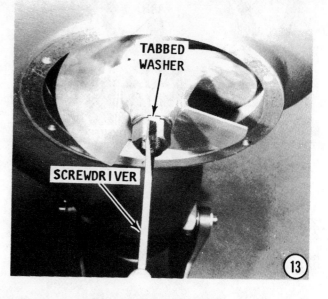

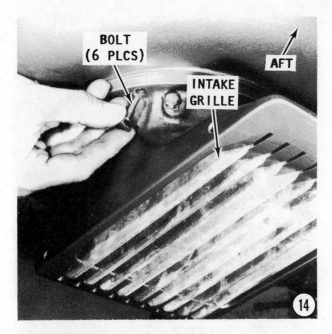

BOLT
(6 PLCS)

AFT

INTAKE
GRILLE

14

JET DRIVE ADJUSTMENTS

Cable Alignment
and Free Play Adjustment

1- Move the shift lever downward into the **FORWARD** position. The leaf spring should snap over on top of the lever to lock it in position.

2- Remove the locknut, washer, and bolt from the threaded end of the shift cable.

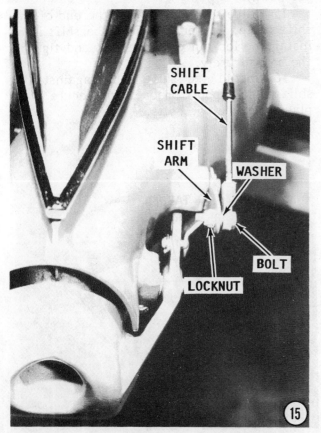

SHIFT
CABLE

SHIFT
ARM

WASHER

BOLT

LOCKNUT

15

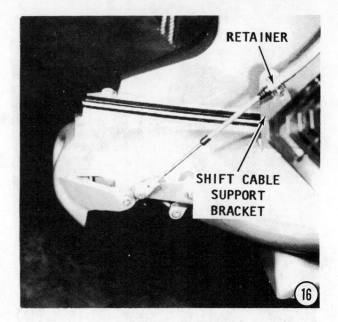

RETAINER

SHIFT CABLE
SUPPORT
BRACKET

16

Push the reverse gate firmly against the rubber pad on the underside of the jet drive housing.

Check to be sure the link between the reverse gate and the shift arm is hooked into the **LOWER** hole on the gate.

Hold the shift arm up until the link rod and shift arm axis form an imaginary

LEAF
SPRING

SHIFT
LEVER

1

straight line, as indicated in the accompanying illustration. Adjust the length of the shift cable by rotating the threaded end, until the cable can be installed back onto the shift arm **WITHOUT** disturbing the imaginary line. Pass the nut through the cable end, washer, and shift arm. Install and tighten the locknut.

Neutral Stop Adjustment
FIRST, THESE WORDS

In the **FORWARD** position, the reverse gate is neatly tucked underneath and clear of the exhaust jet stream.

In the **REVERSE** position, the gate swings up and blocks the jet stream deflecting the water in a forward direction under the jet housing to move the boat sternward.

In the neutral position, the gate assumes a "happy medium" -- a balance between forward and reverse when the powerhead is operating at **IDLE** speed. Actually, the gate is deflecting some water to prevent the boat from moving forward, but not enough volume to move the boat sternward.

WARNING

THE GATE MUST BE PROPERLY ADJUSTED FOR SAFETY OF BOAT AND PASSENGERS. IMPROPER ADJUSTMENT COULD CAUSE THE GATE TO SWING UP TO THE REVERSE POSITION WHILE THE BOAT IS MOVING FORWARD CAUSING

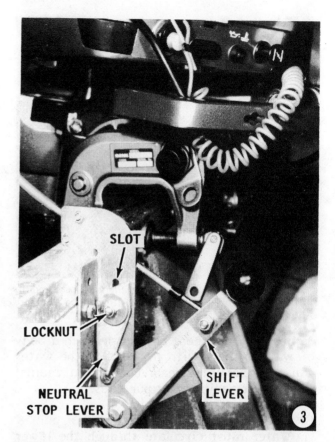

SERIOUS INJURY TO BOAT OR PASSENGERS.

3- Loosen, but do **NOT** remove the locknut on the neutral stop lever. Check to be sure the lever will slide up and down along the slot in the shift lever bracket.

CRITICAL WORDS

The following procedure **MUST** be per-

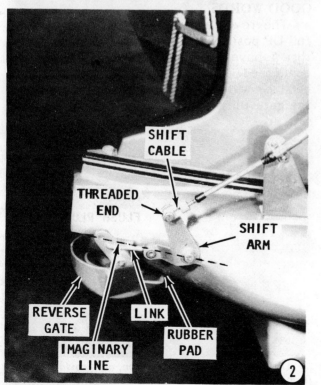

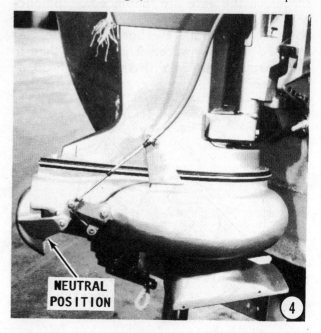

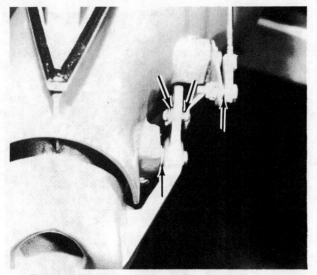

The arrows on this illustration indicate pivot points to be serviced on a regular basis with a good grade of water resistant multi-purpose lubricant.

formed with the boat and jet drive in a body of water. Only with the boat in the water can a proper jet stream be applied against the gate for adjustment purposes.

CAUTION

Water must circulate through the lower unit to the powerhead anytime the power-head is operating to prevent damage to the water pump in the lower unit. Just five seconds without water will damage the water pump impeller.

Start the powerhead and allow it to operate **ONLY** at **IDLE** speed. With the neutral stop lever in the down position, move the shift lever until the jet stream forces on the gate are "balanced". "Balanc-ed" means the water discharged is divided in both directions and the boat moves neither forward nor sternward. The gate is then in the neutral position with the powerhead at idle speed.

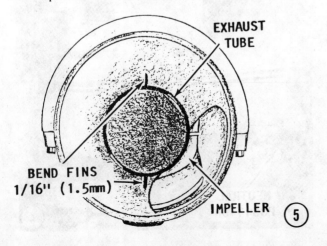

EXHAUST TUBE

BEND FINS 1/16" (1.5mm)

IMPELLER ⑤

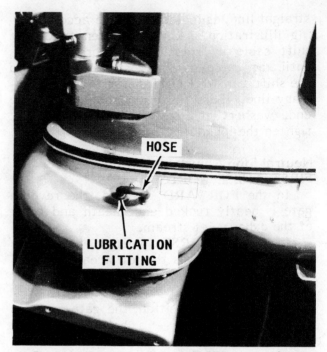

HOSE

LUBRICATION FITTING

Removing the hose is accomplished by deflecting the hose to one side to snap it free of the fitting. Do NOT attempt to "pull" the hose off the fitting.

4- Move the neutral stop lever up against the shift lever until the stop lever barely makes contact with the shift lever. Tighten the locknut to maintain this new adjusted position. Shut down the power-head.

GOOD WORDS

The reverse gate may not swing to the full **UP** position in reverse gear after Steps 1 thru 4 have been performed. Do **NOT** be concerned. This condition is acceptable, because water pressure in reverse will close the gate fully under normal operation.

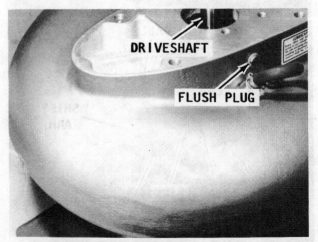

DRIVESHAFT

FLUSH PLUG

All jet drive units, 1987 and on, are provided with a flush plug on the port side of the housing.

Trim Adjustment

5- During operation, if the boat tends to "pull" to port or starboard, the flow fins may be adjusted to correct the condition. These fins are located at the top and bottom of the exhaust tube.

If the boat tends to "pull" to starboard, bend the trailing edge of each fin approximately 1/16" (1.5mm) toward the starboard side of the jet drive. Naturally, if the boat tends to "pull" to port, bend the fins toward the port side.

JET DRIVE LUBRICATION

The bearing and seal housing installed on the jet unit drive shaft is lubricated through an externally mounted zerk fitting. This fitting is located at the top of the jet drive casing on the port side.

To lubricate through the fitting, first push the coupling aside to release the coupling and the hose from the fitting. **DO NOT** attempt to pull on the coupling or the hose. Pushing the coupling aside is the answer.

Pump multi-purpose water resistant marine lubricant into the fitting until the old grease emerges from the hose coupler. Internal passageways are provided through the outer jet drive housing and the bearing and the seal housing to route the lubricant. When the old grease emerges from the hose coupling, there is no doubt the system has been properly lubricated. Wipe off the excess grease, and then snap the coupler back into place over the fitting.

The fitting should be lubricated every ten hours of jet drive operation.

Every 50 hours of jet drive operation,

pump enough new grease into the fitting to replace the old grease. A distinct change in color between the old grease and the new grease will be noted, indicating the unit is filled with the new lubricant.

When the unit is new, a slight discoloration may be expected, as the new seals are "broken in".

If the old grease contains tiny beads of water, the seals are beginning to break down. If the old grease emerging from the hose coupling is a dark dirty grey color, the seals have already broken down and water is attacking the bearings.

Lubricate the gate control linkage pivot points at regular intervals. These points are indicated in the accompanying illustration.

Jet Drive and Flushing Device

Regular flushing of the jet drive will prolong the life of the powerhead, by clearing the cooling system of possible obstructions. A plug is provided for this purpose just above the lubrication hose.

Remove the plug and gasket, install the flush adaptor, connect the garden hose and turn on the water supply.

Start and operate the powerhead at a fast idle for about 15 minutes. Disconnect the flushing adaptor and replace the plug.

SPECIAL WORDS

The procedure just described will only flush the powerhead cooling system, not the jet drive. To flush the jet drive unit, direct a stream of high pressure water through the intake grille.

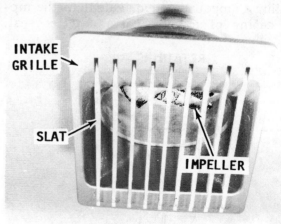

The edges of the impeller should be kept sharp for maximum efficiency and performance.

The clearance between the impeller and the casing core should be checked on a regular basis and should never exceed 1/32" (0.8mm).

8-11 TYPE "H" — COLT — 1990 ONLY

This short section provides instructions to disassemble and assemble the 1.25hp Colt/Junior lower unit, reference the exploded drawing on this page. Some instructions call for removal and discarding of an item. Be sure a replacement is on hand, especially a water pump repair kit.

Disassembling

1- Remove the spark plug lead. Remove the attaching hardware and the propeller.

2- Remove the two screws securing the lower unit to the exhaust housing, illustration **"A"**.

3- **Take Care** not to damage the driveshaft and the water tube and separate the lower unit from the exhaust housing.

4- Remove the fill/drain plug and drain the lower unit of lubricant.

5- Remove the two screws, and then lift the water pump housing up and free of the driveshaft. Remove and discard the water tube grommet. Lift the impeller and slide it free of the driveshaft.

6- Remove the two bearing carrier retaining screws, and then slide the carrier free of the propeller shaft. To start movement of the bearing carrier, rotate the carrier, and then tap the ears with a soft head mallet. Discard the O-ring.

7- If the driveshaft seal and the bearing carrier seal are damaged, obtain OMC Puller Bridge P/N 432127 and OMC Bearing Puller P/N 432130. Use these two tools to remove the seals.

Assembling

1- Apply a light coating of OMC Gasket Sealing Compound, or equivalent to the metal casing of a new bearing carrier seal.

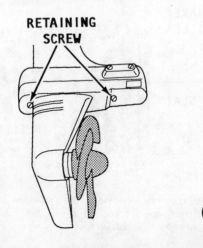

RETAINING SCREW

(A)

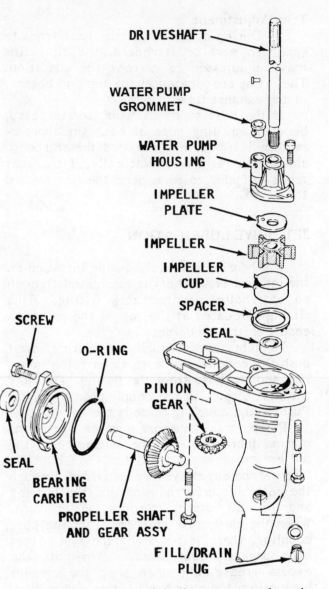

Exploded drawing of the lower unit used on the Model Colt and Junior, with major parts identified.

Install the seal with the lip of the seal facing **INWARD**. Use OMC Seal Installer, P/N 330219, or equivalent.

2- Apply a light coating of OMC Gasket Sealing Compound, or equivalent, to the metal casing of a new driveshaft seal. Install the seal into the lower unit with the lip of the seal facing **DOWN**. Coat a new O-ring with OMC Triple-Guard grease, or equivalent, and then install the ring into the bearing carrier.

3- Insert the pinion gear and the propeller shaft and gear assembly into the lower unit. Now, slide the driveshaft down through the seal and the lower unit. Rotate the driveshaft slightly to allow the splines on the end of the driveshaft to index with the internal splines of the pinion gear.

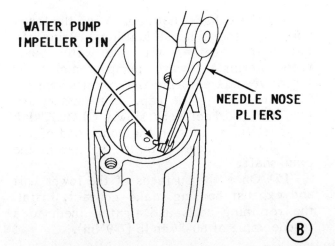

WATER PUMP
IMPELLER PIN

NEEDLE NOSE
PLIERS

B

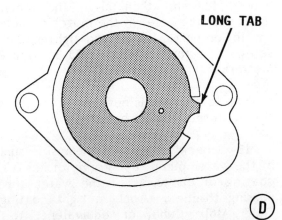

LONG TAB

D

4- Apply a light coating of OMC Gasket Sealing Compound to the threads of the bearing carrier retaining screws, and then install the carrier, with the flat surface at the upper screw hole facing **UP**. Tighten the screws to a torque value of 60-84 in lb (709Nm).

5- With the lower unit fill/drain hole facing **UP**, fill the unit with 1.3 fl oz (38 ml) of OMC Hi-Vis lubricant. The unit is full when lubricant appears at the hole. Install and tighten the fill/drain plug to a torque value of 60-84 in lb (7-9Nm).

6- Using a pair of needle nose pliers, insert the water pump impeller pin into the driveshaft by aligning the **"T"** portion of the pin with the axis of the shaft, illustration **"B"**. Press the pin inward until the pin bottoms against the shaft.

7- Install the plastic pump spacer into the lower unit pump cavity. The side of the spacer with the two small prongs toward the cavity. The prongs must enter the large main water intake hole in the lower unit casting.

8- Apply a very light coating of oil to the impeller blades. Rotate the impeller in a **CLOCKWISE** direction as the impeller is worked into the housing cup, illustration "C". Be sure the blades are seated in the direction shown.

9- Insert the impeller plate into the water pump body with the long tabs extending into the water outlet opening, illustration **"D"**.

10- Apply a light coating of OMC Gasket Compound, or equivalent, to the exterior surface of the impeller cup. Install the cup, with the impeller in place, into the impeller pump body. Check to be sure the tabs on the cup index into the slot in the pump body.

11- Apply a light coating of oil to the outside and inside surfaces of the water pump tube grommet. Force the grommet into the water pump opening, illustration **"E"**.

12- Apply a coating of OMC Gasket Sealing Compound, or equivalent, to the threads of the water pump retaining screws.

WATER PUMP
IMPELLER
BLADE

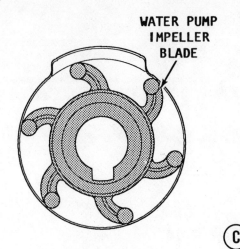

C

WATER PUMP
TUBE GROMMET

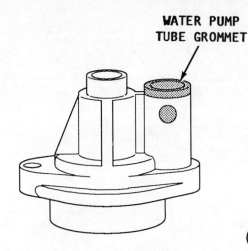

E

Carefully slide the water pump down the driveshaft, and then align the mounting holes with the holes in the matching holes in the lower unit. Install and tighten the two screws to a torque value of 25-35 in lb (2.8-4.0Nm).

The lower unit is now ready for installation with the exhaust housing.

13- Check to be sure the inside surface of the water pump tube grommet has a light coating of oil to assist the water tube in passing through. Apply a light coating of OMC Moly Lube, or equivalent, to the splines at the upper end of the driveshaft. **BE CAREFUL** not to get any lube on the end of the driveshaft. Any lubricant on the end may prevent the driveshaft from fully seating in the powerhead crankshaft.

CRITICAL WORDS

The flywheel may be rotated in the next step to permit the splines of the driveshaft to index with the splines in the crankshaft. The flywheel must be rotated **ONLY** in a **CLOCKWISE** direction to prevent damage to the water pump vanes.

14- Apply a coating of OMC Nut Lock to the threads of the bolts securing the lower unit to the exhaust housing. Align the lower unit with the exhaust housing. Bring the two units together with the water pump tube entering the water pump grommet and the driveshaft entering the exhaust housing. Continue closing the gap. If necessary rotate the flywheel very slowly **CLOCKWISE** to assist the splines on the upper end of the driveshaft to index with the splines in the crankshaft.

15- Once the surfaces of the lower unit and exhaust housing make contact, install the retaining bolts and tighten them to a torque value of 60-84 in lb (7-9Nm).

Closing Tasks

Apply a coating of OMC Triple-Guard grease to the propeller shaft, and then install the propeller. Always use a new drive pin and new cotter pin.

Install the spark plug lead onto the spark plug.

Mount the outboard unit in a test tank or body of water. Start the powerhead and check the completed work.

After the unit has been operated for a short time, shut the powerhead down, remove the lower unit fill/drain plug and check the lubricant level. Add lubricant as required.

9
HAND REWIND STARTERS

9-1 INTRODUCTION

The hand rewind starters installed on the Johnson/Evinrude outboards covered in this manual may be one of three basic designs:

One is a vertical spool type mounted on the side of the powerhead with a drive gear engaging the teeth of the flywheel.

The second design is a disc type which engages the teeth of the flywheel and may be mounted either horizontally or vertically.

The third design is a flat disc type mounted atop the flywheel. This design starter engages a ratchet plate on the flywheel.

UNFORTUNATELY, engineering changes have resulted in several models being used for each design. The only logical and practical method of designating different procedures for each model for each design was to assign a letter designation -- **"A" thru "F"** for each different starter. The outboard models using each type, at press time, are listed under the heading for each starter.

EMERGENCY STARTING

The 50hp powerheads are equipped with an electric cranking motor only. However, cutouts are manufactured into the flywheel to permit the use of a rope for emergency start, provided an NFL linebacker is aboard.

SAFETY WORD

If an emergency rope is used on **ANY** powerhead **NEVER** wrap the rope end around your hand for a better grip. If the powerhead should happen to backfire, the sudden jerk on the rope would severely **INJURE** an individual's hand.

"A" Hand Rewind Starter
Model 5 thru 8 — 1990 Only

The **"A"** starter is a cylinder with a pinion gear arrangement similar to an automotive cranking motor. The unit is mounted vertically on the side of the powerhead. When the starter rope is pulled, a nylon drive gear slides upward and engages the flywheel ring gear. After the powerhead starts, the drive gear automatically disengages and retracts to the "rest" position.

*Typical **"A"** hand rewind starter installation for a Model 5 thru 8 hp powerhead. This starter is similar in principle to an automotive type cranking motor.*

"B" swing arm hand rewind starter mounted on the port side of a Model 4D powerhead.

"B" Hand Rewind Starter
Model 4D Only -- 1990 & On

This **"B"** starter is mounted on the port side of the powerhead and the drive gear works on an axis. As the rope is pulled, a swing arm moves the drive gear upward to engage with the teeth of the flywheel ring gear. A coil spring winds and tightens as the rope unwinds. The spring then coils the rope around a pulley as the rope handle is returned to the control panel.

Typical "C" hand rewind starter mounted flat (horizontally), next to the flywheel on Model 9.9 thru 15 powerheads -- 1990 and on. The drive gear moves upward and engages the ring gear of the flywheel when the starter rope is pulled.

"C" Hand Rewind Starter
Model 9.9 thru 15 -- 1990 & On

This pinion gear starter is a design employing the principles of an automotive-type cranking motor. A nylon pinion gear slides upward and engages the flywheel ring gear as the starter rope is pulled. The pinion gear automatically disengages when the engine starts. The ratio between the pinion gear and the ring gear was selected to provide maximum cranking speed with minimum pulling effort to ensure fast and easy powerhead start.

A lockout pawl linked to the cam follower prevents manual starter engagement if the throttle is advanced beyond the start position.

"D" Hand Rewind Starter
Colt, and Models 3 & 4 -- 1990 & On
Models 5 thru 8 -- 1991 & On
Models 20 thru 30 -- 1990 & On

This design hand rewind starter is usually mounted atop the flywheel with three mounting legs attached to the powerhead. The rewind spring is under tremendous tension when it is wound -- a real tiger in a cage. Therefore, the service procedures **MUST** be closely followed to prevent the spring from unwinding at the wrong time. Such action could cause severe personal injury.

"D" hand rewind starter mounted on an older small horsepower powerhead. The procedures and illustrations in this chapter are valid for the Colt -- 1990, also Model 3 and Model 4 -- both 1991 and on.

"E" Hand Rewind Starter
Model 40 thru 55 — 1990 & On

The "E" starter is mounted atop of the flywheel with three mounting legs. The unit has a large rectangular pawl plate visible on the underside surface.

Because this starter is used on fairly large horsepower powerheads, actual number in the field is rather small. As can be imagined, a very husky fellow is required to crank this size powerhead with sufficient force to achieve engine start. Qualified individuals might be a linebacker for one of the NFL teams.

Typical "F" hand rewind starter mounted on a Model 2.3 and Model 3.3 powerhead — 1991 and on.

"F" Hand Rewind Starter
Model 2.3 & 3.3 — 1991 & On

This hand rewind starter for these small powerheads is a simple coiled spring unit mounted on top of the powerhead under the fuel tank.

OPERATION

Normally, very few problems are encountered with the hand rewind starter. It is strictly a mechanical device to crank the powerhead for starting. The spring will last an incredibly long time, if used properly. The greatest enemy of the spring is the operator.

Three causes contribute to starter failure. Two may be prevented, the third cannot.

The most common problem is the result of the operator pulling the starter rope too far outward. If the operator places one hand on the powerhead and pulls the rope with the other hand, it is physically impossible, in this position, to pull the rope too far. Problems develop when the operator uses both hands to pull on the rope, with no control on how far the rope can be extended. The rope may be broken or the knot released from the starter disc. In either case, the spring rewinds with tremendous speed and in almost all cases travels past its normal rewind position bending the end of the spring in reverse. Therefore, more maintenance work is involved than merely replacing the rope.

Another bad habit, while using the hand rewind starter, is the operator releasing his/her grip on the rope when it is in the extended position, allowing the rope to freely rewind. The operator should **NEVER** release his grip, but hold onto the rope, and thus control the rewind. The owner should always be alert to any wear on the rope and replace it long before the possibility of breaking might occur. If the rope should break, the spring would rewind with incredible speed, the same as if the rope were released, causing damage to the spring and other starter parts.

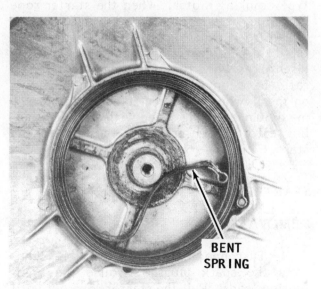

BENT SPRING

The pull rope broke on this "D" hand rewind starter causing the spring to rewind with such incredible speed, the spring actually doubled back in a reverse direction, as shown.

The third cause of spring failure cannot be prevented -- age. As the outboard continues to perform year after year, the age of the spring steel will finally take its toll.

SAFETY WORDS

Depending on the model and the powerhead, from 6 to 12 **FEET** of spring steel length is wound into about a 4 **INCH** diameter. This places the spring under unbelievable tension -- a real **"TIGER IN A CAGE"** -- making it a highly **DANGEROUS** force. Therefore, any time the hand starter is serviced, especially during work on the spring, **SAFETY GLASSES** should be worn and the work performed with the utmost care. The procedures **MUST** be followed exactly as presented for each starter, to prevent possible injury to the worker or others in the area.

Any time the rope is broken, the starter spring will rewind with incredible speed. Such action will cause the spring to rewind past its normal travel and the end of the spring will be bent back out of shape. Therefore, if the rope has been broken, the starter should be completely disassembled and the spring repaired or replaced.

9-2 "A" HAND REWIND STARTER

MODEL 5 THRU 8 — 1990 ONLY

This gear-driven starter is a new design employing the principle of an automotive type cranking motor. When the starter rope is pulled, the starter rotates, and a nylon pinion gear slides upward and engages the flywheel ring gear. The gear automatically disengages when the powerhead starts. The ratio between the pinion gear and the ring on the flywheel has been selected to provide maximum cranking speed with minimum pulling effort to ensure fast, easy powerhead start.

STARTER ROPE REPLACEMENT

REMOVAL

1- Disconnect the high-tension leads from the spark plugs. Ground the high-tension leads. Pull the starter rope out until it is fully extended. Now, allow the rope to retract just a little, until the knot end on the spool is facing the port side of the

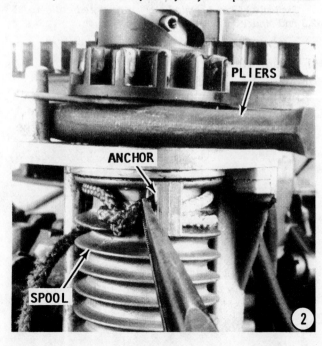

powerhead. Lift the pinion gear to engage the flywheel ring gear. Hold the pinion gear engaged with the ring gear, and at the same time, slide the handles of a pair of pliers under the pinion gear to lock the pinion gear with the ring gear, as shown.

2- Remove the handle from the end of the starter rope. **OBSERVE** how the rope is wound onto the spool and how the rope is secured by a loop formed in a slot in the spool. Remove the rope from the starter spool.

INSTALLATION

Rope Purchase Instructions

The length and diameter of the starter rope required will vary depending on the horsepower size of the model being serviced. Therefore, check the Hand Starter Rope Specifications in the Appendix, and then purchase a quality nylon piece of the

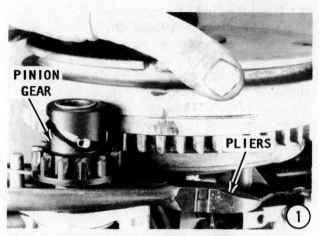

proper length and diameter size. Only with the proper rope, will you be assured of efficient operation following installation.

Each end of the nylon rope should be "fused" by burning them slightly with a very small flame (a match flame will do) to melt the fibers together. After the end fibers have been "fused" and while they are still hot, use a piece of cloth as protection and pull the end out flat to prevent a "glob" from forming.

3- Feed one end of the new rope through the spool anchor, make a loop, and then thread the end back through the hole in the anchor, but **DO NOT** pull it tight at this time, leave a loop.

4- Bring the short end of the rope through the loop just formed.

5- Work the short end back through the anchor, as shown.

6- Now, pull both ends of the rope tight. Feed the rope through the front engine

cowling, and then install the starter rope handle. Pull and hold tension on the rope, and at the same time remove the pliers from under the pinion gear. Allow the starter rope to rewind in a normal manner. After the rope is fully wound onto the starter spool, the rope handle should be up tight against the cowling. If the handle is not up tight, the rope was installed too long or the starter spring is weak and should be replaced.

STARTER REMOVAL

AUTHOR'S NOTE

For photographic clarity, the accompanying pictures were taken servicing a starter from a powerhead removed from the

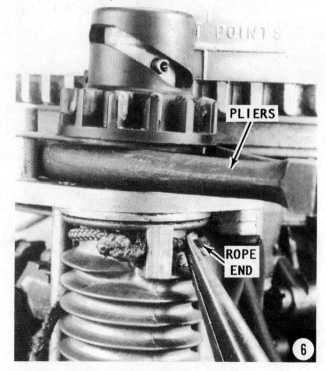

exhaust housing. The hood need only be removed to work on the starter.

1- Pull approximately 3/4 of the starter rope out, and then form a knot in the rope to prevent it from recoiling. Allow the rope to recoil until the knot is tight against the cowling. Remove the rope handle.

2- The rope may be removed now, or later. To remove the rope now, first pull the rope all the way out. Slide the handles of a pair of pliers under the pinion gear to hold the gear engaged with the flywheel. Remove the rope from the starter spool.

3- Grasp the spool firmly. Remove the pliers and allow the spool to slip a little at a time until the spring is completely unwound.

4- Remove the two retaining bolts on top of the starter.

5- Loosen, but **DO NOT** remove the bolts on the bottom and on each side of the starter spool. When the bottom two bolts are loosened, the retainer will separate from the lower cap. Lift the starter from the powerhead. If the spring is still clipped into the lower retainer, release the spring by disengaging the spring tang from the retainer.

BOLT

5

DISASSEMBLING

6- Remove the pin in the pinion gear, and then remove the pinion gear from the collar. Slip the bearing head off the spool. Remove the retainer, installed under the pinion gear, from the starter.

7- Remove the spring retainer (the long tube) from the center of the spool.

8- Pull the starter spring from the spool.

PINION GEAR

BEARING HEAD

6

SPRING RETAINER   7

CLEANING AND INSPECTING

Wash all parts in solvent, and then dry them with compressed air.

Inspect and replace the main spring if it is damaged or worn. Check the bottom end of the spring very carefully to be sure the two tangs (one on the inner and the other on the outer spring) are in good condition with no sign of distortion.

Inspect the bushing in the bottom collar of the starter housing. This bushing was not removed in the disassembling procedures. Feel with a finger for any roughness, burrs, or other evidence of excessive wear or damage. If the bushing is in good condition, it need not be removed.

Check the rope condition. If the rope is frayed or shows any sign of weakness, it should be replaced. There will never be an easier time to replace the rope than while the starter is disassembled.

Inspect the teeth of the pinion gear. The teeth will show some signs of normal wear. A broken tooth or excessive wear on one side of the teeth is justification for replacement.

Inspect the groove through the pinion gear. This is the groove to accommodate the roll pin. Check to be sure the upper part of the pinion gear is not cracked or distorted.

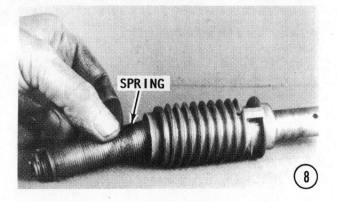

SPRING

8

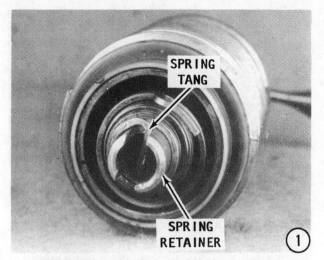

DO NOT lubricate the pinion gear. Oil applied to the pinion gear will attract dirt causing the gear to bind on the spool. Lubricate the upper and lower spool bearing surfaces with just a drop of outboard lubricant. Apply outboard oil to the spring on the pinion gear. **DO NOT** oil the pinion gear bearing or the surfaces of the spring.

ASSEMBLING

GOOD NEWS

Two methods of assembling and installing this starter are presented. The first is the factory suggested procedure and begins with the following steps on this page.

An alternate method is also outlined which many professional mechanics feel is much simplier, easier, and quicker. The alternate method is given on the following pages, Steps 1A thru 5A.

After the starter is assembled and installed on the powerhead, continue the work with Step 7.

Factory Method

1- Install the spring retainer from the bottom side. Slide the spring onto the

spring retainer. Work the spring upward until the inner spring tang engages the slot on the bottom of the retainer.

2- Align the hole in the retainer with the hole in the spool sleeve.

3- Slip the bearing head and pinion gear down over the spool shaft. Install the roll pin through the pinion gear and sleeve.

4- If the lower bushing was removed, install a **NEW** bushing into the bottom collar on the powerhead.

HELPFUL WORD

As an assist to installation, first soak the bushing in hot water for about ten minutes, and then lubricate it with just a drop of outboard oil.

5- The tang on the outer spring must hook into the slot of the lower spring retainer plate. Pull on the outer spring to elongate the spring, and at the same time, lower

SPRING TANG

5

the spring into the spring retainer plate and hook the tang into the slot in the plate. Rotate the spring **CLOCKWISE** to lock the tang in the plate. Hold upward and turn the spring **CLOCKWISE,** and at the same time tighten the two screws in the lower spring retainer plate.

6- Place the starter assembly in position on the powerhead and start the two upper screws through the upper bearing support. Check to be sure the guide is in place in the bottom and top retainers. Tighten the two bottom retainer screws.

Alternate Assembling Method

1A- If the lower bushing was removed, install a **NEW** bushing into the bottom collar on the powerhead.

BOLT

6

BOTTOM COLLAR

BUSHING

1A

HELPFUL WORD

As an assist to installation, first soak the bushing in hot water for about ten minutes, and then lubricate it with just a drop of outboard oil.

2A- Install the spring retainer from the bottom side of the spool. Slip the bearing head and pinion gear down the spool shaft.

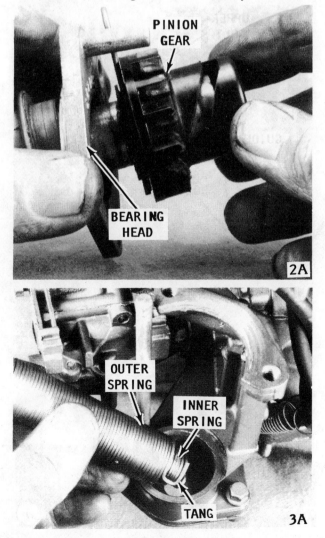

PINION GEAR

BEARING HEAD

2A

OUTER SPRING

INNER SPRING

TANG

3A

4A

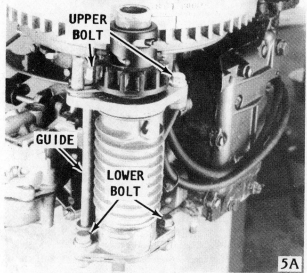

5A

7

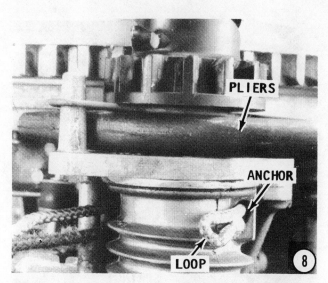

8

Align the holes and install the roll pin through the pinion gear and spool.

3A- Take the spring and lower it into the bottom retainer. Hook the outer spring into the retainer.

4A- Lower the spool assembly down over the spring and engage the spring retainer into the tang of the inner spring.

5A- Start the two upper bolts through the upper bearing suport. Check to be sure the guide is in place in the bottom and top retainers. Tighten the two bottom retainer bolts.

And now the work continues from Step 6

7- Insert a large size screwdriver into the top of the spool, and then rotate the spool, by count, exactly 16-1/2 complete turns.

8- Lift the pinion gear to engage the flywheel ring gear. Hold the pinion gear engaged with the ring gear, and at the same time, slide the handles of a pair of pliers

9

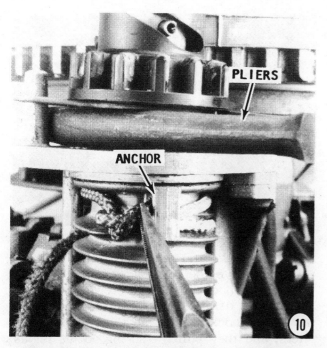

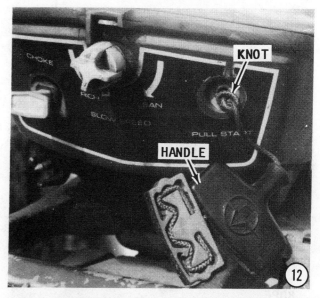

not up tight against the cowling the rope was installed too long and needs to be shortened.

9-3 "B" HAND REWIND STARTER COIL SPRING WITH SWING ARM VERTICAL MOUNT MODEL 4D ONLY — 1990 AND ON

REMOVAL

1- Remove the spark plugs and ground the high tension leads. Pull the starter rope out, and then tie a knot in the rope behind the handle. Allow the rope to rewind until the knot is against the cowling. Untie the knot in the end of the rope, and then remove the handle and the rubber bumper.

2- Remove the knot tied in the rope in Step 1. Allow the rope to **SLOWLY** wind into the starter. Before the rope end passes

under the pinion gear to lock the pinion gear with the ring gear, as shown.

Feed one end of the new rope through the spool anchor, make a loop, and then thread the end back through the hole in the anchor, but **DO NOT** pull it tight at this time, leave a loop.

9- Bring the short end of the rope through the loop just formed.

10- Work the short end back through the anchor, as shown.

11- Now, pull both ends of the rope tight.

12- Feed the rope through the front engine cowling, and then install the starter rope handle. Hold tension on the rope with the handle and at the same time, remove the pliers from underneath the pinion gear. Allow the starter rope to wind onto the spool. After the rope has been wound onto the spool, the starter handle should be up tight against the cowling. If the handle is

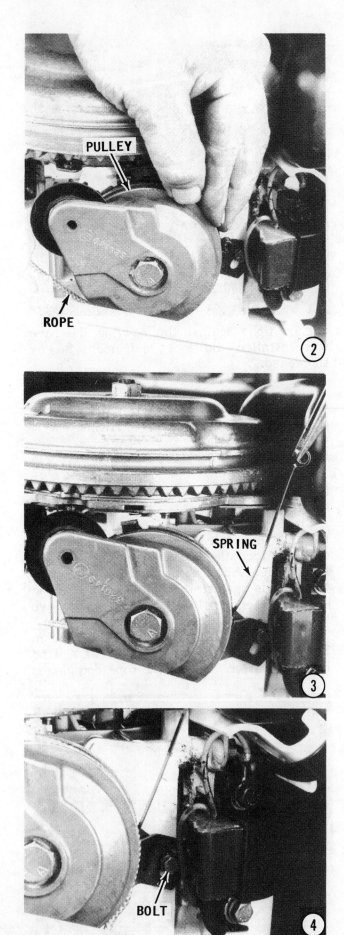

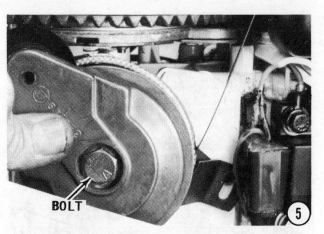

the cowling, firmly grasp the starter pulley, and then allow the starter to unwind.

3- Observe the back side of the starter. Notice the hook of the starter spring protruding out of a hole in the starter. Grasp the spring hook with a pair of needle-nose pliers, and then pull the spring out as far as possible, to relieve tension on the spring.

4- Remove the 3/8" bolt from the bracket between the starter and the exhaust housing.

5- Hold the starter together with one hand, and at the same time **LOOSEN** the large bolt from the center of the starter. **DO NOT** remove this bolt at this time. Remove the starter from the powerhead.

6- If the starter is only removed in order to accomplish other work, install a 3/8" x 16 nut onto the far side of the thru-bolt to hold the starter together and prevent the spring from escaping.

DISASSEMBLING

7- Remove the center bolt, idler gear arm, and the idler gear arm spring.

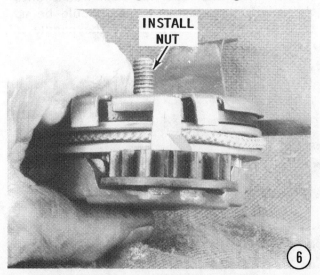

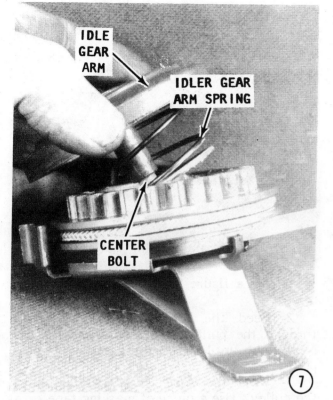

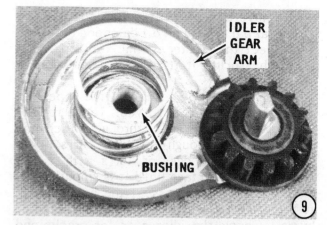

be released from the cup. After the pulley has been removed, notice the position of the spring loop. Remove the rope from the pulley. Notice how the rope unwinds from the pulley **COUNTERCLOCKWISE**.

9- Remove the bushings from the idler gear arm and the bushings installed one on each side of the pulley.

10- Two different methods are suggested to remove the spring from the starter cup. One method involves pulling continuously on the end of the spring that contains the loop. The second method is to simply toss the cup a safe distance onto carpeting or a lawn, allowing the spring to be released instantly from the cup.

If this second method is used, be sure the spring will not cause a threat to any individual in the area when it is released.

CLEANING AND INSPECTING

Wash all parts except the rope in solvent and then blow them dry with compressed air.

SAFETY WORDS

The next step could be dangerous. Removing the pulley from the cup **MUST** be done with care to prevent personal injury.

8- Lift the pulley **SLIGHTLY** and then use a screwdriver and work the spring free of the pulley. **DO NOT** allow the spring to

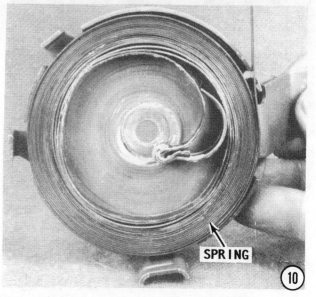

Remove any trace of corrosion and wipe all metal parts with an oil dampened cloth.

Inspect the starter spring end loops. Replace the spring if it is weak, corroded or cracked.

Inspect the rope. Replace the rope if it appears to be weak or frayed. If the rope is frayed, check the hole through which the rope passes for rough edges or burrs. Remove the rough edges or burrs with a file, and polish the surface until it is smooth.

Inspect the dog ears of the pulley gears to be sure they are not worn and are free of burrs. Check the idler gear for cracks and missing teeth.

ASSEMBLING

Rope Purchase Instructions

The length and diameter of the starter rope required will vary depending on the horsepower size of the model being serviced. Therefore, check the Hand Starter Rope Specifications in the Appendix, and then purchase a quality nylon piece of the proper length and diameter size. Only with the proper rope, will you be assured of efficient operation following installation.

Each end of the nylon rope should be "fused" by burning them slightly with a very small flame (a match flame will do) to melt the fibers together. After the end fibers have been "fused" and while they are still hot, use a piece of cloth as protection and pull the end out flat to prevent a "glob" from forming.

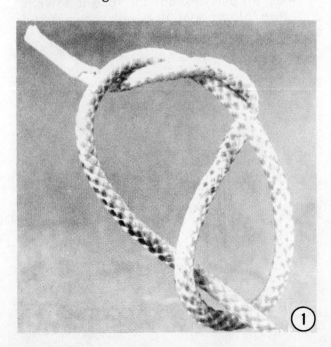

1- Tie a figure **8** knot in one end of the rope.

2- Feed the other end of the rope through the pulley hole, and then pull the rope tight until the knot is seated in the pulley.

3- Wind the rope **CLOCKWISE** around the pulley. Use a piece of masking tape or a rubber band to hold the rope in place in the pulley.

4- Coat the bushings with a light film of OMC Type A lubricant. Insert one bushing into the pulley, another bushing into the idler gear arm, and two more bushings, one on each side of the pulley.

5- If the spring was **NOT** removed from the cup, lower the pulley down over the spring and insert the end of the spring into the spring anchor post of the pulley. If the spring **WAS** removed from the cup, hook the end of the spring into the pulley and then

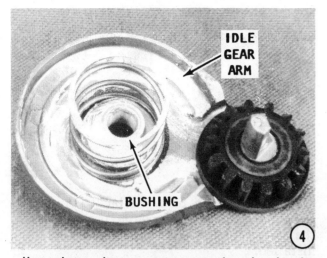

allow the spring to come out the slot in the cup. Turn the pulley and wind the spring into the cup until about 1/2 of the spring length has been wound. Hold the gear and allow it to back off slowly. **DO NOT** allow the spring to rewind quickly. The remainder of the spring will be installed later.

6- Assemble the idler gear with the shoulder against the idler gear arm. Install the idler gear arm and spring to the pulley and cup with the stop on the underside of the idler gear shaft located between the upper stop and the lower stop on the cup and stop assembly.

7- Hold the assembly together and install the assembly onto the powerhead. Thread the shoulder bolt into place first. This bolt will hold the starter assembly together.

8- Install the bolt through the idler arm and into the exhaust manifold. **DO NOT** tighten this bolt at this time. Apply a light coating of OMC Type A lubricant to the portion of the spring extending out of the starter.

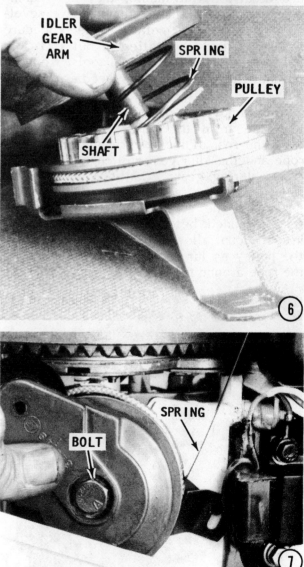

Housing (right) with the spring properly installed and the spring end bent toward the center. The pin in the pulley (left) must index into the loop on the spring end during installation.

BOLT

⑧

SPECIAL WORDS

Special tools are available to lock-in the starter to the flywheel. However, the tools are usually not available; they are expensive; and professional mechanics have developed an alternate method. The procedure will take time and patience, but it is the only way without the special tools. To work without the special tools proceed as follows:

9- Remove the rubber band or the masking tape from the coiled rope. Working from the rear of the powerhead, pull on the rope and the idler gear will engage with the flywheel ring gear. Continue to pull the rope, and at the same time, work the spring down into the cup. If the rope becomes fully extended, before the spring is installed into the cup, allow the rope to rewind onto the pulley as far as possible and then wind the rope around the pulley again. Now, pull

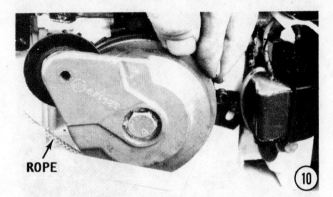

ROPE

⑩

on the rope again from the back side of the rear of the powerhead, and continue to work the spring into the cup until it is completely installed.

10- Ease back on the rope until there is no spring tension on the starter. Thread the rope into the pulley **CLOCKWISE** around the starter. Two, or possibly more, loops may be required to accomplish the task. Use all of the rope in the pulley with the starter in the relaxed position. After all of the rope has been fed into the pulley, grab the end of the rope in front of the starter and pull it out, then feed it through the cowling at the front of the powerhead. Continue to pull the rope until about two feet is extending out through the cowling. Tie a slip knot in the rope.

11- Install the rubber bumper and handle onto the rope. Tie a figure **8** knot in the end of the rope, and then pull the knot into the handle.

SPRING

⑨

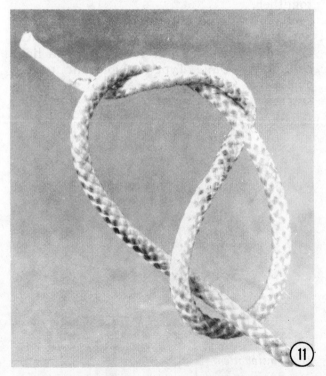

⑪

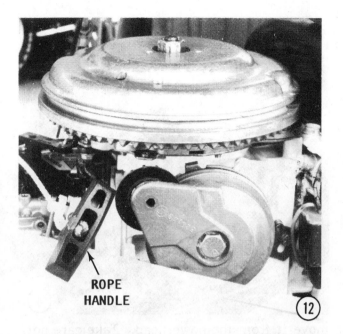

ROPE
HANDLE

⑫

12- Untie the slip knot and ease the rope back into the starter. The starter handle must be up tight against the cowling when the rope is completely rewound on the starter pulley. If the rope is not tight against the cowling, remove the knot and handle from the rope, and then wind the rope around the pulley one complete turn. Tie another knot in the end of the rope as described earlier in this step and then check to be sure the handle is tight against the cowling when the rope is wound onto the pulley.

Starter Adjustment

13- Hold the idler gear arm stop against the cup stop. Fully engage the idler gear teeth with the teeth in the flywheel ring gear. Tighten the cup and stop assembly screw. Tighten the shoulder screw securely.

BOLT

⑬

9-4 "C" HAND REWIND STARTER COIL SPRING WITH SLIDING GEAR HORIZONTAL MOUNT MODEL 9.9 THRU 15 — 1990 AND ON

This pinion gear starter is a design employing the principles of an automotive-type starter motor. A nylon pinion gear slides upward and engages the flywheel ring gear as the starter rope is pulled. The pinion gear automatically disengages when the engine starts. The ratio between the pinion gear and the ring gear was selected to provide maximum cranking speed with minimum pulling effort to ensure fast and easy powerhead start.

A lockout pawl linked to the cam follower prevents manual starter engagement if the throttle is advanced beyond the start position.

REMOVAL

SPECIAL NOTE

If the only work to be performed on the starter is replacement of the rope, perform Steps 1 thru 17, then jump to Page 9-19, Installation (rope). If the starter is to be rebuilt, perform Steps 1 thru 4, then jump to Step 18.

1- Disconnect the high-tension leads from the spark plugs. Ground the high-tension leads. Pull the rope out far enough, and then tie a knot in the rope. Allow the rope to rewind until the knot is against the front of the cowling.

2- Remove the handle rope anchor and then the rope from the handle. Remove the rubber bumper.

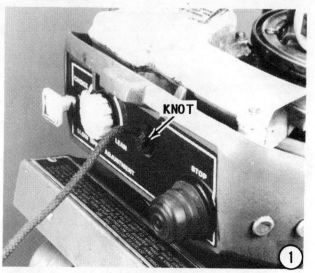

KNOT

①

ROPE

3- Untie the knot in the rope and allow the rope to **SLOWLY** wind onto the pulley and the spring to unwind. When the rope is about to pass through the opening in the cowling, hold the pulley, and then allow your grip to slip to permit the rope to continue winding onto the pulley, and the spring to unwind slowly. **DO NOT** release the grip on the pulley completely. If the spring rewinds too rapidly the spring may be damaged along with other parts.

WARNING
The rewind spring is under tremendous tension and is a potential hazard. Therefore, **SAFETY GLASSES** should be worn and extreme **CARE** exercised to follow the procedures carefully during removal, disassembling, and assembling work with the starter.

4- HOLD the pulley and cup together, and at the same time, loosen the center mounting bolt and back it out of the intake manifold. **DO NOT** remove the center bolt from the starter. Continue to hold the starter **FIRMLY** together and carefully re-

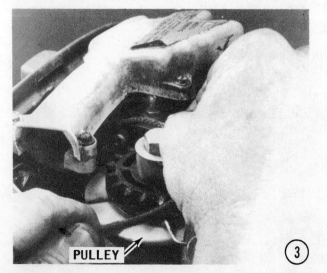

PULLEY

PULLEY AND CUP

move it from the powerhead. Take care not to damage the starter pawl.

5- If the starter is only removed in order to accomplish other work, install a 3/8" x 16 nut onto the far side of the thru-bolt to hold the starter together and prevent the spring from escaping.

STARTER ROPE REPLACEMENT

REMOVAL

6- Clamp the starter in a vise, as shown. Tighten the vise just **SLIGHTLY** onto the cup bushing. Remove the center bolt.

7- Remove the pinion spring and gear from the pulley.

8- **EXERCISE CARE** during this next procedure. Slide a putty knife or other similar flat tool in between the bottom side of the pulley and top side of the spring. Using the tool, work the pulley off the

NUT   THRU-BOLT

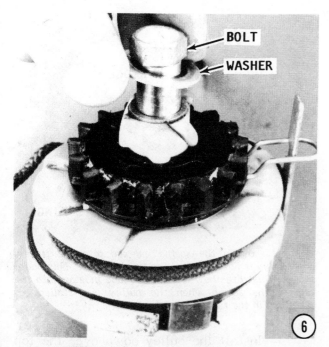

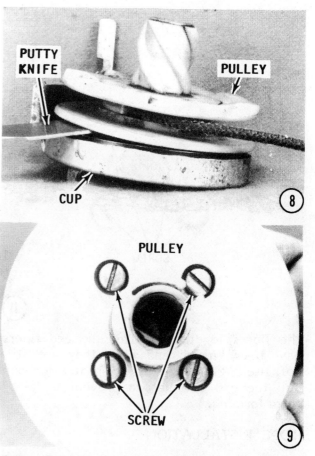

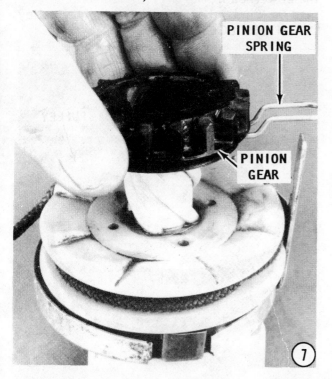

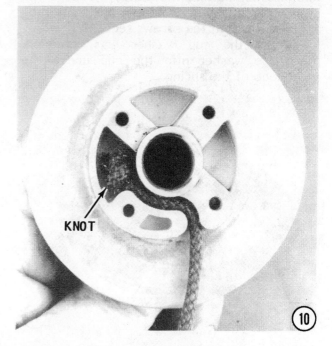

spring **WITHOUT** allowing the spring to escape from the cup.

9- Remove the four screws from the back side of the pulley, and then separate the pulley. Remove the rope.

INSTALLATION

Rope Purchase Instructions

The length and diameter of the starter rope required will vary depending on the horsepower size. of the model being serviced. Therefore, check the Hand Starter Rope Specifications in the Appendix, and then purchase a quality nylon piece of the proper length and diameter size. Only with the proper rope, will you be assured of efficient operation following installation.

Each end of the nylon rope should be "fused" by burning them slightly with a very small flame (a match flame will do) to melt

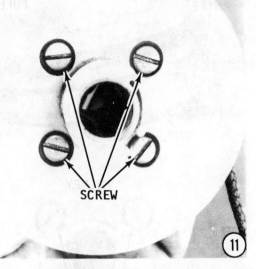

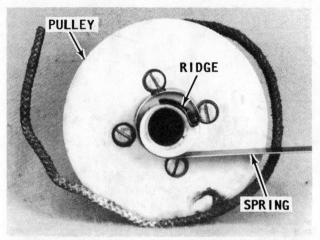

View showing the spring indexed into the pulley. Installation is not performed in this manner. This photo merely illustrates how the pulley ridge indexes into the loop in the spring when the pulley is installed, as described in the text.

the fibers together. After the end fibers have been "fused" and while they are still hot, use a piece of cloth as protection and pull the end out flat to prevent a "glob" from forming.

ROPE INSTALLATION

10- Tie a figure **8** knot as close to the end of the rope as possible, as shown. After the knot has been tied, feed the knot into the recess of the pulley.

11- Position the other half of the pulley over the rope. Rotate the cap slightly to align the four holes in the cap with the holes in the other half of the pulley. Secure the pulley together with the four retaining screws. Tighten the screws securely.

12- If the cup washer was removed, place the washer into the cup under the inner loop of the spring.

13- Install the pulley down over the top of the spring. Check to be sure the end of the spring is engaged in the pulley slot where the rope is installed. **DO NOT** lubricate the pulley or the pinion gear.

14- Install the pinion gear and the pinion spring onto the pulley.

15- Lubricate the mounting screw and washer with OMC outboard oil, and then install the screw through the pulley and cup.

16- Thread a 3/8" x 16 nut onto the mounting bolt to hold the pulley and spring in the cup until the starter is installed on the powerhead.

17- Wind the rope onto the pulley **COUNTERCLOCKWISE.**

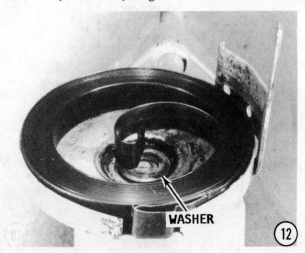

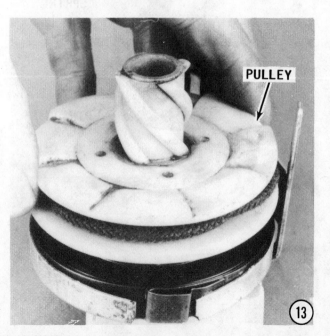

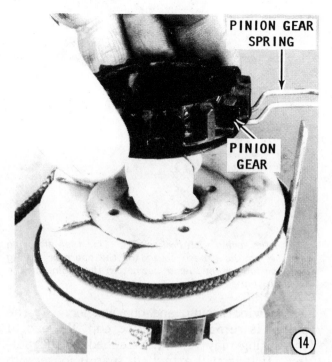

PINION GEAR SPRING

PINION GEAR

NUT

BOLT

ROPE

SPECIAL NOTE

If the only work to be performed is replacement of the starter rope, proceed directly to Step 8, under Starter Installation, to install the starter onto the powerhead.

DISASSEMBLING

18- After the starter has been removed as outlined in Steps 1 thru 4 of this section,

thread a 3/8" x 16 nut onto the center thrubolt. Clamp the starter in a vise with the mounting screw secured between the vise jaws.

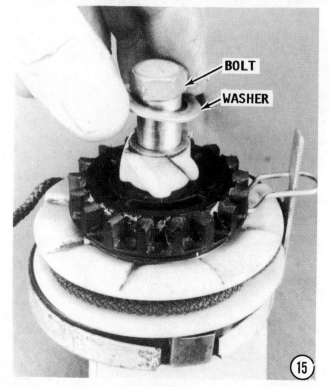

BOLT

WASHER

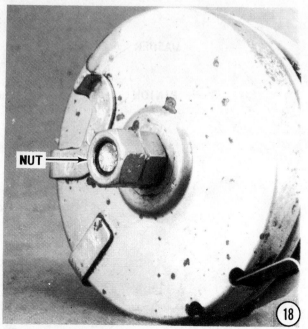

NUT

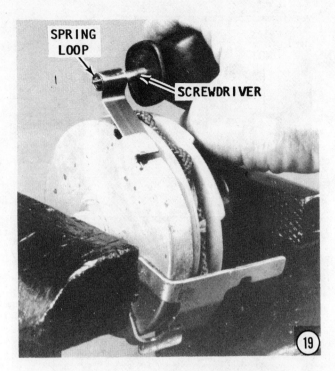

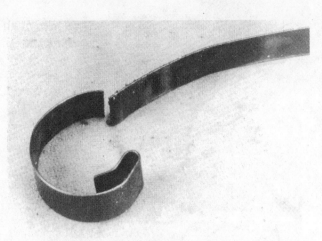

Starter spring with broken end. This type of spring damage may have been caused by the operator pulling the rope too far (one hand was not on the engine) or the rope may have broken.

19- Use a screwdriver to engage the spring loop, and then pull the spring out of the cup until it is completely unwound.

20- Release the starter from the vise. Remove the nut, mounting bolt, and washer. Remove the pinion gear, pinion spring, pulley, rewind spring, and cup washer. As the pulley is removed from the cup, the end of the spring may remain attached to the pulley. If it is still attached, snap it loose with a screwdriver.

ROPE REPLACEMENT

If the rope is to be replaced during the disassembling work, perform Steps 4 thru 16, at the begining of this section.

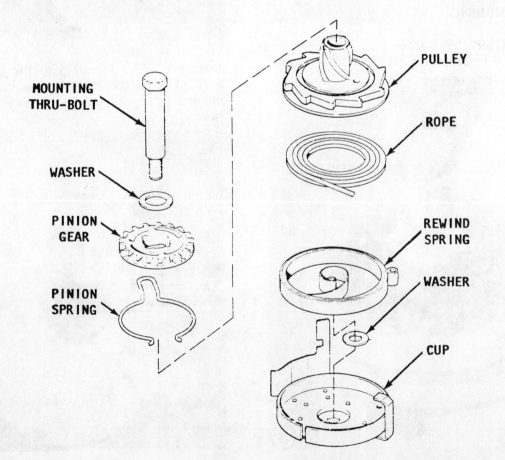

Broken outside end of a spring. This failure was most likely the result of metal fatigue -- age, and cannot be prevented.

CLEANING AND INSPECTING

Wash all parts except the rope in solvent and then blow them dry with compressed air.

Remove any trace of corrosion and wipe all metal parts with an oil dampened cloth.

Inspect the starter spring end loops. Replace the spring if it is weak, corroded or cracked.

Inspect the rope. Replace the rope if it appears to be weak or frayed. If the rope is frayed, check the hole through which the rope passes for rough edges or burrs. Remove the rough edges or burrs with a file, and polish the surface until it is smooth.

Inspect the pinion gear and pulley for wear, chipped or broken teeth. Inspect the

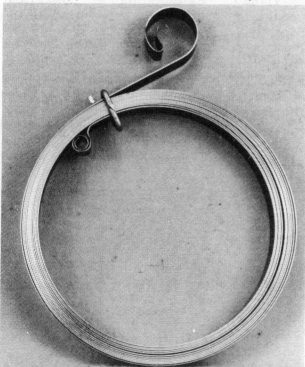

A new spring as it appears when purchased from the marine dealer. Notice the loop on the outside. Prior to installation, the spring must be rewound with the large loop on the inside.

A distorted starter cup no longer fit for service. This type damage was probably caused because the rope broke and the spring rewound with such incredible speed the cup was damaged. The cup material is not capable of withstanding such force from within.

cup for corrosion or damage, such as being warped out of shape.

SPECIAL GOOD WORDS

If the rope is to be replaced during the disassembling work, perform Steps 4 thru 16, at the beginning of this section.

STARTER ASSEMBLING

The following procedures pickup the work after a new rope has been installed according to Steps 4 thru 16 of this section.

1- Coat the inside surface of the cup with OMC Type A Lubricant. Position the rewind spring into the cup, as shown. Place the cup washer into the cup.

2- Install the pulley with the spring loop engaging the pulley, as shown. Check to be sure the spring feeds out of the cup slot.

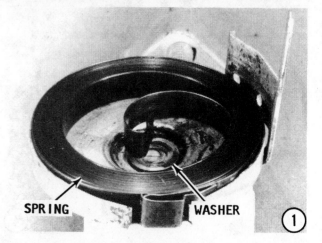

SPRING   WASHER

①

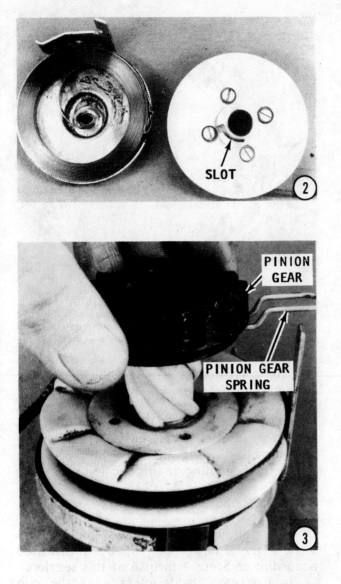

SLOT

②

PINION GEAR

PINION GEAR SPRING

③

BOLT

WASHER

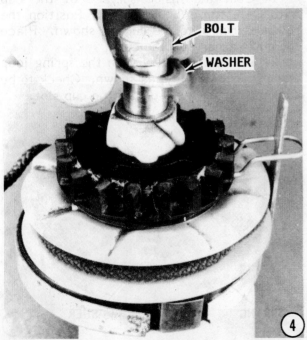

④

3- Install the pinion gear onto the pulley.

4- Coat the threads of the mounting screws with OMC Outboard Oil. Insert the center thru-bolt through the pulley and cup.

5- Thread a 3/8" x 16 nut onto the bolt to hold the parts together.

6- Hold the cup and at the same time wind the spring into the cup by rotating the pulley **COUNTERCLOCKWISE** as viewed from the top of the pulley. As soon as excessive resistance is felt during the winding, feed the spring into the cup through the slot to relieve spring tension. Continue to wind and feed the spring into the cup until the loop on the end of the spring is drawn up tight against the side of the cup.

7- Wind the rope **COUNTERCLOCKWISE** around the pulley. Install the pinion gear spring. Hold the spring in place with a rubberband or piece of string.

STARTER INSTALLATION

8- Hold the starter pulley and cup together and position the assembly in place on the powerhead, as shown. Thread the mounting screw into the manifold and tighten the screw securely.

Thread the rope through the front cowling and pull it all the way out. With the rope fully extended, the starter spring end must be free to extend a minimum of 1/2" from the cup. If the spring is not free to extend 1/2" from the cup, allow the starter rope to fully rewind onto the pulley. After the rope has rewound onto the pulley, release one full turn of the rope from the pulley and make the test again. Repeat the procedure until the spring is free to extend

NUT

BOLT

⑤

a minimum of 1/2" from the cup when the rope is fully extended. With the rope still through the front of the cowling tie a knot in the rope. Allow the rope to rewind until the knot is tight against the cowling.

Install the rubber bumper, handle and anchor onto the rope. Secure the rope by pressing the anchor into the handle. Remove the knot from the rope and allow the rope to rewind onto the pulley. When the rope is fully rewound, the handle should be up tight against the cowling.

STARTER ADJUSTMENT

9- Place the shift lever in the **NEUTRAL** position, and the throttle in the **START** position. Check to see if the lockout pawl clears the highest point on the starter pulley by 0.050" to 0.110" (1.27 to 2.79 mm). This clearance is required to allow powerhead start with the shift mechanism in gear and the throttle in the start position. To

adjust, loosen the nut and rotate the adjusting screw inward or outward until the proper clearance is obtained. Tighten the nut securely to hold the adjustment. To check, the hand starter should crank the powerhead with the shift mechanism in either **FORWARD** or **REVERSE** gear and with the throttle in the **START** position. Move the throttle to the **FAST** position. The starter lockout pawl should prevent the starter from cranking the powerhead.

SPECIAL WORDS

To restart a powerhead from the throttle **FAST** position, the throttle must be moved to the **SLOW** position and then advanced to the **START** position in order to remove backlash from the lockout linkage.

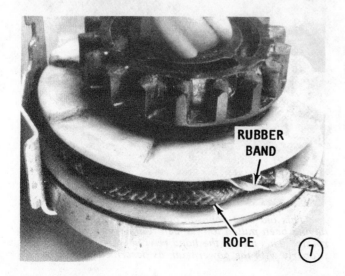

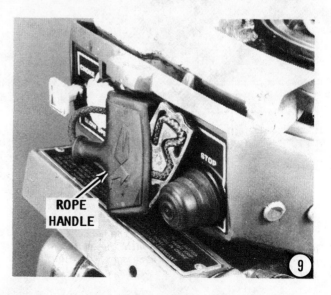

9-5 "D" HAND REWIND STARTER
MOUNTED ATOP FLYWHEEL
WITH TWO NYLON PAWLS
COLT, AND MODELS 3 & 4 — 1990 & ON
MODELS 5 THRU 8 — 1991 & ON
MODELS 20 THRU 30 — 1990 & ON

WARNING

As with other types of hand starters, the rewind spring is a potential hazard. The spring is under tremendous tension when it is wound -- a real tiger in a cage. If the spring should accidentally be released, severe personal injury could result from being struck by the spring with force. Therefore, the service instructions **MUST**, and we say again **MUST**, be followed closely to prevent release of the spring at the wrong time. Such action would be a **BAD SCENE**, a very **BAD SCENE**, because serious personal injury could result.

The starter rope should **NEVER** be released from the extended position. Such action would allow the spring to wind with incredible speed, resulting in serious damage to the starter mechanism.

Any time the rope is broken, the starter spring will rewind with incredible speed. Such action will cause the spring to rewind past its normal travel and the end of the spring will be bent back out of shape. Therefore, if the rope has been broken, the starter must be completely disassembled and the spring repaired or replaced.

Hand Starter Timing

Surprising as it may sound, this starter, mounted on top of the powerhead over the flywheel, can actually be timed to the powerhead. This timing can best be described by using an example.

If two marks were made on the flywheel 180° apart, and matching marks made on the powerhead, then each time the powerhead was shut down, one set of marks on the flywheel would align very closely with one of the marks on the powerhead. What is actually happening, is the powerhead is stopping with either the top piston at TDC (top dead center) or the bottom piston at the TDC position.

Now, assume the powerhead has been operating at idle speed and then suddenly stops for any number of reasons. The problem is corrected and the powerhead is once again ready to be started. Two notches are manufactured into the inside diameter of the flywheel. A single dog on the pulley engages with one of these dogs when the rope is pulled. Now, if it is necessary to pull an excessive amount of rope before the

Closeup view of the pulley and the housing with the arrow on the housing aligned with the marks on the pulley. This alignment is necessary to "time" the starter with the engine.

This old time Johnson outboard illustrates the rope having been pulled too far before the flywheel begins to rotate. This means the hand rewind starter is not timed properly with the powerhead, as described in the text.

dog is able to engage the flywheel, full starting rotation power would not be available in the rope.

Therefore, the starter is timed to engage the starter with the flywheel after the rope has been pulled exactly the same amount each time. This distance is very short to allow as much rope pull as possible to rotate the crankshaft for fast start. This "timing" will assure an adequate amount available for starting **AND** that one of the pistons will return to TDC when the pull is completed, if the powerhead fails to start. If an excessive amount of pull is necessary before the flywheel begins to rotate, the starter was not assembled properly -- the arrow on the pulley was not aligned with the two marks on the starter housing when the spring is relaxed and the rope handle is retracted.

COLT ONLY
PRELIMINARY TASKS

WORDS OF CAUTION

DO NOT ATTEMPT to remove the large slotted head screw from atop the starter housing. Loosening this screw will cause the starter spring to disengage and forceably unwind within the confines of the starter housing.

ALL OTHER MODELS COVERED
IN THIS SECTION

Remove the screw retaining the throttle cable trunnion to the powerhead. Pull the throttle cable down to release it from the armature plate. Remove the one nut and two bolts securing the starter housing to the powerhead.

BOLT (3 PLCS)

①

To remove the throttle cable from the starter housing, remove the two screws on the small retaining bracket across the throttle cable and then twist the cable **COUNTERCLOCKWISE** to release the cable from the throttle knob.

STARTER REMOVAL

1- Remove the attaching bolts securing the three legs of the starter housing to the powerhead. On some smaller horsepower powerheads, the starter housing is attached to the fuel tank with screws. Remove the hand starter and lay it on the bench with the pulley facing toward you.

DISASSEMBLING

2- Pull the rope out enough to tie a knot in the rope. Tie a knot, and then allow the rope to rewind to the knot. Work the rope anchor out of the rubber covered handle, then remove the rope from the anchor. Remove the handle from the rope. Untie the knot in the rope, and then hold the disc pulley, but permit it to turn and thus allow the rope to wind back onto the pulley **SLOWLY**. Continue to allow the spring in the pulley to unwind **SLOWLY** until all tension has been released.

SPECIAL WORDS

This series of photographs depicts a hand rewind starter with a single pawl. If servicing a starter with two pawls, merely repeat Steps 3 and 4 to remove the other

ANCHOR   HANDLE

②

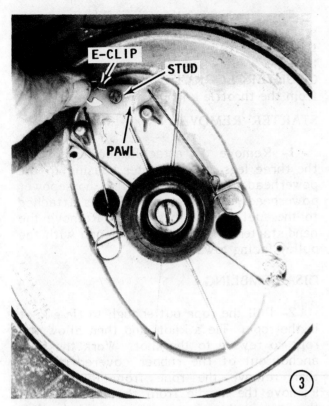

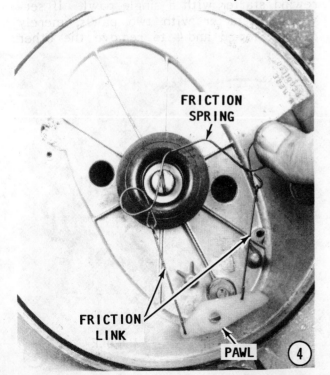

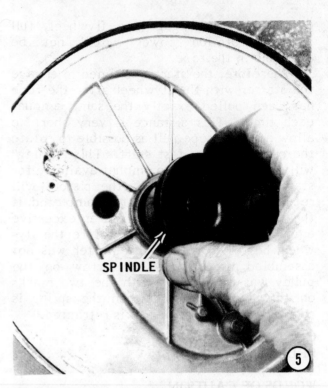

pawl. All other procedural steps are valid for both starters.

3- Remove the E-clip from the nylon pawl. Lift the pawl from the stud.

4- Remove the friction spring and friction link from the pawl.

5- Remove the bolt, lockwasher, and washer from the center of the pulley spin-

dle. Lift the spindle out of the pulley, and at the same time hold the pulley firmly together with the housing.

SPECIAL WORDS FOR TWO PAWL STARTERS

A starter spring shield is installed between the pulley and the spring on these units.

6- Lift the pulley straight up and at the same time work the spring free of the pulley. The spring has a small loop hooked into the pulley. An alternate and safe method is to hold the pulley and the housing together tightly and turn the complete assembly with the legs extending downward in

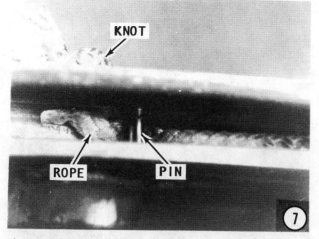

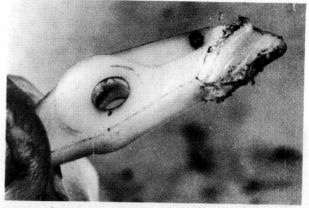

Damaged pawl unfit for further service.

the normal manner. Now, lower the complete assembly to the floor. When the legs make contact with the floor, release your grip. The pulley will fall and the spring will be released from the housing, but the three legs will contain the spring and prevent it from traveling across the room. If the spring was not released from the housing, the only safe method is to again make contact with the three legs on the floor and jar the spring free.

7- Unwind the rope out of the pulley groove. Notice the pin next to the knot in the rope and how the rope feeds **BEHIND** the pin. Pull the knot and the rope out far enough to untie the knot, and then pull the rope free of the pulley.

CLEANING AND INSPECTING

If the rope was broken and the spring is bent backward, as shown in the accompany-

ing illustration, it is a simple matter to bend the spring end back to its normal position. The next illustration clearly shows a spring end properly positioned in the housing.

Wash all parts except the rope in solvent and then blow them dry with compressed air.

Remove any trace of corrosion and wipe all metal parts with an oil dampened cloth.

Inspect the rope. Replace the rope if it appears to be weak or frayed. If the rope is frayed, check the hole through which the rope passes for rough edges or burrs. Remove the rough edges or burrs with a file, and polish the surface until it is smooth.

The rope on this unit broke, causing the spring to rewind with incredible speed. The end of the spring was bent back in the wrong direction.

Knot tied in the end of the starter rope to prevent the rope from being pulled from the starter.

Inspect the starter spring end loops. Replace the spring if it is weak, corroded or cracked. Check the spring pin located at the back side of the pulley to be sure it is straight and solid.

Check the inside surface of the housing and remove any burrs.

Check the condition of the pawl spring to be sure it is not stretched out of shape. The end of the spring should be bent back toward the coil of the spring. Inspect the pawl for wear and that the edges are not rounded.

Check the friction spring and link to be sure they are not distorted.

Inspect the spindle. The spindle must be straight and tight.

SPECIAL WORDS

This series of photographs depicts a hand starter with a single pawl. If servicing a unit with two pawls, merely repeat Steps 5 and 6 to install the other pawl. All other procedural steps are valid for both starters.

STARTER ASSEMBLING

ROPE INSTALLATION

Rope Purchase Instructions

The length and diameter of the starter rope required will vary depending on the horsepower size of the model being serviced. Therefore, check the Hand Starter Rope Specifications in the Appendix, and then purchase a quality nylon piece of the proper length and diameter size. Only with the proper rope, will you be assured of efficient operation following installation.

Each end of the nylon rope should be "fused" by burning them slightly with a very small flame (a match flame will do) to melt

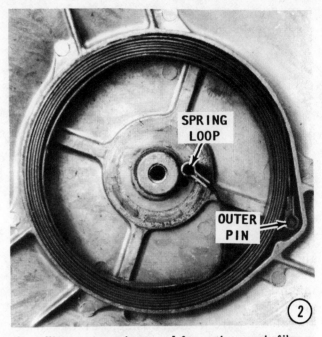

the fibers together. After the end fibers have been "fused" and while they are still hot, use a piece of cloth as protection and pull the end out flat to prevent a "glob" from forming.

1- Tie a figure 8 knot in the end of the rope. Insert one end of the new rope through the hole in the pulley and housing and on the back side of the pin, as shown. Continue to wrap the remainder of the rope COUNTERCLOCKWISE around the pulley.

SAFETY WORD

Wear a good pair of gloves while installing the spring. The spring will develop tension and the edges of the spring steel are

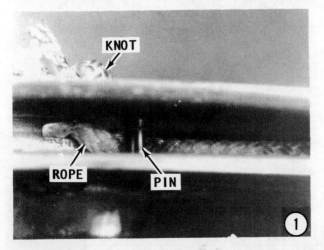

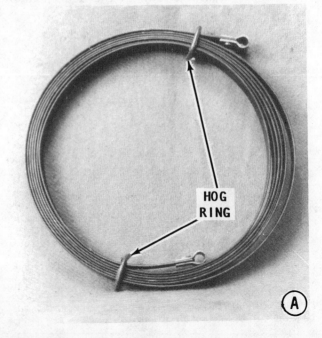

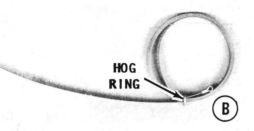

HOG RING

sharp. The gloves will prevent cuts on your hands and fingers.

2- Slide the spring onto the outer pin and then start the spring from the outside edge of the housing and insert it into the housing **COUNTERCLOCKWISE,** as shown in the accompanying illustration. Notice the small hump in the housing. This hump prevents the spring from being wound in the wrong direction. Work the first turn into the housing, and then hold the spring down with one hand and continue to wind the spring into the housing. Patience and time are required to work the spring completely into the housing. After the last portion is in place, bend the end of the spring towards the center of the housing. This position will allow the pulley pin to align with the loop in the end of the spring, when the pulley is installed.

Alternate Method

An alternate method of installing a **NEW** spring into the housing with less risk of personal injury is presented with accompanying illustrations.

HOG RING

C

A- Remove **ONLY** the hog ring next to the end of the outside wrap of the spring.

B- Pull on the outside end of the spring. As the spring is pulled, the inside diameter will get smaller and smaller, as shown. When the diameter is a bit smaller than the inside diameter of the starter housing, wrap the entire free end of the spring around the coiled portion.

C- CAREFULLY lower the coiled spring into the starter housing with the loop on the free end of the spring indexed over the peg in the housing and the spring feeding **COUNTERCLOCKWISE,** as shown. Remove the second hog ring **WITHOUT** allowing the spring to escape from the housing. An easy and safe method is to cut the hog ring with a pair of "dikes". Bend the inside end of the spring toward the center of the housing to permit the pin in the pulley to index into the loop.

SPECIAL WORDS FOR TWO PAWL STARTERS

A starter spring shield is installed between the pulley and the spring on these units. Place the shield against the underside of the pulley aligning the holes.

3- Lower the pulley down over the top of the spring with the pulley pin indexing into the loop in the end of the spring. In the accompanying illustration, notice the call-out for the boss on the backside of the pulley. The pin is located directly under the boss. The boss can, therefore, be a guide during pulley installation.

4- If servicing a model 20 or 30hp: Install the bushing and shim material onto the pulley followed by the square friction plate. Then install the spindle, friction ring

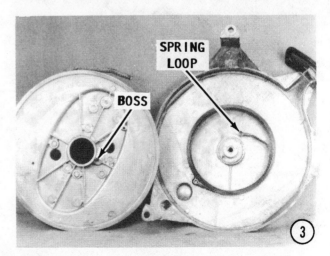

BOSS SPRING LOOP

3

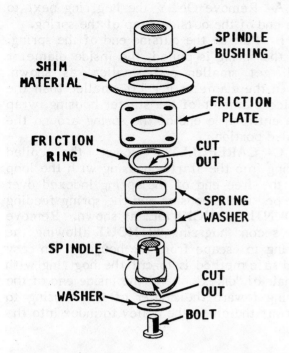

SPINDLE BUSHING

SHIM MATERIAL

FRICTION PLATE

FRICTION RING

CUT OUT

SPRING WASHER

SPINDLE

CUT OUT

WASHER

BOLT

Arrangement of parts comprising the spindle for Models 20hp and 30hp.

and two wavy washers, indexing the friction ring with the spindle, flat-spot-to-flat-spot, as shown in the accompanying illustration. Install the washer and bolt and tighten the bolt to a torque value of 10 ft lb (13.6Nm).

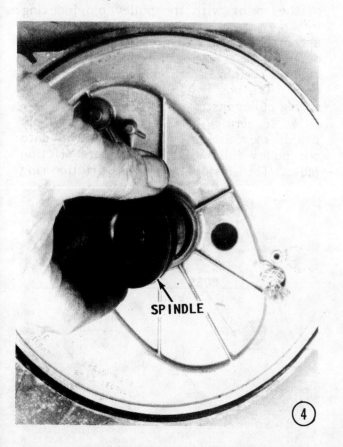

SPINDLE

4

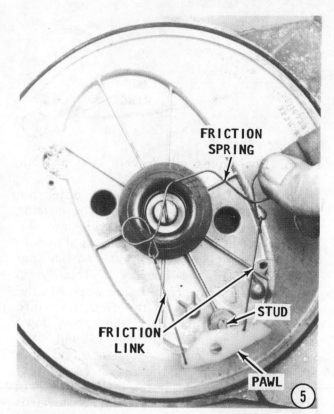

FRICTION SPRING

FRICTION LINK

STUD

PAWL

5

All other models: Lower the spindle assembly down through pulley. Place the washer and lockwasher inside the spindle housing and then install the bolt through the washer into the housing. Tighten the bolt securely. Check to be sure the pulley will rotate smoothly and does not bind on the

PAWL

E-CLIP

6

spindle.   Rotate   the   pulley   slightly **COUNTERCLOCKWISE** and then release it to be sure there is proper engagement with the spring and the pulley has good spring tension.

5- Install the friction spring and link and the nylon pawl onto the starter hub. The friction spring fits into a groove in the spindle. Pull just a little on the pawl and set it over the stud on the flywheel pulley.

6- Snap the E-clip over the top of the pawl to secure it in place.

7- Rotate the pulley three complete revolutions, and then work the rope out through the hole in the pulley and the housing. Pull on the rope until a couple of feet are exposed. Tie a knot in the rope and allow the rope to rewind until the knot is tight against the housing. Work the end of the rope into the handle anchor. Secure the rope in place by pushing the anchor into the rubber handle.

Pull on the rope enough to untie the knot, and then allow the rope to slowly recoil into the pulley.

8- Lay the starter on its back with the pulley facing toward you. Notice the imprint on the pulley: **ARROW HERE ROPE-RECOILED.** Also notice the arrow on the housing. On Johnson powerheads, when the rope is fully coiled (the starter pulley completely wound) the arrow must fall between the marks on the pulley. Further up on the pulley you will notice the letter **E** (for Evinrude). On the Evinrude powerheads, the arrow must fall between the two marks on

the pulley. If the arrow is not properly aligned the starter rope is not the proper length. It is either too long or too short. The arrow must align properly for the starter to be timed with the powerhead. See the beginning of this section.

9- With the starter still on its back, pull the rope with quick movements, and at the same time check the pawl to be sure it moves toward the center of the pulley. Release the rope slowly and check to be sure the pawl returns to its original position.

STARTER INSTALLATION

Position the starter over the flywheel with the three legs aligned over the holes in the powerhead for the retaining bolts. Install the retaining bolts and tighten them to a torque value of 6.8 ft lb (8Nm).

Throttle Cable Installation
Models Equipped with Remote Control
or Tiller Handle

Insert the throttle cable through the cable guide in the housing. Rotate the cable **CLOCKWISE** to engage the cable with the throttle knob. Place the small retaining bracket across the cable and secure it to the housing with two screws.

To adjust the throttle cable, back off the toothed idle stop wheel away from the throttle knob until the throttle cable can be snapped back into place on the armature plate. Push the throttle knob toward the starter housing while pushing the armature plate **CLOCKWISE** as far as possible. Adjust the position of the throttle cable trunnion to allow the cable to slide into the side retaining bracket and then tighten the bracket retaining screw.

9-6 "E" HAND REWIND STARTER
WITH LARGE PAWL PLATE
MODEL 40 THRU 55 — 1990 AND ON

This design hand rewind starter is mounted atop the flywheel with three mounting legs attached to the powerhead. Very few are produced because of the strength required to crank the powerhead.

STARTER REMOVAL

1- Remove the screw retaining the lockout cable clamp to the housing, and then remove the lockout slide. Remove the attaching bolts securing the three starter housing legs to the powerhead. Remove the two screws securing the starter handle bracket to the powerhead. Remove the hand starter and lay it on the bench upside down with the pulley and legs facing up.

DISASSEMBLING

2- Pull the rope out enough to tie a knot in the rope. Tie a knot, and then allow the rope to rewind to the knot. Work the rope anchor out of the rubber covered handle, and then remove the rope from the anchor. Remove the handle from the rope. Untie the knot in the rope, but hold and prevent the disc pulley from rotating. Now, slack the hold on the pulley, and permit it to turn thus winding the rope back onto the pulley **SLOWLY**. Continue to allow the spring in the pulley to unwind **SLOWLY** until all tension has been released.

3- Remove the bolt retaining the lockout lever, the lockout lever spring, and the washer to the housing.

Place one hand under the housing to prevent all the large pawl plate and other small parts from flying loose. Keep one finger over the center bolt.

Remove the large center nut and washer from atop the starter housing. Carefully turn the starter housing over without disturbing any parts.

WARNING

THE REWIND SPRING IS A POTENTIAL HAZARD. The spring is under tremendous tension when it is wound — a real **"tiger"** in a cage! If the spring should accidentally be released, severe personal injury could result from being struck by the spring with force. Therefore, the following steps **MUST** be performed with care to prevent personal injury to self and others in the area.

4- Lift off the center bolt and washer, and then lift off the large pawl plate and the plate return spring from the side of the starter housing.

Remove the center spring. Use a screwdriver to pry out the pulley lockring from under the spring. Lift off the friction plate and the spring washer beneath the plate.

5- Remove the starter pawl and spring washer from the pawl pivot post.

6- Lift the pulley straight up and at the same time work the spring free of the pulley. The spring has a small loop hooked into the pulley. An alternate and safe method is to hold the pulley and the housing together tightly and turn the complete assembly over with the legs extending downward in the normal manner. Now, lower the complete assembly to the floor. When the legs make contact with the floor, release your grip. The pulley will fall and the spring will be released from the housing almost instantly and with considerable **FORCE**. However, the three legs will contain the spring and prevent it from lashing out causing possible injury to self or others in the area. If the spring was not released from the housing as just described, the only safe method is to again jar the three legs on the floor and dislodge the spring.

7- Unwind the rope out of the pulley groove.

CLEANING AND INSPECTING

If the rope was broken bending the spring backward, it is a simple matter to bend the spring end back to its normal position.

Wash all parts except the rope in solvent and then blow them dry with compressed air. Remove any trace of corrosion and wipe all metal parts with an oil dampened cloth.

Inspect the rope. Replace the rope if it appears to be weak or frayed. If the rope is frayed, check the hole through which the rope passes for rough edges or burrs. Remove the rough edges or burrs with a file, and polish the surface until it is smooth.

Inspect the starter spring end loops. Replace the spring if it is weak, corroded or cracked.

Check the inside surface of the housing and remove any burrs.

Check the condition of the pawl spring to be sure it is not stretched out of shape.

Inspect the pawl for wear and that the edges are not rounded.

Check the friction spring to be sure they are not distorted.

Rope Purchase Instructions

Purchase a quality piece of nylon rope 96 1/2" (245cm) long. Only with the proper rope, can efficient operation be ensured following installation.

Each end of the nylon rope should be "fused" by burning the last half inch slightly with a very small flame (a match flame will do) to melt the fibers together. After the end fibers have been "fused" and while they are still hot, use a piece of cloth as protection and pull the end out flat to prevent a "glob" from forming.

Tie a figure **8** knot in the end of the rope and keep the rope close at hand for installation in Step 4.

STARTER ASSEMBLING

SAFETY WORDS

Wear a good pair of gloves while installing the spring. The spring will develop tension and the edges of the spring steel are sharp. The gloves will prevent cuts on hands and fingers.

It is **STRONGLY** recommended a pair of safety goggles or a face shield be worn while the spring is being installed. As the work progresses a "tiger" is being forced

into a cage. If the spring is accidentally released, it will lash out with tremendous ferocity and very likely could cause personal injury to the installer or other persons nearby.

1- Slide the spring onto the outer pin and then start the spring from the outside edge of the housing and insert it into the housing **COUNTERCLOCKWISE**. Work the first turn into the housing, and then hold the spring down with one hand and continue to wind the spring into the housing. Patience and time are required to work the spring completely into the housing. After the last portion is in place, bend the end of the spring towards the center of the housing. This position will allow the long slot in the center hub to align with the loop in the end of the spring, when the pulley is installed.

2- Lower the pulley down over the top of the spring with the long slot in the center hub indexing into the loop in the end of the spring.

Install the friction plate spring washer and friction plate over the pulley hub. Position the lockring over the plate and tap around the circumference of the ring until it is well seated into the groove of the hub. Place the starter housing spring over the lockring.

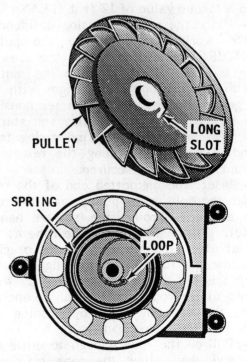

The long slot at the pulley center MUST engage the inner hook of the rewind spring for the starter to function correctly.

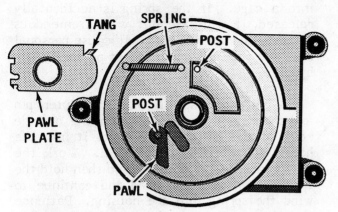

The pawl plate return spring is positioned and attached between a post on the pulley and a tang on the pawl plate.

Apply a coat of OMC Triple-Guard grease onto the pawl pivot post before installing the spring washer and pawl onto the post.

Hook one end of the pawl plate return spring over the pulley post. Hook the other end of the spring onto the pawl plate tang. Slide the pawl plate into position to align with the pulley hub. Place the washer over the pawl plate and install the center bolt.

Place one hand over the entire pawl plate and keep one finger over the center bolt to keep all the parts together while the assembly is inverted.

3- Place the washer over the end of the installed center bolt. Install and tighten the nut to a torque value of 12 ft lb (13.6Nm).

4- Place the starter housing upsidedown on the workbench. Rotate the pulley **COUNTERCLOCKWISE** until the rewind spring is under tension. Stop the motion when the pulley rope notch aligns with the rope guide mounted on the starter housing. Keep a tight hold on the pulley and starter housing to prevent the rewind spring from unwinding while performing the remaining work until the rope is secured.

5- Insert the unknotted end of the rope through the hole in the pulley, past the rope guide and out through the starter handle bracket. Tie a slip knot in the rope at the point at which it emerges from the bracket and allow the rope to rewind slightly until the knot is tight against the housing. Work the end of the rope into the handle anchor. Secure the rope in place by pushing the anchor into the rubber handle.

6- Pull on the rope enough to untie the knot, and then allow the rope to slowly recoil into the pulley.

7- With the starter still on its back, pull

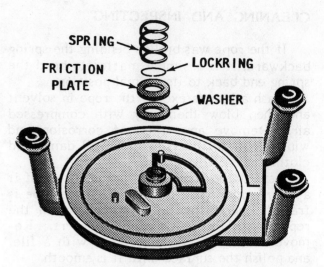

The friction plate spring washer and friction plate are secured with the pulley lockring. The starter housing spring rests on top of the lockring.

the rope with quick movements, and at the same time check the pawl to be sure it extends outward when the rope is pulled out.

Release the rope slowly and check to be sure the pawl retracts and returns to its original position.

8- Install the starter lockout lever, spring, and washer onto the housing. Tighten the bolt securely.

STARTER INSTALLATION

9- Position the starter over the flywheel with the three legs aligned over the holes in the powerhead for the retaining bolts. Install the retaining bolts and tighten them to a torque value of 6.8 ft lb (8Nm). Install and tighten the two starter handle bracket screws to the same torque value.

Position the lockout slide in the channel of the housing. Lightly secure the lockout cable to the starter housing using the cable clamp and bolt. Check to be sure the lower unit is in **NEUTRAL** gear and adjust the lockout cable to align the slide with the lockout lever. Tighten the cable clamp securely to hold this adjustment.

Shift the lower unit into **FORWARD** and then **REVERSE** gear and make a check of the starter lockout system. The starter **MUST** be **LOCKED,** and unable to rotate when the lower unit is in any gear other than **NEUTRAL.**

If the starter fails this test, loosen the cable clamp and adjust the position of the slide until no motion of the starter is possible when the lower unit is in **FORWARD** or **REVERSE** gear.

9-7 "F" HAND REWIND STARTER
MODEL 2.3 AND 3.3 — 1990 AND ON

This hand rewind starter for these very small powerheads is a simple coiled spring unit mounted on top of the powerhead just ahead of the fuel tank.

REMOVAL AND DISASSEMBLING

1- Disconnect the high tension lead from the spark plug. Remove the cowling from both sides of the powerhead.

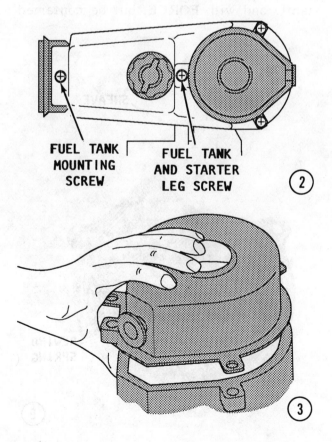

FUEL TANK MOUNTING SCREW

FUEL TANK AND STARTER LEG SCREW

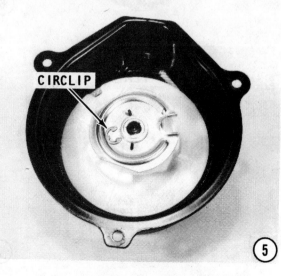

2- Remove the two screws securing the fuel tank to the powerhead. The forward screw also secures the aft leg of the rewind starter. Remove the other to screws securing the other two legs of the starter.

3- Lift the fuel tank slightly and remove the rewind starter assembly from the powerhead.

4- Turn the starter housing over with the sheave facing up. Pull on the handle to gain some slack in the rope; hold the sheave from rewinding; and then remove the handle. Two types of handle arrangement are used, as shown in the illustration.

After the handle is removed, ease the grip on the sheave and allow the sheave to completely rewind with the rope inside the sheave.

5- Snap the Circlip out of the groove in the center shaft.

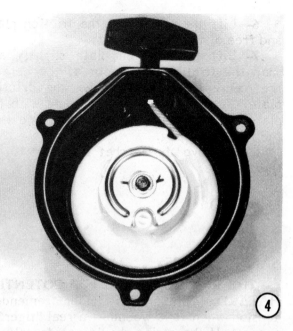

CIRCLIP

6- Lift the spacer off the friction plate and free of the center shaft.

7- Lift the friction plate slightly and snap the friction spring out of the hole in the center shaft. Remove the friction plate. The return spring will come with the plate. Remove the spring cover, and then the friction spring.

8- Remove the ratchet from the top of the sheave.

SAFETY WORDS

THE REWIND SPRING IS A POTENTIAL HAZARD. The spring is under tremendous tension when it is wound -- a real **"tiger"** in a cage! If the spring should accidentally be released, severe personal injury could result from being struck by the spring with force. Therefore, the following two steps **MUST** be performed with care to prevent personal injury to self and others in the area.

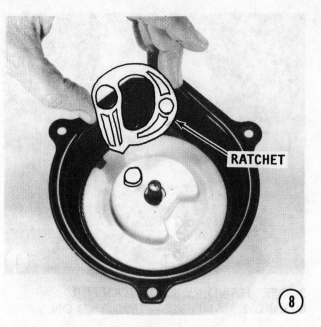

RATCHET

⑧

9- Very **CAREFULLY** "rock" the sheave and at the same time lift the sheave about 1/2" (13mm). The spring will disengage from the sheave and remain in the housing, as shown.

10- Now, **SLOWLY** turn the housing over and **GENTLY** place it on the floor with the spring facing the floor. Tap the top of the housing with a mallet and the spring will fall free of the housing and unwind almost instantly and with **FORCE**, but be contained

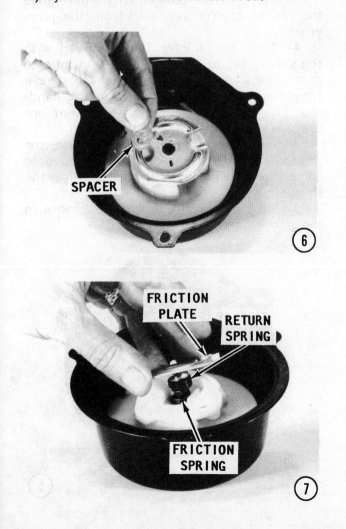

SPACER

⑥

FRICTION PLATE

RETURN SPRING

FRICTION SPRING

⑦

SHEAVE

REWIND SPRING

⑨

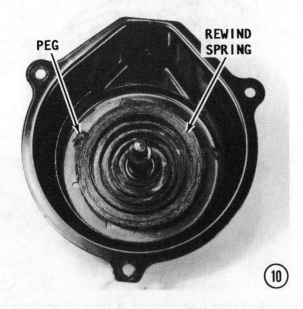

PEG   REWIND SPRING

(10)

Remove any trace of corrosion and wipe all metal parts with an oil dampened cloth.

Inspect the rope. Replace the rope if it appears to be weak or frayed. If the rope is frayed, check the holes through which the rope passes for rough edges or burrs. Remove the rough edges or burrs with a file and polish the surface until it is smooth.

Inspect the starter spring end hooks. Replace the spring if it is weak, corroded or cracked. Inspect the tab on the spring retainer plate. This tab is inserted into the inner loop of the spring. Therefore, be sure it is straight and solid. Inspect the inside surface of the sheave rewind recess for grooves or roughness. Grooves may cause erratic rewinding of the starter rope.

within the housing. Tilt the container with the opening **AWAY** from you. The spring will be released from the housing and unwind rapidly. Turn the housing over and unhook the end of the spring from the peg in the housing.

11- Untie the knot in the end of the old starter rope and pull the rope free of the sheave.

CLEANING AND INSPECTING

Wash all parts except the rope and the handle in solvent, and then blow them dry with compressed air.

ASSEMBLING AND INSTALLATION TYPE "F" REWIND STARTER

SAFETY WORDS

Wear a good pair of gloves while winding and installing the spring. The spring will develop tension and the edges of the spring steel are extremely sharp. The gloves will prevent cuts to the hands and fingers.

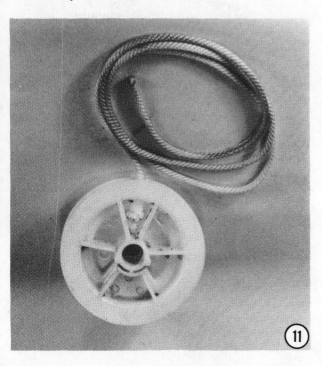

(11)

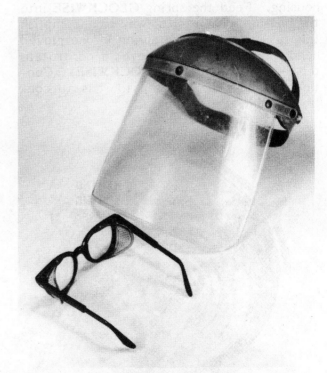

*Protect your eyes with a face mask or safety glasses while working with the rewind spring, especially a used one. The spring is a real **"tiger in a cage"**, almost 13' (4 m), of spring steel wound into less than 4" (about 10cm).*

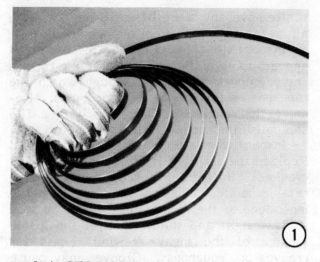

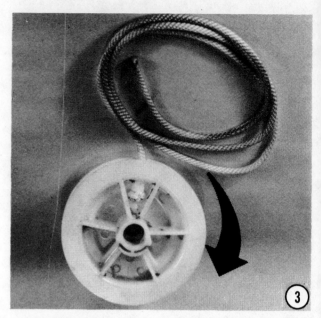

It is **STRONGLY** recommended a pair of safety goggles or a face shield be worn while the spring is being installed. As the work progresses a **"tiger"** is being forced into a cage, over 14' (4.3m) of spring steel wound into about 4" (10.2cm) circumference. If the spring is accidentally released, it will lash out with tremendous ferocity and very likely could cause personal injury to the installer or other persons nearby.

1- Hold the spring in a coil in one gloved hand, as shown.

2- With the other hand, hook the looped end of the spring over the peg in the housing. Feed the spring **CLOCKWISE** into the housing. The easiest way to accomplish this task is to hold the spring in one gloved hand and with the other gloved hand, rotate the housing **COUNTERCLOCKWISE.** Continue working the spring until it is all confined within the housing.

3- Feed one end of a new rope through the hole in the sheave, and then tie a figure "8" knot in the end of the rope. Pull on the rope until the knot is confined inside the sheave recess. Wind the entire pull rope **CLOCKWISE** around the sheave.

4- Lower the sheave down over the center shaft of the housing. As the sheave goes into the housing the hook on the lower face of the sheave **MUST** index into the loop end of the spring. In the illustration, the sheave is turned over to expose the hook to view. Once the hook is indexed into the spring, the sheave may be fully seated in the housing.

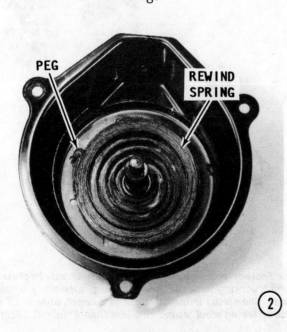

PEG

REWIND SPRING

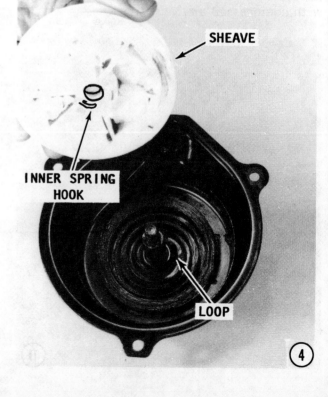

SHEAVE

INNER SPRING HOOK

LOOP

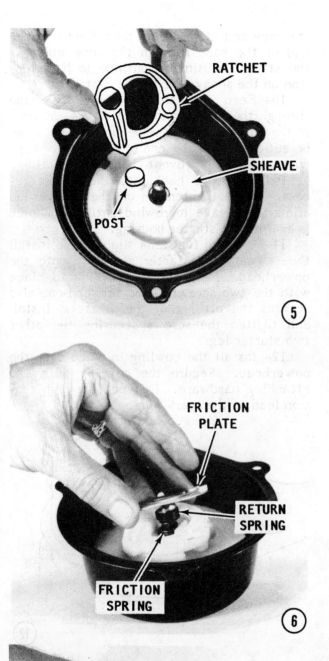

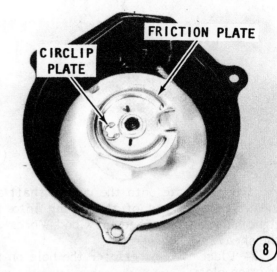

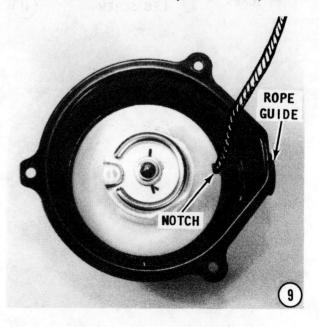

5- Position the ratchet in place on the sheave, with the flat side of the ratchet against the sheave and the round hole indexed over the post.

6- Slide the friction spring down the center shaft. This spring serves as a spacer and exerts an upward pressure on the friction plate. Install the spring cover on top of the spring.

GOOD WORDS

Two holes are located in the sheave on the same side of the center shaft. With the holes on the side of the shaft facing you, one end of the friction spring must index into the right hole.

Hook one end of the return spring into the slot of the friction plate. Now, lower

10

the friction plate onto the center shaft and index the free end of the spring into the right hole, as described in "Good Words", in the sheave.

7- Place the spacer over the hole on the friction plate.

8- Push down on the friction plate, and then snap the Circlip into the groove on the center shaft to secure the plate and associated parts in place.

9- Place the pull rope into the notch in the sheave. With the rope in the notch, rotate the sheave **THREE** complete turns **COUNTERCLOCKWISE.** Hold tension on

the rope and at the same time feed the free end of the rope through the rope guide in the starter housing. Continue to hold tension on the sheave for the next step.

10- Feed the free end of the rope through the handle and tie a figure "8" knot in the end. Excess rope beyond the knot can be cut off. Burn the end of a non-fiber rope with a match to prevent it from unravelling.

Pull the rope back into the handle recess. Relax the tension on the sheave and allow the sheave to rewind until the handle is against the starter housing.

11- Lift the fuel tank slightly and install the rewind hand starter assembly onto the powerhead. Secure the fuel tank in place with the two screws -- the forward one also secures the aft leg of the starter. Install and tighten the screws securing the other two starter legs.

12- Install the cowling in place over the powerhead. Secure the cowling with the attaching hardware. Connect the high tension lead to the spark plug.

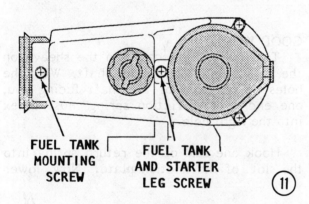

FUEL TANK
MOUNTING
SCREW

FUEL TANK
AND STARTER
LEG SCREW

11

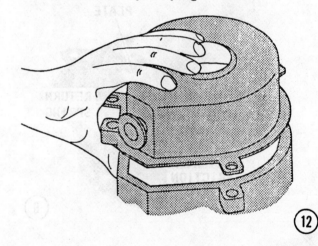

12

10
TRIM/TILT SYSTEM

10-1 SYSTEM DESCRIPTION

The power trim/tilt unit is a hydraulic/-mechanical unit mounted between the stern brackets on the larger outboard units covered in this manual. The system consists of a single cylinder, hydraulic pump, electric motor to drive the pump, and a manifold containing a hydraulic reservoir, and the necessary valves, check valves, and relief valves to make it all function properly.

The trim/tilt unit pivots on a thrust rod through the stern brackets. The working end of the cylinder piston is attached to the outboard swivel bracket. The unit is controlled by a remote set of switches -- one for the **UP** direction and the other for **DOWN**.

Cylinder Action

When the **UP** switch is activated, current is provided to the electric motor and the pump rotates in a **COUNTERCLOCKWISE** direction, relative to the motor end. The proper valves open and close; hydraulic pressure is applied to the bottom of the cylinder; the piston (ram) extends; and the outboard unit moves upward. The first 15° of movement is considered the trim range. If the **UP** button is held depressed, the outboard may be raised another 50° -- the tilt range. Once the full up position -- 65° -- is reached, a relief valve is mechanically opened to protect the system from excessive hydraulic pressure.

When the **DOWN** switch is activated, current is provided to the other circuit of the motor and the pump rotates in a **CLOCKWISE** direction, relative to the motor end. Again, the proper valves open and close; hydraulic pressure is directed to the top of the cylinder; the piston (ram) retracts; and the outboard is lowered.

If the outboard is to remain in the full tilt position for any length of time or if the unit is raised for trailering to and from the water, the weight of the outboard must be supported through a set of trailering locks. Once the locks are engaged, the piston must be retracted a little -- until the locks are fully seated in the stern brackets.

Overload Protection

The electric motor has a built-in thermal overload protection device. This device shuts off power to the motor in the event of an overload in the circuit. The device cools

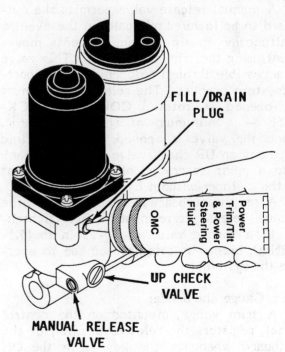

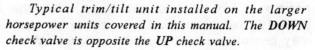

*Typical trim/tilt unit installed on the larger horsepower units covered in this manual. The **DOWN** check valve is opposite the **UP** check valve.*

in approximately one minute and the switch closes to allow current to flow once again to the electric motor.

Impact Relief

If the outboard should strike an underwater object, extreme pressure develops almost instantly above the cylinder. Under such a condition, a series of five impact valves and one of the **UP** relief valves open almost instantly to allow hydraulic fluid to pass through as the cylinder rod -- ram -- extends. The fluid is routed back through the system to the reservoir.

Following an accidental impact, the trim/tilt system must be operated in the **DOWN** direction to return the outboard to the "before impact" position.

Thermal Expansion Relief

As might be expected, if outside ambient temperatures increase significantly, the hydraulic fluid in the system will expand. Expansion will cause an increase in hydraulic pressure. When internal pressure reaches about 8000 psi (55,000 kPa), an expansion relief valve built into the manual release valve will vent fluid directly into the reservoir.

Manual Release Valve

A manual release valve permits the outboard to be lowered manually in the event a malfunction in the system prevents movement using the trim/tilt system. This valve is accessible through an opening in the portside stern bracket. The release valve must be opened -- rotated **COUNTERCLOCKWISE** -- a minimum of three full turns. Once the valve is opened, hydraulic fluid bypasses an **UP** check valve and flows back into a pump reservoir, allowing movement of the cylinder without hydraulic pressure.

After the outboard has been lowered, the manual release valve must be fully tightened to a torque value of 45-55 in lb (5.1-6.2Nm), before an attempt is made to operate the system.

Trim Gauge and Sender

A trim gauge, mounted on the control panel, registers the relative position of the outboard whenever the key is in the **ON** position. A sending unit is located on the port side swivel bracket. An eccentric cam may be rotated to adjust the sender. Moving the cam outward will increase needle movement and movement of the cam inward will decrease movement. Adjustment is made with the outboard unit in the full tilt position and the trailering locks engaged.

Supporting Components

Many of the following components are located internally and are not accessible except during a complete disassembly of the hydraulic unit. Because the trim/tilt unit is almost a completely closed system -- contaminates can only enter through the fill opening -- seldom do internal seals, O-rings, valves, etc., fail causing problems.

Because the trim/tilt unit is classed as a "high pressure" unit, total disassembly and service to the system would best be left to a shop properly equipped with the proper test equipment and trained personnel with the expertise to work on high pressure hydraulic systems.

Pump Control Piston

This very little piston moves by hydraulic pressure and controls the **UP** and **DOWN** check valves. This piston along with the **UP** and **DOWN** check valves control direction of hydraulic fluid and therefore cylinder movement.

UP Check Valve

This check valve is opened by pump pressure or by the pump control piston. When opened by pump pressure, hydraulic fluid is routed to the bottom of the cylinder to move the piston (ram) **UPWARD**.

When the check valve is opened by the pump control piston low pressure hydraulic fluid is routed back to the pump as the piston moves **DOWNWARD**.

DOWN Check Valve

This valve, like the up check valve, is opened by hydraulic pressure or by the pump control piston. If the valve is opened by hydraulic pressure fluid is routed to the top of the cylinder to move the piston **DOWNWARD**. When the valve is opened by the pump control valve, low pressure is routed back to the pump as the piston moves **UPWARD**.

DOWN Relief Valve

When the cylinder is fully retracted and internal pressures reach about 1600 psi (11,032 kPa), this valve opens as a unit protection against extreme operating pressures.

UP Relief Valve

When the cylinder is fully extended, this valve opens mechanically to protect the unit from extreme operating pressures.

Filter Valve

Permits filtered hydraulic fluid to move from the reservoir to the top of the cylinder when the outboard is lowered manually, through use of the manual release valve.

10-2 FUNCTIONAL OPERATION

Flow Diagram

In order to gain some appreciation of the complexity of the trim/tilt system, the accompanying functional flow diagrams are included. The diagrams will assist the read-er in following the flow of hydraulic fluid for the UP direction and for the DOWN direction.

UP Operation

The UP button is depressed: Current is released to the electric motor; the pump rotates COUNTERCLOCKWISE; hydraulic pressure builds almost instantly; the UP check valve opens; the pump control piston mechanically opens the DOWN check valve.

Hydraulic pressure is routed to the bottom of the cylinder; the piston (ram) begins to move upward; hydraulic pressure on top of the cylinder is low; hydraulic fluid is routed from the top of the cylinder through the open DOWN check valve back to the pump. To compensate for the volume difference when the cylinder moves upward, additional fluid is drawn into the pump from the reservoir.

Once the piston is fully extended, the UP relief valve is mechanically opened as a protection for the unit from extreme operating pressure.

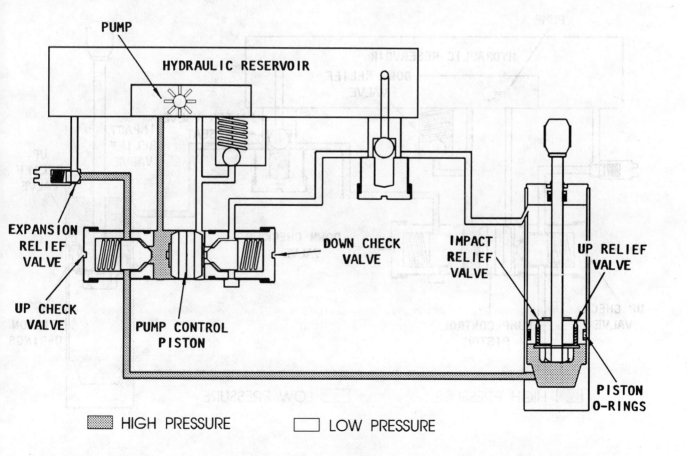

HIGH PRESSURE   LOW PRESSURE

Functional diagram to depict hydraulic fluid flow, mainly the high-pressure side, to move the outboard upward. This diagram will be most helpful when following the UP operation of the system presented in the text on this page.

DOWN Operation

The **DOWN** button is depressed: Current is released to the other circuit of the electric motor; the pump rotates **CLOCKWISE**; hydraulic pressure builds almost instantly; the down check valve opens; the pump control piston mechanically opens the **UP** check valve; hydraulic pressure is routed to the top of the cylinder; the piston (ram) moves down -- retracts; the outboard is lowered.

In this case the low fluid pressure on the bottom of the cylinder is routed through the open **UP** check valve and back to the pump.

To compensate for the volume difference caused by the piston rod, the excess fluid is through the **DOWN** relief valve and into the reservoir.

Once the piston (ram) is fully retracted, pump pressure is routed through the **DOWN** relief valve as protection of the unit from extreme operating pressure.

10-3 TROUBLESHOOTING

Troubleshooting **MUST** be done **BEFORE** the system is opened in order to isolate the problem to one area. Always attempt to proceed with troubleshooting in a definite, orderly, and directed manner. The "shot in the dark" approach will only result in wasted time, incorrect diagnosis, replacement of unnecessary parts, and frustration.

The following procedures are presented in a logical sequence to check the electrical and mechanical components, with the most prevalent, easiest, and less costly items to be checked listed first.

PRELIMINARY CHECKS AND INSPECTION

The following items are logical areas causing problems with the trim/tilt system.

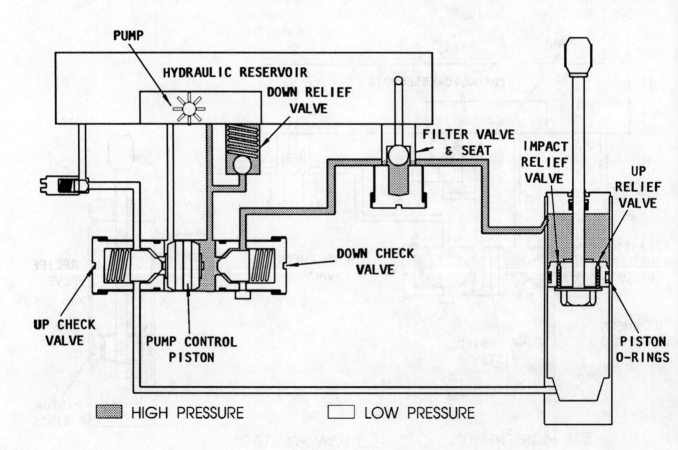

Functional diagram to depict hydraulic fluid flow, mainly the high-pressure side, to move the outboard downward. This diagram will be most helpful when following the DOWN operation of the system presented in the text on this page.

All may be checked and corrected without the use of special tools or equipment, with the exception of a torque wrench.

1- Check to be sure the battery is up to a full charge and the terminal connections are clean and tight. Make a visual inspection of all exposed wiring for an open circuit or other damage that might cause a problem.

2- Check the hydraulic fluid level at the reservoir fill plug **AFTER** all air has been removed from the unit and when the piston (ram) has been fully extended. The outboard unit must be in the vertical position. The fluid level should be even with the bottom of the fill hole with the motor full tilted up. Top off the reservoir to the plug level.

Operate the motor and then again check the oil level. Attempt to cycle the unit several times and again check the level when the piston is fully extended. The system should be cycled through at least five complete movements to ensure all air has been purged from the system.

3- Add **ONLY** OMC Power Trim/Tilt Fluid or GM "Dexron II" Automatic Transmission Fluid as required. Total capacity of the system is 11.7 fl. oz. (345 mL).

4- Make an external (outside) inspection of the system for damage or signs of a fluid leak.

5- Seat the manual release valve by tightening the screw to a torque value of 45 to 55 in lbs (5.1 to 6.2 Nm).

6- Inspect the stern brackets for signs of binding with the swivel brackets in the thrust rod area.

7- Trailering the boat with the motor in the full tilt position and unsupported can cause a hydraulic "lock-up". To relieve such a "lock-up", loosen the cylinder end cap 1/4 turn with a socket type spanner wrench. Operate the unit down then up slightly. Tighten the end cap.

ELECTRICAL TROUBLESHOOTING

As mentioned earlier, the trim/tilt unit is an almost completely closed system -- contaminats can only enter through the hydraulic fluid fill opening. Therefore, if problems are encountered, the most logical areas to check are electrical. The following areas are listed for the most common and least expensive components first.

CRITICAL WORDS

The location and orientation of the **UP** and **DOWN** relays in the relay bracket will not be the same for all powerheads. **HOWEVER** the relay terminals will always be connected to the electrical system and to the trim/tilt electric motor using the following color code:

TERMINAL	CIRCUIT	COLOR CODE
S	**UP** Switch	Blue/White
S	**DOWN** Switch	Green/White
B+	Battery -- POS (+)	Red
B-	Battery - NEG (-)	Black
M	Motor	**UP** -- Blue
M	Motor	**DOWN** -- Green

OHMMETER DAMAGE

To guard against almost instant meter damage, **NEVER** apply an ohmmeter to a circuit if voltage is present.

When performing any of the tests outlined in this section, refer to the wiring diagram on Page 10-6 and the lettered illustrations **"A"** thru **"G"**, mentioned in the text.

Trim Gauge

1- Obtain a voltmeter. Turn the key switch to the **ON** position. At the trim gauge, check for voltage between the **"I"** terminal and the **"G"** terminal, see wiring diagram, Page 10-6.

If the meter indicates voltage is not present, proceed directly to test No. 4.

If the meter indicates voltage is present, continue with the next test, No. 2.

2- With the key switch still in the **ON** position, check for voltage between the trim gauge terminal **"G"** and the **"A"** terminal at the connector just outside the remote control box.

If voltage is present -- check the condition of the accessory harness Purple lead.

If voltage is not present -- proceed with the next test No. 3.

3- With the key switch still in the **ON** position, make contact with the Black meter lead to a good ground and with the red meter lead to the **"A"** key switch terminal.

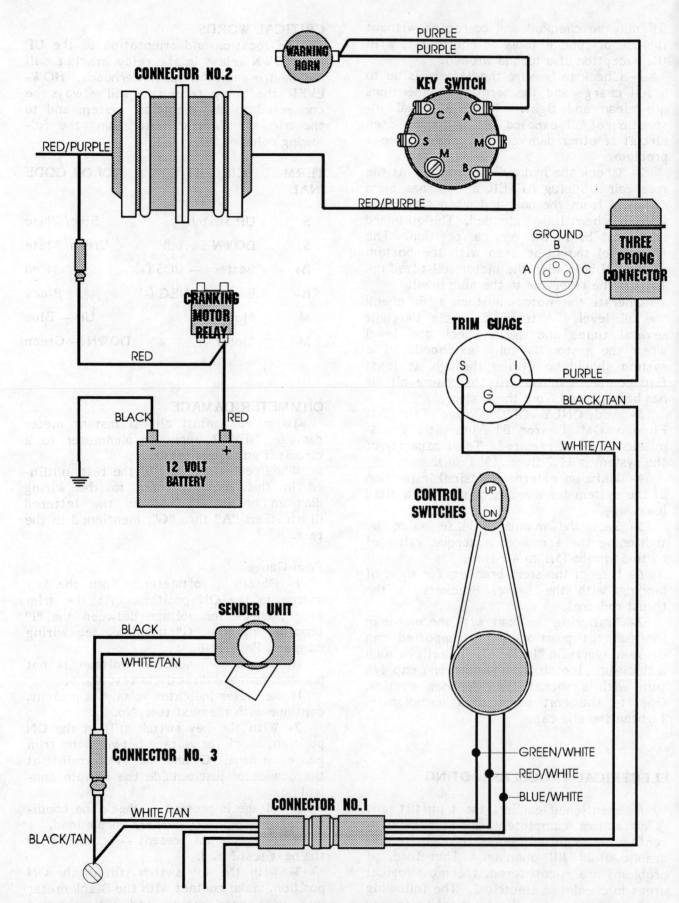

Functional schematic diagram of only the trim/tilt electrical circuits. This illustration to be used in conjunction with the troubleshooting tests presented on the previous and following pages.

If the meter indicates voltage is present, check the Purple remote control leads.

If the meter indicates voltage is not present, check the condition of the key switch and the powerhead harness 20-amp fuse.

4- Obtain an ohmmeter. Set the ohmmeter to the low ohms scale. Turn the key switch to the **OFF** position. Remove the trim harness leads from the "S" and "G" terminals on the trim gauge. Connect the meter leads between the disconnected leads. Test results should be as follows:

Outboard unit fully **DOWN** -- about 88 ohms.

Outboard unit fully **UP** -- about 1 ohm.

If the test result is as given, replace the trim gauge.

If the test results are not as given, proceed with the next test -- No. 5.

5- Perform the same test listed in No. 4 at connector No. 3. If test results are as given in test No. 4, check the condition of the trim harness at Connector No. 1.

If the test results are not as listed in test No. 4, replace the trim/tilt sending unit.

Power Supply

Check to be sure the battery has a full charge. Turn the key switch to the **OFF** position and actually remove the key. Obtain a voltmeter and connect the Black meter lead to a good powerhead ground. Check for voltage on the battery side of the cranking motor solenoid.

If voltage is not present check the battery again, the cable connections, and the wiring from the battery to the solenoid.

If voltage is present, continue with the testing.

SPECIAL WORDS

The UP circuit and **DOWN** circuit tests are all made from the relay side of the trim/tilt relay bracket. Remove both relays from the bracket. Refer to the illustrations "A" and "B" supporting the text for these tests.

An assistant is almost necessary to depress the buttons on the control handle while checking the various terminals at the powerhead.

UP Circuit
(Upper Half of Relay Bracket)

1- Connect the Black voltmeter lead to a good ground on the powerhead. Make contact with the Red meter lead to the Red terminal in the upper half of the relay bracket. If the meter indicates voltage is present, continue with the testing. If the meter indicates voltage is not present check for voltage at the terminal strip. If voltage is present at the strip, check out the wire.

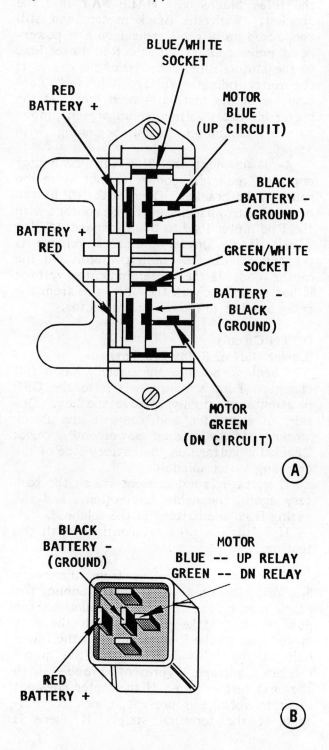

2- Have the assistant depress the **UP** trim button on the control handle and make contact with the Red meter lead to the Blue/White relay bracket terminal. The meter should indicate voltage. If the meter indicates voltage is present, continue with the testing. If the meter indicates voltage is not present, check the trim switch and/or the red/purple wire.

3- Install one of the relays into the upper -- Blue -- half of the bracket. Push the relay blades only **HALF WAY** into the bracket. With the Black meter lead still connected to a good ground on the power-head, make contact with the Red meter lead to the Motor (Blue) terminal of the relay. If the meter indicates voltage is present, continue with the testing -- next step. If the meter indicates voltage is not present, move to the ohmmeter tests, beginning with Step 9.

4- Disconnect the Blue wire connector and the Green wire connector, but leave the relay in the bracket. Depress the **UP** button on the control handle and make contact with the Red meter lead to the disconnected Blue wire. If the meter indicates voltage is present, check the electric motor and the motor lead. If the meter indicates voltage is not present, check the Blue wire from the relay bracket to the motor connector.

DOWN Circuit
(Lower Half of Relay Bracket)
Check to be sure the battery has a full charge. Turn the key switch to the **OFF** position and actually remove the key. Obtain a voltmeter and connect the Black meter lead to a good powerhead ground. Check for voltage on the battery side of the cranking motor solenoid.

If voltage is not present check the battery again, the cable connections, and the wiring from the battery to the solenoid.

If voltage is present, continue with the testing.

5- Remove the relays from the relay bracket. Obtain a voltmeter. Connect the Black meter lead to a good ground on the powerhead. Make contact with the Red meter lead to the Red terminal in the lower half of the relay bracket. If the meter indicates voltage is present, proceed with the next test --No. 6. If the meter does not indicate voltage is present, check for voltage at the terminal strip. If there is

voltage at the terminal strip, check the lead between the relay bracket and the terminal strip.

6- Have the assistant depress the **DOWN** button on the control handle and make contact with the Red meter lead to the Green/-White terminal in the relay bracket.

If the meter indicates voltage, proceed with the next test -- No. 7. If the meter indicates voltage is not present, check the trim switch and the Red/Purple wire to the switch.

7- Install one of the relays into the down -- lower half -- Green portion -- of the relay terminal. Push the relay in only **HALF WAY**. With the Black meter lead still connected to a good ground on the power-head, make contact with the Red meter lead to the Motor -- Green --terminal. If the meter indicates voltage is present, proceed with the next test -- No. 8. If the meter indicates voltage is not present, move to test No. 9.

8- Leave the relay half way into the bracket and disconnect the trim motor Blue and Green power wire connector. Have the assistant depress the **DOWN** button on the control handle and make contact with the Red meter lead to the Green wire. If the meter indicates voltage, check the electric motor and the motor lead. If the meter indicates voltage is not present, check the Green wire from the relay bracket to the motor connector.

SPECIAL WORDS
Before proceeding with the following tests, remove the relays from the sockets in the relay bracket, to prevent damaging a test light or ohmmeter during the tests.

9- Obtain a continuity test light or an ohmmeter. Set the meter to the high ohms scale or the continuity scale. Connect the Black meter lead to the center -- ground -- terminal, illustration "C". Make contact

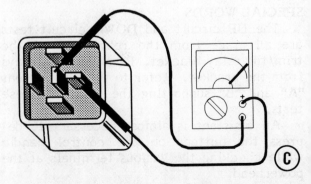

Ⓒ

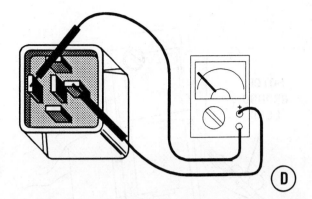

(D)

with the Red meter lead to the Motor terminal. The meter must indicate continuity.

10- Connect the Red meter lead to the Motor terminal, illustration **"D"**. Now, make contact with the Black meter lead to the B+ Battery terminal. The meter must indicate continuity.

CRITICAL WORDS

To prevent immediate damage, never connect an ohmmeter to an electrical circuit where voltage is present.

11- Set the ohmmeter to indicate 50-100 ohms. Make contact with the two meter leads to the two **S** terminals, illustration **"E"**. The meter should indicate 70-100 ohms.

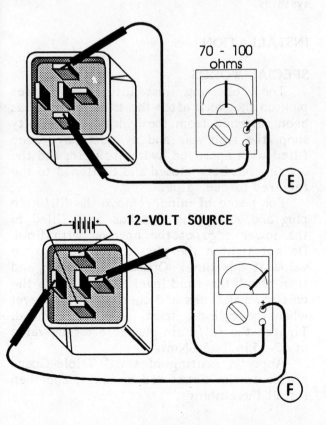

70 - 100 ohms

(E)

12-VOLT SOURCE

(F)

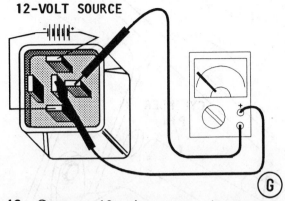

12-VOLT SOURCE

(G)

12- Connect 12-volts across the Battery (+) and Battery (-) terminals, illustration **"F"**. Set the ohmmeter to the high ohms scale.

13- Leave the 12-volt source connected across the Battery terminals, as in the previous test. Make contact with the black meter lead to the center "ground" terminal and the red meter lead to the motor lead, illustration **"G"**. The meter must indicate **NO** continuity.

If the results are not satisfactory for these tests -- replace the relay.

10-4 TRIM TILT UNIT SERVICE

REMOVAL

Raise the outboard unit, manually if necessary, by first opening the manual release valve **COUNTERCLOCKWISE** at least three full turns. Once the outboard is in the full up position, engage the trailering locks to secure the outboard in a safe manner. Disconnect the Blue and Green electrical leads from the pump motor connector housing.

1- Remove the spring clip from the cylinder pin.

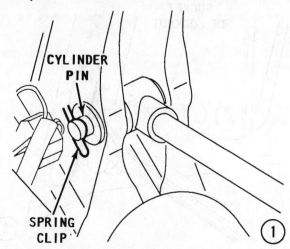

CYLINDER PIN

SPRING CLIP

(1)

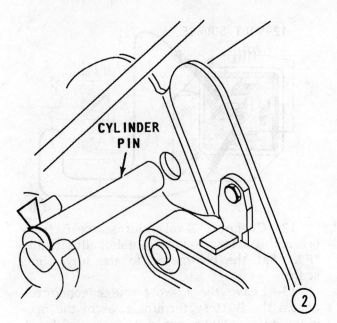

CYLINDER PIN

2

2- Withdraw the cylinder pin from the same side of the stern brackets as the spring clip was removed.

3- Remove one of the 3/4" locknuts from the angle adjustment rod, and then remove the rod from the stern brackets.

4- Remove the trim/tilt unit from the stern brackets far enough to rotate the unit and disconnect the ground lead from the pump motor mounting screw.

The electric motor can be separated from the hydraulic unit through the four attaching screws. Remove and discard the O-ring.

Testing and repair of the electric motor is covered in the next section -- this Chapter.

NOW, THESE WORDS

In the majority of cases, service of the system through the procedures covered thus

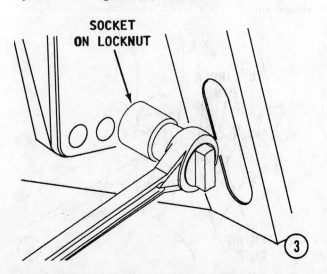

SOCKET ON LOCKNUT

3

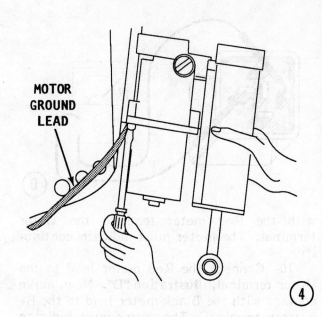

MOTOR GROUND LEAD

4

far will solve any rare problems encountered with the trim/tilt system.

The trim/tilt components comprise what is considered a "closed" system. The only route for entry of foreign material is through the fill opening.

Because the trim/tilt unit is classed as a "high pressure" unit, further disassembly and service to the system would best be left to a shop properly equipped with the proper test equipment and trained personnel with the expertise to work on high pressure hydraulic systems.

INSTALLATION

SPECIAL WORDS

The following installation procedures pick up the work after the trim/tilt unit has been returned from the hydraulic speciality shop; the reservoir and cylinder have been filled with approved hydraulic fluid; and the opening has been closed and tightened to the required torque value.

For peace of mind, remove the fill/drain plug and verify the unit has been filled to the lower edge of the opening with fluid. Do not tighten the plug to the full torque value at this time. Operate the motor, and then check the fluid level again. Cycle the unit several times and check the fluid level when the cylinder (ram) is fully extended. Tighten the fill/drain plug to a torque value of 45-55 in lb (5-6Nm).

Apply a coating of OMC Triple-Guard grease to the thrust rod bushings, and then install the bushings.

Connect the ground lead to the electric motor with the mounting screw. Apply a coating of OMC Triple-Guard grease to the angle adjustment rod, and then install the rod. Secure the rod with the locknuts. Tighten the locknuts to a torque value of 20-25 ft lb (27-34Nm).

Stretch a new O-ring into place at the base of the electric motor.

Place the electric motor in position on the manifold and secure it with the four attaching screws and lockwashers. Tighten the screws to a torque value of 35-52 in lb (4-6Nm).

10-5 ELECTRIC MOTOR TESTING AND REPAIR

Motor Testing

The condition of the motor can be bench tested with a current draw test on a no load test.

On a no load test, the motor should have a maximum current draw of 18 amps at a minimum of 8500 rpm at 12-volts.

To make the test, connect the black wire to negative and the Green/White wire (**DOWN**) to positive. The motor shaft from the drive end should turn in a **COUNTER-CLOCKWISE** direction. Repeat the test with the Blue/White (**UP**) wire to positive and the motor shaft should turn in a **CLOCKWISE** direction.

If the motor fails either of the tests it must be serviced or replaced.

Motor Repair

The following procedures pick up the work after the motor has been removed.

Remove and discard the O-ring. Remove the thru bolts and discard the seals on the bolts.

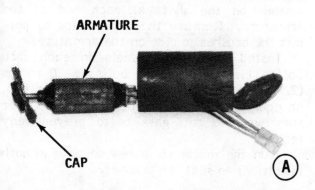

ARMATURE

CAP

(A)

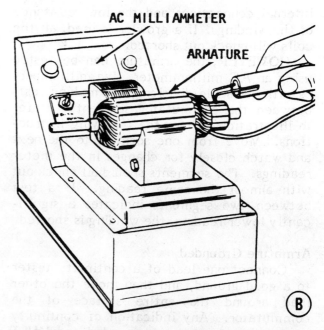

AC MILLIAMMETER

ARMATURE

(B)

Refer to illustration **"A"** for the following instructions. Remove the drive end cap from the motor. Discard the gasket. Removing the seal exercising **CARE** not to scratch the casting surfaces to ensure the new seal will seat properly. Discard the old seal.

Remove the armature from the motor housing. **TAKE CARE** not to lose the fiber washer on each end of the shaft. Tip the end cap free of the motor housing. Discard the springs and the end cap gasket.

SPECIAL WORD

The end cap is serviced as a complete assembly.

CLEANING AND INSPECTION

Clean all parts with a dry cloth. **DO NOT** clean either head in solvent, because the solvent will remove the lubricating oils in the armature shaft bushings. **DO NOT** clean the armature in solvent, because the solvent will leave traces of oil residue on the commutator segments. Oil will cause arcing between the commutator and the brushes.

Brush Replacement

A new brush head may be purchased from the Local OMC dealer as a complete assembly with new brushes installed.

Armature Shorted

The armature **CANNOT** be checked in the usual manner on a growler, because the internal connections and the low resistance of the windings. If a growler is used, all the coils will check out shorted.

HOWEVER, the armature can be tested using an AC milliammeter, five milliamperes with 100 scale divisions, and making tests between the commutator segments. Refer to illustration "B" for the following instructions. Move from one segment to the next and watch closely for changes in the meter readings. The segments should all check out with almost the same reading. If a test between two segments indicates a significantly lower reading, the winding is shorted.

Armature Grounded

Connect one lead of a continuity tester to a good ground, and then move the other lead around the entire surface of the commutator. Any indication of continuity means the armature is grounded and **MUST** be replaced. If the commutator segments are dirty or show signs of wear (roughness), clean between the bars, and then true it in a lathe. **NEVER** undercut the mica because the brushes are harder than the insulation.

After turning the armature, the insulation between the segments **MUST** be under cut to a depth of 1/32" (0.79 mm). The undercut **MUST** be flat at the bottom and should extend the full width of each insulated groove and beyond the brush contact in both directions. This will prevent the segment insulation from being smeared over the commutator as the segments wear.

After undercutting, the commutator should be sanded to remove the ridges left during the undercutting. Now, clean the commutator thoroughly to remove any metal chips or sanding grit.

Again perform the shorting and grounding tests on the armature.

End Head Bushings

Side play of each end head on the armature should be carefully checked. Any side play indicates bearing wear and the end head **MUST** be replaced, because the bearings are not serviced separately. If the heads with worn bearings are returned to service, the armature will rub against the pole shoes, or the armature shaft may actually bind.

To replace the commutator end head, first cut the lead connecting it to the field

coil as close to the end head as possible. Next, solder the new end lead to the brush holder.

Field Coils

The field coils are series-wound and are **NOT** grounded to the frame. To test the field coils for a short, make contact with one probe of a test light to a good ground on the frame. Make contact with the other probe to the Blue/White or Green/White tilt motor lead. If the test light comes on, the field coil is grounded and **MUST** be replaced. The field coils are only available as a complete field coils and frame assembly.

ASSEMBLING

Press a **NEW** seal into the end cap with the lip and seal spring facing **UP,** toward the tool. Use a flat ended bar. Continue to press the seal into place until the seal is below the chamber in the end cap.

Install a new gasket onto the end cap. Split the end of the **NEW** brush lead **CAREFULLY** to 1/8" (3.18 mm) to fit over the piece of brush lead left on the motor head. Fit the end of the new split brush leads over the old brush lead ends on the motor head, and then twist them slightly.

Hold the new leads with a pair of pliers and solder them using resin core solder. The pliers will act as a heat barrier, preventing the solder from spreading and the rest of the brush leads will remain flexible.

Install a **NEW** end cap gasket. Slide the new gasket over the wires and the end head into position.

Use two paper clips bent to make brush and spring holders. Install the new brush springs and insert the brushes into position using the paper clip as a tool to hold the brushes retracted in place.

Install the armature into the motor housing and into the motor head, with a fiber washer on the shaft at each end of the armature. Remove the paper clips to permit the brushes to ride on the armature.

Install new washer seals on each thru bolt. Tighten the thru bolts to 20 in lbs (2.5 Nm). Apply OMC Black Neoprene Dip, or equivalent, over the bolt heads and over the two end cap gaskets to prevent any leakage.

Run the motor for a few seconds in both directions to seat the brushes.

II
MAINTENANCE

11-1 INTRODUCTION

The authors estimate 75% of engine repair work can be directly or indirectly attributed to lack of proper care for the engine. This is especially true of care during the off-season period. There is no way on this green earth for a mechanical engine, particularly an outboard motor, to be left sitting idle for an extended period of time, say for six months, and then be ready for instant satisfactory service.

Imagine, if you will, leaving your automobile for six months, and then expecting to turn the key, have it roar to life, and be able to drive off in the same manner as a daily occurrence.

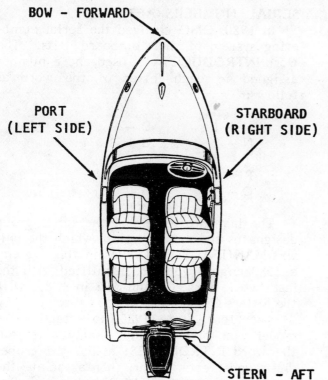

Common terminology used throughout the world for reference designation on boats. These are the terms used in this book.

It is critical for an outboard engine to be run at least once a month, preferably, in the water. At the same time, the shift mechanism should be operated through the full range several times and the steering operated from hard-over to hard-over.

Only through a regular maintenance program can the owner expect to receive long life and satisfactory performance at minimum cost.

Many times, if an outboard is not performing properly, the owner will "nurse" it through the season with good intentions of working on the unit once it is no longer being used. As with many New Year's resolutions, the good intentions are not completed and the outboard may lie for many months before the work is begun or the unit is taken to the marine shop for repair.

Imagine, if you will, the cause of the problem being a blown head gasket. And let us assume water has found its way into a cylinder. This water, allowed to remain over a long period of time, will do considerably more damage than it would have if the unit had been disassembled and the repair work performed immediately. **THEREFORE**, if an outboard is not functioning properly, **DO NOT** stow it away with promises to get at it when you get time, because the work and expense will only get worse, the longer corrective action is postponed. In the example of the blown head gasket, a relatively simple and inexpensive repair job could very well develop into major overhaul and rebuild work.

Outboards On Sail Boats

Owners of sail boats pride themselves in their ability to use the wind to clear a harbor or for movement from Port A to Port B, or maybe just for a day sail on a lake. The outboard is carried only as a last resort -- in case the wind fails completely, or in an emergency situation.

As a result, the outboard is stowed below, usually in a very poorly ventilated area, and subjected to moisture, stale air --in short, an excellent enviroment for "sweating" and corrosion.

If the owner could just take the time about once every month or two, to pull out his outboard, clean it up, and give it a short run, not only would he have "peace of mind" knowing it **WILL** start in an emergency, but also his maintenance cost will be drastically reduced.

Chapter Coverage

The material presented in this chapter is divided into five general areas.

1- General information every boat owner should know.

2- Maintenance tasks that should be performed periodically to keep the boat operating at minimum cost.

3- Care necessary to maintain the appearance of the boat and to give the owner that "Pride of Ownership" look.

4- Winter storage practices to minimize damage during the off-season when the boat is not in use.

5- Preseason preparation work that should be performed to ensure satisfactory performance the first time it is put in service.

In nautical terms, the front of the boat is the **bow** and the direction is **forward**; the rear is the **stern** and the direction is **aft**; the right side, when facing forward, is the **starboard** side; and the left side is the **port** side. All directional references in this manual use this terminology. Therefore, the direction from which an item is viewed is of

no consequence, because **starboard** and **port** **NEVER** change no matter where the individual is located or in which direction he may be looking.

11-2 ENGINE SERIAL NUMBERS

The engine serial numbers are the manufacturer's key to engine changes. These numbers identify the year of manufacture, the qualified horsepower rating, and the parts book identification. If any correspondence or parts are required, the engine model number **MUST** be used or proper identification is not possible. The accompanying illustrations will be very helpful in locating the engine identification tag for the various models.

The model number establishes the year in which the engine was produced and not necessarily the year of first installation.

On some model engines, the serial number and model number were stamped on a plate mounted between the two swivel brackets underneath the hood.

On other model engines, the plate is mounted on the port side of the engine on the front or side of the swivel bracket. The hp and rpm range will also be found on the plate.

SERIAL NUMBERS AFTER 1979

In 1980, OMC changed the serial numbering system of their outboard units. The word **INTRODUCES** was used and a number assigned to each letter of the word as follows:

I -- 1	D -- 6
N -- 2	U -- 7
T -- 3	C -- 8
R -- 4	E -- 9
O -- 5	S -- 0

The last two letters of the serial group designates the model year in which the unit was **MANUFACTURED**. Using this system, a 1980 engine would be identified with the last two letters "C" and "S"; in 1985, with the letters C and O, etc.

Therefore, since 1980, to establish the model year of an engine, simply write out the word INTRODUCES; assign the proper digits under each letter; then associate the letters on the engine with the corresponding letters of "INTRODUCES" and the model year is established.

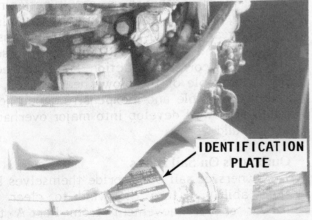

IDENTIFICATION PLATE

Manufacturer's identification plate installed between the transom brackets.

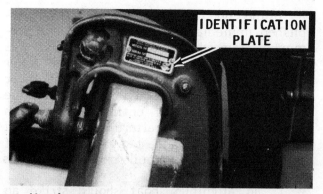

Manufacturer's identification plate installed on the port side of the transom bracket.

11-3 FIBERGLASS HULLS

Fiberglass reinforced plastic hulls are tough, durable, and highly resistant to impact. However, like any other material they can be damaged. One of the advantages of this type of construction is the relative ease with which it may be repaired. Because of its break characteristics, and the simple techniques used in restoration, these hulls have gained popularity throughout the world. From the most congested urban marina, to isolated lakes in wilderness areas, to the severe cold of far off northern seas, and in sunny tropic remote rivers of primitive islands or continents, fiberglass boats can be found performing their daily task with a minimum of maintenance.

A fiberglass hull has almost no internal stresses. Therefore, when the hull is broken or stove-in, it retains its true form. It will not dent to take an out-of-shape set. When the hull sustains a severe blow, the impact will be either absorbed by deflection of the laminated panel or the blow will result in a

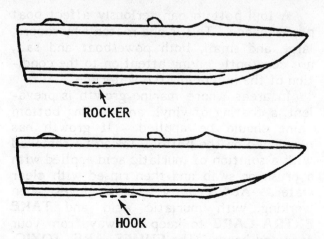

ROCKER

HOOK

Simple drawing to illustrate two types of possible damage to the hull. Such injury to the boat will affect the boat's performance and subtract from the owner's enjoyment.

A new fiberglass boat and trailer outfit ready for an owner and a power package.

definite, localized break. In addition to hull damage, bulkheads, stringers, and other stiffening structures attached to the hull, may also be affected and therefore, should be checked. Repairs are usually confined to the general area of the rupture.

11-4 ALUMINUM HULLS

Aluminum boats have become popular in recent years because they are so lightweight and may be carried with ease atop an automobile or other vehicle. These aluminum

An aluminum boat ready for an engine. The owner of this type boat will probably carry it atop his vehicle and be saved the expense and trouble of trailering to the water.

MARINE GROWTH

CLEAN BOTTOM

A boat and outboard used in salt water. Notice the marine growth on the lower unit and the anti-fouling bottom paint on the hull which prevented the marine growth.

craft are available in sizes ranging from small 8-foot prams to twin-hulled pontoon houseboats or swimming "rafts" in excess of 30 feet. Naturally, the large units cannot be carried atop a vehicle.

One of the advantages of an aluminum hull is the easy maintenance program required, and the ability of the material to resist corrosion.

As an added protection against marine growth, the below the waterline area may be painted with an anti-fouling paint. Bottom paint sold for use on a wooden or fiberglass hull is **NOT** suitable. At the time of purchase, check to be sure the paint contains the chemical properties required for an aluminum surface. The label should clearly indicate the intended use is specifically for aluminum.

If the aluminum hull does not have anti-fouling paint but requires cleaning to remove marine growth, one method is to rub the hull with a gunny sack just as soon as the boat is removed from the water and while it is still wet. The roughness of the sack is fairly effective in cleaning the surface of marine growth, including crustaceans (barnacles for instance) that have attached themselves to the hull. As soon as the rubdown has been completed the hull should be washed with high-pressure fresh water.

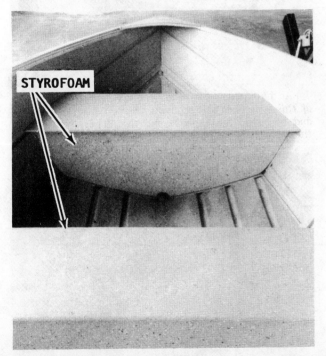

Aluminum boat with the wooden seat removed exposing the Styrofoam blocks for flotation. The seat should be removed at least once each season and the blocks thoroughly dried.

If the rubdown and wash was not accomplished immediately after the boat was removed from the water and the hull was allowed to dry, it will be necessary to cover the hull with wet blankets, gunny sacks or other suitable material and to continue soaking the covering until the growth is loosened. An easy alternate method, of course, is to return the boat to the water, if possible, and then too pull it out after it has been allowed to soak.

If an aluminum boat should strike an underwater object resulting in damage to the hull and a leak develops, the only emergency action possible is to make an attempt to reduce the amount of water being taken on by stuffing any type of available material into the opening until the boat is returned to shore. The aluminum cannot be repaired while it is wet. Repair of a damaged hull must be performed by a shop equipped for heliarc welding and other aluminum work.

Styrofoam blocks are installed under the seats of all aluminum boats. The foam blocks are designed for flotation to prevent the boat from sinking even if it should fill with water. Once each season, the wooden seat should be removed and the foam allowed to dry. Some manufacturers enclose the foam blocks in plastic bags prior to installation to protect them from moisture and loss of their flotation ability. New blocks may be purchased in a wide range of sizes. If new blocks are obtained, make an attempt to enclose the block in some form of plastic covering, then seal the package before installing it under the seat.

11-5 BELOW WATERLINE SERVICE

A foul bottom can seriously affect boat performance. This is one reason why racers, large and small, both powerboat and sail, are constantly giving attention to the condition of the hull below the waterline.

In areas where marine growth is prevalent, a coating of vinyl, anti-fouling bottom paint should be applied. If growth has developed on the bottom, it can be removed with a solution of muriatic acid applied with a brush or swab and then rinsed with clear water. **ALWAYS** use rubber gloves when working with muriatic acid and **TAKE EXTRA CARE** to keep it away from your face and hands. The **FUMES ARE TOXIC.** Therefore, work in a well-ventilated area, or if outside, keep your face on the windward side of the work.

Barnacles have a nasty habit of making their home on the bottom of boats which have not been treated with anti-fouling paint. Actually they will not harm the fiberglass hull, but can develop into a major nuisance.

If barnacles or other crustaceans have attached themselves to the hull, extra work will be required to bring the bottom back to a satisfactory condition. First, if practical, put the boat into a body of fresh water and allow it to remain for a few days. A large percentage of the growth can be removed in this manner. If this remedy is not possible, wash the bottom thoroughly with a high-pressure fresh water source and use a scraper. Small particles of hard shell may still hold fast. These can be removed with sandpaper.

11-6 SUBMERGED ENGINE SERVICE

A submerged engine is always the result of an unforeseen accident. Once the engine is recovered, special care and service procedures **MUST** be closely followed in order to return the unit to satisfactory performance.

NEVER, again we say **NEVER** allow an engine that has been submerged to stand more than a couple hours before following the procedures outlined in this section and making every effort to get it running. Such delay will result in serious internal damage. If all efforts fail and the engine cannot be started after the following procedures have been performed, the engine should be disassembled, cleaned, assembled, using new gaskets, seals, and O-rings, and then started as soon as possible.

Submerged engine treatment is divided into three unique problem areas: submersion in salt water; submerged engine while running; and a submerged engine in fresh water, including special instructions.

Crankshaft from a submerged two-cylinder engine recovered from salt water. In a very short time the crank was severely damaged by corrosion.

Rod bearing and cages badly damaged by salt water corrosion.

The most critical of these three circumstances is the engine submerged in salt water, with submersion while running a close second.

Salt Water Submersion

NEVER attempt to start the engine after it has been recovered. This action will only result in additional parts being damaged and the cost of restoring the engine increased considerably. If the engine was submerged in salt water the complete unit **MUST** be disassembled, cleaned, and assembled with new gaskets, O-rings, and seals. The corrosive effect of salt water can only be eliminated by the complete job being properly performed.

Cleaner to restore an engine recovered from fresh water after it has been submerged.

Submerged While Running
Special Instructions

If the engine was running when it was submerged, the chances of internal engine damage is greatly increased. After the engine has been recovered, remove the spark plugs and attempt to rotate the flywheel with the rewind starter. On larger horsepower engines without a rewind starter, use a socket wrench on the flywheel nut. If the attempt to rotate the flywheel fails, the chances of serious internal damage, such as, bent connecting rod, bent crankshaft, or damaged cylinder, is greatly increased. If all attempts to rotate the flywheel fail, the powerhead must be completely disassembled.

Submerged Engine — Fresh Water
SPECIAL WORD

As an aid to performing the restoration work, the following steps are numbered and should be followed in sequence. However, illustrations are not included with the procedural steps because the work involved is general in nature.

1- Recover the engine as quickly as possible.

2- Remove the cowling and the spark plugs.

3- Remove the carburetor. To rebuild the carburetor, see Chapter 4.

4- Flush the outside of the engine with fresh water to remove silt, mud, sand, weeds, and other debris. **DO NOT** attempt to start the engine if sand has entered the powerhead. Such action will only result in serious damage to powerhead components. Sand in the powerhead means the unit must be disassembled.

5- Remove as much water as possible from the powerhead. Most of the water can be eliminated by first holding the engine in a horizontal position with the spark plug holes **DOWN,** and then cranking the engine with the rewind starter or with a socket wrench on the flywheel nut.

6- Alcohol will absorb water. Therefore, pour alcohol into the carburetor throat and again crank the engine.

7- Lay the engine in a horizontal position, and then roll it over until the spark plug openings are facing **UPWARD.** Pour alcohol into the spark plug openings and again crank the engine.

8- Roll the engine in the horizontal position until the spark plug openings are again facing **DOWN.** Pour engine oil into the carburetor throat and, at the same time, crank the engine to distribute oil throughout the crankcase.

9- With the engine still in a horizontal position, roll it over until the spark plug holes are again facing **UPWARD.** Pour approximately one teaspoon of engine oil into each spark plug opening. Crank the engine to distribute the oil in the cylinders.

10- Install the spark plugs and tighten them to the torque value given in the Appendix. Connect the high-tension leads to the spark plugs.

11- Install the carburetor onto the engine with a **NEW** gasket on the intake manifold.

12- Mount the engine in a test tank or body of water.

CAUTION: Water must circulate through the lower unit to the engine any time the engine is run to prevent damage to the water pump in the lower unit. Just five seconds without water will damage the water pump.

Obtain **FRESH** fuel and attempt to start the engine. If the engine will start, allow it to run for approximately an hour to eliminate any water remaining in the engine.

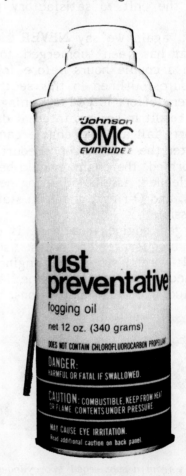

Rust preventative to be sprayed inside the engine in preparation for storage, as explained in the text.

13- If the engine fails to start, determine the cause, electrical or fuel, correct the problem, and again attempt to get it running. **NEVER** allow an engine to remain unstarted for more than a couple hours without following the procedures in this section and attempting to start it. If attempts to start the engine fail, the unit should be disassembled, cleaned, assembled, using new gaskets, seals, and O-rings, just as soon as possible.

11-7 WINTER STORAGE

Taking extra time to store the boat properly at the end of each season, will increase the chances of satisfactory service for the next season. **REMEMBER**, idleness is the greatest enemy of an outboard motor. The unit should be run on a monthly basis. The boat steering and shifting mechanism should also be worked through complete cycles several times each month. The owner who spends a small amount of time involved in such maintenance will be rewarded by satisfactory performance, and greatly reduced maintenance expense for parts and labor.

ALWAYS remove the drain plug and position the boat with the bow higher than the stern. This will allow any rain water

Before storing the outboard for the off-season, the powerhead should be operated, following the procedures listed in the text.

and melted snow to drain from the boat and prevent "trailer sinking". This term is used to describe a boat that has filled with rain water and ruined the interior, because the plug was not removed or the bow was not high enough to allow the water to drain properly.

Proper storage for the engine involves adequate protection of the unit from physical damage, rust, corrosion, and dirt.

The following steps provide an adequate maintenance program for storing the unit at the end of a season.

1- Remove the cowling. Start the powerhead and allow it to warm to operating temperature.

CAUTION: Water must circulate through the lower unit to the engine any time the engine is run to prevent damage to the water pump in the lower unit. Just five seconds without water will damage the water pump.

Disconnect the fuel line from the engine and allow the unit to run at **LOW** rpm and, at the same time, inject about 4 ounces of rust preventative spray through each carburetor throat. Allow the engine to run until it shuts down from lack of fuel.

2- Drain the fuel tank and the fuel lines. Pour approximately one qt. of benzol (benzene) into the fuel tank, and then rinse the tank and pickup filter with the benzol. Drain the tank. Store the fuel tank in a cool dry area with the vent **OPEN** to allow air to circulate through the tank. **DO NOT** store the fuel tank on bare concrete. Place the tank to allow air to circulate around it. If

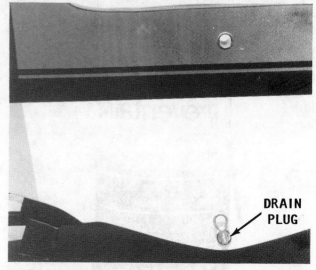

DRAIN PLUG

Drain plug removed from the transom to allow rain and melted snow to drain from the boat. Failure to remove this plug during long periods of storage can cause "beaucoup" problems.

Standard OMC fuel tank. During periods of storage the tank should be empty and the cap "cracked" open to allow the tank to "breathe".

the fuel tank containing fuel is to be stored for more than a month, a commercial additive such as Sta-Bil should be added to the fuel. This type of additive will maintain the fuel in a "fresh" condition for up to a full year.

3- Clean the carburetor fuel filter/s with benzol, see Chapter 4, Carburetor Repair Section.

OMC fuel conditioner added to the fuel will keep it fresh for up to one full year.

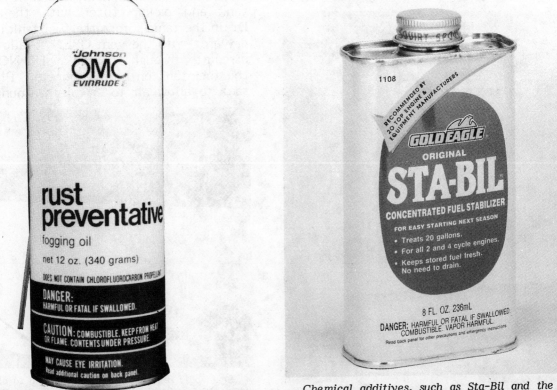

Rust preventative to be used when preparing the engine for long periods of non-use and storage.

Chemical additives, such as Sta-Bil and the OMC fuel conditioner at the top of the page will prevent fuel from "souring" for up to twelve months.

4- Drain, and then fill the lower unit with OMC Lower Unit Gear Lubricant, as outlined in Section 11-8.

5- Lubricate the throttle and shift linkage. Lubricate the swivel pin and the tilt tube with Multi-purpose Lubricant, or equivalent.

Clean the engine thoroughly. Coat the powerhead with Corrosion and Rust Preventative spray. Install the cowling and then apply a thin film of fresh engine oil to all painted surfaces.

Remove the propeller. Apply anti-sieze or a waterproof sealer to the propeller shaft, and then install the propeller back in position.

FINAL WORDS: Be sure all drain holes in the gear housing are open and free of obstruction. Check to be sure the FLUSH plug has been removed to allow all water to drain. Trapped water could freeze, expand, and cause expensive castings to crack.

ALWAYS store the engine off the boat with the lower unit below the powerhead to prevent any water from being trapped inside. The ideal storage position for an outboard is to hang it with the lower unit down. If hanging is not practical, lay the outboard on its back. This will place the lower unit below the powerhead.

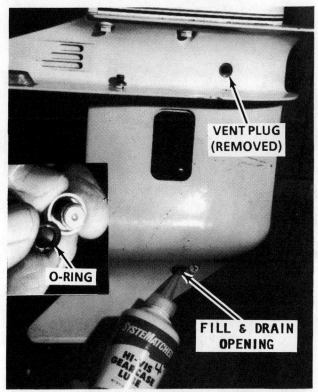

The vent plug MUST be removed when adding or draining lubricant from the lower unit. Both plugs have an O-ring, as shown in the insert.

Propeller with grooves worn into the hub.

11-8 LOWER UNIT SERVICE

PROPELLERS

With Shear Pin

The propeller should be checked regularly to be sure all the blades are in good condition. If any of the blades become bent or nicked, such damage will set up vibrations in the motor. Remove and inspect the propeller. Use a file to trim nicks and burrs. TAKE CARE not to remove any more material than is absolutely necessary. For a complete check, take the propeller to your marine dealer where the proper equipment and knowledgeable mechanics are available to perform a proper job at modest cost.

Inspect the propeller shaft to be sure it is still true and not bent. If the shaft is not perfectly true, it should be replaced.

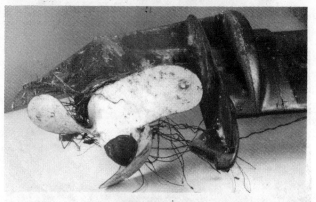

Sea weed entangled in the propeller. The propeller should be removed frequently and any foreign material, especially fish line removed before it can cause damage.

With Exhaust

Propellers with the exhaust passing through the hub **MUST** be removed more frequently than the standard propeller. Removal after each weekend use or outing is not considered excessive. These propellers do not have a shear pin. The shaft and propeller have splines which **MUST** be coated with an anti-corrosion lubricant prior to installation as a aid to removal the next time the propeller is pulled. Even with the lubricant applied to the shaft splines, the propeller may be difficult to remove.
The propeller with the exhaust hub is more expensive than the standard propeller and therefore, the cost of rebuilding the unit, if the hub is damaged, is justified.

A replaceable diffuser ring on the backside of the propeller disperses the exhaust away from the propeller blades. If the ring becomes broken or damaged "ventilation" would be created pulling the exhaust gases back into the negative pressure area behind the propeller. This condition would create considerable air bubbles and reduce the effectiveness of the propeller.

Standard Propeller Removal

On some model engines, the shear pin is installed behind the propeller. First, pull the cotter key, and then remove the propeller nut, drive pin, and washer. Because the drive pin is not a tight fit, the propeller is able to move on the pin and cause burrs on

the hole. These burrs may make removing the propeller difficult. To overcome this problem, the propeller hub has two grooves running the full length of the hub. Hold the shaft from turning, and then rotate the propeller 1/4 turn to position the grooves over the drive pin holes. The propeller can then be pulled straight off the shaft. After the propeller has been removed, file the drive pin holes on both sides of the shaft to remove the burrs.

Standard Propeller Installation

On some model engines, the shear pin is installed behind the propeller. On such units the propeller shaft should be coated with Perfect Seal, and then the propeller installed. After the propeller is on the shaft, install the washer, shear pin, propeller nut, and finally a **NEW** cotter pin.

Propeller With Exhaust Removal

First, disconnect the high tension leads to the spark plugs to prevent accidental engine start. Next, pull the cotter pin from the propeller nut. Wedge a piece of wood between one of the propeller blades and the cavitation plate to prevent the propeller from rotating. Back off the castellated propeller nut. Pull the propeller straight off

Propeller installation using a spacer, propeller nut, and a cotter pin.

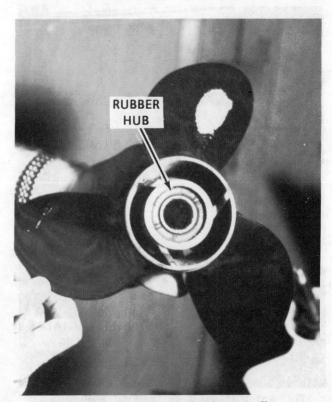

A rubber hub installed inside the propeller serves the same function as a "shear pin".

the shaft. It may be necessary to carefully tap on the front side of the propeller with a soft headed mallet to jar it loose. If the propeller appears to be "frozen" to the shaft, see Chapter 8 for special removal instructions. The thrust washer does not have to be removed unless it appears damaged.

Propeller Exhaust Installation

First, check to be sure the high tension leads have been disconnected from the spark plugs to prevent accidental engine start. Install the thrust washer onto the propeller shaft, if it was removed. Coat the splines of the driveshaft with anti-corrosion lubricant. The lubricant **MUST** be applied to the shaft **EACH** time the propeller is installed to prevent it from "freezing" to the shaft. The propeller may "freeze" to the shaft in a short time in fresh water and much sooner in salt water.

Slide the propeller onto the shaft with the splines on the shaft indexed with the splines in the propeller hub. Force the propeller onto the shaft until it is tight against

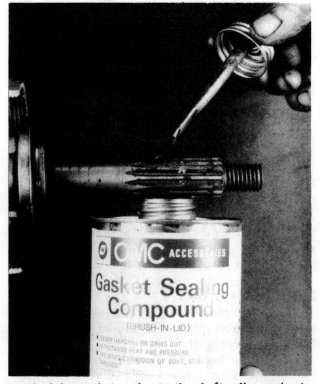

Applying gasket sealer to the shaft splines prior to installing a "prop exhaust" propeller.

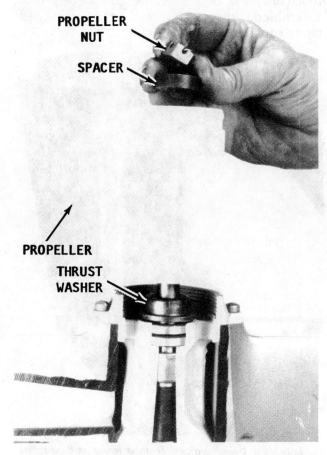

PROPELLER NUT

SPACER

PROPELLER

THRUST WASHER

Installing a "prop exhaust" propeller. The thrust washer is installed first, then the propeller, spacer, nut and finally the cotter pin through the propeller nut.

OMC anti-corrosion lubricant that should be applied to the propeller shaft before the propeller is installed. Such lubricant on the shaft will not only fight corrosion, but assist in propeller removal.

the thrust washer. If the propeller cannot be moved tight against the thrust washer, the splines in the hub or on the shaft are dirty and must be cleaned.

After the propeller is in place, wedge a piece of wood between one of the blades and the cavitation plate to prevent the propeller from rotating. Thread the castellated nut onto the shaft and bring it up tight against the propeller. Insert the cotter pin through the nut and propeller shaft. If the hole in the shaft does not align with one of the holes in the nut, **TIGHTEN** the nut until one of the holes is aligned. **NEVER** loosen the nut to align the holes. Install the cotter pin, remove the piece of wood, and connect the high tension leads to the spark plugs.

Draining Lower Unit

Remove the **VENT** plug just above the anti-cavitation plate **FIRST**, and then the **FILL** plug from the gear housing. **NEVER** remove the vent or filler plug when the drive unit is hot. Expanded lubricant would be released through the plug hole.

CRITICAL WORD

The Phillips screw securing the shift fork in place is located very close to the drain plug. If the wrong screw is removed,

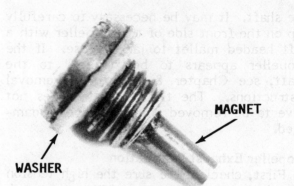

Late model lower unit drain plugs have a magnet to attract metal particles in the lubricant before they cause damage. If an unusual amount of particles are discovered, disassembly and search should be made to discover the source.

BAD NEWS, VERY BAD NEWS. The lower unit will have to be disassembled in order to return the shift fork to its proper location.

Allow the gear lubricant to drain into the container. As the lubricant drains, catch some with your fingers, from time-to-time, and rub it between your thumb and finger to determine if any metal particles are present. If metal is detected in the lubricant, the unit must be completely disassembled, inspected, and the damaged parts replaced.

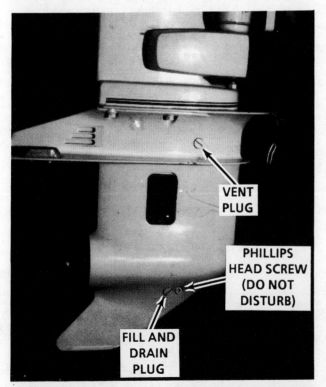

Care MUST be exercised NOT to remove the Phillips head screw next to the fill/drain plug by mistake. Such an error would be BAD NEWS indeed, because this screw secures the shift fork in place.

Lubricant being drained from a lower unit without shift capabilities. Therefore, the Phillips screw mentioned in the illustration in the left column is not present.

If the lubricant appears milky brown, or if large amounts of lubricant must be added to bring the lubricant up to the full mark, a thorough check should be made to determine the cause of the loss.

Filling Lower Unit

Add only OMC lower unit lubricant. Lubricant, for engines covered in this manual, is as follows: Use **ONLY** Type C, now known as Premium Blend Gearcase Lube, in all electric shift models. Use either Premium Blend Lube or the OMC Hi-Vis Gearcase Lube for all other engines. **NEVER** use regular automotive-type grease in the lower unit because it expands and foams too much. Lower units do not have provisions to accommodate such expansion.

The gearcase lubricant should be changed twice each year or season. If the lubricant is purchased in a large container, say the one-gallon size, a considerable savings can be realized. What is not used this season will be used in the next or the one after. A small inexpensive pump can be

LOWER UNIT LUBRICANT CAPACITIES

MODEL SIZE	YEAR	CAPACITY OUNCES
Colt	1990	1.2
2.3	1991 & On	3
3.3	1991 & On	3
3	1990 & On	2.7
4	1990 & On	2.7
4D	1990 & On	11
6	1990 & On	11
8	1990 & On	11
9.9	1990 & On	9
15	1990 & On	9
20	1990 & On	11
25	1990 & On	11
30	1990 & On	11
40	1990 & On	16.4
50	1990 & On	16.4

purchased to move the lubricant from the large container to the lower unit.

Position the drive unit approximately vertical and without a list to either port or starboard. Insert the lubricant tube into the **FILL/DRAIN** hole at the bottom plug hole, and inject lubricant until the excess begins to come out the **VENT** hole. Install the **VENT** and **FILL** plugs with **NEW** gaskets.

The two types of lubricant used for Johnson/Evinrude engines. The text clearly identifies the lubricant to be used on the different model engines covered in this manual.

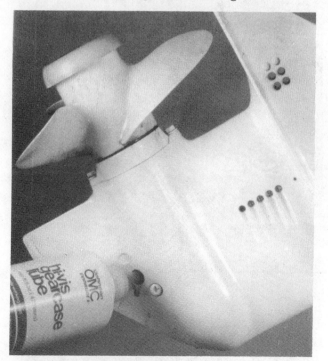

Filling a lower unit with a "prop exhaust" propeller. The lower unit should be "topped off" as outlined in the text.

Filling a non-shifting lower unit with lubricant.

After the lower plug is in place, remove the vent plug again and using a squirt-type oil can, add lubricant through this vent hole. A squirt-type oil can must be used to allow the trapped air in the lower unit to escape at the same time the final lubricant is added. Once the unit is completely full, install and tighten the vent plug.

Using a squirt can to "top off" the lubricant in the lower unit, as explained in the text.

Check to be sure the vent and drain plug gaskets are properly positioned to prevent water from entering the housing.

See lower unit capacity list on the previous page.

11-9 BATTERY STORAGE

Remove the batteries from the boat and keep them charged during the storage period. Clean the batteries thoroughly of any dirt or corrosion, and then charge them to full specific gravity reading. After they are fully charged, store them in a clean cool dry place where they will not be damaged or knocked over.

NEVER store the battery with anything on top of it or cover the battery in such a manner as to prevent air from circulating around the filler caps. All batteries, both new and old, will discharge during periods of storage, more so if they are hot than if they remain cool. Therefore, the electrolyte

A check of the electrolyte in the battery should be on the maintenance schedule for any boat. A hydrometer reading of 1.300 or in the green band, indicates the battery is in satisfactory condition. If the reading is 1.150 or in the red band, the battery needs to be charged.

level and the specific gravity should be checked at regular intervals. A drop in the specific gravity reading is cause to charge them back to a full reading.

In cold climates, **EXERCISE CARE** in selecting the battery storage area. A fully-charged battery will freeze at about 60 degrees below zero. A discharged battery, almost dead, will have ice forming at about 19 degrees above zero.

11-10 PRESEASON PREPARATION

Satisfactory performance and maximum enjoyment can be realized if a little time is spent in preparing the engine for service at the beginning of the season. Assuming the unit has been properly stored, as outlined in Section 11-7, a minimum amount of work is required to prepare the engine for use.

The following steps outline an adequate and logical sequence of tasks to be performed before using the engine the first time in a new season.

1- Lubricate the engine according to the manufacturer's recommendations. Remove, clean, inspect, adjust, and install the spark plugs with new gaskets if they require gaskets. Make a thorough check of the ignition system. This check should include: the points, coil, condenser, condition of the wiring, and the battery electrolyte level and charge.

2- If a built-in fuel tank is installed, take time to check the tank and all of the

Today, numerous type spark plugs are available for service. ALWAYS check with the local OMC dealer to be sure you are purchasing the proper plugs for the engine being serviced.

fuel lines, fittings, couplings, valves, and the flexible tank fill and vent. Turn on the fuel supply valve at the tank. If the fuel was not drained at the end of the previous season, make a careful inspection for gum formation. If a six-gallon fuel tank is used, take the same action. When gasoline is allowed to stand for long periods of time, particularly in the presence of copper, gummy deposits form. This gum can clog the filters, lines, and passageways in the carburetor. See Chapter 4, Fuel System Service.

3- Check the oil level in the lower unit by first removing the vent screw on the port side just above the anti-cavitation plate. Insert a short piece of wire into the hole and check the level. Fill the lower unit according to procedures outlined in Section 11-8.

4- Close all water drains. Check and replace any defective water hoses. Check to be sure the connections do not leak.

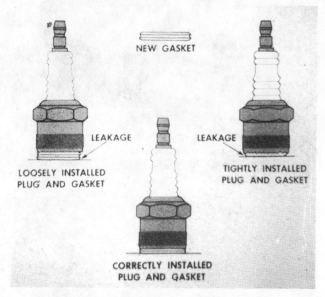

Correct and incorrect spark plug installation. The plugs MUST be installed properly and tightened to the proper torque value for satisfactory performance.

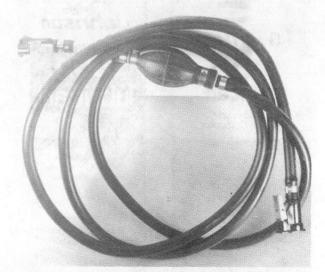

Typical fuel hose with squeeze bulb. The hose and bulb must remain flexible. The O-rings MUST prevent fuel leakage.

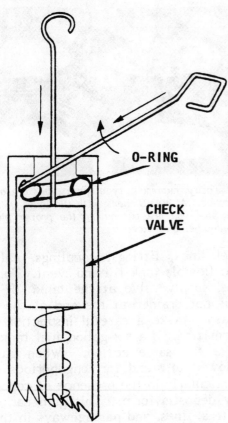

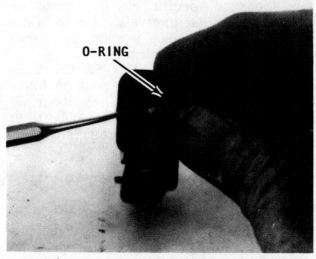

Using a punch to depress the check ball while installing an oiled O-ring.

Method of removing an O-ring from a connector. The connector is also replaceable.

Replace any spring-type hose clamps, if they have lost their tension, or if they have distorted the water hose, with band-type clamps.

5- The engine can be run with the lower unit in water to flush it. If this is not

practical, a flush attachment may be used. This unit is attached to the water pick-up in the lower unit. Attach a garden hose, turn on the water, allow the water to flow into the engine for awhile, and then run the engine.

CAUTION: Water must circulate through the lower unit to the engine any time the engine is run to prevent damage to the water pump in the lower unit. Just five seconds without water will damage the water pump.

Adding OMC oil to the fuel. Only a high grade oil should be added to the fuel to ensure proper lubrication.

OMC lubricants for Johnson/Evinrude outboards.

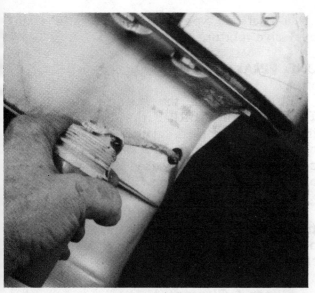

Using a squirt can to "top off" the lubricant in a lower unit, as described in the text.

Check the idle exhaust port for water discharge. At idle speed, only a fine water mist will be visible. Check for leaks. Check operation of the thermostat. After the engine has reached operating temperature, tighten the cylinder head bolts to the torque value given in the Specifications in the Appendix.

The battery should be located near the engine and well secured to prevent even the slightest amount of movement. The battery, including the terminals MUST be kept clean for maximum performance.

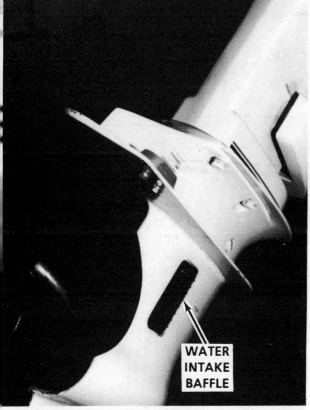

WATER INTAKE BAFFLE

The water intake baffles on both sides of the lower unit MUST face forward and the openings kept clear. If the baffles are installed backwards or the openings become clogged, the powerhead would quickly overheat.

An OMC degreaser widely used for cleaning the boat and engine.

6- Check the electrolyte level in the batteries and the voltage for a full charge. Clean and inspect the battery terminals and cable connections. **TAKE TIME** to check the polarity, if a new battery is being installed. Cover the cable connections with grease or special protective compound as a prevention to corrosion formation. Check all electrical wiring and grounding circuits.

7- Check all electrical parts on the engine and electrical fixture or connections in the lower portions of the hull inside the boat to be sure they are not of a type that could cause ignition of an explosive atmosphere. Rubber caps help keep spark insulators clean and reduce the possibility of arcing. Starters, generators, distributors, alternators, electric fuel pumps, voltage regulators, and high-tension wiring harnesses should be

of a marine type that cannot cause an explosive mixture to ignite.

ONE FINAL WORD

Before putting the boat in the water, **TAKE TIME** to **VERIFY** the drain plugs are installed. Countless number of boating excursions have had a very sad beginning because the boat was eased into the water only to have water begin filling the inside.

Keep the gas tank full, the trim tab trimmed, the fuel pump pumping, the spark plugs sparking, and the pistons, well, keep them working too.

Joan & Clarence

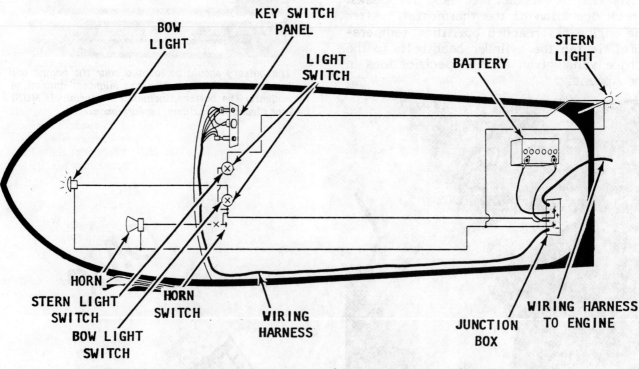

Principle electrical points requiring a careful check at the start of each season.

APPENDIX

METRIC CONVERSION CHART

LINEAR

inches	X 25.4	= millimetres (mm)
feet	X 0.3048	= metres (m)
yards	X 0.9144	= metres (m)
miles	X 1.6093	= kilometres (km)
inches	X 2.54	= centimetres (cm)

AREA

inches^2	X 645.16	= millimetres^2 (mm^2)
inches^2	X 6.452	= centimetres^2 (cm^2)
feet^2	X 0.0929	= metres^2 (m^2)
yards^2	X 0.8361	= metres^2 (m^2)
acres	X 0.4047	= hectares (10^4 m^2) (ha)
miles^2	X 2.590	= kilometres^2 (km^2)

VOLUME

inches^3	X 16387	= millimetres^3 (mm^3)
inches^3	X 16.387	= centimetres^3 (cm^3)
inches^3	X 0.01639	= litres (l)
quarts	X 0.94635	= litres (l)
gallons	X 3.7854	= litres (l)
feet^3	X 28.317	= litres (l)
feet^3	X 0.02832	= metres^3 (m^3)
fluid oz	X 29.60	= millilitres (ml)
yards^3	X 0.7646	= metres^3 (m^3)

MASS

ounces (av)	X 28.35	= grams (g)
pounds (av)	X 0.4536	= kilograms (kg)
tons (2000 lb)	X 907.18	= kilograms (kg)
tons (2000 lb)	X 0.90718	= metric tons (t)

FORCE

ounces - f (av)	X 0.278	= newtons (N)
pounds - f (av)	X 4.448	= newtons (N)
kilograms - f	X 9.807	= newtons (N)

ACCELERATION

feet/sec^2	X 0.3048	= metres/sec^2 (m/S^2)
inches/sec^2	X 0.0254	= metres/sec^2 (m/s^2)

ENERGY OR WORK (watt-second - joule - newton-metre)

foot-pounds	X 1.3558	= joules (j)
calories	X 4.187	= joules (j)
Btu	X 1055	= joules (j)
watt-hours	X 3500	= joules (j)
kilowatt - hrs	X 3.600	= megajoules (MJ)

FUEL ECONOMY AND FUEL CONSUMPTION

miles/gal	X 0.42514	= kilometres/litre (km/l)

Note:
235.2/(mi/gal) = litres/100km
235.2/(litres/100 km) = mi/gal

LIGHT

footcandles	X 10.76	= lumens/metre^2 (lm/m^2)

PRESSURE OR STRESS (newton/sq metre - pascal)

inches HG (60 F)	X 3.377	= kilopascals (kPa)
pounds/sq in	X 6.895	= kilopascals (kPa)
inches H$_2$O (60 F)	X 0.2488	= kilopascals (kPa)
bars	X 100	= kilopascals (kPa)
pounds/sq ft	X 47.88	= pascals (Pa)

POWER

horsepower	X 0.746	= kilowatts (kW)
ft-lbf/min	X 0.0226	= watts (W)

TORQUE

pound-inches	X 0.11299	= newton-metres (N·m)
pound-feet	X 1.3558	= newton-metres (N·m)

VELOCITY

miles/hour	X 1.6093	= kilometres/hour (km/h)
feet/sec	X 0.3048	= metres/sec (m/s)
kilometres/hr	X 0.27778	= metres/sec (m/s)
miles/hour	X 0.4470	= metres/sec (m/s)

TEMPERATURE

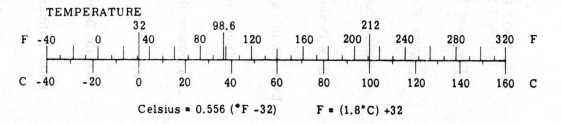

Celsius = 0.556 (°F -32)   F = (1.8°C) +32

TORQUE VALUE LISTING

SCREW OR BOLT LOCATION	COLT	2.3 & 3.3	3-4D	5 thru 8	9.9 thru 15	20 thru 30	40 thru 50
Flywheel Nut	22-25 ft lb (30-34Nm)	29-33 ft lb (40-45Nm)	30-40 ft lb (40-54Nm)	40-50 ft lb (54-70Nm)	45-50 ft lb (60-70Nm)	100-105 ft lb (135-140Nm)	100-105 ft lb (135-140Nm)
Spark Plug	18-21 ft lb (24-27Nm)	18-21 ft lb (24-27Nm)	18-21 ft lb (24-27Nm)	18-21 ft lb (24-27Nm)	18-21 ft lb (24-27Nm)	18-21 ft lb (24-27Nm)	18-21 ft lb (24-27Nm)
Water Pump	25-35 in lb (3-4Nm)	62-89 in lb (7-12Nm)	See Note 1	60-84 in lb (7-9Nm)	60-80 in lb (7-9Nm)	60-80 in lb (7-9Nm)	60-84 in lb (7-9Nm)
Fill & Drain Plugs	60-84 in lb (7-9Nm)	31-40 in lb (3.5-4.5Nm)	See Note 2	60-84 in lb (7-9Nm)	60-80 in lb (7-9Nm)	60-80 in lb (7-9Nm)	60-84 in lb (7-9Nm)
Lwr. Crankcase Head	60-84 in lb (7-9Nm)	60-84 in lb (7-9Nm)	---	---	60-84 in lb (7-9Nm)	60-84 in lb (7-9Nm)	60-84 in lb (7-9Nm)
Cylinder Head	60-84 in lb (7-9Nm)	60-84 in lb (7-9Nm)	60-84 in lb (7-9Nm)	144-168 in lb (16-19Nm)	216-240 in lb (25-27Nm)	216-240 in lb (25-27Nm)	18-20 ft lb (25-27Nm)
Connecting Rod Cap	60-70 in lb (7-7.5Nm)	---	60-70 in lb (7-7.5Nm)	60-70 in lb (7-7.5Nm)	60-70 in lb (7-7.5Nm)	30-32 ft lb (40-43Nm)	30-32 ft lb (40-43Nm)
Pwr. Head Retain.	60-84 in lb (7-9Nm)	60-84 in lb (7-9Nm)	60-84 in lb (7-9Nm)	60-84 in lb (7-9Nm)	60-84 in lb (7-9Nm)	192-216 in lb (22-24Nm)	216-240 in lb (25-27Nm)
Main Bearing	90-120 in lb (10-14Nm)	90-120 in lb (10-14Nm)	60-84 in lb (7-9Nm)	See Note 3	144-168 in lb (16-19Nm)	168-192 in lb (19-22Nm)	18-20 ft lb (25-27Nm)
Inner Exhaust Tube	---	---	60-84 in lb (7-9Nm)	60-84 in lb (7-9Nm)	60-84 in lb (7-9Nm)	96-120 in lb (11-14Nm)	---
Cyl. Head Cover	---	---	---	---	60-84 in lb (7-9Nm)	60-84 in lb (7-9Nm)	---

Component					
Exhaust Cover	60-84 in lb (7-9Nm)	60-84 in lb (7-9Nm)	95-130 in lb (11-14Nm)	48-84 in lb (5-9Nm)	48-84 in lb (5-9Nm)
Bypass Cover	---	---	60-84 in lb (7-9Nm)	60-84 in lb (7-9Nm)	---
Intake Manifold	60-84 in lb (7-9Nm)	60-84 in lb (7-9Nm)	60-84 in lb (7-9Nm)	60-84 in lb (7-9Nm)	60-84 in lb (7-9Nm)
Gearcase Head	60-84 in lb (7-9Nm)	60-84 in lb (7-9Nm)	---	---	---

Note 1 Model 3hp, 25-35 in lb (3-4Nm).
Model 4D, 60-84 in lb (7-9Nm).

Note 2 Model 3hp, 40-50 in lb (4.5-5.6Nm).
Model 4D, 60-84 in lb (7-9Nm).

Note 3 Model 5-8hp, 1990 only, 144-168 in lb (16-19Nm).
Model 5-8hp, 1991 & On, 60-84 in lb (7-9Nm).

STANDARD BOLTS AND NUTS

Torque value for standard sizes not listed above.

Bolt Size	In./Lbs	Ft/Lbs	Newton Meters
No. 6	7-10	---	0.8-1.2
No. 8	15-22	---	1.6-2.4
No. 10	25-35	2-3	2.8-4.0
No. 12	35-40	3-4	4.0-4.6
1/4"	60-80	5-7	7-9
5/16"	120-140	10-12	14-16
3/8"	220-240	18-20	24-27
7/16"	340-360	28-30	38-40

POWERHEAD SPECIFICATIONS

MODEL	NO. CYL	YEAR	CU. IN. DISPL.	STROKE (INCHES)	BORE STD. (INCHES)	BORE OVERSIZE	PISTON DIA. STANDARD (INCHES) NOTE 1	PISTON RING GROOVE SLIDE CLEAR. BOTH (MAX.)	PISTON RING END GAP BOTH
Colt	1	1990	2.65	1.374	1.565	Note 2	1.5620-1.5625	0.004	0.015-0.025
2.3	1	1991 & On	4.47	1.69	1.89	Note 2	1.8868-1.8873	0.0026	0.0059-0.0138
3.3	1	1991 & On	4.47	1.69	1.89	Note 2	1.8868-1.8873	0.0026	0.0059-0.0138
3	2	1990 & On	5.29	1.374	1.565	Note 2	1.5625-1.5631	0.004	0.005-0.015
4	2	1990 & On	5.29	1.374	1.565	Note 2	1.5625-1.5631	0.004	0.005-0.015
4D	2	1990 & On	5.28	1.374	1.565	Note 2	1.5625-1.5631	0.004	0.005-0.015
6	2	1990 & On	10	1.70	1.937	Note 2	1.9345-1.9355	0.004	0.005-0.015
8	2	1990 & On	10	1.70	1.937	Note 2	1.9345-1.9355	0.004	0.005-0.015
9.9	2	1990 & On	13.2	1.76	2.188	Note 2	2.1845-2.1850	0.004	0.005-0.015
15	2	1990 & On	13.2	1.76	2.188	Note 2	2.1845-2.1850	0.004	0.005-0.015
20	2	1990 & On	31.8	2.25	3.00	Note 2	Note 3	0.004	0.007-0.017
25	2	1990 & On	31.8	2.25	3.00	Note 2	Note 3	0.004	0.007-0.017
30	2	1990 & On	31.8	2.25	3.00	Note 2	Note 3	0.004	0.007-0.017
40	2	1990 & On	44.99	2.82	3.187	Note 2	3.1831	0.004	0.007-0.017
48	2	1990 & On	44.99	2.82	3.187	Note 2	3.1831	0.004	0.007-0.017
50	2	1990 & On	44.99	2.82	3.187	Note 2	3.1831	0.004	0.007-0.016

Note 1 Piston diameter measurements to be taken 1/8" to 1/4" from bottom edge of piston skirt.
Note 2 When necessary to bore oversize, add piston oversize dimension to standard bore dimension.
Note 3 Manufacturer has used different vendors to supply pistons. Therefore, individual must visually identify the piston to know the appropriate dimension. See small diagrams on this page.
Zollner piston: 1990 -- 2.99656 in. (76.11mm), in line with piston pin, 1/8 to 1/4" from skirt edge.
Art piston: 1990 -- 2.99690 in. (76.12mm), in line with piston pin, 1/8 to 1/4" above skirt edge.
Zollner piston: 1991 & on -- 2.9956 in. in line with piston pin, 1/8 to 1/4" above skirt edge.
Art & Rightway piston: 1991 & on -- 2.9969 in. in line with piston pin, 1/8 to 1/4" above skirt edge.

TUNE-UP SPECIFICATIONS

MODEL	NO. CYL.	YEAR	CU. IN. DISPL.	GEAR RATIO	WOT THROTTLE RPM	IDLE RPM IN GEAR	SPARK PLUG CHAMP	PLUG GAP (IN.)	POINT GAP (IN.)	CARB. TYPE	CARB. FLOAT DROP	IGN. TYPE
Colt	1	1990	2.65	12:25	4500-5500	625-675	RJ6C	0.030	0.014	"A"	Note 1	Magneto
2.3	1	1991 & On	4.47	13:34	4200-5200	1100-1300	QL77JC4	0.030	0.014	"C"	Note 1	Magneto
3.3	1	1991 & On	4.47	13:34	4300-5000	1100-1300	QL77JC4	0.030	0.020	"C"	Note 1	Magneto
3	2	1990 & On	5.29	12:25	4500-5500	700-800	QL77JC4	0.030	N/A	"A"	1-1/8 to 1-1/2	CDII
4	2	1990 & On	5.29	12:25	4500-5500	700-800	QL77JC4	0.030	N/A	"A"	1-1/8 to 1-1/2	CDII
4D	2	1990 & On	5.28	13:29	4500-5500	600-650	QL77JC4	0.030	N/A	"A"	1-1/8 to 1-1/2	UFI
6	2	1990 & On	10	13:29	4500-5500	650-700	QL77JC4	0.030	N/A	"A"	1 to 1-3/8	UFI
8	2	1990 & On	10	13:29	5000-6000	650-700	QL77JC4	0.030	N/A	"A"	1 to 1-3/8	UFI
9.9	2	1990 & On	13.2	12:29	5500-6500	650-700	QL77JC4	0.030	N/A	"A"	1-1/8 to 1-1/2	UFI
15	2	1990 & On	13.2	12:29	5500-6500	650-700	QL77JC4	0.030	N/A	"A"	1-1/8 to 1-1/2	UFI
20	2	1990 & On	31.8	13:28	4500-5500	650-700	QL77JC4	0.030	N/A	"B"	1 to 1-3/8	UFI
25	2	1990 & On	31.8	13:28	4500-5500	650-700	QL77JC4	0.030	N/A	"B"	1-1/8 to 1-1/2	UFI
30	2	1990 & On	31.8	13:28	5200-5800	650-700	QL77JC4	0.030	N/A	"B"	1-1/8 to 1-1/2	UFI
40	2	1990 & On	44.99	12:29	4500-5500	725-775	QL78C	0.030	N/A	"B"	1-1/8 to 1-1/2	UFI
48	2	1990 & On	44.99	12:29	4500-5500	725-775	QL78C	0.030	N/A	"B"	1-1/8 to 1-1/2	UFI
50	2	1990 & On	44.99	12:29	4500-5500	725-775	QL78C	0.030	N/A	"B"	1-1/8 to 1-1/2	UFI

Note 1 Distance from top of float hinge to top of gasket should be 0.090 in (2.3mm).

LOWER UNIT
LUBRICANT CAPACITY

MODEL SIZE	YEAR	CAPACITY OUNCES
Colt	1990	1.2
2.3	1991 & On	3
3.3	1991 & On	3
3	1990 & On	2.7
4	1990 & On	2.7
4D	1990 & On	11
6	1990 & On	11
8	1990 & On	11
9.9	1990 & On	9
15	1990 & On	9
20	1990 & On	11
25	1990 & On	11
30	1990 & On	11
40	1990 & On	16.4
50	1990 & On	16.4

WIRING COLOR CODE

At press time, OMC was using the following color code system. We list the codes as "always", but add -- "subject to change".

Red -- always a "hot" lead when outboard cables are connected to a battery.

Purple -- always a hot lead when key switch is turned to the **ON** or **START** position.

Black -- always a ground lead.

Tan -- always a warning system -- very low oil or excessive temperature.

Gray -- always a tachometer lead.

Black/Yellow -- always the "Kill" circuit.

Yellow/Red -- always the start circuit.

Purple/White -- always the primer solenoid lead.

Blue/Green, Blue/White, or Green/White -- always the trim/tilt circuits. Lead with Green is **DOWN** circuit; lead with Blue is **UP** circuit.

Yellow, Yellow/Blue, and/or Yellow/Gray -- always the charging circuit.

OIL/FUEL MIXTURE

and

BREAK-IN PROCEDURE
(New or Rebuilt Powerhead)

Acceptable Products

Always use OMC or BIA certified TC-W oil lubricant. Check the container label for the certification information. **NEVER** use automotive oils, premixed fuel of unknown oil quantity, or premixed fuel richer than 50:1.

With Standard Fuel Tank

Use 25:1 fuel/oil mixture (16 fl oz -- 473 ml per 3-gallons of fuel), for the first 12 gallons of gasoline used, then use 50:1 mixture.

With AccuMix or VRO System

Use 50:1 fuel/oil mixture (8 fl oz -- 236 ml --per 3 gallons of fuel), in **ADDITION** to the AccuMix or VRO system for the first 12 gallons of fuel consumed.

Initial Powerhead Operation

Operate the powerhead at a fast idle with the unit in gear for the first 20 minutes. Check the fuel, exhaust, and water systems for leaks. Observe the water stream which should be visible at the rear -- starboard side -- of the powerhead. The stream will verify water is circulating through the powerhead.

First Hour

During first hour of operation, do not exceed 1/2 throttle. Actually it is best to vary powerhead rpm every 10 to 15 minutes.

Second Hour

During the second hour of operation, increase powerhead rpm to 3/4 throttle and increase rpm to full throttle for short periods -- just a couple minutes. Change powerhead rpm every 10 to 15 minutes.

Next Eight Hours

Avoid full throttle for long periods of time. Change powerhead rpm every 10 to 15 minutes.

Check VRO System

Check the VRO reservoir to verify oil has been used during this break-in period, before changing to clear fuel.

After Ten Hours

Allow powerhead to cool, and then check torque value on the cylinder head bolts.

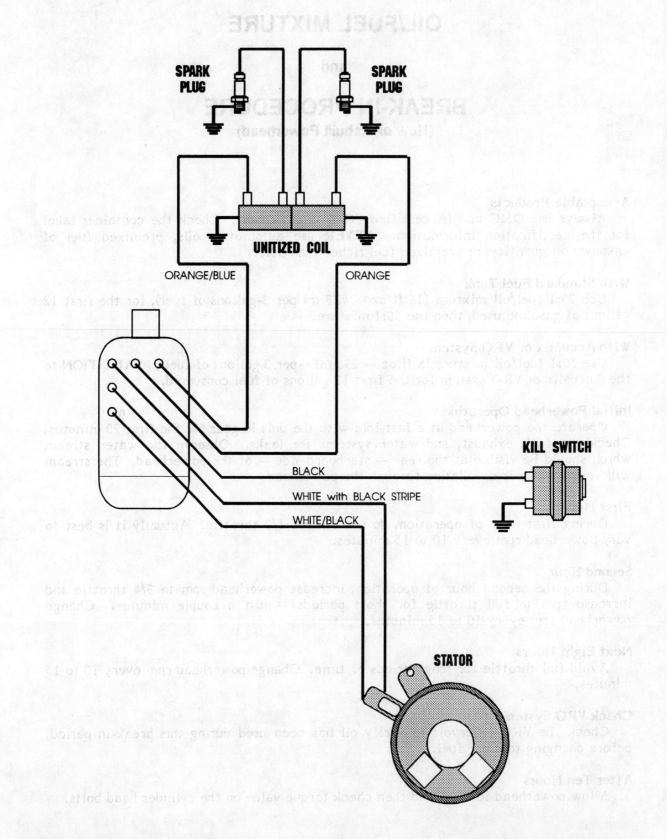

Wire identification for a simple 2-cylinder powerhead with: CDII ignition having unitized coils in a single sealed unit.

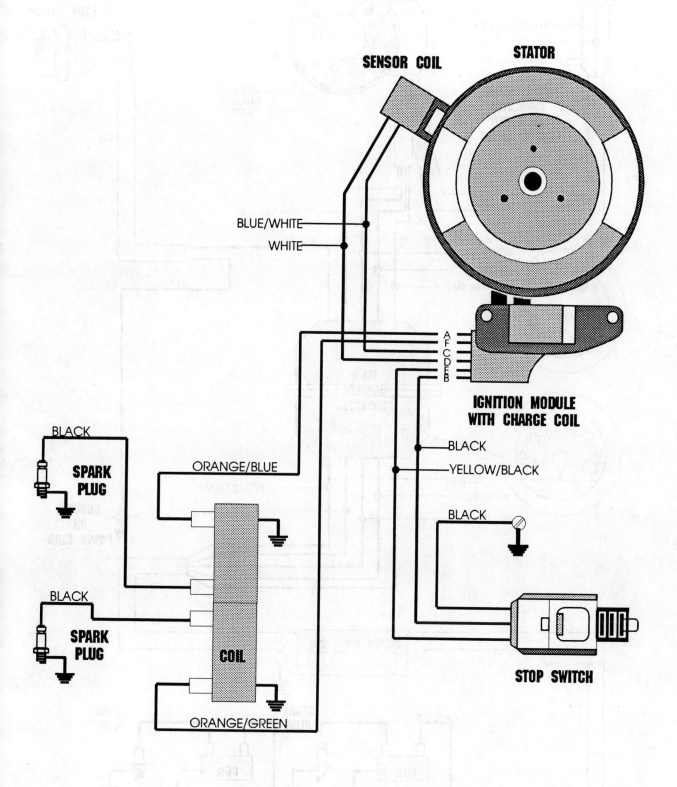

SENSOR COIL

STATOR

BLUE/WHITE

WHITE

IGNITION MODULE
WITH CHARGE COIL

A
F
C
D
E
B

BLACK

SPARK
PLUG

ORANGE/BLUE

BLACK

SPARK
PLUG

COIL

ORANGE/GREEN

BLACK

YELLOW/BLACK

BLACK

STOP SWITCH

*Wire identification for the Model 3 and Model 4 powerhead with: ignition module; unitized coils;
and charge coil are an integral part of the ignition module.*

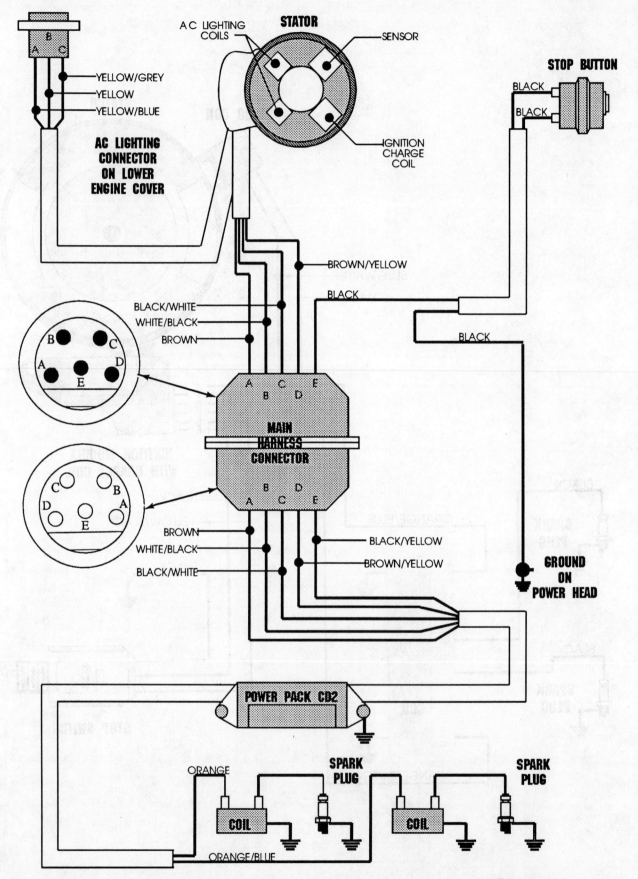

Wire identification for a typical 2-cylinder powerhead with: manual start; CDII ignition; power pack; separate ignition coils; and AC lighting coils, but no battery charging capability.

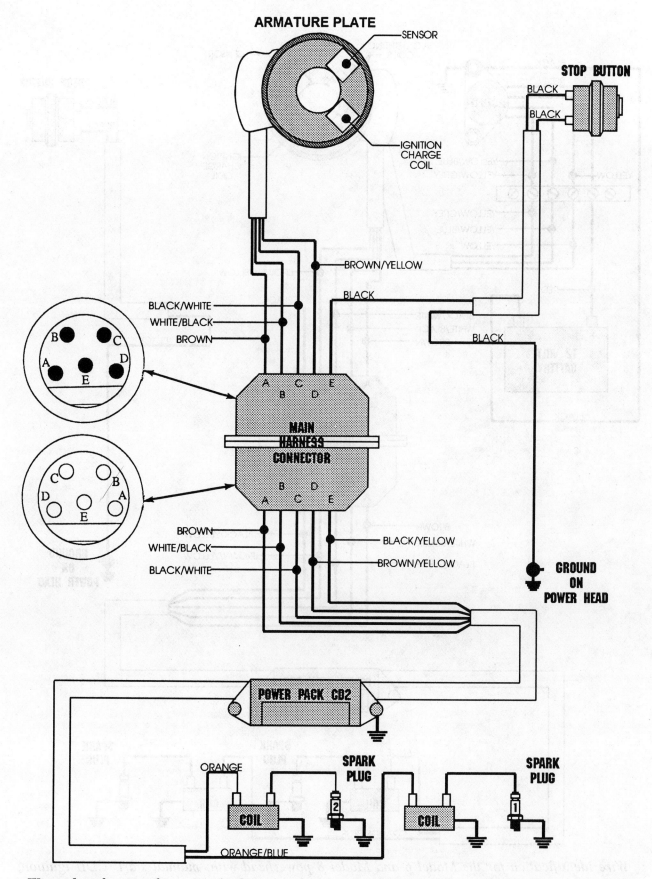

Wire identification for a 2-cylinder powerhead with: CDII ignition; power pack; and separate individual ignition coils.

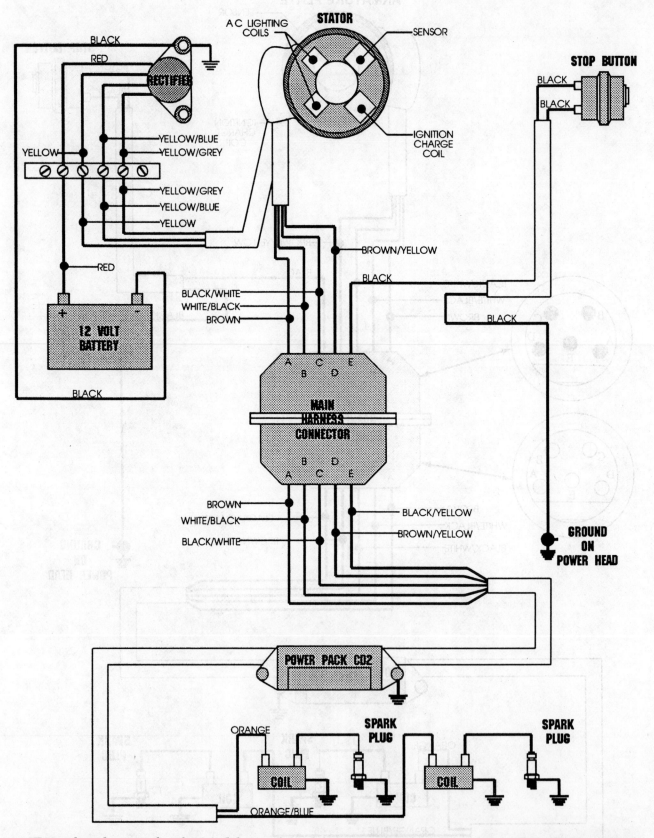

Wire identification for the Model 6 and Model 8 powerhead with: manual start; CDII ignition; power pack; AC lighting coils and rectifier, providing battery charging capabiltiy; and separate individual ignition coils.

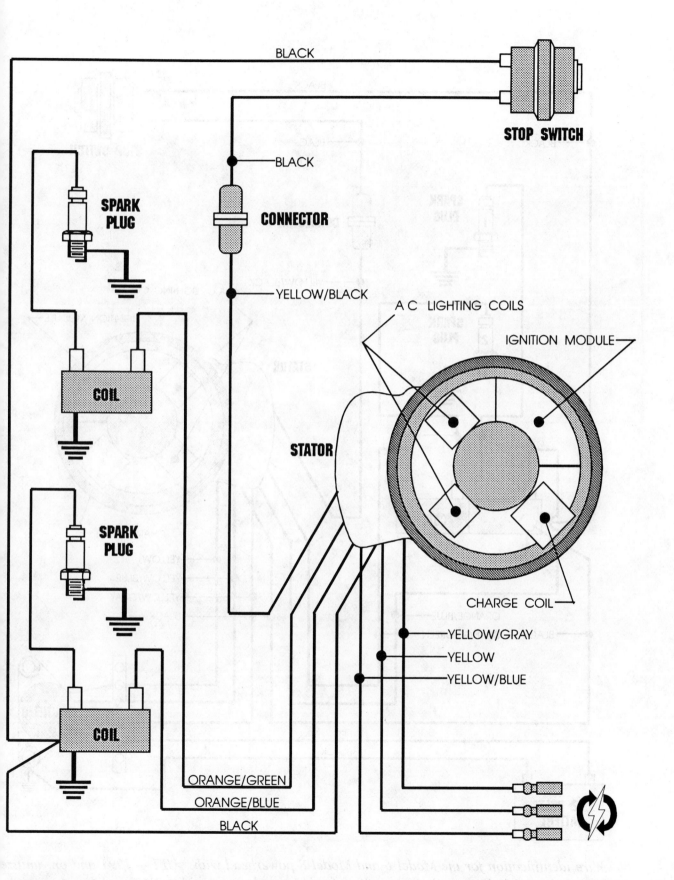

Wire identification for the Model 6 and Model 8 powerhead with: manual start; UFI; AC lighting ut no battery charging capability; separate ignition coils.

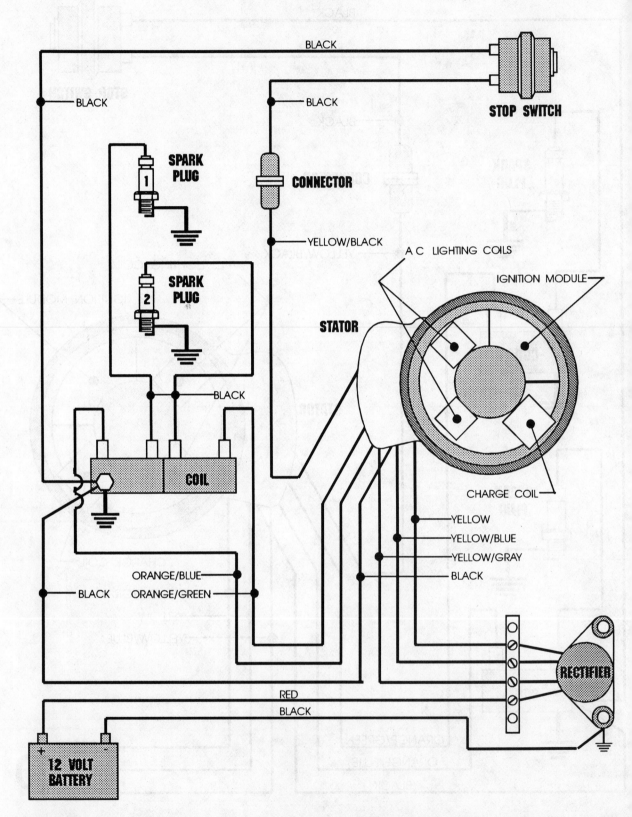

Wire identification for the Model 6 and Model 8 powerhead with: UFI -- 1991 and on; unitize ignition coils; AC lighting coils and rectifier for battery charging. The 1990 models have separa ignition coils.

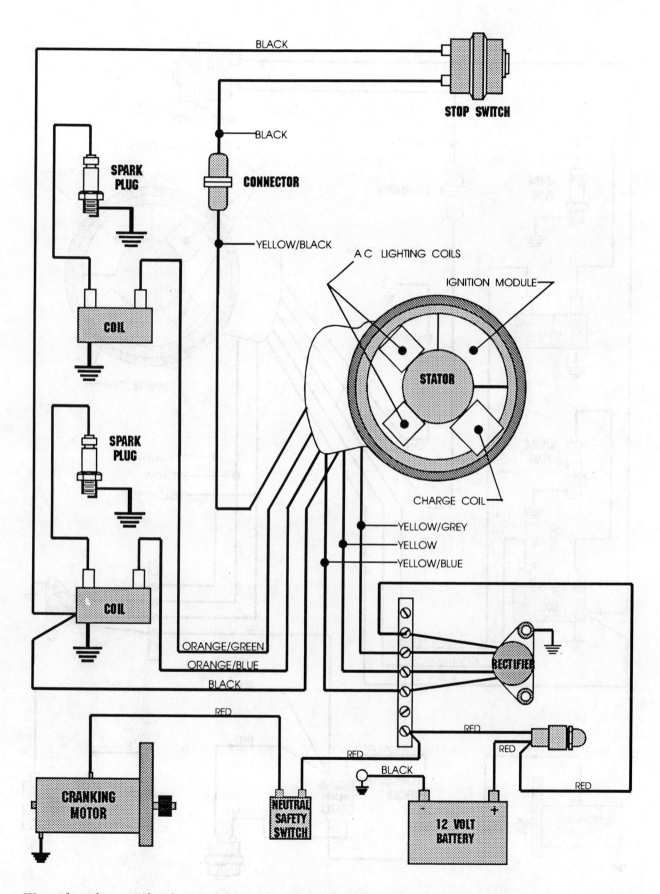

Wire identification for the Model 9.9 and Model 15 with: UFI; electric start; AC lighting coils and rectifier for battery charging; and separate ignition coils.

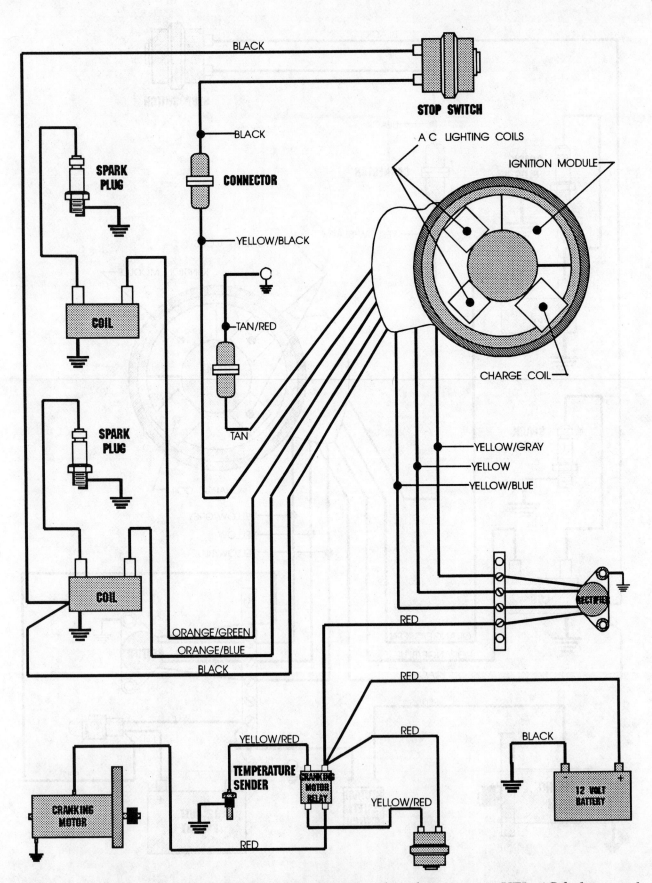

Wire identification for the Model 20, 25, and 30TE with: electric start; UFI; AC lighting with rectifier for battery charging; and separate ignition coils.

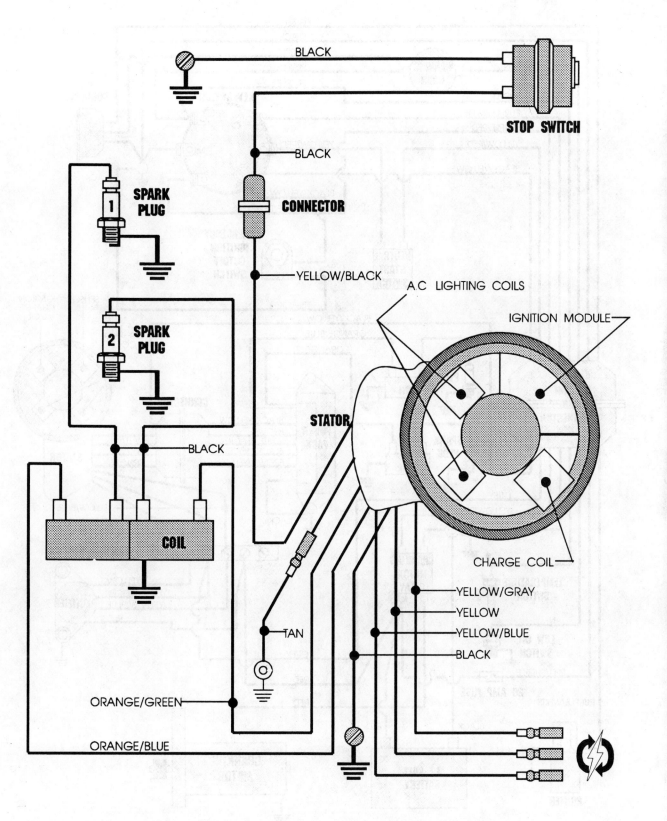

Wire identification for the Model 40 and Model 50 powerhead with: manual start; UFI; AC lighting -- no battery charging capability; unitized ignition coils -- both coils encased in a single sealed unit.

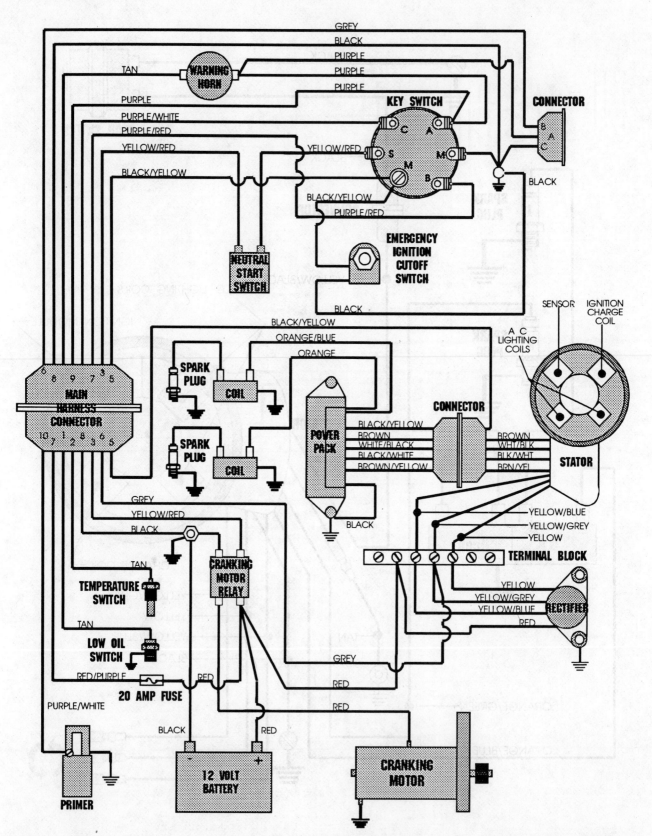

Wire identification for the Model 40 and Model 50 powerhead with: electric start; UFI; AC lighting with rectifier for battery charging; power pack; separate individual ignition coils; and remote control.

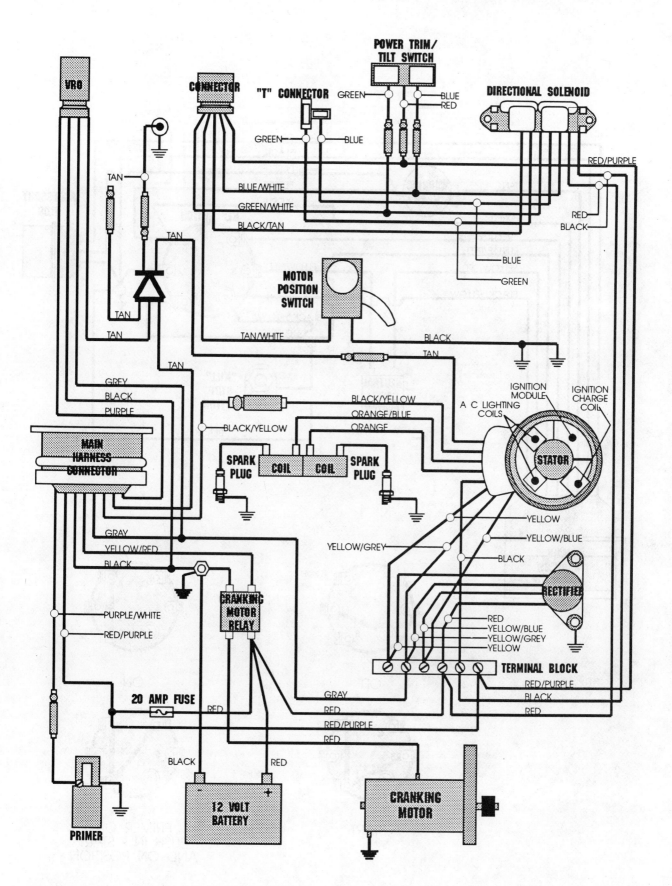

Wire identification for the Model 40 and Model 50 powerhead with: electric start; UFI; AC lighting with rectifier for battery charging; power trim/tilt; and remote control.

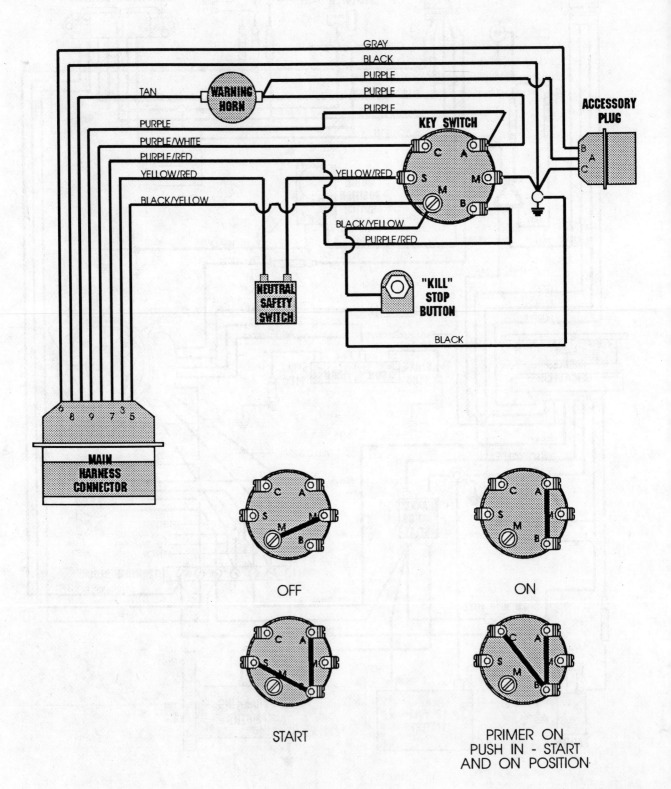

Wire identification for a remote control unit without the optional power trim/tilt.